THE FUNGAL SPORE: MORPHOGENETIC CONTROLS

Proceedings of the Third International Fungal Spore Symposium (3IFSS) held 18-22 August 1980 at Gwatt near Thun, Switzerland

THE FUNGAL SPORE: MORPHOGENETIC CONTROLS

Edited by

G. Turian

Département de biologie végétale
Université de Genève,
Genève, Switzerland

and

H.R. Hohl

Institut für Pflanzenbiologie
Universität Zürich,
Zürich, Switzerland

1981 · ACADEMIC PRESS

A Subsidiary of Harcourt Brace Jovanovich, Publishers
London · New York · Toronto · Sydney · San Francisco

ACADEMIC PRESS INC. (LONDON) LTD.
24–28 Oval Road,
London NW1

United States Edition published by
ACADEMIC PRESS INC.
111 Fifth Avenue
New York, New York 10003

British Library Cataloguing in Publication Data
International Fungal Spore Symposium, *3rd, Gwatt, 1980*
The fungal spore.
1. Fungi – Spores – Congresses
I. Turian, G II. Hohl, H R
589'.2'0446 QK601 80-49681

ISBN 0-12-703680-0

Printed in Great Britain by
Whitstable Litho Ltd., Whitstable, Kent

LIST OF CONTRIBUTORS

Dr. A. AL-RAWI, Department of Applied Microbiology, Strathclyde University, Glasgow G1 1XW, Scotland

Dr. J.G. ANDERSON, Department of Applied Microbiology, Strathclyde University, Glasgow G1 1XW, Scotland

Dr. N. ARPIN, Département de Biologie végétale, Université de Lyon, 43 Bd du 11 novembre 1918, F-69622 Villeurbanne France

Dr. W.L. AUSTIN, Department of Botany, Howard University, P.O. Box 83, Washington, D.C. 20069, USA

Dr. D.E. AXELROD, Waksman Institute of Microbiology, Rutgers University, New Brunswick, N.J. 08854, USA

Dr. S. BARTNICKI-GARCIA, Department of Plant Pathology, University of California, Riverside, CA 92521, USA

Dr. G.W. BEAKES, Department of Plant Biology, University of Newcastle, Newcastle-upon-Tyne NE1 7RU, U.K.

Dr. A. BECKETT, Department of Botany, Bristol University, Bristol BS8 1UG, U.K.

Dr. M.L. BOUILLANT, Laboratoire de Mycochimie, Département de Biologie végétale, Université de Lyon 1, 43 Bd du 11 novembre 1918, F-69622 Villeurbanne, France

Dr. R. BRAMBL, 304 Stakman Hall, University of Minnesota, Saint Paul, MN 55108, USA

Dr. S. BRODY, Biology Department, University of California, La Jolla, CA 92037, USA

Dr. N.J. BUTNIK, Waksman Institute of Microbiology, Rutgers University, New Brunswick, N.J. 08854, USA

Dr. S. P. CHAMPE, Waksman Institute of Microbiology, Rutgers University, New Brunswick, N.J. 08854, USA

Dr. G.T. COLE, Department of Botany, University of Texas, Austin, TX 78712, USA

Dr. D.A. COTTER, Department of Biology, University of Windsor, Windsor, Ontario, Canada N9B 3P4

Dr. K.R. DAHLBERG, Department of Plant Pathology, University of Nebraska, Lincoln, NE 68583, USA

Dr. S.K. DUTTA, Department of Botany, Howard University, P.O. Box 83, Washington, D.C. 20059, USA

Dr. U.M. EDEN, Institute of Genetics, University of Copen-

hagen, Øster Farimagsgade 2A, DK-1353 Copenhagen, Denmark
Dr. L. FREDERICK, Department of Botany, Howard University, P.O. Box 83, Washington, D.C. 20059, USA
Dr. M.S. FULLER, Department of Botany, University of Georgia, Athens, GA 30602, USA
Dr. B. FURCH, Botanisches Institut der Universität, Olshausenstrasse 40-60, D-2300 Kiel, FRG
Dr. G.W. GOODAY, Department of Microbiology, Marischal College, University of Aberdeen, Aberdeen AB9 1AS, Scotland
Dr. D. GOTTLIEB, N 519 Turner Hall, Department of Plant Physiology, University of Illinois, Urbana, ILL 61801, USA
Dr. T.M. HAMMILL, Department of Biology, University of New York, Oswego, N.Y. 13126, USA
Dr. L.E. HAWKER, Department of Botany, The University, Bristol BS8 1UG, U.K.
Dr. H. HEGNAUER, Institut für Pflanzenbiologie, Universität Zürich, Zollikerstrasse 107, CH-8008 Zurich, Switzerland
Dr. W. HESS, 245 WIDB, Department of Botany and Range Science, Brigham Young University, Provo, UT 84602, USA
Dr. H.R. HOHL, Institut für Pflanzenbiologie, Universität Zürich, Zollikerstrasse 107, CH-8008 Zurich, Switzerland
Dr. B.-F. HUANG, Boyce Thompson Institute, Tower Road, Ithaca, N.Y. 14853, USA
Dr. B. KRISTANSEN, Department of Applied Microbiology, Strathclyde University, Glasgow G1 1XW, Scotland
Dr. M.B. KURTZ, Waksman Institute of Microbiology, Rutgers University, P.O. Box 759, Piscataway, N.J. 08854, USA
Dr. L. LANGE, Danish Government Institute of Seed Pathology, Ryvangs Allé 78, DK-Hellerup, Copenhagen, Denmark
Dr. J.S. LOVETT, Department of Biological Sciences, Purdue University, West Lafayette, IN 47907, USA
Dr. V. MACKO, Boyce Thompson Institute, Ithaca, N.Y. 14853, USA
Dr. M.F. MADELIN, Department of Botany, The University, Bristol BS8 1UG, U.K.
Dr. L. NYHLEN, Institut für Pflanzenbiologie, Universität Zürich, Zollikerstrasse 107, CH-8008 Zurich, Switzerland
Dr. G. OLAH, Université Laval, Faculté des Sciences, Cité Universitaire, Québec, P.Q., Canada G1K 7P4
Dr. L.W. OLSON, Institute of Genetics, University of Copenhagen, Øster Farimagsgade 2A, DK-1353 Copenhagen, Denmark
Dr. L.M. POPE, Department of Botany, University of Texas, Austin, TX 78712, USA

Dr. D. RAST, Institut für Pflanzenbiologie, Universität Zürich, Zollikerstrasse 107, CH-8008 Zurich, Switzerland
Dr. O. REISINGER, Université de Nancy 1, Laboratoire de Microbiologie, Case Officielle No 140, F-54037, Nancy Cedex France
Dr. G.M. RUSSO, Department of Plant Pathology, University of Nebraska, Lincoln, NE 68583, USA
Dr. W.D. SIKKEMA, Department of Biological Sciences, Purdue University, West Lafayette, IN 47907, USA
Dr. J.E. SMITH, Department of Applied Microbiology, Strathclyde University, Glasgow G1 1XW, Scotland
Dr. R.C. STAPLES, Boyce Thompson Institute, Tower Road, Ithaca, N.Y. 14853, USA
Dr. H. STUSSI, Institut für Pflanzenbiologie, Universität Zürich, Zollikerstrasse 107, CH-8008 Zurich, Switzerland
Dr. A.S. SUSSMAN, Department of Botany, University of Michigan, Ann Arbor, MI 48189, USA
Dr. G. TURIAN, Laboratoire de Microbiologie générale, Université de Genève, 3 Place de l'Université, CH-1211 Geneva 4, Switzerland
Dr. J.L. VAN ETTEN, Department of Plant Pathology, University of Nebraska, Lincoln, NE 68583, USA
Dr. D.J. WEBER, 285 Widtsoe, Brigham Young University, Provo, UT 84602, USA
Dr. J.D. WEETE, Department of Botany, Auburn University, Auburn, AL 36830, USA
Dr. D. WOLF, Biochemisches Institut, Universität Freiburg, D-7800 Freiburg, FRG
Dr. A.G. YAHYA, Department of Applied Microbiology, Strathclyde University, Glasgow G1 1XW, Scotland
Dr. L.N. YAGER, Waksman Institute of Microbiology, Rutgers University, New Brunswick, N.J. 08854, USA

PREFACE

The Second International Fungal Spore Symposium was held in Provo, Utah, USA in 1974. There, emphasis was put on questions of spore dormancy and germination. For the present Third International Fungal Spore Symposium we aimed at events accompanying spore formation without, however, neglecting recent advances in the field of spore activation and germination. Again, a major goal of the Symposium was to bring together leading authorities to assess the present state of knowledge on the ultrastructure, physiology, biochemistry and genetics of fungal spores.

It should be remembered, though, that our common denominator, the spore, represents a very diverse entity. Not only do spores differ greatly in shape and mode of formation but also functionally there is a wide gap between asexual spores and those of sexual origin which are the result of recombination mechanisms initiated by the complex interplay of mating factors. The one and all-embracing feature of spores may be recognized in their representing the end product of the transition from a vegetative, relatively undifferentiated growth condition to a reproductive, differentiated structural entity, followed by a reversal of this state during germination.

The questions raised and the problems posed by these series of transitional events are wide open and presently under study at different levels of organization. These studies cover the entire range from gene level to environmental parameters. Considered in its fullest context, spore research would encompass almost all events leading from one spore generation to another and we might have entitled the present volume 'From Spore to Spore'. For practical reasons the topics covered were restricted to those directly connected with initiation, formation and germination of spores of the major fungal groups.

Relatively simple environmental triggers such as elevated temperature (heat shock) or nutritional deficiencies (starvation) may interrupt the vegetative stage and channel the organism into developmental and reproductive pathways. Research with microcycling fungi should help to answer the question of how much (or how little) vegetative growth is needed to render the organism competent for the morphogenetic switches.

Initial and basic controls of these morphogenetic processes must be based on the informational content of the organism's DNA which forms the *sine qua non* of reproducing a 'true to type' spore. Expression of gene information is conveyed by spore specific mRNAs which call into action controls at the translational and perhaps posttranslational levels to sustain synthesis of specific spore proteins, pigments and structural characteristics.

Germination, often the result of specific triggering mechanisms, represents the return to a vegetative condition, implying processes of dedifferentiation and a switch to growth which is in essence a synthesis of protoplasm and cell walls. While in some lower forms and also in early stages of spore germination in higher fungi, growth occurs in an unpolar, multidirectional fashion, in filamentous fungi the germinating spore quickly redifferentiates into an apically growing, polarized germ tube. This polarized growth pattern is maintained as long as it is enforced by the genetic potential and the environmental conditions. Eventually, morphogenetic controls will again become operative thereby imposing a return to sporogenesis. This brief outline may suffice to indicate the scope of the present volume on 'The Fungal Spore : Morphogenetic Controls'.

We would like to thank the various institutions that financially supported our meeting : Swiss National Science Foundation, Swiss Academy of Natural Sciences, Fonds Académique (Genève), Interpharma (Basel), Association Suisse des Fabricants de Cigarettes (Fribourg), University of Geneva and University of Zürich. The Symposium was sponsored by the Swiss Microbiological Society and the Swiss Botanical Society. Many thanks are due to the members of the organizing committee and the technical assistants whose untiring efforts contributed so much to the success of the meeting. They are : Erika Rüegg, Barbara Duchoud, Rosmarie Honegger, Ursula Heiniger, Marianne Hohl, Jakob Bodenmann, Mukti Ojha, Jürg Jehli, Jean Zuber, René Gees, Françoise Grange, Katja Iselin, Mehrbanou Michéa, The Can Ton-That and Bernhard Häne. We are also indebted to Academic Press for publishing the proceedings and for their cooperation and help during the editing of this volume.

June 1981 G. Turian and H.R. Hohl

CONTENTS

III. NUCLEAR CONTROLS

IV. INFORMATIONAL EXPRESSION

V. ENVIRONMENTAL CONTROLS

VI. BIOCHEMICAL EVENTS AND CONTROLS

HISTORICAL INTRODUCTION

Lilian E. Hawker, Emeritus Professor of Mycology

Department of Botany, The University
Bristol, England.
BS8 1UG

INTRODUCTION

The Colston Research Society, a group of eminent citizens of Bristol makes an annual grant to finance a symposium on a branch of knowledge chosen by the University and held in one of its Halls of Residence. In 1966 I was asked to arrange a symposium on a mycological subject of my choice. After consultation my mycological colleagues in the Department of Botany, and I decided to adopt Dr. Madelin's suggestion that the title should be "The Fungus Spore" and that we should attempt a wide outline coverage of the subject including initiation, maturation, dispersal, dormancy, germination and the importance of fungus spores in agriculture, medicine and industry. Our guests and contributors entered into the enterprise with enthusiasm and expert knowledge. The results were published as *Colston Papers* No. 18. Later it became clear that we had initiated a continuing study. Drs. Hess and Weber arranged a second symposium in Utah U.S.A., in 1974. By then it was no longer possible to cover all aspects of the subject and the organisers wisely limited the scope of the symposium to detailed studies of the mature spore, its dormancy and germination. This second symposium was both enjoyable and stimulating. It is fitting that the present third symposium on the fungus spore should be limited to another particular aspect of the subject. Since the 1974 meeting in Utah, great advances have been made in the study of morphogenetic controls, particularly of spore initiation and development but also of activation and germination. Our present meeting is to discuss this morphogenetic control.

In a historical introduction to this fascinating and fast growing subject it is obviously not possible to cover more than a fraction of the vast literature on the development of fungi which now exists. Accordingly an attempt will be made to indicate the general principles which emerged from earlier work on the subject and to illustrate this with selected examples to form a basis for appreciating the accounts of advanced specialist topics which are to follow.

FACTORS INFLUENCING REPRODUCTION IN FUNGI

During the first two decades of the present century little was known about factors influencing growth of fungi and still less about the events leading to the initiation of reproduction.

Apart from a few isolated papers the study of the physiology of fungi began with the pioneer work of William Brown at the Imperial College of Science in London, the results of which were published in a series of papers (Brown 1922, 1923, 1925, Brown and Horne, 1926 etc.) Brown was primarily a plant pathologist but he realised that his studies on the physiology of parasitism and the control of plant disease must be soundly based on knowledge of the factors controlling growth and reproduction of the parasite.

Brown's early work on fungal physiology was done with a species of *Fusarium* isolated from apples and then known as *F. fructigenum*, later as *F. lateritium*. This fungus grew well on most of the natural media then in general use, such as potato extract and malt extract.

Brown set out to determine the minimum nutritional requirements for growth and sporulation. From a chemical analysis of potato extract he attempted to produce a solution of defined chemicals reproducing as closely as possible the composition of the natural extract. Brown's next move was to leave out various components until he was finally left with a simple medium (containing glucose, asparagine, tri-potassium phosphate and magnesium sulphate) on which the fungus grew well. With the aid of this simple medium he studied the response of the fungus to such external factors as temperature, light, pH of the medium and aeration. It was soon evident that conditions sufficient to allow good mycelial growth were not necessarily favourable to the formation of conidia and might even inhibit spore formation. Brown and his students then attempted to grow a range of different fungi on simple synthetic media and it was soon clear that these fungi differed widely in their requirements for growth and sporulation and that fungi producing more than one type of spore needed specific conditions to produce each of these.

It was the writer's good fortune to enter Brown's laboratory in 1932 and to have the benefit of his guidance for 13 years. The starting point of the writer's researches on reproduction in fungi was a plate of a synthetic medium (Medium A: containing 0.5% glucose, KNO_3, KH_2PO_4 and $MgSO_4$) inoculated with a Pyrenomycete, *Melanospora destruens* (later shown to be a strain of *Sordaria*, *S. destruens*; Hawker, 1951). On this plate was a small contaminant colony. The *Sordaria* grew only sparsely and produced no perithecia in pure culture in medium A, but around the contaminant colony was a ring of perithecia containing viable ascospores. A number of fungi from diverse groups were then grown in mixed culture with *S. destruens* and many of them induced the latter to form perithecia. This phenomenon could have been due to one or more of a number of external factors including a mechanical barrier to growth and changes in the medium brought about by the second fungus, such as removal of nutrients, changes in pH, or the production of staling substances or of stimulating substances (growth substances). Preliminary experiments indicated that the most likely cause was the production of substances by the contaminant colony which had a stimulating effect on perithecial production by the *Sordaria* (Asthana and Hawker, 1936). Dilution of the medium was also important.

It must be remembered that in the 1930's the study of plant growth substances (auxins) was only just beginning, that the chemical nature of common vitamins was largely unknown and that pure preparations of these were not commercially available.

Nielsen (1931) had already shown that culture medium in which *Rhizopus suinus* had grown induced increased growth both of the oat coleoptile and a species of *Aspergillus*. By fractionation of the culture filtrate he showed that the factor influencing *Aspergillus* was distinct from that causing extension of the oat coleoptile (heteroauxin). Schopfer (1931, 1932) showed that growth of *Phycomyces* spp required a supply of aneurin (Vitamin B_1, now termed thiamin). Buston and his collaborators (Buston and Pramanik, 1931a, 1931b, Buston and Kasinathan, 1933) reported that *Nematospora gossypii* (the cotton boll rot fungus) which was unable to grow on Brown's glucose-asparagine medium, grew well with the addition of an extract of lentils to the medium. Later Kögl and Fries (1937) working with the same fungus extracted from large quantities of yeast or eggs three substances (inositol, vitamin B_1 and an unknown substance which they termed biotin). An external supply of all three was essential for growth of *N. gossypii*.

The lentil extract or its ether insoluble fraction used by Buston proved to be sufficient to permit formation of perithecia by *S. destruens* [1] when added in small quantities to medium A. Through the kindness of Kögl and Fries the writer was provided with a small sample of biotin and was able to show that *S. destruens* did not need an external supply of inositol, showed improved growth when biotin was added to medium A and needed the further addition of vitamin B_1 for the formation of perithecia (Hawker, 1938). Later it was shown that the pyrimidine component of thiamin was as active as the whole molecule (Hawker,1939).

Experiments with other fungi by a number of workers (listed by Lilly and Barnett, 1951, and others) soon showed that requirements varied not only between species but often also between strains of the same species. Experiments with mixed cultures showed that while all fungi required a range of growth substances for growth and sporulation they varied in their ability to synthesize these from simple chemicals. Some like *F. lateritium* could synthesise all the growth substances they needed; others like *S. destruens* were able to synthesize only small quantities of particular ones or were totally unable to synthesize one or more essential substance(s).

Parallel with these studies on the effects of growth substances were others on those of the nature and concentration of major nutrients.

It has long been recognised that a dilute medium is often more favourable to sporulation than a concentrated one. The classic experiments of Klebs (1898, 1899, 1900) with *Saprolegnia* spp and Claussen (1912) with *Pyronema confluens* showed that transfer from a medium rich in nutrients to a more dilute one induced the formation of sporangia and apothecia respectively. It is now generally accepted that such a transfer of a well fed mycelium to a medium poor in one or more essential nutrient is likely to induce sporulation.

A detailed study was made of the effects of different carbon sources at different concentrations on the production of perithecia by *S. destruens* (Hawker, 1939, 1947).

With glucose or fructose as the source of carbon, perithecia were most numerous at a sugar concentration of 0.5% and were absent at concentrations of 1.0% and higher.

The concentrations of some other carbohydrates optimal for production of perithecia were higher than that with glucose. Thus sucrose, starch and lactose were optimal at concentrations of 10.0%, 2.0% and 1.0% respectively. However, if in a liquid medium a total amount of glucose larger than 0.5% were given in small increments, fruiting increased but was still inferior to that on 10.0% sucrose,

It was concluded and supported by other experiments that fruiting was best on those more complex carbohydrates which were broken down to hexoses at a rate giving an optimal concentration of these over a relatively long period (Hawker, 1948). Since other methods of maintaining such a supply did not give such good results as sucrose, it was concluded that some intermediate stage in the inversion of sucrose stimulated the initiation of fruiting.

It had already been shown by Buston *et al.*(1953) working with *Chaetomium globosum* and by Miller *et al.*(1955) with the yeast *Saccharomyces cerevisiae* that small quantities of certain hexose phosphates favoured spore production. Perithecial formation by *S. destruens* also proved to be responsive to small quantities of glucose - 1 - phosphate or fructose - 1: 6 - diphosphate. It was concluded that the favourable effect of sucrose and some other complex carbohydrates was due to a combination of the rate of hydrolysation and the formation of particular phosphoric esters (Hawker, 1948).

The complete inhibition of formation of perithecia of *S. destruens* by high concentrations of glucose could be prevented by increasing the amount of thiamin in the medium and it was shown (Hawker, 1944) that the concentration of glucose optimal for fruiting increases with increase in the concentration of thiamin. The favourable effects of thiamin were found to be linked with an increased rate of respiration. Respiration was also greater on a sucrose rather than on a glucose medium. It was demonstrated that with *S. destruens* and some other fungi, growing on culture media, respiration reached a peak and then fell and that initiation of sporulation closely followed the fall from the peak.

The nature and concentration of nitrogen sources and the carbon/nitrogen ratio also influenced sporulation of many fungi but here the situation is complicated by the effect of toxic staling substances and changes in pH resulting from the utilization of particular nitrogen compounds.

Many trace elements also influence sporulation as has been shown particularly with *Aspergillus niger* beginning with the work of Steinberg in 1919.

Physical factors, such as light, are also important in inducing sporulation of a number of fungi. The effects were reviewed by Carlile (1965) and more recently there is a suggestion that light may act by inducing the formation by the fungus of spore-promoting substances (Leach, 1965).

Thus by the time (1966) of the first of these "Fungus Spore" symposia the general factors controlling sporulation were known. In addition, a number of specific studies in depth on particular fungi or groups of fungi have added further information on the external and internal factors causing the change from the vegative to the reproductive state. Such studies as those of Lilly and his co-workers (Lilly, 1966) with *Phytophthora* spp; Cantino (1956) with *Blastocladiella emersonii*; Emerson (Emerson and Fox, 1940) with *Allomyces* spp; Raper in a series of papers on *Achlya* spp (Raper, 1939, 1951) and Gooday (1973, 1974) with the biosynthesis of trisporic acid and the role of this substance in differentiation in the Mucorales, show that the study of morphogenesis in fungi is indeed a fertile field.

FACTORS INFLUENCING SPORE GERMINATION

Fungus spores vary greatly in their mode of formation, their genetic history, their ability to survive adverse conditions in a dormant state and their mode of germination. Different types of spore respond differently to particular external factors but, in general, conditions leading to germination are more exacting than those permitting vegetative growth (Brown, 1922). The physiology of spore germination was summarised by Gottlieb in 1950 and he also discussed biosynthetic processes in germinating spores at the first "Fungus Spore" symposium (Gottlieb, 1966) and at the second one (Gottlieb, 1976). Many aspects of the subject were discussed by other contributors to the Utah Symposium. The report of this symposium gives a sufficient basis for appreciation of further work to be reported later in the present meeting.

TARGETS FOR FUTURE WORK

A historical introduction must not trespass on present day studies. Hence this paper has largely ignored work in progress or done during the 70's. However, during that decade the tools available to the investigator have become increasingly sophisticated, making possible experiments which could not have been done 10 years ago and holding out the prospect of the eventual elucidation of the internal control of morphogenesis.

Two broad lines of investigation present themselves. In one, use may be made of improved micro-chemical techniques; in the other, the study of changes taking place in ultrastructure is facilitated by advances in the techniques of electron microscopy. Much has been done recently along both these lines but much more remains to be done. It is an exciting prospect.

FOOTNOTE

[1] *S. destruens* proved to be particularly suitable for this study since an estimate of "perithecial frequency" could readily be made by counting the number of fruit bodies at arbitrally chosen points on a Petri dish culture.

References

Asthana, R.P. and Hawker, L.E.(1936) The influence of certain fungi on the sporulation of *Melanospora destruens* Shear, and of some other Ascomycetes. *Ann. Bot* 50, 325-44

Brown, W. (1922) On the germination and growth of fungi at various temperatures and in various concentrations of oxygen and of carbon dioxide. *Ann. Bot.* 36, 257-83.

Brown, W. (1923) Experiments on the growth of fungi in culture media. *Ann. Bot.* 37, 105-29.

Brown, W. (1925) Studies in the genus *Fusarium II.* An analysis of factors which determine the growth-forms of certain strains. *Ann. Bot.* 39, 373-408.

Brown, W. and Horne, A.S. (1926) Studies on the genus *Fusarium III.* An analysis of factors which determine certain microscopic features of *Fusarium* strains. *Ann. Bot.* 40, 203-21.

Buston, H.W. and Pramanik, B.K. (1931a). The accessory factor necessary for the growth of *Nematospora gossypii I.* The chemical nature of the accessory factor. *Biochem. J.* 25, 1656-20.

Buston, H.W. and Pramanik, B.K. (1931b). The accessory factor necessary for the growth of *Nematospora gossypii II.* The relation of the factor to "Bios". *Biochem. J.* 25, 1671-3.

Buston, H.W. and Kasinathan, S. (1933). The accessory factor necessary for the growth of *Nematospora gossypii III. Biochem J.* 27, 1859-68.

Buston, H.W., Jabbar, A. and Etheridge, D.E. (1953). The influence of hexose phosphates, calcium and jute extract on the formation of perithecia by *Chaetomium globosum. J. Gen. Microbiol.* 8, 302-6.

Cantino, E.C. (1956) The relation between cellular metabolism and morphogenesis in *Blastocladiella. Mycologia* 48, 225-40.

Carlile, M.J. (1965) The photobiology of fungi. *Ann. Rev. Plant Physiol.* 16, 104-8.

Claussen, P. (1912) Zur Entwicklungsgeschichte der Ascomyceten *(Pyronema confluens). Z. Bot.* 4, 1.64.

Emerson, R. and Fox, D.L. (1940) γ - Carotene in the sexual phase of the aquatic fungus *Allomyces*. *Proc. Roy. Soc. London B.* 128, 275-93.
Gooday, G.W. (1973) Differentiation in the Mucorales in Microbial Differentiation. (Ed, J.M. Ashworth and J.E. Smith). Symposium 23, *Soc. Gen. Microbiol. C.U.P. Cambridge, England,* 269-93.
Gooday, G.W. (1974) Fungal Sex Hormones. *Ann. Rev. Biochem.* 35-49.
Gottlieb, D. (1950) The physiology of spore germination in fungi. *Bot. Rev.* 16, 229-57.
Gottlieb, D. (1964) Germination of fungus spores. *Endeavour* 23., 85-9.
Gottlieb, D. (1966) Biosynthetic processes in germinating spores, in The Fungus Spore (Ed. M.F. Madelin), *Colston Papers* 18 Lond. 217-234.
Hawker, L.E. (1936) The effect of certain accessory substances on the sporulation of *Melanospora destruens* and of some other fungi. *Ann. Bot.* 50, 699-718.
Hawker, L.E. (1938) Effect of growth substances on growth and fruiting of *Melanospora destruens*. *Nature (Lond.)* 142 p. 1038.
Hawker, L.E. (1939a) The influence of various sources of carbon on the formation of perithecia by *Melanospora destruens* Shear in the presence of accessory growth factors. *Ann. Bot. N.S.* 3, 455-68.
Hawker, L.E. (1939b) The nature of the accessory growth substances influencing growth and fruiting of *Melanospora destruens* Shear and of some other fungi. *Ann. Bot. N.S.* 3. 657-76.
Hawker, L.E. (1944) The effect of Vitamin B_1 on the utilization of glucose by *Melanospora destruens* Shear *Ann. Bot. N.S.* 8. 79.90.
Hawker, L.E. (1947) Further experiments on growth and fruiting of *Melanospora destruens*. Shear. in the presence of various carbohydrates with special reference to the effects of glucose and sucrose. *Ann. Bot. N.S.* 11. 245-59.
Hawker, L.E. (1948) Stimulation of the formation of perithecia of *Melanospora destruens* Shear by small quantities of certain phosphoric esters of glucose and fructose. *Ann. Bot. N.S.* 12, 77-9.
Hawker, L.E. (1950) Physiology of Fungi. University of London Press. London, England. 360 pp.
Hawker, L.E. (1951) Morphological and physiological studies on *Sordaria destruens* (Shear). Comb. nov. *(syn Melanospora destruens)*, *Sordaria fimicola,* and *Melanospora*

zamii. Trans. Brit. Mycol Soc. 34, 174-86
Hawker, L.E. (1957) The Physiology of Reproduction in Fungi. Camb. Univ. Press., Cambridge, England, 128 pp.
Hawker, L.E. (1974) The effects of some environmental factors on the ultrastructure of fungi. *Kavaka* 2, 19-24.
Klebs, G. (1898,1899, 1900) Zur Physiologie der Fortpflanzung einiger Pilze. *J. Wiss Bot.* 32, 1-70; 33, 513-93; 35, 80-203.
Kögl, F. and Fries, N. (1937) Über den Einfluss von Biotin, Aneurin und Meso-Inosit auf das Wachstum verschiedenen Pilzarten. *Z. physiol. Chem.* 249. 23-110.
Leach, C.M. (1965) Ultraviolet-absorbing substances associated with light-induced sporulation in fungi. *Canad. J. Bot.* 43, 187-200.
Lilly, V.G. (1966) The effects of sterols and light on the production and germination of *Phytophthora* spores, in The Fungus Spore (Ed. M.F. Madelin) *Colston Papers* 18, London, England, 388 pp.
Lilly, V.G. and Barnett, H.L. (1951) Physiology of Fungi. McGraw Hill, New York.
Miller, J.J., Calvin, J. and Tremaine, J.H. (1955). A study of factors influencing sporulation of *Saccharomyces cerevisiae. Canad. J. Microbiol.* 1. 560-73.
Nielsen, N. (1931) The effect of rhizopin on the production of dry matter of *Aspergillus niger*, C.R. Lab. Carlsberg. *Sér Physiol* 19, 1.10.
Raper, J.R. (1939) Role of hormones in the sexual reaction of heterothallic *Achlyas. Science* 89, 321-2.
Raper, J.R. (1951) Sexual hormones in *Achlya. Amer. Sci.* 39, 110-20.
Schopfer, W.H. (1931) Etude de l'influence des extraits de levures et des concentrés de vitamine B sur la sexualité d'un champignon. *C.R. Soc. Phys. His. Nat. Genève* 48, 105-7.
Schopfer, W.H. (1934) Les vitamines crystallisées B comme hormones de croissance chez un microorganisme *(Phycomyces) Arch. Mikrobiol.* 5, 502-10.
Steinberg, R.A. (1919) A study of some factors in the chemical stimulation of the growth of *Aspergillus niger. Amer. J. Bot.* 30, 37-48.
Steinberg, R.A. (1936) Some effects of heavy metals essential for the nutrition of *Aspergillus niger* upon its growth. *Amer. J. Bot.* 23, 227-31.

I. ULTRASTRUCTURAL MORPHOGENESIS

LOWER FUNGI

ULTRASTRUCTURAL MORPHOGENESIS IN LOWER FUNGI: DISCUSSANT'S INTRODUCTION

Melvin S. Fuller

Department of Botany, University of Georgia, Athens, Georgia 30602

The use of electron microscopes in the study of "lower" fungi has come a long way since 1951 when Manton and her colleagues (Manton *et al*, 1951) first looked at the spores of *Saprolegnia* and, subsequently (Manton *et al*, 1952), other aquatic fungi. Following a long period of cut them up and look, electron microscopy has become more exact with the acquisition of qualitative and quantitative data being required in studies today. We insist that our students, especially those planning careers in experimental mycology and with interests in fungal morphogenesis, be thoroughly acquainted with the cell biological, biochemical, and molecular aspects of fungal cells. While we can make progress working in special areas of our science, the excitement and rapid progress comes when we combine the structural and functional in our work. De Bary's laboratories at Strasbourg were described by Farlow (see Sparrow, 1978) as looking more like chemical laboratories than botanical ones. Today this has become the norm for the study of fungi and deBary, where he still alive, would have used the occasion of the Third International Fungus Spore Symposium (3IFSS) to take refuge, accompanied by his family, along the shores of beautiful Lake Thun. As we review our own discoveries we should remember that the success of deBary and his students was because of his insistence upon starting with the germination of the spore of the fungus and tracing its development from spore to spore. Where deBary may have had a bottle of 40% alcohol and pieces of split pith, we have a battery of chemicals, techniques and instrumentation available to us that allows better resolution by several orders of magnitude. If we pursue our studies with the care exercised by deBary much excitement lies ahead. In his talk that follows Dr. Hess will give an overview of what we know about the ultrastructure of fungal cells. I hope, in the few minutes available to me, to review some of the progress we have made with aquatic fungi and to point to areas needing study before the Fourth International Fungus Spore Symposium. Hopefully, some of these results will be forthcoming in the sessions to follow.

When I reported (Fuller, 1966) upon the structure of zoospores in uniflagellate fungi at 1IFSS, it was already apparent that spore structure correlated well with other parameters used in determining relationships among the aquatic fungi. In the ensuing years, we, and primarily others; most important of whom are Barr (1978), Powell (1978) and Lange and Olson (1979), have looked at many more spores. The structure of zoospores has become extremely important to fungal systematics and Barr (personal communication) tells me that he intends, on the basis of zoospore ultrastructure, to make two orders out of the present Chytridiales. This is satisfying to one who has been a part of the work on spore ultrastructure yet I must add a note of caution that we not, because of the ease with which we can now look at spore structure, place undue emphasis on this one phase in the life cycle of these fungi.

Less can be said about the functional nature of the structures and specializations we see in fungal motile cells. In spite of many careful studies (see Lovett, 1975), we still cannot say why the ribosomes of the motile cells in many of these fungi are physically separated from other cell components. How and for what reasons do the mitochondria fuse during the differentiation of motile cells in the Blastocladiales? Attempts to isolate and determine the function of basal bodies and their associated rootlet structures (L. P. Gauriloff, personal communication) have been promising but much remains to be done by persons interested in structure-function relationships. Powell (personal communication) is attempting to determine the function of microbody-lipid complexes, a consistent feature of chytridiomycete zoospores. I look forward to someone's determining the function of the rumposome, a name I presented in jest at 1IFSS, and have had to live with for the ensuing 14 years. While I certainly do not question the existence of contractile vacuoles in the zoospores of biflagellates, and probably in the uniflagellates, we need more critical methods of separating the structure and function of contractile vacuoles from other vacuoles and membrane-bounded structures in zoospores. We know much about the levels of nucleic acid and protein synthesis in motile cells (Lovett, 1975), yet very few studies (see Suberkropp and Cantino, 1973) have been made of other biochemical events occurring in these cells. Although Held *et al* (1969) gave a convincing case for the lack of an aerobic system in several fermentative water molds, the structural data I provided for that paper is insufficient and unconvincing in 1980. Certainly, a student interested in structure-function relationships would find fertile ground for study among this group of water molds. What about the obligately anaerobic rumen chytrids (see Orpin, 1977)?

When the zoospore stops swimming, it quickly retracts or loses its flagellum and the cell that had been specialized for motility begins making nucleic acids, protein and other essential structural compounds (Lovett, 1975). The biochemists and molecular biologists are considerably ahead of the morphologists in terms of events occurring during encystment and spore germination. Recently, Barstow and Pommerville (1980) have examined cell wall formation in encysting zoospores of *Allomyces* and the possible role therein of gamma particles. Now that most of the aquatic fungi can be grown under controlled conditions (see Fuller, 1978), we need to examine the fate and function of other organelles present in the zoospore but not in the growing cells.

As growth proceeds and the fungus increases in diameter or develops a mycelial habit, many questions have been asked and some answers provided (see Grove, 1978; Gooday, 1978; and Gull, 1978), but much remains for a person interested in morphogenetic events occurring during the growth phase. While septum form and development is extremely important in the higher fungi (Gull, 1978), few studies of septa have been made among the aquatics. Are they all complete and associated with differentiation? I doubt it! More than any other studies that come to mind, and whatever the final outcome, research that beautifully exemplifies what may be gained from collaboration between morphologists and biochemists is that of Bartnicki-Garcia (see later chapter), Bracker, and Ruiz-Herrera with the chitosome.

Mitotic studies of aquatic fungi have proven valuable with respect to the relationships of these fungi and the general evolution and function of the mitotic apparatus in eukaryotes (Fuller, 1976; Heath, 1980a). I contend that the sequence of events occurring during mitosis is extremely useful in determining relationships and that there is a need for more studies of mitosis, particularly in the Monoblepharidales and Hyphochytriales. There is an urgent need for studies of the prophase events occurring during mitosis in the aquatic fungi. All too often the sequence we are presented with jumps from duplicated centrioles, that may or may not have migrated to opposite poles of the prophase nucleus, to metaphase figures with a fully developed spindle. Only in the studies of Heath (1980b) with biflagellates do we gain some insight into how the chromosomes become associated with the spindle during prophase. Again, synchronized cultures (see Olson and Fuller, 1971) offer the best material for such studies and we are presently having limited success with *Allomyces* and *Rhizidiomyces*.

The lower fungi offer considerable opportunity for examining the events occurring during the sexual process. Outstanding work on sexual hormones has been done with these fungi (Kochert, 1978), but there is a need for more work on the morphology and physiology of sexuality in lower fungi. Jeff Pommerville and I (1976) described the fusion of anisogametes in *Allomyces* and Pommerville (1978) continues experimental work with this system. Dr. Beakes, from whom we shall hear more later in this program, is moving rapidly with description of sexual events in oogamous forms. We can examine events ultrastructurally that would be missed with the light microscope. The synaptonemal complex is diagnostic for meiosis and can be a powerful technique for examining resting structures of many aquatics to determine the site of meiosis. Olson and his colleagues (Olson and Reichle, 1978) have used this technique extensively with blastocladiaceous fungi.

The production of zoospores and their subsequent release is what, for me, makes the aquatics so exciting. In 1978, many of my colleagues joined me in producing a book (Fuller, 1978) and workshop on *Lower Fungi in the Laboratory* to honor the late Ralph Emerson. Representatives of all of the major groups of aquatics can be grown axenically and induced to produce zoospores synchronously in the laboratory. This allows for concommitant studies of the structural and functional events occurring during spore formation and release. The modern demonstration of mini- and midicycles, possible in most of these groups, allows for studying spore differentiation in sporangia making one or a few spores (Hennessy and Cantino, 1972; Barstow and Lovett, 1974).

In *Blastocladiella* (Lovett, 1975), we know a good deal about the structural and molecular events associated with spore differentiation. This is not true for many other aquatics. On the basis of very few studies, namely one by Renaud and Swift (1964), we assume that flagellum formation and cleavage occur simultaneously. Yet, studies by McNitt (1974) and others (Fuller, 1977) suggest that there may be different sequences. Our own observations on *Chytridium* support this and more complete descriptions of spore formation in other genera are in order. Other structural changes that occur during spore formation are equally interesting. We (Taylor and Fuller, 1980) have completed structural studies of the Golgi in *Chytridium* during zoospore differentiation. This readily grown and easily controlled fungus (Gauriloff and Fuller, 1979) has considerable potential for studying functional and structural events occurring during spore differentiation.

Let me address briefly a few structural specializations that occur during spore differentiation. One organelle in blastocladiaceous zoospores is extremely well known (Myers and Cantino, 1974). Barstow (Barstow, 1979; Barstow and Lovett, 1975) has carefully documented its ontogeny in three genera and is pursuing its function. Mitochondrial fusions are common in the Blastocladiales with *Blastocladiella* and *Catenaria* having a single, larger mitochondrion. This fusion occurs late in spore development and needs to be more carefully documented. Nuclear caps such as those in the Blastocladiales, and ribosomal aggregations variously delimited by membranes as found in the Chytridiales, have been looked at only briefly with respect to development. What is the origin of the cap membranes? When and where does the cap form? Interestingly, in the *Chytridium* we are studying, the "cap" is not yet formed when the spores are in the vesicle just a few seconds prior to swimming, yet we know that the spores have "caps".

Let me turn briefly to spore release mechanisms. Structural data is slowly accumulating on the pores (opercula, papilla, or variations thereof) through which the spores are released (see Taylor and Fuller, 1980 for references). We (Taylor and Fuller, 1980) have looked carefully at operculum development in *Chytridium*. In only one case, *Allomyces* studied by Youatt (1976) has anything approaching the composition of the material associated with the release apparatus been reported. We shall hear more of the blastocladiaceous papilla during this meeting (see paper by Lovett). What about those fungi, e.g., *Rhizophydium patellarium*, that lack a papilla or operculum and in which large portions of the wall are dissolved?

Finally, I speak of a most exciting and little studied process, i.e., the exit of zoospores of uncleaved prezoospore masses into the discharge medium. Why are so few people studying this process? Clearly, water relations are involved (Hoch and Mitchell, 1973; Gisi *et al*, 1979). Were we all turned off to this subject by our courses in plant physiology? In 1977 (Fuller), I expressed a strong belief that contractile proteins are involved in protoplast release by *Rhizidiomyces* sp. and *Pythium* spp. Our own attempts to isolate such proteins from *Rhizidiomyces* have been frustrated by poor synchrony and our inexperience with the biochemical techniques involved. I hope that someone present or a person reading this later will find a challenge in the process of zoospore release since so many basic questions about the process are unanswered.

While I have touched upon just a few of the exciting problems that are being explored in the aquatic fungi, I hope I have set the stage for the speakers to follow and pointed to some problems that we need to be closer to solving by 4IFSS.

REFERENCES

Barr, D.J.S. (1978), Taxonomy and phylogeny of chytrids, *Biosystems* 10, 153-165.

Barstow, W.E. (1979), Ultrastructure of the formation of γ particles during zoosporogenesis in *Allomyces macrogynus* and *Catenaria anguillulae*, *Exper. Mycol.* 3, 28-41.

Barstow, W.E. and Lovett, J.S. (1974), Ultrastructure of a reduced developmental cycle (minicycle) in *Blastocladiella emersonii*, *Exper. Mycol.* 2, 145-155.

Barstow, W.E. and Lovett, J.S. (1975), Formation of gamma particles during zoosporogenesis in *Blastocladiella emersonii*, *Mycologia* 67, 518-529.

Barstow, W.E. and Pommerville, J. (1980), The ultrastructure of cell wall formation and gamma particles during encystment of *Allomyces macrogynus* zoospores, *Arch. Microbiol.* in the press.

Fuller, M.S. (1966), Structure of the uniflagellate zoospores of aquatic phycomycetes. *In* "The Fungus Spore" (Ed. M.F. Madelin), pp 67-84. Butterworth's, London.

Fuller, M.S. (1976), Mitosis in fungi, *Intern. Rev. Cytol.* 45, 113-153.

Fuller, M.S. (1977), The zoospore, hallmark of the aquatic fungi, *Mycologia* 69, 1-20.

Fuller, M.S. (Ed. 1978), "Lower Fungi in the Laboratory", pp i-ix, 1-213. Palfrey Contributions in Botany No 1. Department of Botany, University of Georgia, Athens.

Gauriloff, L.P. and Fuller, M.S. (1979), Morphological synchrony in axenic cultures of *Chytridium confervae*, a promising developmental system, *Exp. Mycol.* 3, 3-15.

Gisi, U., Hemmes, D.E., and Zentmyer, G.A. (1979), Origin and significance of the discharge vesicle in *Phytophthora*, *Exper. Mycol.* 3, 321-339.

Gooday, G.W. (1978), The enzymology of hyphal growth. *In* "The Filamentous Fungi" (Ed. J.E. Smith and D.R. Berry) Vol. 3, Developmental Mycology, pp 51-77. Halsted, New York.

Grove, S.N. (1978), The cytology of hyphal tip growth. *In* "The Filamentous Fungi" (Ed. J.E. Smith and D.R. Berry) Vol. 3, Developmental Mycology, pp 28-50. Halsted, New York.

Gull, K. (1978), Form and function of septa in filamentous fungi. *In* "The Filamentous Fungi" (Ed. J.E. Smith and D.R. Berry) Vol. 3, Developmental Mycology, pp 78-93. Halsted, New York.

Heath, I.B. (1980a), Variant mitoses in lower eukaryotes: Indicators of the evolution of mitosis? *Intern. Rev. Cytol.* 64, 1-80.

Heath, I.B. (1980b), The behavior of kinetochores during mitosis in the fungus *Saprolegnia ferax*, *J. Cell Biol.* 84, 531-546.

Held, A.A., Emerson, R., Fuller, M.S., and Gleason, F.H. (1969), *Blastocladia* and *Aqualinderella*: fermentative water molds with high carbon dioxide optima, *Science* 165, 706-709.

Hennessy, S.W. and Cantino, E.C. (1972), Lagphase sporogenesis in *Blastocladiella emersonii*: induced formation of unispored plantlets, *Mycologia* 64, 1066-1087.

Hoch, H.C. and Mitchell, J.E. (1973), The effects of osmotic water potentials on *Aphanomyces euteiches* during zoosporogenesis, *Canad. J. Bot.* 51, 413-420.

Kochert, G.D. (1978), Sexual pheromones in algae and fungi, *Ann. Rev. Plant Physiol.* 29, 461-486.

Lange, L. and Olson, L.W. (1979), The uniflagellate phycomycete zoospore, *Dansk Bot. Arkiv* 33, 1-95.

Lovett, J.S. (1975), Growth and differentiation of the water mold *Blastocladiella emersonii*: cytodifferentiation and the role of ribonucleic acid and protein synthesis, *Bacteriol. Rev.* 39, 345-404.

Manton, I., Clarke, B., and Greenwood, A.D. (1951), Observations with the electron microscope on a species of *Saprolegnia*, *J. Exp. Bot.* 2, 321-331.

Manton, I., Clarke, B., Greenwood, A.D., Flint, E.A. (1952), Further observations on the structure of plant cilia by a combination of visual and electron microscopy, *J. Exp. Bot.* 3, 204-215.

McNitt, R. (1974), Zoosporogenesis in *Phlyctochytrium irregulare*, *Cytobiologie* 9, 290-306.

Myers, R.B. and Cantino, E.C. (1974), The gamma particle. A study of cell-organelle interactions in the development of the water mold *Blastocladiella emersonii*. *In*

"Monographs in Developmental Biology" (Ed. A. Wolsky) Vol. 8, pp 1-117. Karger, Basel.

Olson, L.W. and Fuller, M.S. (1971), Leucine-lysine synchronization of *Allomyces* germlings, *Arch. Mikrobiol.* 78, 76-91.

Olson, L.W. and Reichle, R.E. (1978), Synaptonemal complex formation and meiosis in the resting sporangium of *Blastocladiella emersonii*, *Protoplasma* 97, 261-273.

Orpin, C.G. (1977), On the induction of zoosporogenesis in the rumen phycomycetes *Neocallimastix frontalis*, *Piromonas communis and Sphaeromonas communis*, *J. Gen. Microbiol.* 101, 181-189.

Pommerville, J. (1978), Analysis of gamete and zygote motility in *Allomyces*, *Expl. Cell Res.* 113, 161-172.

Pommerville, J. and Fuller, M.S. (1976), The cytology of the gametes and fertilization of *Allomyces macrogynus*, *Arch. Microbiol.* 109, 21-30.

Powell, M.J. (1978), Phylogenetic implications of the microbody-lipid globule complex in zoosporic fungi, *Biosystems* 10, 167-181.

Renaud, F.L. and Swift, A. (1964), The development of basal bodies and flagella in *Allomyces arbuscula*, *J. Cell Biol.* 23, 339-354.

Sparrow, F.K. (1978), Professor Anton deBary, *Mycologia* 70, 222-252.

Suberkropp, K.F. and Cantino, E.C. (1973), Utilization of endogenous reserves by swimming zoospores of *Blastocladiella emersonii*, *Arch. Mikrobiol.* 89, 205-221.

Taylor, J.W. and Fuller, M.S. (1980), The Golgi apparatus and development of the zoospore discharge apparatus in the fungus, *Chytridium confervae*, *Exp. Mycol.* 4, in the press.

Youatt, J. (1976), Sporangium formation in *Allomyces* throughout the growth cycle, *Trans. Br. mycol. Soc.* 67, 159-161.

FUNGAL ORGANELLES

AND OTHER CELL STRUCTURES

W. M. Hess

Department of Botany and Range Science
Brigham Young University, Provo, Utah, USA

An organelle is normally defined as a subcellular structure which is bounded by a limiting membrane. However, organelles may also be defined as structural differentiation of the cell or a structure which contains specific enzymes or are responsible for a particular function. The term implies a sub-cellular "organ" or structure which performs a function. Perhaps a limiting membrane should not be necessary in order to use the term since ribosomes are vital to the cell and are not membrane-bound. On the other hand, a membrane invagination would be a structure, but not an organelle.

CELL STRUCTURES OF FUNGI

At the second spore symposium many fungal organelles or structures were discussed. The organelles which are present in all fungal spores were discussed and included endoplasmic reticulum (ER), nuclei, mitochondria, ribosomes, and vesicles. Other organelles and structures such as lomasomes and vacuoles, microbodies, the apical corpuscle, unidentified bodies, side bodies, stüben bodies and symphyomicrobodies were also discussed. Since the last symposium some of these organelles and structures have been investigated in much more detail and a few structures have been described which were not previously known.

Phycomycete organelles and cell structures

The phycomycetous fungi in general have a much broader array of recognizable cell organelles and structures than the non-phycomycetous fungi. Dictyosomes are generally readily evident, microbodies are often reported, centrioles, basal bodies and microtubules are prominent, some spores have flagella, smooth and rough ER is commonly seen in abundance, many kinds of vesicles are evident, and there are many specialized structures such as rumposomes, gamma particles and an array of storage structures.

Membranes. Freeze fracture is a powerful tool to study membranes and has been used in some studies to demonstrate dynamics of membranes. Pinto da Silva and Nogueira (1977) used freeze fracture to demonstrate the dynamics of fusion of the peripheral vesicle of *Phytophthora palmivora* with the plasmalemma. This fusion occurred at several sites and resulted in the formation of a hemi-micelle which provided continuity between the inner half of the plasmalemma and the outer half of the peripheral vesicle membrane.

The use of freeze fracture has made it possible to better understand membrane interactions and membrane structure. As an example, Takeo (1974) demonstrated "rod-like" invaginations in the plasmalemma of a dormant sporangiospore of a mucor, and Sjöstrand (1979) recently demonstrated that there is no unique location of the fracture plane relative to membranes. He indicated that the layer that appears after etching is a layer of contamination formed by solutes freed by sublimation of water and the most common type of fracture face of membranes he has studied exposes the surfaces of membranes rather than their interior.

The plasmalemma of many kinds of spores elaborate to form plasmalemmal invaginations projecting into the cytoplasm. Sometimes they appear as a pocket containing a number of vesicles and tubules which may reflect the plane sectioned, and may be called paramural bodies. Khan (1976a) supported the hypothesis that lomasomes are paramural bodies that contain vesicles of cytoplasmic origin, and suggested that vesicles of plasmalemmal origin should be called vesicular plasmalemmasomes.

Plasmalemmasomes may be associated with vesicles in septum formation during sporangial development (Khan, 1977), they may have a variety of appearances in other fungal spores (Beakes, 1980a) or similar structures may be referred to as structures that resemble plasmalemmasomes (Jeffries and Young, 1976). In addition, fungi may also contain plasmodesmata. O'Donnell et al. (1976) briefly reviewed plasmodesmata of mucoraceous fungi.

Endoplasmic reticulum and dictyosomes. Endoplasmic reticulum and the Golgi apparatus are common features of phycomycetous fungal spores (Beakes, 1980b; Hemmes and Ribeiro, 1977; Powell, 1979). Smooth and rough ER may be seen in the same spore in some instances (Barr and Hadland-Hartmann, 1977), and ribosomes are readily evident during many stages of development. In addition to presence in other areas of the cytoplasm cisternae of ER are commonly adjacent to microbodies (Heath, 1977; Powell, 1979), and sometimes nuclei (Lunney and Bland, 1976b), and ER increased during sporangial development in *Albugo* (Khan, 1976b). The perinuclear vesicles and smooth surfaced cisternae also differentiated into a well-developed Golgi apparatus which continued to secrete until sporangia matured. Perhaps this is a typical example.

In many instances dictyosomes are distinct (Barr and Hadland-Hartmann, 1977; Beakes, 1980a) and may also be seen in close association with nuclei (Lunney and Bland, 1976a). Perhaps they could be visualized in most phycomycetous fungal spores if they were carefully looked for. During some developmental stages perhaps they are only one or a few membrane profiles.

Vesicles and vacuoles. Possibly one of the most interesting and least understood areas of investigation with fungal spores is vesicles and vacuoles. In many instances it is not clear whether the sub-cellular body is a vesicle or a vacuole. In some instances the differences are evident. When they are small and numerous and the function appears to be obvious, such as with wall vesicles (Beakes and Gay, 1978b) or when they are large and related to a specific cellular process such as the abscission vacuole of zygomycetes (Benny and Aldrich, 1975) the terms vesicles and vacuoles are appropriate. However, when vesicles and vacuoles appear to be similar in size and appear to be distinguished by whether the content is "crystalline" or "granular" (Clerk, 1974) it becomes evident that there is a need to be able to define these terms and standardize the terminology. Perphaps histochemistry and x-ray microanalysis could be used to help elucidate these problems.

As examples it would be interesting to know the differences in chemical nature and specific function of central vacuoles, peripheral vacuoles, storage bodies, granular bodies and central globules (Ruben and Stanghellini, 1978) or reserve globules, lipid vesicles, reserve vesicles, and vacuoles (Tewari and Skoropad, 1977). Secretory vesicles, multivesicular bodies, fibrillar vesicles, (Khan, 1976b) crystalline vesicles with lipid, granular vesicles with protein (Bimpong and Hickman, 1975) dense body vesicles, finger-print vacuoles (Traquair and McKeen, 1980) and the various types of vesicles which have been tabulated (Beakes, 1980a; Lunney and Bland, 1976b) need to be better understood.

Microbodies. Although it is a common practice for investigators to label microbodies on micrographs, not all of these labeled structures are investigated for enzyme activity since few investigators use histochemistry to elucidate function. Also, it is difficult to demonstrate the presence of catalase, the microbody enzyme, in fungi (Beakes, 1980b; Maxwell et al., 1977), although Powell (1979) demonstrated cytochemical localization of catalase activity in zoosporangia of *Entophlyctis*. She studied three dimensional reconstruction from serial sections and demonstrated that the morphology of microbodies and spatial association with other organelles varied during development and suggested that microbodies originated as local dilations in tubular ER, as lateral buds from opposite sides of ER cisternae and as fragments from elongate microbodies.

Maxwell et al. (1977) reviewed microbodies of fungi and stated that they are apparently present in all eucaryotes, including fungi and the terms glyoxysome and peroxisome refer to microbodies with demonstrated metabolic functions. These authors pointed out that cytosome, stüben body, and symphyomicrobody are examples of other terms used for microbodies. Traquair and McKeen (1980) used the term peroxisome-like body, and Lunney and Bland (1976b) stated that they were also called parastrasomes and x-bodies.

It was also pointed out by Maxwell et al. (1977) that the enzyme content seems to be profoundly affected by different conditions such as the developmental stage of the fungus and the carbon source upon which it is grown, and although function is not entirely known they do have a role during sporulation of fungi and during colonization of host tissue.

Lipid bodies. Lipids in the cytoplasm of spores are normally called lipid bodies, lipid droplets, lipid vesicles (Tewari and Skoropad, 1977) and lipid-like bodies (Baker et al., 1977; Hemmes and Bartnicki-Garcia, 1975) and they vary in morphological appearance depending upon the composition of the lipid, fixation procedures, methods of staining and embedding, and photographic techniques. They may be difficult to distinguish from vacuoles (Khan, 1975) or other cellular components (Baker, 1979), and may appear to have multilayered shells in freeze-etch replicas (Takeo, 1974).

Lipid bodies may be surrounded by or be in juxtaposition to membranous cell structures including microbodies, which in turn may be surrounded or partially surrounded by another membranous complex (Chong and Barr, 1974), they may be partially surrounded by mitochondria (Powell, 1979), they may be within crystalline vesicles (Bimpong and Hickman, 1975) depending upon interpretation, or more commonly they are free in the cytoplasm.

Distribution of neutral lipids and dense body granules during oospore and oosphere maturation with *Saprolegnia* was studied by Beakes and Gay (1978a). The lipids dispersed in the cytoplasm, they became centrally concentrated in the cytoplasm, they then again dispersed in the cytoplasm, then they moved toward the periphery. A similar phenomenon was observed by Traquair and McKeen (1980) in their investigations with oogonia and oospores of *Aphanomyces*.

In a later investigation with *Saprolegnia* oospheres Beakes (1980b) demonstrated a steady accumulation of lipid and dense-body (lipo-phospho glucan complexes) reserves, preceded by a marked increase in lipid-associated microbodies (glyoxysomes) which appeared to develop from smooth ER. The neutral lipid fraction was partially used before germ tubes emerged, indicating that the lipid is a source of energy.

In zoospores of *Entophlyctis* (Powell, 1976) ER occurred on the surface of lipid bodies toward the exterior, and the microbody, subtended by mitochondria, was appressed to the opposite surface of the lipid body. As the zoospore swam and encysted, lipid bodies coalesced and the microbody-lipid globule became disorganized. After the lipid body coalesced, the microbody-lipid complex regained its order and several microbodies were clustered adjacent to a single lipid body. The microbodies persisted and were found on all sides of the lipid body. It was suggested that the microbody-lipid complex appears to be a "survival package" until substrate can be utilized.

It was shown by Barr and Hadland-Hartmann (1978) that zoospores of *Rhizophydium* and *Chytridium* have a lipid globule complex compartmentalized into the core of the zoospore by a double membrane system, and 2-5 microtubules connect one side of the kinetosome to the rumposome on the lipid globule surface. This anchors the lipid globule in a lateral-posterior position in the zoospore. These authors briefly reviewed *Chytridium-Rhizophydium* zoospore structure and indicated that in spite of variations in the structure of the lipid globule complex a microbody and mitochondrion are associated with the lipid globule. This indicates that the lipid globule is a source of energy.

Mitochondria. Mitochondria are associated with microbodies and ER which are spatially associated with nuclei and dictyosomes in developing sporangia (Powell, 1979). In serial section reconstructions of zoospores the several mitochondria seen in a single section may be parts of a single complex mitochondrion associated with lipid bodies, ER, and the nucleus (Lange and Olson, 1976a). These authors stated that there are three arrangements of mitochondria in zoospores of uniflagellate phycomycetes; a single mitochondrion usually associated with the kinetosome located in the posterior end of the zoospore, numerous small mitochondria randomly distributed along with a larger mitochondrion associated with the kinetosome, and a random distribution of mitochondria with no mitochondrion-kinetosome association. Lunney and Bland (1976b) indicated that mitochondria in mature swimming zoospores of *Pythium* are larger than in immature zoospores. In zoospores mitochondria may vary from 1 to more than 30 in various species of the genus *Rhizophydium* (Barr and Hadland-Hartmann, 1978).

In sporangia of *Albugo* (Khan, 1976b) numbers of mitochondria and amounts of ER increased during maturation, and near the end of sporangial maturation the number of mitochondria and the amount of ER decreased. In oospores of *Aphanomyces* (Traquair and McKeen, 1980) peripheral mitochondria are associated with fingerprint vacuoles, lomasomas, and wall vesicles and it was suggested that the dense-body vacuoles or fingerprint vacuoles are a source of energy for oospore-wall synthesis. Beakes (1980b) discussed mitochondrial number, appearance associated with function and distribution in the oospore during maturation, along with the changes evident in other cellular organelles and structures, and summarized his observations in a diagrammatic table. It is obvious that mitochondria are involved in a complex and well-controlled process in fungal spores.

Nuclei. In many ultrastructural investigations of fungal spores of phycomycetous fungi, nuclei are labeled for identification, an unusual observation such as an inclusion (Hemmes and Bartnicki-Garcia, 1975) may be noted, but little other attention is normally given to them. However, in studies with some fungi number and location are important (Khan, 1976b) and structures associated with nuclei in some spores such as bars, pairing of axial elements, synaptonemal complexes, centriole, centriolar cap, (Traquair and McKeen, 1980) and microtubules (Khan 1976b) are discussed, although microtubules not associated with nuclei and flagella are also reported (Powell, 1979).

Flagella. Zoospores have interesting and complex ultrastructure, particularly when the complexity of flagella, and structures associated with flagella, are studied. A flagellum may consist of a rhizoplast (Lange and Olson, 1976b) composed of fine fibrils, running lengthwise from the kinetosome and connected to a functional or non-functional centriole. The kinetosome may have a terminal plate and props may radiate from the kinetosome (Barr and Hartmann, 1977). In addition, mastigonemes which form within cisternae of the perinuclear continuum and move into the groove region may be dumped to the periphery of a zoospore for subsequent flagellar attachment and they may also be located adjacent to mitochondria (Lunney & Bland, 1976a and b).

Non-phycomycete organelles and cell structures

For the most part non-phycomycetous fungal spores have a relatively simple cell structure compared to phycomycete spores except for the myxomycetes, which have some interestin and complex stages in spore development and spore germination, although relatively few ultrastructural studies of myxomycete spores have been done. Some myxomycetes also have interesting and apparently unique organelles or structures such as digestive vacuoles and doughnut-shaped structures bounded by a thick unit membrane (Farley et al., 1975).

Membranes. Spines of rust teliospores, in at least some genera, form in invaginations of the plasmalemma and are subsequently exserted when a secondary wall is laid down beneath them (Bennell and Henderson, 1978; Bennell et al., 1978; Mims and Thurston, 1979).

Invaginations are prominent features of the plasma membranes of most fungal spores. Sekiguchi et al. (1975a) studied conidiation in *Penicillium* with freeze fracture and thin sectioning techniques and reported that the plasma membranes of immature conidia formed complex depressions which were round, ring-shaped, or irregular "crack-shaped" linear structures. The round depressions were also seen in heat-treated conidia before and after germination, the "ring-shaped" depressions were rarely seen in germinating conidia and the "crack-shaped" structures were only seen during germination. Mature conidia had much more developed indentations and particulate elements associated with them. Mature spores had trench-like depressions (Sekiguchi et al., 1975b) which were visible in freeze-etch replicas and thin sections. Invaginations vary in appearance depending upon the organism examined (Caputo et al., 1977; Casperson and Purz, 1974).

In investigations with germinating sugar cane smut teliospores Paine and Hess (1980) reported that invaginations averaged 418 nm in width when spores were not fixed or glycerol treated, 732 nm when spores were fixed without glycerol treatment, and 1189 when spores were fixed and glycerol treated, and germ tubes of spores not fixed or glycerol treated had wrinkled membranes at the tip or along the entire germ tube. This indicates that treatment before freeze fracture can significantly affect the morphological appearance of the membrane.

Both invaginations and lomasomes may be reported for some fungal spores (Hawes, 1980) and vesicles and invaginations associated with the plasma membrane may be visualized with freeze-etch replicas (Steele and Miller, 1974). As was mentioned for phycomycetous fungi, lomasomes (Hanlin, 1976), lomasome-like structures (Garrison and Boyd, 1977) paramural bodies (Beckett, 1979; Hackett and Chen, 1976; Jones, 1976) and plasmalemmasomes (Madelin and Ogunsanya, 1979) are also common features of non-phycomycetous fungi. Myelin figures in the cytoplasm and microtubules not associated with nuclear division have also been reported (Carroll and Carroll, 1974), although Tewari and Skoropad, (1975) reported that with macroconidia of *Fusarium* routinely fixed, conidia had myelin figures which were not seen in replicas of conidia freeze-etched without the use of chemical fixation and cryoprotection.

Mesosomes (Hardisson and Manzanal, 1976) and intracytoplasmic membrane systems (Soina, et al., 1977) have been reported for some species of *Streptomyces* and strains of *Fusidium* (Lechevalier et al., 1977), and complex membranous configurations have been reported for the shitake mushroom *Lentinus* (Nakai, 1978). Mesosomes are distinct from multivesiculate structures reported for some fungal spores (Hackett and Chen, 1976). Although mesosomes appear to be involved in cell division, no function is presently known for multivesiculate structures.

Microbodies and Woronin bodies. Microbodies (Gull and Newsam, 1975; Mims, 1977; Ton That and Turian, 1978) and Woronin bodies (Ellis, 1978; Hanlin, 1976; Jones, 1976) have been reported for non-phycomycetous fungi, and Maxwell et al. (1977) stated that Woronin bodies, found near septa of certain Ascomycotina resemble the inclusions observed in the microbodies of the same organisms and are surrounded by similar membranes. These authors indicated that Woronin bodies appear to arise by pinching off of inclusions from microbodies, that the matrix of Woronin bodies and microbody inclusions is not extracted by lipid solvents, and both are digested by proteases.

Lipids. Lipid bodies may be small and numerous (Kollmorgen et al., 1979; Madelin and Ogunsanya, 1979), arranged primarily along the cell periphery (Tewari and Skoropad, 1975), spores may have low or high lipid content from "low and high reserve" spores respectively (Van Eck, 1978), they may be in layers (Hohl and Streit, 1975) associated with ER (Garrison and Lane, 1974), or structures which appear to be lipid bodies may be labeled as osmiophilic bodies (Jones, 1976). Staining methods are now available to help identify lipid bodies both with light microscopy and electron microscopy (Wigglesworth, 1976).

Lipid bodies may be enclosed in vacuoles (Mims, 1977), surrounded by half-unit membranes (Grayson and Lacy, 1975), enclosed in paired unit membranes (Stevenson and Becker, 1979; Clough et al., 1977) or may have a myelin-like appearance (Carroll and Carroll, 1974; Hassan and Littlefield, 1979). Gardner and Hess (1977) reported a lamellar substructure with 6 to 10 nm center-to-center spacing in lipids of *Tilletia* teliospores which was reflected in the irregular fracture patterns visible with freeze-etch replicas. It is likely that the differences in appearance of lipid bodies are reflected by chemical composition (Gunasekaran et al., 1974; Johnson et al., 1976).

Other organelles and structures. Most spores have ER although the amount varies with the developmental stage, the type of spore, and the metabolic state of the spore. In some instances abundant smooth-membraned ER is seen (Gull and Newsam, 1975), although normally only a few profiles are evident (Beckett, 1979). Unusual membranes are also seen, such as those reported by Harder (1976) which he called a "crochet-like" pattern.

Few non-phycomycetous fungal spores appear to have dictyosomes, although some have an impressive Golgi apparatus (Ellis and Griffiths, 1976). Madelin and Ogunsanya (1979) reported compact orderly stacks of membranes, sometimes in each locule of *Botryodiplodia* conidia. Perhaps these stacks of membranes function as dictyosomes, since it is likely that dictyosomes consist of only one or two, or at best, a few membrane profiles in most non-phycomycetous spores.

Mitochondria vary significantly in shape, size, density, and appearance, but are normally easily recognized (Beckett, 1976a), although they may also be associated with other cell organelles (Carroll and Carroll, 1974); and nuclei normally receive little attention unless studies are involved with nuclear division when other structures associated with nuclei are also mentioned, such as microtubules (Stiers, 1976) and the spindle-pole body (Mims, et al., 1976; Poon and Day, 1976).

For most organelles and structures in fungal spores there are suggested or demonstrated functions. Vesicles or vacuoles with electron dense inclusions (Beckett, 1976b) have been evident in micrographs for many years, but have received little attention. Recently White and Brown (1979) used energy-dispersive X-ray microanalysis to analyze similar structures in a vesicular-arbuscular mycorrhizal fungus and concluded that it was high in concentrations of phosphorus and calcium. They suggested that the inclusions were polyphosphate granules and indicated that there were two other types of cellular inclusions, one vacuolar and one cytoplasmic, that could not be identified by X-ray microanalysis, but could be easily mistaken for phosphorous granules.

POSSIBLE APPROACHES TO INVESTIGATIONS

Although significant amounts of new information have been obtained during the last two decades it is likely that an even greater accumulation of new information, particularly as related to function, will be possible in the next two decades. The availability of new technology and the sophistication of the technology becoming available will make this possible. Hess (1978, 1979) briefly reviewed examples of investigative procedures for fungal spores.

X-ray microanalysis

There are two methods which are commonly used for X-ray microanalysis of biological tissue: energy-dispersive and wavelength-dispersive (microprobe). Wavelength-dispersive requires scanning of several crystals through a range of angles to obtain analyses for several elements. Each setting requires additional analysis, which is likely to damage biological specimens. With energy-dispersive analysis several energies can be simultaneously processed. Relatively brief reviews of X-ray microanalysis in biology were published by Zs.-Nagy et al. (1977), Echlin (1977) and Duncumb (1979).

Wavelength-dispersive analysis has been used to study fungal structures. Jones et al. (1977) used these procedures to determine that phosphorus, sulphur, calcium, chlorine and magnesium were present in hyphal and spore walls of *Cunninghamella*. Also, it was reported by Ho (1979) that oogonial walls of species of *Phytophthora* had similar emission spectra with a characteristic calcium peak, and mycorrhizal roots fixed with glutaraldehyde-ferric chloride had electron-dense particles adjacent to external hyphae which bound iron (Holley and Petterson, 1979). Histochemical localization of heavy metals greatly facilitates analysis.

Energy-dispersive analysis has also been used in several studies which have involved fungi. Doonan et al. (1979) developed methods for analysis by using air-dried *Aureobasidium* blastospores which were unfixed and nonembedded, and demonstrated three types of polyphosphate bodies, based upon peak-to-background ratios of X-rays. Whole spore analysis of untreated spores gives much more consistent data than analysis when spores are treated or processed for thin sectioning (Hess, unpublished data). Thin sectioning-histochemical procedures need to be worked out so consistent data can be obtained.

For energy-dispersive analysis many attempts have been made to obtain reliable data with diffusible substances. It was pointed out by Appleton (1979) that qualitative and quantitative information can be obtained, at a resolution which is roughly comparable to the resolution of autoradiography, with frozen section procedures and it was suggested that cryoprotectants should not be used because they cause diffusible substances to diffuse.

In a number of studies precipitation with antimonate has been used. However, at least in some instances, antimonate does not precipitate with sodium, and calcium is trapped only after serious dislocation (Van Iren et al., 1979), which indicates that for other than whole spores which are not pre-treated it will be difficult to obtain reliable information.

Air-dried conidia and conidiophores of *Erysiphe* were analyzed by Kunoh and Ishizaki (1980). These investigators were able to get prominent peaks for elements which were not detected in chemically fixed specimens. Similar results were reported by Hess (1979) with teliospores of *Tilletia*. By using non-treated whole spores it was possible to use elemental composition to distinguish between species of *Tilletia* which are morphologically identical.

It appears that critical point drying promotes the redistribution of normal intracellular elements, but that rapid freezing preserves normal distribution of mobile elements such as sodium and calcium (Parrish et al., 1979). Perhaps the recently developed cryochambers (available from SEM supply sources) will help to elucidate these problems since they will, in theory, make it possible to rapidly freeze tissue, fracture the tissue while it is frozen, and shadow and analyze the tissue while it is frozen.

Scanning electron microscopy

In addition to cryochambers, mentioned above, examples of other SEM procedures are to freeze-fracture fungal structures in liquid nitrogen, with or without cryoprotectants, critical point dry and examine the structures in the SEM (O'Donnell and Hooper, 1977; Flegler and Hooper, 1978). The charging effect can be reduced by use of the methods of Kubotsu and Ueda (1980) which involve successive exposure of a specimen to osmium tetroxide and hydrazine hydrate vapors. Freeze-etch replicas can also be examined with SEM (Nei et al., 1974). Although SEM resolution is relatively limited it provides a means to examine large replicas where relatively low magnifications are suitable.

Freeze-fracture

Freeze-fracture procedures have made it possible to study dormant fungal spores which could not be examined by conventional thin sectioning procedures until special procedures were developed (Gardner et al., 1975), at least for spores where fixation solutions and embedding resins do not penetrate due to impervious spore walls. Freeze-fracture technology is now much better understood (Rash and Hudson, 1979), and although not yet applied to fungal spores, freeze-fracture autoradiography is also possible (Rix et al., 1976; Schiller et al., 1979). Recent developments with high resolution scintautoradiography (Fischer, 1979), and better freezing methods (Frederik and Busing, 1979), should help to make freeze-fracture autoradiography an even more effective procedure.

Although fungal spores, particularly dormant spores, are relatively easy to freeze-fracture the availability of propane jet freezing (Moor et al., 1976) or the new cryo-jet freezing device (Balzers, Balzers, Lichenstein) should make it possible to get even more reproducibility of information in freeze fracture replicas. In addition, the use of stereology, aids in interpretation (Allen et al., 1978; Hess and Allen, 1978) and provides a means to determine quantities in three-dimensional sample space which can be verified by other methods (Underwood, 1979).

Summary

For many fungal organelles and sub-cellular structures there are a number of different terms which are used, and lucid definitions for many terms are not available. For example, there is a significant disparity in perception of vesicles and vacuoles. There is a need to have lucid definitions for all organelles and sub-cellular structures and a plea for international acceptance of the terms. In addition, there is a need to develop standard histochemical, analytical or other procedures so cell organelles and sub-cellular structures can be readily identified by standard procedures. It is recognized that much careful research will be necessary to accomplish these suggested objectives, but if these objectives can be accomplished it will significantly simplify the publication of investigations with fungal spores and will help to reduce confusion.

REFERENCES

Allen, J. V., Gardner, J. S. and Hess, W. M. (1978). Pseudoscopic illusions in stereo views of freeze-etch replicas. *Protoplasma* 93, 473-475.

Appleton, T. C. (1979). The localization of diffusible substances. Experience in the electron microscope. *Journal of Histochemistry and Cytochemistry* 27, 1518-1519.

Baker, K. K. (1979). Ultrastructural development of merosporangia in the mycoparasite *Piptocephalis indica*. *Canadian Journal of Botany* 57, 2460-2465.

Baker, K. L., Hooper, G. R. and Beneke, E. S. (1977). Ultrastructural development of merosporangia in the mycoparasite *Syncephalis sphaerica* (Mucorales). *Canadian Journal of Botany* 55, 2077-2215.

Barr, D. J. S. and Hadland-Harmann, V. E. (1977). Zoospore ultrastructure of *Olpidium cucurbitacearum* (Chytridiales). *Canadian Journal of Botany* 55, 3063-3074.

Barr, D. J. S. and Hadland-Hartmann, V. E. (1978). Zoospore ultrastructure in genus *Rhizophydium* (Chytridiales). *Canadian Journal of Botany* 56, 2380-2404.

Barr, D. J. S. and Hartmann, V. E. (1977). Zoospore ultrastructure of *Olpidium brassicae* and *Rhizophlyctis rosea*. *Canadian Journal of Botany* 55, 1221-1235.

Beakes, G. W. (1980a). Electron microscopic study of oospore maturation and germination in an emasculate isolate of *Saprolegnia ferax*. 2. Wall differentiation. *Canadian Journal of Botany* 58, 195-208.

Beakes, G. W. (1980b). Electron microscopic study of oospore maturation and germination in an emasculate isolate of *Saprolegnia ferax*. 3. Changes in organelle status and associations. *Canadian Journal of Botany* 58, 209-227.

Beakes, G. W. and Gay, J. L. (1978a). Light and electron microscopy of oospore maturation in *Saprolegnia furcata*. 1. Cytoplasmic changes. *Transactions of the British Mycological Society* 71, 11-24.

Beakes, G. W.and Gay, J. L. (1978b). Light and electron microscopy of oospore maturation in *Saprolegnia furcata*. 1. Cytoplasmic changes. *Transactions of the British Mycological Society* 71, 25-35.

Beckett, A. (1976a). Ultrastructural studies on exogenously dormant ascospores of *Daldinia concentrica*. *Canadian Journal of Botany* 54, 689-697.

Beckett, A. (1976b). Ultrastructural studies on germinating ascospores of *Daldinia concentrica*. *Canadian Journal of Botany* 54, 698-705.

Beckett, A. (1979). Ultrastructure and development of the ascospore germ slit in *Xylaria longipes*. *Transactions of the British Mycological Society* 72, 269-276.

Bennell, A. P. and Henderson, D. M. (1978). Urediniospore and teliospore development in *Tranzschelia* (Uredinales). *Transactions of the British Mycological Society* 71, 271-278.

Bennell, A. P., Henderson, D. M. and Prentice, H. T. (1978). The teliospores of *Puccinia smyrnii* (Uredinales). *Grana* 17, 17-27.

Benny, G. L. and Aldrich, H. C. (1975). Ultrastructural observations on septal and merosporangial ontogeny in *Linderina pennispora* (Kickxellales; zygomycetes). *Canadian Journal of Botany* 53, 2325-2335.

Bimpong, C. E. and Hickman, C. J. (1975). Ultrastructural and cytochemical studies of zoospores, cysts, and germinating cysts of *Phytophthora palmivora*. *Canadian Journal of Botany* 53, 1310-1327.

Caputo, R., Innocenti, M., and Shimono, M. (1977). Cell wall and plasma membrane of dermatophytes. A freeze-fracture study. *Journal of Ultrastructure Research* 59, 149-157.

Carroll, G. C. and Carroll, F. E. (1974). The fine structure of conidium development in *Phialocephala dimorphospora*. *Canadian Journal of Botany* 52, 2119-2128.

Casperson, G., and Purz, H. J. (1974). Gefrierätzung an den Hefen *Saccharomyces*, *Torulopsis* und *Rhodotorula*. *Zeitschrift für Allgemeine Mikrobiologie* 14, 645-654.

Chong, J. and Barr, D. J. S. (1974). Ultrastructure of the zoospores of *Entophlyctis confervaeglomeratae*, *Rhizophydium patellarium* and *Catenaria anguillulae*. *Canadian Journal of Botany* 52, 1197-1204.

Clerk, G. C. (1974). Ultrastructure of wall swellings of germinating sporangia of *Phytophthora palmivora* (Butl.) Butl. *Annals of Botany* 38, 1103-1106.

Clough, K. S., Schneider, E. F., and Seaman, W. L. (1977). Cellular bodies in developing chlamydospores of *Thielaviopsis basicola*. *Canadian Journal of Microbiology* 23, 325-330.

Doonan, B. B., Crang, R. E., Jensen, T. E. and Baxter, M. (1979). In situ X-ray energy dispersive microanalysis of polyphosphate bodies in *Aureobasidium pullulans*. *Journal of Ultrastructure Research* 69, 232-238.

Duncumb, P. (1979). X-ray microanalysis. *Journal of Microscopy* 117, 165-174.

Echlin, P. (1977). Microanalysis in biology and medicine. *Nature* 270, 102-103.

Ellis, D. H. (1978). Fine structure of conidia in *Trichocladium opacum* (syn. *Torula allii*). *Transactions of the British Mycological Society* 70, 67-71.

Ellis, D. H. and Griffiths, D. A. (1976). The fine structure of conidial development in the genus *Torula*. III. *T. graminis* Desm. *Canadian Journal of Microbiology* 22, 858-866.

Farley, F. E., Crang, R. E. and Brent, M. M. (1975). Ultrastructure of the germinating spores of *Physarella oblonga* F. *alba*. *Transactions of the American Microscopical Society* 94, 136-141.

Fischer, H. A. (1979). High resolution scintiautoradiography. *Journal of Histochemistry and Cytochemistry* 27, 1527-1528.

Flegler, S. L. and Hooper, G. R. (1978). Ultrastructure of *Cyathus stercoreus*. *Mycologia* 70, 1181-1190.

Frederik, P. M., and Busing, W. M. (1979). Freezing of tissue-limits for the autoradiographic localization of diffusable substances. *Journal of Histochemistry and Cytochemistry* 27, 1520-1523.

Gardner, J. S., Allen, J. V. and Hess, W. M. (1975). Fixation of dormant *Tilletia* teliospores for thin sectioning. *Stain Technology* 50, 347-350.

Gardner, J. S. and Hess, W. M. (1977). Ultrastructure of lipid bodies in *Tilletia caries* teliospores. *Journal of Bacteriology* 131, 662-671.

Garrison, R. G. and Boyd, K. S. (1977). The fine structure of microconidial germination and vegetative cells of *Histoplasma capsulatum*. *Annales de Microbiologie* 128, 135-149.

Garrison, R. G. and Lane, J. W. (1974). The fine structure of the microconidium of *Blastomyces dermatitidis*. *Mycopathologia et Mycologia applicata* 52, 93-100.

Grayson, R. L. and Lacy, M. L. (1975). Development and nuclear history of the teliospores of *Urocystis colchici*. *Phytopathology* 65, 994-999.

Gull, K. and Newsam, R. J. (1975). Ultrastructural organization of cystidia in basidiomycete *Agrocybe praecox*. *Journal of General Microbiology* 91, 74-78.

Gunasekaran, M., Hess, W. M. and Weber, D. J. (1974). Lipids and ultrastructure of *Phymatotrichum omnivorum*. *Transactions of the British Mycological Society* 63, 519-525.

Hackett, C. J. and Chen, K. (1976). Ultrastructure of developing ascospores in *Sordaria brevicollis*. *Journal of Bacteriology* 126, 883-894.

Hanlin, R. T. (1976). Phialide and conidium development in *Aspergillus clavatus*. *American Journal of Botany* 63, 144-155.

Harder, D. E. (1976). Electron microscopy of urediospore formation in *Puccinia coronata avenae* and *P. graminis avenae*. *Canadian Journal of Botany* 54, 1010-1019.

Hardisson, C. and Manzanal, M. B. (1976). Ultrastructural studies of sporulation in *Streptomyces*. *Journal of Bacteriology* 127, 1443-1454.

Hassan, Z. M., and Littlefield, L. J. (1979). Ontogeny of the uredium of *Melampsora lini*. *Canadian Journal of Botany* 57, 639-649.

Hawes, C. R. (1980). Conidium germination in the Chalara state of *Ceratocystis adiposa*: a light and electron microscope study. *Transactions of the British Mycological Society* 74, 321-328.

Heath, I. B. (1977). Morphology and ontogeny of microbodies in the oomycete fungus *Sapromyces elongatus*. *Canadian Journal of Botany* 55, 556-562.

Hemmes, D. E. and Bartnicki-Garcia, S. (1975). Electron microscopy of gametangial interaction and oospore development in *Phytophthora capsici*. *Archives of Microbiology* 103, 91-112.

Hemmes, D. E. and Ribeiro, O. K. (1977). Electron microscopy of early gametangial interaction in *Phytophthora megasperma* var. *sojae*. *Canadian Journal of Botany* 55, 436-447.

Hess, W. M. (1978). Ultrastructure of dried spores. *In* Dry Biological Sytems. (Ed. J. H. Crowe and J. S. Clegg). pp. 75-96. Academic Press, New York.

Hess, W. M. (1979). Electron microscopy and plant sciences emphasis upon fungal spores, *In* Proceedings of the Electron Microscopy Society of America (Ed. G. W. Bailey) pp. 168-171. Claitor's Publishing Div., Baton Rouge.

Hess, W. M. and Allen, J. V. (1978). The practice and use of stereo pair photography in TEM. Examples of interpretation of freeze etch replicas. *Norelco Reporter* 25, 26-33.

Ho, H. H. (1979). Electron microscope studies of oogonial wall and elemental composition of oogonia in *Phytophthora*. *Mycopathologia* 68, 17-21.

Hohl, H. R. and Streit, W. (1975). Ultrastructure of ascus, ascospore and ascocarp in *Neurospora lineolata*. *Mycologia* 67, 367-381.

Holley, J. D. and Peterson, R. L. (1979). Development of a vesicular-arbuscular mycorrhiza in bean roots. *Canadian Journal of Botany* 57, 1960-1978.

Jeffries, P. and Young, T. W. K. (1976). Physiology and fine structure of sporangiospore germination in *Piptocephalis unispora* prior to infection. *Archives of Microbiology* 107, 99-107.

Johnson, D., Weber, D. J., and Hess, W. M. (1976). Lipids from conidia of *Erysiphe graminis tritici* (Powdery mildew). *Transactions of the British Mycological Society* 66, 35-43.

Jones, D., McHardy, W. J., Farmer, V. C. and Jones, Y. (1977). Electron probe microanalysis of cell walls of *Cunninghamella echinulata*. *Transactions of the British Mycological Society* 69, 71-75.

Jones, J. P. (1976). Ultrastructure of conidium ontogeny in *Phoma pomorum*, *Microsphaeropsis olivaceum*, and *Coniothyrium fuckelii*. *Canadian Journal of Botany* 54, 831-851.

Khan, S. R. (1975). Wall structure and germination of spores in *Cunninghamella echinulata*. *Journal of General Microbiology* 90, 115-124.

Khan, S. R. (1976a). Paramural bodies of *Albugo candida*. *Archives of Microbiology* 109, 271-275.

Khan, S. R. (1976b). Ultrastructural changes in maturing sporangia of *Albugo candida*. *Annals of Botany* 40, 1285-1292.

Khan, S. R. (1977). Light and electron microscopic observations of sporangium formation *Albugo candida* (Peronosporales; oomycetes). *Canadian Journal of Botany* 55, 730-739.

Kollmorgen, J. F., Hess, W. M. and Trione, E. J. (1979). Ultrastructure of primary sporidia of a wheat-bunt fungus, *Tilletia caries*, during ontogeny and mating. *Protoplasma* 99, 189-202.

Kubotsu, A. and Ueda, M. (1980). A new conductive treatment of the specimen for scanning electron microscopy. *Journal of Electron Microscopy* 29, 45-53.

Kunoh, H. and Ishizaki, H. (1980). X-ray microanalysis of air-dried conidia and conidiophores of *Erysiphe graminis hordei*. *Canadian Journal of Botany* 58, 133-135.

Lange, L. and Olson, L. W. (1976a). The zoospore of *Olpidium brassicae*. *Protoplasma* 90, 33-45.

Lange, L. and Olson, L. W. (1976b). The flagellar apparatus and striated rhizoplast of the zoospore of *Olpidium brassicae*. *Protoplasma* 89, 339-351.

Lechevalier, H., Lechevalier, M. P., Handley, D. A., Ghosh, B. K. and Carmichael, J. W. (1977). Strains of *Fusidia* which can be mistaken for actinomycetes. *Mycologia* 69, 8195.

Lunney, C. Z., and Bland, C. E. (1976a). An ultrastructural study of zoosporogenesis in *Pythium proliferum* de Bary. *Protoplasma* 88, 85-100.

Lunney, C. Z. and Bland, C. E. (1976b). Ultrastructural observations of mature and encysting zoospores of *Pythium proliferum* de Bary. *Protoplasma* 90, 119-137.

Madelin, M. F. and Ogunsanya, O. C. (1979). The fine structure of conidia of *Botryodiplodia ricinicola* with observation on effects of chilling. *Annals of Botany* 44, 417-425.

Maxwell, D. P., Armentrout, V. N. and Graves, Jr., L. B. (1977). Microbodies in plant pathogenic fungi. *Annual Review of Phytopathology* 15, 119-134.

Mims, C. W. (1977). Fine structure of basidiospores of the cedar-apple rust fungus *Gymnosporangium juniperi - virginianae*. *Canadian Journal of Botany* 55, 1057-1063.

Mims, C. W., Seabury, F. and Thurston, E. L. (1976). An ultrastructural study of spermatium formation in the rust fungus *Gymnosporangium juniperi - virginianae*. *American Journal of Botany* 63, 997-1002.

Mims, C. W. and Thurston, L. (1979). Ultrastructure of teliospore formation in the rust fungus *Puccinia podophylli*. *Canadian Journal of Botany* 57, 2533-2538.

Moor, H., Kistler, J. and Müeller, M. (1976). Freezing in a propane jet. *Experientia* 32, 805.

Nakai, Y. (1978). The mode of cytoplasmic separation between a basidiospore and a sterigma in shiitake mushroom, *Lentinus edodes*. *Mushroom Science* 10, 191-199.

Nei, T., Yotsumoto, H., Hasegawa, Y. and Hasegawa, M. (1974). Letter: freeze-etching in scanning electron microscopy. *Journal of Electron Microscopy* (Tokyo) 23, 137-138.

O'Donnell, K. L. and Hooper, G. R. (1977). Cryofracturing as a technique for the study of fungal structures in the scanning electron microscope. *Mycologia* 69, 309-320.

O'Donnell, K. L., Hooper, G. R. and Fields, W. G. (1976). Zygosporogenesis in *Phycomyces blakesleeanus*. *Canadian Journal of Botany* 54, 2573-2586.

Paine, W. A. and Hess, W. M. (1980). Ultrastructure of germinating sugarcane smut, *Ustilago scitaminea*, teliospores. *Botanical Society of America* (published abstracts) Publication 185 p. 85.

Parrish, W. B., Bronn, D. G., Wright, J. R. and Minton, J. P. (1979). Elemental analysis of biological material in the fresh-frozen state. *Ohio Journal of Science* 79, 271-273.

Pinto da Silva, P. and Nogueira, M. L. (1977). Membrane fusion during secretion. A hypothesis based on electron microscope observation of *Phytophthora palmivora* zoospores during encystment. *Journal of Cell Biology* 73, 161-181.

Poon, N. H. and Day, A. W. (1976). Somatic nuclear division in the sporidia of *Ustilago violacea*. IV. Microtubules and the spindle-pole body. *Canadian Journal of Botany* 22, 507-522.

Powell, M. J. (1976). Ultrastructure and isolation of glyoxysomes (microbodies) in zoospores of the fungus *Entophlyctis*. *Protoplasma* 89, 1-27.

Powell, M. J. (1979). The structure of microbodies and their associations with other organelles in zoosporangia of *Entophlyctis variabilis*. *Protoplasma* 98, 177-198.

Rash, J. E. and Hudson, C. S. (Eds.) (1979). Freeze Fracture; Methods, Artifacts, and Interpretations. Raven Press, New York.

Rix, E., Schiller, A. and Taugner, R. (1976). Freeze-fracture-autoradiography. *Histochemistry* 50, 91-101.

Ruben, D. M. and Stanghellini, M. E. (1978). Ultrastructure of oospore germination in *Pythium aphanidermatum*. *American Journal of Botany* 65, 491-501.

Schiller, A., Taugner, R. and Rix, E. (1979). Freeze-fracture autoradiography. Progress towards a routine technique. *Journal of Histochemistry and Cytochemistry* 27, 1514-1515.

Sekiguchi, J., Gaucher, G. M., and Costerton, J. W. (1975a). Microcycle conidiation in *Penicillium urticae*: An ultrastructural investigation of conidiogenesis. *Canadian Journal of Microbiology* 21, 2069-2083.

Sekiguchi, J., Gaucher, G. M., and Costerton, J. W. (1975b) Microcycle conidiation in *Penicillium urticae*: An ultrastructural investigation of spherical spore growth. *Canadian Journal of Microbiology* 21, 2048-2058.

Sjöstrand, F. S. (1979). The interpretation of pictures of freeze-fractured biological material. *Journal of Ultrastructure Research* 69, 378-420.

Soina, V. S., Sokolov, A. A. and Agre, N. S. (1977). Electron-microscopic study of spore formation of *Actinomadura verrucosospora*. *Microbiology* 46, 576-580.

Steele, S. D. and Miller, J. J. (1974). Ultrastructural changes in germinating ascospores of *Saccharomyces cerevisiae*. *Canadian Journal of Microbiology* 20, 929-933.

Stevenson, I. L. and Becker, S. A. W. E. (1979). The fine structure of mature and germinating chlamydospores of *Fusarium oxysporum*. *Canadian Journal of Microbiology* 25, 808-817.

Stiers, D. L. (1976). The fine structure of ascospore formation in *Ceratocystis fimbriata*. *Canadian Journal of Botany* 54, 1714-1723.

Takeo, K. (1974). Ultrastructure of polymorphic mucor as observed by means of freeze-etching. III. Dormant sporangiospore. *Archives of Microbiology* 99, 99-107.

Tewari, J. P. and Skoropad, W. P. (1975). Fine structure of the macroconidia of *Fusarium solani*. *Canadian Journal of Botany* 53, 2134-2146.

Tewari, J. P. and Skoropad, W. P. (1977). Ultrastructure of oospore development in *Albugo candida* on rapeseed. *Canadian Journal of Botany* 55, 2348-2357.

Ton That, T. C. and Turian, G. (1978). Ultrastructural study of microcyclic macroconidiation in *Neurospora crassa*. *Archives of Microbiology* 116, 279-288.

Traquair, J. A., and McKeen, W. E. (1980). Electron microscopy of developing *Aphanomyces oogonia* and oospores. *Mycologia* 72, 378-394.

Underwood, E. E. (1979). Quantification of microstructures by stereological analysis. *Journal of histochemistry and cytochemistry* 27, 1536-1537.

Van Eck, W. H. (1978). Lipid body content and persistence of chlamydospores of *Fusarium solani* in soil. *Canadian Journal of Microbiology* 24, 65-69.

Van Iren, F., Van Essen-Joolen, L., Van der Duyn Shouten, P., Boers-Van der Sluijs, P., and deBruijn, W. C. (1979). Sodium and calcium localization in cells and tissues by precipitation with antimonate. A quantitative study. *Histochemistry* 63, 273-294.

White, J. A. and M. F. Brown. (1979). Ultrastructure and X-ray analysis of phosphorus granules in a vesicular-arbuscular mycorrhizal fungus. *Canadian Journal of Botany* 57, 2812-2818.

Wigglesworth, V. B. (1976). Lipid staining for the electron microscope: A new method. *Journal of Cell Science* 19, 425-437.

Zs.-Nagy, I., Pieri, C., Giuli, C., Bertoni-Freddari, C. and Zs.-Nagy, V. (1977). Energy dispersive X-ray microanalysis of the electrolytes in biological bulk specimen. Specimen preparation, beam penetration, and quantitative analysis. *Journal of Ultrastructure Research* 58, 22-23.

ZOOSPOROGENESIS
MODEL SYSTEMS - PROBLEMS - POSSIBLE APPROACHES

L.W. Olson*, U.M. Edén*, L. Lange**

**Institute of Genetics, University of Copenhagen, Øster Farimagsgade 2A, DK-1353 Copenhagen K, Denmark*

***Institute of Seed Pathology for Developing Countries, Ryvangs Allé 78, DK-2900 Hellerup, Denmark*

INTRODUCTION

In 1966 at the 18th Colston symposium on the fungal spore two papers were presented which are key papers in the historical development of our knowledge of the aquatic fungal spore and its development. Professor Fuller (1966) in his paper "Structure of the uniflagellate zoospores of aquatic phycomycetes" described the general types of uniflagellate zoospore ultrastructural organization, as known at that time. And Gay and Greenwood (1966) described "Structural aspects of zoospore production in *Saprolegnia ferax* with particular reference to the cell and vacuolar membranes". These two papers set that stage for the many studies of spore structure and formation which have followed.

In order to appreciate the process of zoosporogenesis we must first examine the products of spore formation. Since Fuller (1966) the spores of a large number of the members of both the uniflagellate and biflagellate fungi have been examined. However, at the present time the ultrastructure of spores of the uniflagellate fungi is better understood than that of the biflagellate fungi. A number of review articles on the ultrastructural organization of the uniflagellate fungal spore have been published (Fuller 1977, Barr 1978, Powell 1978, Lange and Olson 1979). The authors of these papers have chosen different combinations of ultrastructural characters which they believe best typify certain types of spore organization.

Lange and Olson (1979) analysed the ultrastructural organization of 50 species of uniflagellate *Phycomycetes* representing 14 of 17 families of uniflagellate *Phycomycetes*.

They described 5 major ultrastructural zoospore types. The simplest organization of organelles in a uniflagellate fungal spore is found among the *Chytridiales*. However, when these spores which have an apparently low level of organization in light microscopic observations are studied from serial section reconstructions there is an amazing uniformity in organellar arrangement, and it would be incorrect to envision that it is mere chance that controls the cleavage of the cytoplasm during zoosporogenesis. The most complex spore organization is found in the spores of the *Blastocladiales* where the ribosomes are sequestered in a nuclear cap. There is a large basal mitochondrion which partially surrounds the kinetosome, and a complex arrangement of lipid bodies, microbodies, membrane and a portion of the basal mitochondrion form the so-called side body.

In this treatment of zoosporogenesis more than perhaps a fair share is devoted to the formation of spores in the *Blastocladiales*. The reasons for this are several: the members of the *Blastocladiales* lend themselves to synchronous culturing; a great deal of information is now available on the physiology and biochemistry of spore formation particularly on RNA and protein synthesis primarily from the work of Professor Lovett and his students (Lovett 1975); and the cytology of spore formation in the *Blastocladiales* has been extensively studied.

Whether we consider spore formation in a uniflagellate or a biflagellate fungus, during zoosporogenesis each cell must come to contain all the necessary elements for survival. In nature, and in the laboratory this process occurs with what may be considered "boring regularity". However, when one considers just how a fungus goes about forming its spores much less doing it so perfectly each time, we really know very little. At present it is virtually impossible to correlate the cytology of zoosporogenesis with specific biochemical aspects of the process. In the first portion of the discussion of zoosporogenesis will be presented a short review of the cytological aspects of spore formation which will hopefully convey some of the amazing complexity of the process, and to introduce some new observations on zoosporogenesis, review very briefly what is known of the physiology and biochemistry of spore formation, and then to examine two well established model systems which have been used with great success in cytological as well as biochemical studies of zoosporogenesis - model systems using *Blastocladiella* and *Achlya*. In addition to these model systems other systems which possibly can be developed for studies of spore formation in the aquatic fungi, will be described. At the conclusion of this presentation some "speculation" on what might be termed unorthodox approaches to examining zoosporogenesis, approaches which may be of use in the future, will be presented.

SPORE CLEAVAGE

There appears to be two basic systems by which the sporangial cytoplasm is divided into spores. In the biflagellate fungi is found what, conceptually at least, is the simpler of the two types. In *Saprolegnia* a central vacuole is formed at the center of the sporangium which then proceeds to expand between the uninucleate blocks of cytoplasm thus cutting-out the spore initials, and the central vacuole membrane finally fuses with the plasmalemma of the sporangium (Gay and Greenwood 1966, Gay *et al.* 1971). As noted by Heath (1976) the sporangium itself does not expand appreciably during cleavage and thus the vacuole must expand at the expense of the cytoplasm. The expansion of the central vacuole is presumed to be due to an increase in the osmotic potential of the vacuole. With the expansion of the central vacuole there is a concomitant synthesis of additional membrane. Gay and Greenwood (1966) and Gay *et al.* (1971) have suggested that both the increase in osmotic potential of the central vacuole and the synthesis of additional membrane is mediated by the so-called "dense bodies" found in the sporangium cytoplasm. These dense bodies in *Saprolegnia* are observed only after the induction of sporulation and they do appear to be rich in phospholipids (Heath 1976). Gay *et al.* (1971) and Heath (1976) have suggested that some of this phospholipid is incorporated into the expanding vacuolar membrane while some is degraded into smaller molecules which would lead to an increase in the osmotic pressure of the vacuole.

The second mode of cytoplasmic cleavage, in contrast to the expanding central vacuole model described above, is by means of the progressive fusion of small vesicles. This type of spore cleavage has been observed in diverse fungal groups, all species of uniflagellate fungi which have been examined to date, biflagellate fungi, in species of *Phytophthora* (Hohl and Hamamoto 1967, Elsner *et al.* 1970) and *Thraustotheca* (Heath 1976). In *Phytophthora parasitica*, *P. infestans*, and *Harpochytrium* (Hohl and Hamamoto 1967, Elsner *et al.* 1970, and Travland and Whisler 1971) the small vesicles which form the cleavage membranes, appear to be derived from a Golgi apparatus. In *Blastocladiella emersonii* (Lessie and Lovett 1968, Lovett 1975) the origin of the cleavage vesicles remains obscure. In the resting sporangia of *Physoderma maydis* the cleavage vesicles appear to originate from the endoplasmic reticulum and the Golgi apparatus.

The obvious question to be asked at this point is just how does but one nucleus and the associated spore organelles come to be packaged together. Unfortunately there is no definitive answer at present. Heath and Greenwood (1971) have suggested that the cytoplasmic microtubules which are formed around the nucleus of the spore initial in some way may separate the nuclei so that the cleavage vesicles come to pass through relatively "weaker" areas of the cytoplasm. In the uniflagellate fungi the mechanism of cleavage vesicle formation is assumed to be much more complicated and the microtubules are formed only after the fusion of the cleavage vesicles (Heath 1976). A possible analogy to the cytoplasmic microtubular array observed in biflagellate zoosporogenesis in the uniflagellate fungi will be discussed later.

Zoosporogenesis in the blastocladialian sporangium

Zoosporogenesis has been investigated in a number of the uniflagellate fungi but the model system of spore cleavage best understood is that found in the blastocladialian sporangium. Spore formation has been examined cytologically in *B. emersonii* (Lessie and Lovett 1968, Lovett 1975) and in *Allomyces* by a variety of authors (Blondel and Turian 1960, Renaud and Swift 1964, Moore 1968, Barron and Hill 1974, Olson 1980) and in general the accounts given are in agreement. However, there are still certain additional observations which should be incorporated into a description of sporogenesis in this group of organisms.

The following electron micrographs are of zoosporogenesis in *Allomyces catenoides* zoosporangia. In Fig. 1 is seen a nucleus in close association with the sporangium wall. At the start of spore formation the nuclei lie around the periphery of the cytoplasm and are surrounded by numerous lipid bodies, the so-called lipid crown stage. An array of microtubules is seen to flare out from the functional as well as the non-functional kinetosome (Figs. 1-3). These microtubules extend for some distance from the kinetosome out into the sporangia cytoplasm. In the sporangial cytoplasm are found three types of electron dense bodies: an electron dense membrane bound inclusion of uniform size (Figs. 1-2, double arrow), an electron dense membrane bound inclusion which is larger than the former and observed to vary in size (Fig. 1, open arrow), and a membrane bound inclusion which is composed of concentric electron dense layers (Fig. 2). One of the

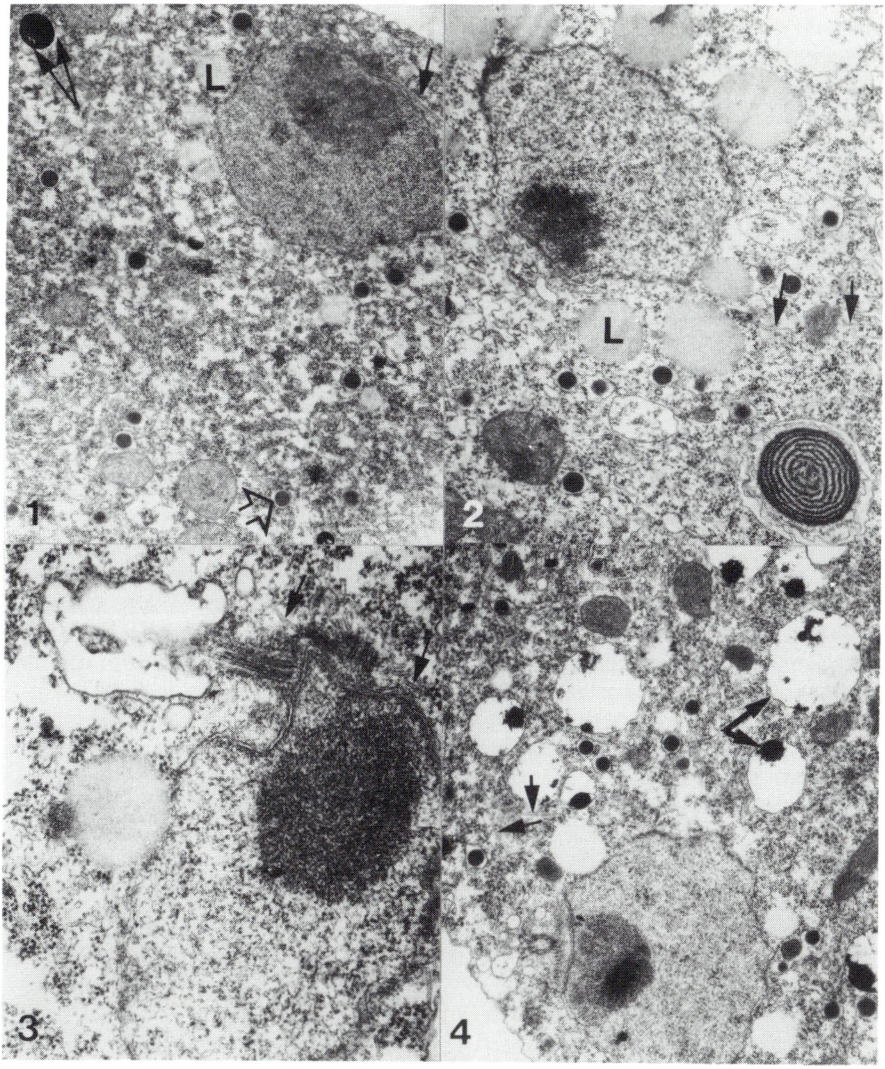

Figs. 1-2. Sporangium nuclei at start of sporogenesis. L = lipid. Arrows = cytoplasmic microtubules which extend from the kinetosomes. Double arrows = membrane electron dense body which will be involved in cleavage. Open arrow = electron dense body not involved in sporogenesis. X 13,000, 16,000. Fig. 3. Early flagellum formation. Arrows = cytoplasmic microtubules. X 32,000. Fig. 4. Mobilization of electron dense material = double arrows. Single arrows = cytoplasmic microtubules. X 13,000.

first events of sporogenesis is the mobilization of the larger electron dense inclusion and the concentric granule (Fig. 4, open arrows). The diameter of the vesicle surrounding the electron dense body increases as the electron dense material decreases in size and the outer layers of the concentric granule appear to be digested. As the electron dense material disappears from the inclusions, small vesicles appear in the cytoplasm. It is interpreted that the electron dense material is analogous to the dense bodies observed in the biflagellate fungi during sporogenesis, and it is these dense bodies that give rise to the primary cleavage vesicles. As noted by Heath (1976) until the chemical nature of these electron dense bodies is determined, the putative role of these structures in vesicle formation must remain but a hypothesis.

The development of the flagellum is the next stage in sporogenesis. The flagellum microtubules extend from the distal end of the functional kinetosome and they appear to extend by tip growth. The flagellum microtubules elongate into a membrane bound evagination. In Fig. 5 the elongating flagellum membrane is formed by the deposition of vesicles at the junction of the kinetosome with the elongating microtubules. One surface of the vesicle forms the flagellum membrane, the other surface forms the plasmalemma. In Figs. 5-6 thin connections between the flagellum membrane and the plasmalemma (arrows) suggest that the final dissolution of the two unused sides of the vesicle takes time, and that the sides of the vesicle are finally incorporated into either the flagellum or the plasmalemma.

Once the flagellum has been formed the final cleavage of the sporangial cytoplasm takes place. The progressive extension of the cleavage furrow appears to proceed by the incorporation of small vesicles or flattened cisternae along the length of the cleavage furrow (Fig. 6) and not by a systematic incorporation of vesicles along the leading edge of the cleavage furrow.

Renaud and Swift (1964) concluded that the plasmamembrane was the source of membrane material which would be incorporated into the flagellum membrane. They suggested that as the kinetosomes are near the cell membrane they induce the formation of vesicles which would surround the axoneme. However, as stated by Lessie and Lovett (1968) and confirmed by the present investigation the almost simultaneous appearance of flagellar cleavage vesicles throughout the cytoplasm argues against the idea that the plasmalemma is involved in cleavage vesicle formation. Cytoplasmic microtubules can at this stage of sporogenesis still be seen to flare out into

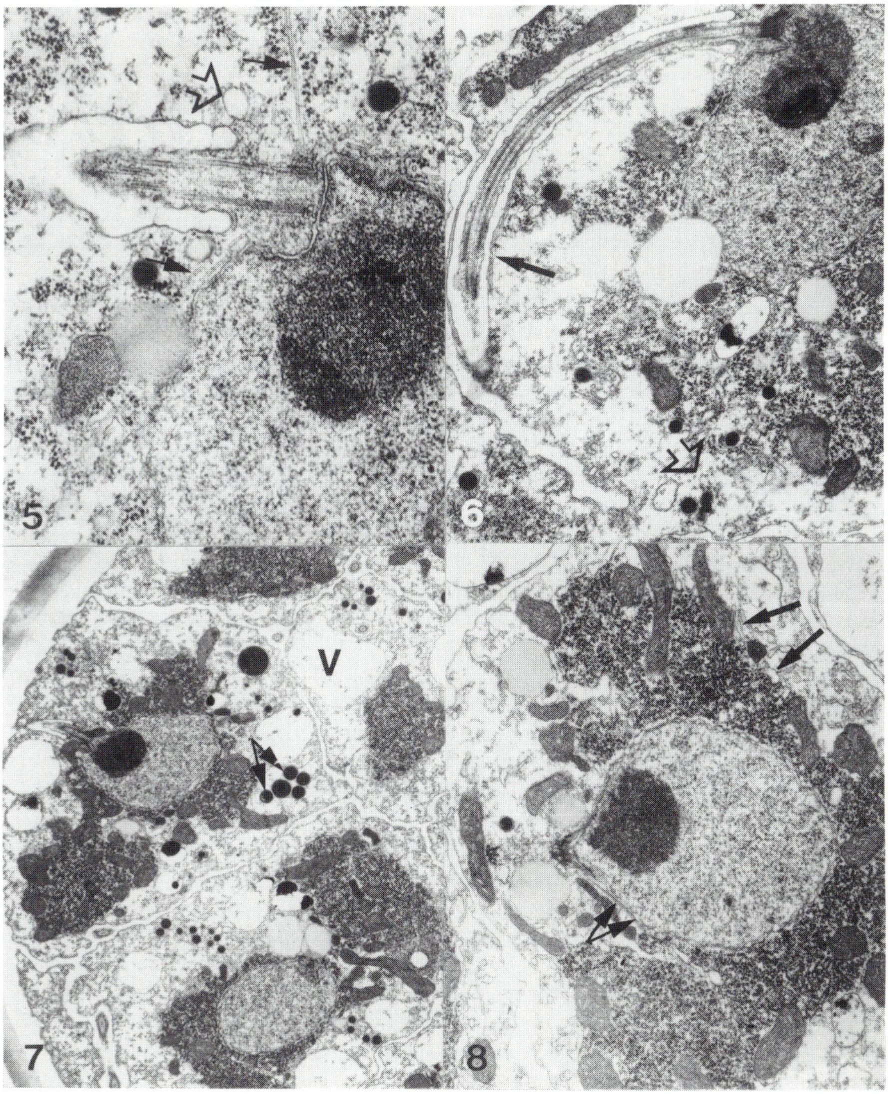

Figs. 5-6. Flagellum formation. Fig. 5. Single arrows = cytoplasmic microtubules. Open arrow = vesicle fusion. Fig. 6. Single arrow = vesicle fusion. Open arrow = bundle of microtubules between spores at cleavage. X 33,000, 12,000. Figs. 7-8. Organellar organization - formation of nuclear cap. Fig. 7. Double arrows = organization of gamma bodies. V = vacuoles. Fig. 8. Double arrows = cytoplasmic microtubules. Single arrows = vesicle forming nuclear cap membrane. X 6,000, 12,000.

the cytoplasm (Fig. 6). In areas between cleavage furrows an aggregate of cytoplasmic microtubules is frequently seen (Fig. 6). It would appear that the cytoplasmic microtubules are actively involved in the determination of the volume of cytoplasm which will be incorporated into each spore.

As cleavage of the sporangial cytoplasm is completed the assembly of the cellular ribosomes to form a nuclear cap takes place (Figs. 6-8), and there is a progressive fusion of the smaller membrane bound electron dense bodies originally observed to be dispersed throughout the sporangial cytoplasm (Fig. 7). It is these electron dense bodies, which have not been active during sporogenesis, which form the gamma bodies of the fully differentiated spore.

The nuclear cap of the spore seen in Fig. 7 has begun to take on its final shape although there are still mitochondria embedded in it. The system of 9 groups of triplet cytoplasmic microtubules which surround the nucleus and nuclear cap appear to be formed at this time (Fig. 8). As this is one of the last stages in spore formation observed prior to spore release in *Allomyces*, it is of interest that the spore still has not formed the large basal mitochondrion, and the lipid bodies-microbody-mitochondrion association of the side body complex has not been established.

In Fig. 8 is seen the final assembly of the nuclear cap membrane. The membrane develops by the fusion of small vesicles. Large vesicles-vacuoles remain in the spore cytoplasm and presumably arise from the fusion of the degraded electron dense bodies observed at the start of spore cleavage (Fig. 7).

The account of sporogenesis given above corresponds in all essential details with the account given by Lessie and Lovett (1968) and Lovett (1975) for zoospore formation in *B. emersonii*, though the array of cytoplasmic microtubules which flares out from the functional and non-functional kinetosomes have not been commented on previously. This apparent oversight would appear to be due to problems in preserving these microtubules. In preparations which are not correctly fixed the microtubules are absent. We would interpret these microtubules as Heath (1976) has done for biflagellate zoosporogenesis. The cytoplasmic microtubules serve to delineate the boundaries of the spore initial and in some way establish cytoplasmic domains through which the flagellum and/or cleavage furrows cannot traverse. Similar arrays of cytoplasmic microtubules radiating from the functional kinetosome have been observed during spore formation in the posteriorly uniflagellate chytrids *Chytridium olla*, *Phlyctochytrium pustulare*, *P. semiglobiferum* and in the anteriorly uniflagellate hyphochytridiomycete, *Anisolpidium* sp. (Lange and Olson unpublished).

In *Allomyces* sporogenesis it appears that the cytoplasmic microtubules which are present during spore cleavage are not the same microtubules which form the array of microtubules which surround the nucleus and nuclear cap of the spore, as these microtubules form during the last stages of spore assembly. It must be determined whether the "skeletal microtubules" observed in the spores of the *Chytridiales* (Barr 1978) are the same microtubules which appear during spore cleavage or whether they also are formed only during the last stages of spore organization. Partial serial section reconstructions of cytoplasmic microtubule arrays observed during sporogenesis in *Allomyces* support the suggestion that they serve to set the boundaries of the spore initial. Complete serial section reconstructions of a large area of the sporangium are required to determine how the cytoplasmic microtubules of one array relate to the surrounding arrays. The physiological evidence that cytoplasmic microtubules are involved in the delineation of the spore initial will be described later.

The morphological similarities between the dense bodies observed during spore formation in the biflagellate fungus *Saprolegnia* (Gay and Greenwood 1966, Gay *et al.* 1971) and the dense bodies observed in *Allomyces* suggest that in spite of the taxonomic distance between these two organisms the dense bodies may serve the same function, the generation of the cleavage vesicles. The mobilization of similar electron dense bodies during the early stages of sporogenesis has been observed in the chytrids *Chytridium olla*, *Phlyctochytrium pustulare*, *P. semiglobiferum*, and the hyphochytridiomycete *Anisolpidium* sp. (Lange and Olson unpublished). On the basis of these observations it would appear that the occurrence of similar electron dense bodies in a diverse group of uniflagellate fungi is the rule.

Molecular events during zoosporogenesis

The molecular events which occur during zoosporogenesis have been described for spore formation in *Achlya*, *Blastocladiella*, and *Allomyces*. Griffin and Breuker (1969) observed that in *Achlya* all stages in differentiation before spore release are sensitive to actinomycin D. As actinomycin D inhibits mRNA synthesis, Griffin and Breuker (1969) concluded that continued mRNA synthesis throughout zoosporogenesis is required and they observed that cycloheximide, an inhibitor of protein synthesis, could inhibit all stages of differentiation. Timberlake *et al.* (1973) refined the system for the induction of zoosporangia formation in *Achlya* so that virtually all of the hyphal cytoplasm became incorporated into the zoosporangium. They also reported that continued mRNA synthesis and protein synthesis are required for zoospore formation, and that more than 90% of the hyphal tip proteins were degraded and recycled during zoosporogenesis. *Achlya* appears to be an organism in which starvation induces zoosporogenesis with a concomitant turnover in RNA and protein.

A second pattern of molecular events is observed in *Allomyces* and *Blastocladiella*. Burke *et al.* (1972) reported that in *Allomyces* all stages of zoosporogenesis were insensitive to actinomycin D and thus concluded that all necessary mRNA required for zoosporogenesis must have been synthesized during the vegetative stage. However, Fähnrich (1974a) and Stumm and Croes (1975) noted that in the differentiation of gametangia there was a continued synthesis of mRNA and protein molecules with a short life time and that it was gametogenesis itself which was sensitive to cycloheximide and totally insensitive to actinomycin D.

Fähnrich (1974b) noted that in *Allomyces* rRNA synthesis is very sensitive to 5-fluorouracil prior to gametogenesis but this compound had no effect on gamete differentiation. Fähnrich (1974a,b, 1976) concluded that spore formation is dependent on mRNA present in the gametangia prior to gametogenesis, protein synthesis on cytoplasmic ribosomes is completed 20-30 min after induction of gametogenesis and that rRNA synthesis appears not to be a prerequisite for the differentiation of gametes. Thus it would appear that there are two patterns of molecular events during sporogenesis. In *Achlya* continued mRNA synthesis and protein synthesis take place during zoosporogenesis while in *Allomyces* and *Blastocladiella* the synthesis of the necessary mRNA for spore differentiation takes place early in the process of sporangium/gametangium differentiation and is presumably stored as long lived mRNA, and protein synthesis occurs throughout sporangium differentiation and sporogenesis.

Among the uniflagellate fungi, the *Chytridiales*, are found a spectrum of growth forms, the monocentric chytrids to the highly complex polycentric forms. It would be of interest to know if there is a variation in the molecular events of sporogenesis among the group or whether they all are of the blastocladialian form. When additional members of the posteriorly uniflagellate fungi (the chytrids), anteriorly uniflagellate fungi (the hyphochytrids) and biflagellate fungi have been studied perhaps general trends of molecular activity during zoosporogenesis will become apparent.

MODEL SYSTEMS

Synchronization of zoosporogenesis, and sporogenesis in general, can be achieved by the transfer of the fungus from a state of balanced nutrition to starvation or near starvation conditions. This method of inducing sporangia-sporogenesis in aquatic fungi was first reported by Klebs (1900) and has been used by mycologists ever since. Emerson (1958) in his classic paper on "Mycological organization" described a variety of techniques which can be used to induce sporogenesis, all based on "Klebs' principle". Fuller organized in 1978 a workshop on the lower fungi in honor of the late Prof. R. Emerson and the Festschrift which was based on this workshop has greatly increased the number of fungi which can be synchronously induced to form spores.

A series of what can be called model systems of zoosporogenesis has evolved for both the uniflagellate and biflagellate fungi and it is from these few systems that the majority of our knowledge of the molecular and cytological events of sporogenesis comes. The following model systems for studying sporogenesis will be briefly described: *Blastocladiella emersonii* - normal cycle, microcyclic-minicyclic sporogenesis, *Achlya* - zoosporogenesis, *Allomyces* - zoosporogenesis, cystocyclic *Allomyces*, and chytrid systems.

Blastocladiella emersonii - normal cycle

The ease with which *B. emersonii* can be cultured and synchronized is based on the pioneering work of Prof. E. Cantino and his students, for descriptions of synchronous culturing of *B. emersonii* see Goldstein and Cantino (1962) and Murphy and Lovett (1966). The basic principle of this culture technique is that after 15½ hrs of growth on a rich broth medium, zoospore formation can be induced in late exponential growth phase plants by exchanging the culture medium with a dilute inorganic salts solution.

Prof. Lovett and his students have described in detail the molecular and cytological events of zoosporogenesis in *B. emersonii* (Lessie and Lovett 1968, Murphy and Lovett 1966, Lovett 1975, Adelman and Lovett 1972, 1974a,b, Barstow and Lovett 1972, 1973, 1974, 1975). The major steps are the following: 150 min after the exchange of the growth medium for a dilute salts solution which starts zoosporogenesis a cross wall is formed separating the sporangium from the rhizoidal system and a discharge papillum is also formed; 20 min later cleavage vesicles begin to form the axonemal sheath and the cytoplasmic cleavage furrows; 190 min after medium exchange the ribosomes begin to aggregate and 20 min later the zoospore is completely differentiated (see Lovett 1975, Fig. 1). 210-230 min after the medium exchange zoospore discharge takes place. The synchronous culturing of *B. emersonii* has provided the system by which all others will have to be compared. As large numbers of plants can be cultured at one time, the problem of having sufficient material for biochemical analysis is overcome. But the culturing of *B. emersonii* does require time, and as liquid culture is used stringent sterile technique is required.

Microcyclic-minicyclic sporogenesis

Hennessy and Cantino (1972) have shown that very young germlings of *B. emersonii* can be induced to form zoospores if they are shifted from growth medium (Difco Cantino PYG) to a dilute salts solution. They observed that these young germlings, as young as 30 min after the initial inoculation to growth medium, would form a single zoospore. Hennessy and Cantino (1972) termed this microcyclic sporogenesis "lag phase sporogenesis". The term microcyclic sporogenesis would appear more appropriate as it is used to describe the immediate recapitulation of sporogenesis following spore germination (Lovett 1975). Microcyclic sporogenesis demonstrates the degree of physiological control which can be used in the manipulation of the development of this organism.

When 1½ hr old germlings are induced to differentiate zoospores an increasing percentage of the sporangia produce two spores. The zoospores which are released from these sporangia are ultrastructurally similar to the "normal" zoospore and display normal viability. Lovett (1975) reports that spinner flask cultures could be induced to undergo zoosporogenesis at any stage in the life cycle. Lodi and Sonneborn (1974) and Silverman and Epstein (1975) have also induced zoosporogenesis after 6 hrs of growth.

Lovett (1975, see Figs. 28-29) notes that germlings induced to undergo zoosporogenesis after 3 hrs of growth are useful in that they form two zoospores, minicyclic sporogenesis, in contrast to the microcyclic sporangia where a single zoospore is formed. The 3 hr germlings, after induction to undergo zoosporogenesis complete all the normal events associated with zoosporangium formation and zoosporogenesis including cytokinesis. As Lovett (1975) notes this minicycle in *Blastocladiella* "seems to provide a very convenient and rapid system in which to study developmental problems. It only requires 7.5 hr to complete the cycle from spore to spore at 25°C, adequate cell densities can be used and the small size permits preparation of electron micrographs of thin sections through entire cells at reasonably high magnifications." Youatt (1976) has demonstrated that microcyclic-minicyclic zoosporogenesis can also be induced in *Allomyces*. Youatt reported that throughout the growth cycle of *Allomyces* zoosporogenesis could be induced by the transfer of plants to nitrogen deficient medium (Youatt 1976, see Figs. 1-4).

Achlya - zoosporogenesis

Among the biflagellate fungi greatest success in obtaining synchronous sporangium differentiation and sporogenesis has been with *Achlya bisexualis*. Timberlake *et al.* (1973) described a modification of the technique of Griffin and Breuker (1969) by which sporangia are differentiated synchronously and the vast majority of the cytoplasm of the hyphae is incorporated into the sporangium. The nearly complete incorporation of the hyphal cytoplasm into the sporangium means that in radioactive labeling experiments the spurious labeling which takes place in the hyphal cytoplasm will not overshadow the molecular events taking place in the differentiating sporangium.

The fungus is grown in a peptone-glucose-yeast extract broth. The flasks (250 ml) are inoculated with 2×10^5 spores and incubated at 30°C on a rotary shaker at 150 rpm for 16 hrs. The fungus develops as a uniform suspension and does not form mycelial balls. The uniformity of the suspension is doubtless part of the key to the synchrony of differentiation and the large amount of the hyphal cytoplasm incorporated into the zoosporangium.

Synchronous sporulation is achieved by washing the hyphal suspension with 0.5 mM $CaCl_2$, suspending the hyphae in 0.5 mM $CaCl_2$ and incubating the suspension on a rotary shaker. The synchronization of zoosporangia differentiation and zoosporogenesis achieved by Timberlake *et al.* (1973) system is impressive. It would be of interest to know whether other members of the biflagellate (Oomycetes) can be synchronized in this way.

Allomyces - zoosporogenesis

The aquatic phycomycete *Allomyces* spp. is closely related to *B. emersonii* but is a hyphal forming fungus in contrast to *B. emersonii* which has a monocentric growth form. A variety of techniques have evolved for the synchronization of sporogenesis in this fungus.

When the fungus develops in a rich growth broth the hyphal

tips can be induced to form sporangia-gametangia by merely transferring the hyphae to a dilute salts solution (Morrison 1978, Barstow 1978, Aronson and Sandstedt 1978). As in *B. emersonii* a septum forms at the base of the sporangium followed by papillum formation, sporogenesis, and spore release, and the process is reasonably synchronous. The primary disadvantage of this technique in studies of sporogenesis is that only the hyphal tips are converted into sporangia-gametangia. Because of the presence of hyphal cytoplasm it is difficult to perform labeling experiments or to isolate RNA from such cultures.

Allomyces can also be grown on solid substrate (Difco Emerson's YpSs agar) and will form copious numbers of sporangia-gametangia. In our laboratory we have worked for a number of years on the synchronization of sporangium germination in *Allomyces* (Olson 1974, Olson and Rønne 1975, Olson and Borkhardt 1978, Olson 1980). Basically the technique is quite simple. Sporangia or gametangia are formed in a semi-synchronous fashion if the surface of the agar is spread with a dense spore suspension. The subsequent crowding of the plants induces a rapid conversion of the hyphal tips into sporangia or gametangia after one or two dichotomous branchings of the hyphae. The sporangia or gametangia of *Allomyces* enter a state of rest at the crown lipid stage. At the crown lipid stage the nuclei assume a peripheral location against the sporangium wall. The sporangia or gametangia will remain at this stage until they are either induced to germinate or eventually die. As the sporangia-gametangia of *Allomyces* are deciduous they can be easily removed from the agar surface by scraping the agar surface with a scalpel. After washing the sporangia they are resuspended in 10-15 ml of dilute salts solution to give a final concentration of 1-5 x 10^4/ml and incubated as sedentary cultures in 9 cm Petri dishes. Sporogenesis in the aquatic fungi can be strongly inhibited or stopped by decreased oxygen tension thus it is important either to aerate the cultures during sporogenesis or to ensure that the depth of the culture is not too great. After the sporangia-gametangia are placed in dilute salts solution spore release occurs after 60-120 min depending on the stage of the life cycle used. The advantage of this system is that it is possible to obtain excellent synchrony of sporogenesis, the disadvantage of the system at the present time is that it is difficult to work with large volumes of germinating sporangia-gametangia.

Cystocyclic Allomyces

Within the genus *Allomyces* is found a subgenus *Cystogenes* where the meiospores released from the thick walled resistant sporangium (meiosporangium) rapidly encyst to form a cyst which then forms a single celled gametangium, undergoes one or two nuclear divisions and after papillum dissolution releases four gametes which fuse in pairs, form a zygote, and then develops into a hyphae forming sporophyte (Teter 1944).

Recent work (Olson 1980) has shown that *Allomyces neomoniliformis* belonging to the subgenus *Cystogenes* could be a suitable organism for studies of sporangium-spore formation. The sporophytic generation of this fungus forms resting sporangia (meiosporangia) in copious numbers and these sporangia can be induced to germinate synchronously. The meiospores swim or crawl by amoeboid motion from the resting sporangium and rapidly encyst. Copious numbers of these cysts can be harvested from a germinating meiospore suspension by mixing the sporangia-cysts on a vortex mixer and then filtering the cysts from the suspension through a fine nylon mesh filter (pore size 30 μm). Within 90 min the single celled gametangium is formed with a single papillum. The gametangial cytoplasm is cleaved into four motile gametes which are then released. The entire process from the induction of meiosporangium germination to zygote formation takes 180 min. The obvious advantages of this system are its synchrony, rapidity and the cyst gametangia are small so that the entire gametangium can be studied at relatively large enlargements in the electron microscope.

Chytrid systems

The chytrids *(Chytridiales)* are certainly the most diverse and taxonomically complicated group of the aquatic fungi (Sparrow 1969). From studies of the ultrastructure of the spores of this group (Barr 1978, Powell 1978, Lange and Olson 1979) it would appear that they represent a most inhomogeneous grouping of organisms. With the publication of Fuller's workshop on "Lower fungi in the laboratory" a variety of chytrids can now be induced to "perform" on schedule for teaching and research *(Rhizophydium sphaerocarpum*, *R. sphaerotheca*, *R.* sp., *Phlyctochytrium kniepii*, *Chytriomyces hyalinus*, *Rhizophlyctis rosea*, *Chytridium confervae*, *Entophlyctis variabilis*, *Cladochytrium replicatum*, *Nowakowskiella elegans)*.

The synchronization of chytrid development and sporogenesis has reached its highest point, at least to date, by Gauriloff (1978a,b,c). Gauriloff (1978a,b,c) has used a variety of techniques to obtain synchronously developing chytrid cultures. The basic theme is that the fungus is grown through one or two sporangium-sporogenesis cycles in a semi-synchronous fashion in broth culture. The final spore suspension is then diluted with a dilute nutrient broth. The spore suspension is poured into a Petri dish containing a number of sterile coverglasses. The spores encyst and attach to the surface of the coverglass and synchronously form sporangia. The coverglasses can be mounted to observe sporangium development or after transfer to dilute salts solution to observe sporogenesis and spore release. The advantage of this technique is the tight synchrony which can be imposed, the relative large numbers of organisms which can be cultured synchronously and the ease with which the organism can be studied microscopically. With the use of plastic coverslips it would be possible to fix the fungi in situ for electron microscopy. It is to be hoped that the ultrastructural aspects of sporogenesis in a variety of these synchronized chytrids, both monocentric and polycentric forms, will be investigated. The taxonomic diversity of the chytrids and their presumed evolutionary distance from the *Blastocladiales* and the biflagellate fungi may show that what we observe in these latter groups of organisms is not quite the same in the *Chytridiales*.

UNORTHODOX APPROACHES TO ZOOSPOROGENESIS

In nature and in the laboratory the cycle of sporangium formation, zoosporogenesis and spore release occurs with surprising exactness. In an attempt to understand the process of zoosporogenesis I would like to examine several alternative experimental approaches.

The exactness of zoosporogenesis is impressive, until one begins to manipulate the process. Every aquatic mycologist will, for example, be familiar with the abnormal spores (multinucleate-multiflagellate spores) released from sporangia which have been covered by a coverglass where sporogenesis takes place under reduced oxygen tension. Extremely little is known about the effect of oxygen on zoosporogenesis and this could be a fruitful area for research.

Phosphate effect

However, induction of zoosporogenesis under controlled gas mixtures requires apparatus which is not always available in a cytologist's laboratory. By chance when we attempted to control the pH of the sporulation medium with dilute Sørensen's phosphate buffer (10 mM pH 7.0) we observed that the gametes were even more abnormal than those released under conditions of reduced oxygen tension.

For some reason dilute Sørensen's phosphate buffer has profound, and from the fungus' point of view, deleterious affects. After the primary cleavage vesicles have formed and coalesced to form the axonemal membrane the process of cleavage furrow formation appears to go amuck! The electron dense membrane bound bodies observed in the sporangium cytoplasm appear to be mobilized but instead of forming small vesicles which will later fuse to form cleavage furrows, these electron dense bodies appear to give rise to extensive sheets of membrane which accumulates in myelin figure-like configurations between the spores as they are cleaved (Figs. 9-10). How these sheets of membranes are actually formed remains somewhat of a mystery, the fact that they are formed is all too obvious. And how these myelin-like configurations perturb the process of spore cleavage is also unknown. The obvious questions as to why abnormal spores are formed during zoosporogenesis in dilute Sørensen's phosphate buffer cannot be answered at present, but it does provide fertile ground for some speculation.

Gay and Greenwood 1966, Gay *et al.* (1971) and Heath (1976) have suggested that in *Saprolegnia*, at least, the electron dense body is the source of vesicles which fuse with the central vacuole membrane which ultimately cleaves the sporangial cytoplasm. Heath (1976) suggested that phospholipase A and B and a diesterase would be the enzymes to be expected to function during zoospore cleavage. But as he correctly notes "at present this hypothesis rests entirely on circumstantial evidence and awaits detailed autoradiographic and enzyme localization investigations for confirmation." But perhaps of more immediate concern, is it possible to isolate these membrane bound organelles, do they contain phospholipases, and if these enzymes are present, is their activity affected by the addition of phosphate? These are some of the possible problems which should be investigated.

Colchicine - benomyl effect

From the previous discussion of spore cleavage in *Allomyces* it is clear that there are arrays of microtubules which radiate from the functional and non-functional kinetosome at the time when the cleavage vesicles are forming. Heath (1976) suggested that the microtubule arrays found around the nuclei in some biflagellate fungi (Oomycetes) might serve to delimit and direct the cleavage furrows as the central vacuole expands during zoosporogenesis. The apparent absence of micro-

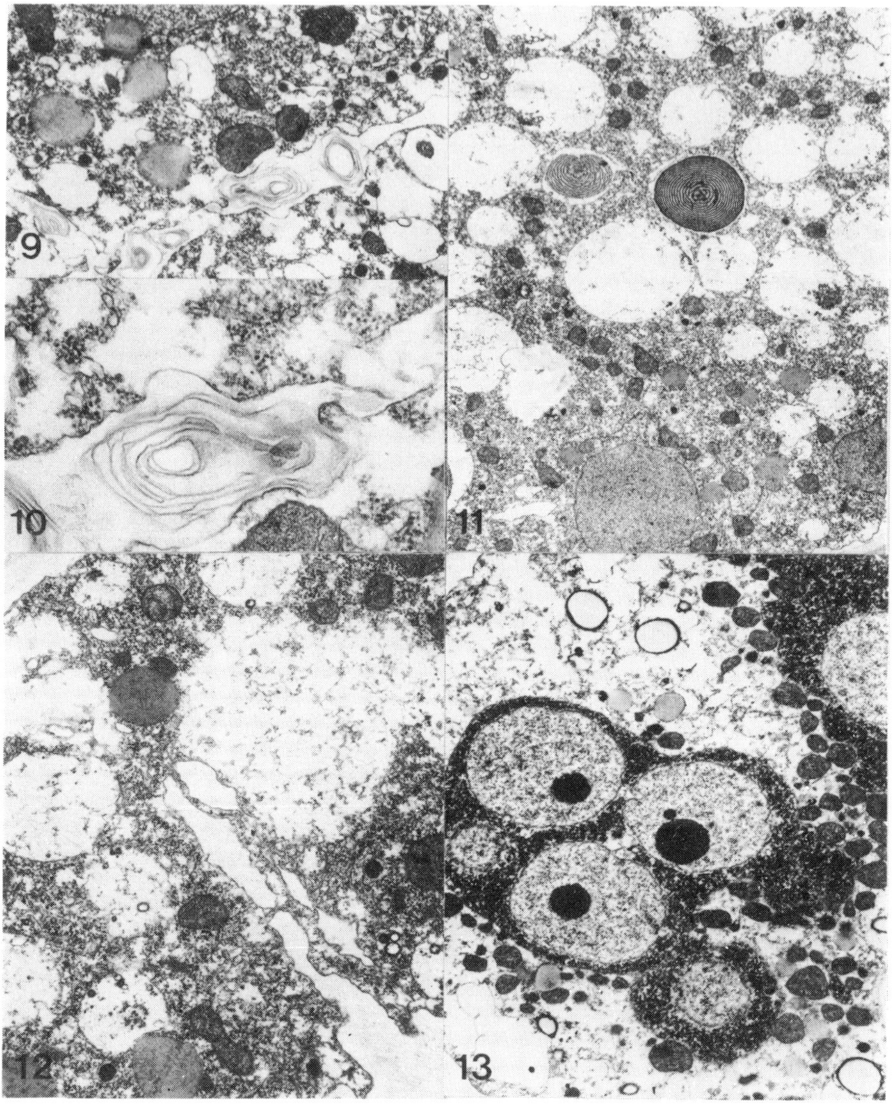

Figs. 9-10. Membrane accumulation between cleaved spores, phosphate buffer, 10 mM, pH 7,0. X 12,000, 36,000. Figs. 11-12. (Colchicine 2 mg/ml). Vacuoles after mobilization of electron dense content. Fig. 12. Fusion and collapse of vesicles to form a cleavage furrow. X 6,000, 12,000. Fig. 13. Multinucleate spore (colchicine 2 mg/ml). X 13,000.

tubules radiating from the kinetosomal region prior to spore cleavage in previous ultrastructural descriptions of zoosporogenesis in the uniflagellate fungi may be a matter of oversight, lack of serial sections, or simply that the fixation image was not sufficient to observe these minute structures.

The cytoplasmic microtubules observed in *Allomyces* sporangia during zoosporogenesis can be envisioned as functioning as suggested by Heath (1976). The cytoplasmic microtubules flare out from the kinetosomes which are in intimate contact with the nuclear membrane. These microtubules encapsulate a certain minimum volume of cytoplasm with the nucleus and its associated kinetosomes being on the periphery of this spherical mass of cytoplasm. The cytoplasmic microtubules establish the boundary of the cytoplasm which will eventually be packaged in the fully differentiated spore. According to this model the flagellum, as it grows, would not pass through an area of cytoplasm "protected" by the cytoplasmic microtubules. Whether this is true or not will only be determined by extensive serial section reconstructions of sporangia during zoosporogenesis. Once the flagellum has formed, the fusion of the cleavage vesicles to form the cleavage furrows-spore membrane would then take place in those regions of the cytoplasm between the cytoplasmic domains "protected" by the kinetosomal associated cytoplasmic microtubules. The organelles to be incorporated into the spore would be those which happened to fall within the cytoplasmic domain controlled by each array of cytoplasmic microtubules.

If this hypothesis contains any truth it would be expected that those agents which interfere with the polymerization of microtubules would interfere with the orderly division of the cytoplasm into spores. When *A. macrogynus* zoosporangia (48 hr old) are induced to form spores in a solution of colchicine (2,4,6,8 mg/ml) or benomyl (100 μg/ml) the spores which are released from the sporangia are abnormal. The spores which emerge are frequently multiflagellate and multinucleate. Some spores are extremely large and multiflagellate and the flagella are observed to extend from the spore in one area or the flagella may be associated in groups (Figs. 13-15). The spores are observed to swim in a highly erratic fashion. Frequently uniflagellate spores are observed which are much smaller than normal and the nuclear cap appears to be greatly reduced in size. In high concentrations of colchicine (8 mg/ml) the spores are frequently aflagellate (85-95%).

During zoosporogenesis in the presence of colchicine or benomyl no microtubules are observed in the sporangial cytoplasm. The electron dense material of the membrane bound in-

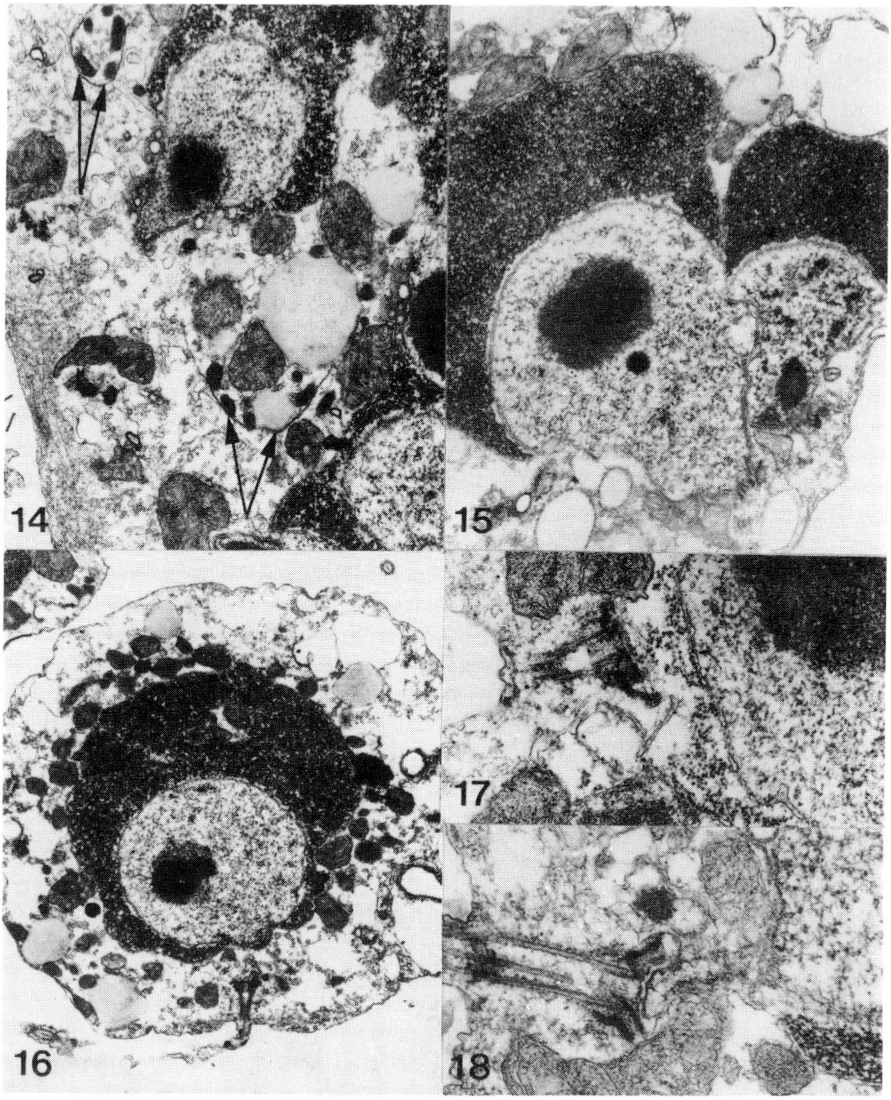

Fig. 14. Multinucleate spore (colchicine 2 mg/ml). Double arrows = side body complex. X 13,000. Fig. 15. Binucleate spore, separate nuclear caps. X 15,000. Figs. 16-18. Inhibition of cytoplasmic microtubule polymerization (colchicine 8 mg/ml). Contact between the nucleus and kinetosome may be lost or tenuous. X 10,000, 30,000, 38,000.

clusions disappears and large vacuoles are left in the cytoplasm (Figs. 11-12). During cleavage of the sporangial cytoplasm these vacuoles appear to collapse, coalesce and form the cleavage furrows (Fig. 12). Although some cleavage vesicle-like structures are seen in the cytoplasm there does not appear to be an orderly inclusion of cleavage vesicles into the cleavage furrow membrane.

Figs. 13-14 are of hexa- and tetra-nucleate spores. The multiple nuclei may be enclosed by a common nuclear cap (Figs. 13-14) or each nucleus may have its own (Fig. 15). The side body complex does not necessarily come to be adjacent to the plasmalemma (Fig. 14), and the nuclei can assume intimate contact (Fig. 15). At present there is no evidence for nuclear fusion in these abnormal spores.

The cytoplasmic microtubules which normally surround the nucleus and nuclear cap are not formed (Figs. 16-18) and the nucleus may become detached from the proximal end of the functional kinetosome (Figs. 16-17) or only a tenuous connection between the kinetosome and the nucleus is maintained.

Inhibition of the polymerization of cytoplasmic microtubules during zoosporogenesis leads to abnormal spore formation. The formation of large vacuoles during zoosporogenesis in solutions of colchicine or benomyl and the fusion of these vacuoles to form the cleavage furrows suggests that there is some sort of regulatory control between cytoplasmic microtubule formation and cleavage vesicle formation and fusion.

The possible use of temperature sensitive mutants of sporogenesis

For the characterization of the molecular events which occur during zoosporogenesis it would be desirable to have a system by which the process of zoosporogenesis could be stopped and started at will. The question which should be asked is, is it possible to isolate temperature sensitive cell cycle mutants or sporulation mutants. Within the last 10 years it has been possible to isolate temperature sensitive cell cycle mutants from a number of organisms: *Saccharomyces cerevisiae* (Hartwell 1974, 1978, Pringle 1978, Simchen 1978); *Schizosaccharomyces pombe* (Nurse *et al.* 1976, Thuriaux *et al.* 1978); *Aspergillus nidulans* (Morris *et al.* 1977); and the slime mold *Physarum polycephalum* (Gingold *et al.* 1976, Wheals *et al.* 1976).

At the present time the isolation of sporulation mutants in the aquatic fungi is but a possibility but one which should be investigated. Among the aquatic fungi, *Allomyces (Euallomyces)* and/or *Blastocladiella emersonii* would appear to be the organisms of choice. The haplontic-diplontic division of the *Allomyces (Euallomyces)* life cycle into separate generations would suggest that either *Allomyces macrogynus* or *A. arbuscula* would be suitable. Recently it has been possible to isolate a number of Mendelian auxotrophic mutants as well as numerous Mendelian sexual mutants (Olson and Nielsen 1980, Olson *et al.* 1980, Nielsen and Olson 1980). The isolation, characterization, and mapping of these mutants demonstrates that *Allomyces* can be used as an organism for genetic studies.

The life cycle of *B. emersonii* would appear to be primarily haplontic and thus especially favorable for the isolation of recessive mutants (Olson and Reichle 1978). The thick walled resting sporangia undergo meiosis at germination and release haploid meiospores. The thin walled zoosporangia appear to also be haploid and the reestablishment of the diploid portion of the life cycle appears to occur when the thin walled zoosporangium is converted into a resting sporangium. In *B. emersonii* sexuality has not been established, thus temperature sensitive sporulation mutants could not be crossed and analysed genetically. However, for an analysis of sporulation this is not really a requirement.

In the following discussion a possible scheme for the isolation of temperature sensitive sporulation mutants in *A. macrogynus* and in *B. emersonii* will be outlined. The system which suggests itself is to take a suspension of meiospores from either organism and irradiate them with X-rays. In our experience with *Allomyces* UV light can and does induce mutations, but as the organism is a coenocyte a back mutation to wild type or a spontaneous suppressor mutation will invariably lead to the loss of the mutant strain. For this reason the mutations of choice would be large deletion mutations which lead to a stable phenotype and which cannot revert to wild type.

In the case of *Allomyces* the irradiated meiospore suspension is plated on freshly prepared agar and sufficient inoculum is plated so that the gametophytic plants which develop will immediately release gametes and selfing will occur. The diploid sporophytes would then be homozygous for the presumed mutation. After 48 hrs of growth the zoosporangia are scraped from the surface of the agar, suspended in dilute salts solution and then filtered through a coarse sterile nylon filter to remove the hyphal fragments from the zoosporangia. The zoosporangia suspension is incubated at 35°C (the restrictive temperature). Zoospore release begins after 50-60 min at 35°C and the incubation of the zoosporangia suspension is for 2 hrs, at which time one would expect that all sporangia that will release spores have done so. The zoospore-zoosporangium suspension is then filtered through a fine mesh sterile nylon filter (pore size 30 μm) which will retain the sporangia but allow the spores to pass through. The sporangia are then washed throughly with a large volume of sterile dilute salts solution to remove those spores which are retained on the filter. The zoosporangia are then resuspended in fresh dilute salts solution by aseptically transferring the nylon filter to a fresh dish of dilute salts. The zoosporangia are then incubated at room temperature (the permissive temperature) for 2 hrs, after which the suspension is again filtered through a sterile fine mesh nylon filter to remove the sporangia from the spore suspension. The spore suspension is then concentrated by low speed centrifugation and the spores are plated on nutrient agar.

The colonies which develop on the nutrient agar are presumed to be derived from zoosporangia which could not complete zoosporogenesis at the restrictive temperature (35°C) but could at the permissive temperature (23°C). In order to test the colonies which develop on the nutrient agar it is necessary to replica plate the colonies. A replica plating device which was developed for replicating *Allomyces* colonies (Olson *et al.* 1980) can equally well be used with *B. emersonii* plants. The device basically consists of a plastic plate which contains 32 wells and a replica head which consists of a plastic disc fitted with 32 steel pins which correspond to the holes in the well plate.

The zoosporangia of a colony to be replica plated are scraped from the surface of the agar and suspended in a few drops of dilute salts solution in a well of the well plate. When each well of the plate has been inoculated, the well plate is incubated at the restrictive temperature (35°C) for 3 hrs, after which time the well plate is replicated to nutrient agar. After 3 hrs incubation at the permissive temperature (23°C) the well plate is again replicated. Both plates are incubated at 35°C for 1-2 days at which time the results are scored. A temperature sensitive sporulation mutant will be found not to grow on the plate from the replication of the sporangium suspension incubated at the restrictive temperature, while growth will be found on the plate from the replica of the suspension incubated at the permissive temperature. Wild type colonies will grow on both of the replica plates. Because of the uncertaintity of the number of colonies which must be replica plated to ensure that mutants are isolated it might be useful to repeat the selection process 2-3 times to enrich for the temperature sensitive mutants. If mutants are isolated from *Allomyces* an allele test must be made to determine whether the mutants are merely descendents of the original mutant or the mutants are non-allelic. If one were to attempt to isolate temperature sensitive sporulation mutants in *B. emersonii* the only basic change in the procedure would be the elimination of the step at which the gametophytic colonies are selfed. Whether it will prove possible to isolate temperature sensitive sporulation mutants remains to be seen. The value of having a "bank" of such mutants in biochemical as well as cytological studies of sporogenesis would be great. It remains to be seen just how much work will be necessary to finally isolate such mutants.

ACKNOWLEDGEMENTS

Portions of this work were supported by funds from the Danish Natural Science Research Council.

REFERENCES

Adelman, T.G. and Lovett, J.S. (1972). Synthesis of ribosomal protein without de novo ribosome production during differentiation in *Blastocladiella emersonii*, *Biochemical and Biophysical Research Communications* 49, 1174-1182.

Adelman, T.G. and Lovett, J.S. (1974a). Evidence for a ribosome-associated translation inhibitor during differentiation of *Blastocladiella emersonii*, *Biochimica et Biophysica Acta* 335, 236-245.

Adelman, T.G. and Lovett, J.S. (1974b). Ribosome function in vitro and in vivo during the life cycle of *Blastocladiella emersonii*, *Biochimica et Biophysica Acta* 349, 240-249.

Aronson, J.M. and Sandstedt, V. (1978). *Allomyces macrogynus*. *In* "Lower Fungi in the Laboratory" (Ed. M.S. Fuller). pp. 51-52. University of Georgia, Athens.

Barr, D.J.S. (1978). Taxonomy and phylogeny of chytrids, *Biosystems* 10, 153-165.

Barron, J.L. and Hill, E.P. (1974). Ultrastructure of zoosporogenesis in *Allomyces macrogynus*, *Journal of General Microbiology* 80, 319-327.

Barstow, W.E. (1978). *Allomyces macrogynus*. *In* "Lower Fungi in the Laboratory" (Ed. M.S. Fuller). pp. 49-50. University of Georgia, Athens.

Barstow, W.E. and Lovett, J.S. (1972). Developmental changes in a "macrotubular" system during growth and differentiation of the fungus *Blastocladiella emersonii*, Abstract *American Journal of Botany* 59 (Suppl.), 665.

Barstow, W.E. and Lovett, J.S. (1973). An ultrastructural study of early sporulation in *Blastocladiella emersonii*, Abstract *American Journal of Botany* 60 (Suppl.), 19.

Barstow, W.E. and Lovett, J.S. (1974). Apical vesicles and microtubules in rhizoids of *Blastocladiella emersonii*: effects of actinomycin D and cycloheximide on development during germination, *Protoplasma* 82, 103-117.

Barstow, W.E. and Lovett, J.S. (1975). Formation of gamma particles during zoosporogenesis in *Blastocladiella emersonii*, *Mycologia* 67, 518-529.

Blondel, B. and Turian, G. (1960). Relation between basophilia and fine structure of cytoplasm in the fungus *Allomyces macrogynus* Em, *Journal of Biophysical and Biochemical Cytology* 7, 127-134.

Burke, D.J., Seale, T.W. and McCarthy, B.J. (1972). Protein and ribonucleic acid synthesis during the diploid life cycle of *Allomyces arbuscula*, *Journal of Bacteriology* 110, 1065-1072.

Elsner, P.R., Vandermolen, G.E., Horton, J.C. and Bowen, C.C. (1970). Fine structure of *Phytophthora infestans* during sporangial differentiation and germination, *Phytopathology* 60, 1765-1772.

Emerson, R. (1958). Mycological organization, *Mycologia* 50, 588-621.

Fähnrich, P. (1974a). Untersuchungen zur Entwicklung des Phycomyceten *Allomyces arbuscula*. I. Einfluss von Inhibitoren der Protein und Nucleinsäuresynthese auf die Differenzierung von Gametangien, *Archives of Microbiology* 98, 85-92.

Fähnrich, P. (1974b). Untersuchungen zur Entwicklung des Phycomyceten *Allomyces arbuscula*. II. Einfluss von Inhibitoren der Protein und Nucleinsäuresynthese auf die Gametogenese, *Archives of Microbiology* 99, 147-153.

Fähnrich, P. (1976). Inhibition of sporogenesis in *Allomyces arbuscula*, *Naturwissenschaften* 63, 247.

Fuller, M.S. (1966). Structure of the uniflagellate zoospores of aquatic phycomycetes. *In* "The Fungus Spore" Proceedings of the Symposium of the Colston Research Society (Ed. M.F. Madelin). Vol. 18, pp. 67-84. Butterworths, London.

Fuller, M.S. (1977). The zoospore, hallmark of the aquatic fungi, *Mycologia* 69, 1-20.

Gauriloff, L.P. (1978a). *Rhizophydium sphaerotheca*. *In* "Lower Fungi in the Laboratory" (Ed. M.S. Fuller). pp. 9-10. University of Georgia, Athens.

Gauriloff, L.P. (1978b). *Rhizophydium sp*. *In* "Lower Fungi in the Laboratory" (Ed. M.S. Fuller). pp. 11-12. University of Georgia, Athens.

Gauriloff, L.P. (1978c). *Chytridium confervae*. *In* "Lower Fungi in the Laboratory" (Ed. M.S. Fuller). pp. 21-22. University of Georgia, Athens.

Gay, J.L. and Greenwood, A.D. (1966). Structural aspects of zoospore production in *Saprolegnia ferax* with particular reference to the cell and vacuolar membranes. *In* "The Fungus Spore" Proceedings of the Symposium of the Colston Research Society (Ed. M.F. Madelin). Vol. 18, pp. 95-108. Butterworths, London.

Gay, J.L., Greenwood, A.D. and Heath, I.B. (1971). The formation and behaviour of vacuoles (vesicles) during oosphere development and zoospore germination in *Saprolegnia*, *Journal of General Microbiology* 65, 233-241.

Gingold, E.C., Grant, W.D., Wheals, A.E. and Wren, M. (1976). Temperature sensitive mutants of the slime mold *Physarum polycephalum* II. Mutants of the plasmodial phase, *Molecular and General Genetics* 149, 115-119.

Goldstein, A. and Cantino, E.C. (1962). Light-stimulated polysaccharide and protein synthesis by synchronized, single generations of *Blastocladiella emersonii*, *Journal of General Microbiology* 28, 689-699.

Griffin, D.H. and Breuker, C. (1969). Ribonucleic acid synthesis during the differentiation of sporangia in the watermold *Achlya*, *Journal of Bacteriology* 98, 689-696.

Hartwell, L.H. (1974). *Saccharomyces cerevisiae* cell cycle, *Bacteriological Reviews* 38, 164-198.

Hartwell, L.H. (1978). Cell division from a genetic perspective, *Journal of Cell Biology* 77, 627-637.

Heath, I.B. (1976). Ultrastructure of freshwater Phycomycetes. *In* "Recent Advances in Aquatic Mycology" (Ed. G. Jones). pp. 603-650. London.

Heath, I.B. and Greenwood, A.D. (1971). Ultrastructural observations on the kinetosomes and Golgi bodies during the asexual life cycle of *Saprolegnia*, *Zeitschrift für Zellforschung und mikroskopische Anatomie* 112, 371-389.

Hennessy, S.W. and Cantino, E.C. (1972). Lag-phase sporogenesis in *Blastocladiella emersonii:* induced formation of unispored plantlets, *Mycologia* 64, 1066-1087.

Hohl, H.R. and Hamamoto, S.T. (1967). Ultrastructural changes during zoospore formation in *Phytophthora parasitica*, *American Journal of Botany* 54, 1131-1139.

Klebs, G. (1900). Zur Physiologie der Fortpflanzung einiger Pilze lll, *Jahrbücher für wissenschaftliche Botanik* 32, 80-203.

Lange, L. and Olson, L.W. (1979). The uniflagellate phycomycete zoospore, *Dansk Botanisk Arkiv* 33 (2), 7-95.

Lessie, P.E. and Lovett, J.S. (1968). Ultrastructural changes during sporangium formation and zoospore differentiation in *Blastocladiella emersonii*, *American Journal of Botany* 55, 220-236.

Lodi, W.R. and Sonneborn, D.R. (1974). Protein degradation and protease activity during the life cycle of *Blastocladiella emersonii*, *Journal of Bacteriology* 117, 1035-1042.

Lovett, J.S. (1975). Growth and differentiation of the water mold *Blastocladiella emersonii:* Cytodifferentiation and the role of ribonucleic acid and protein synthesis, *Bacteriological Reviews* 39, 345-404.

Moore, R.T. (1968). Fine structure of mycota. 13. Zoospore and nuclear cap formation in *Allomyces*. *In* "Mycological Studies Honoring John N. Couch" (Ed. W.J. Koch). pp. 147-165.

Morris, N.R., Felden, R.A., Gealt, M.A., Nardi, R.V., Schier-Neiss, G. and Sanders, M.M. (1977). The *Aspergillus* nucleus: Histones, chromatin, and tubulin. *In* "Genetics and Physiology of *Aspergillus*" (Ed. J.E. Smith and J.A. Pateman). pp. 267-279. Academic Press, New York, London and San Francisco.

Morrison, P.J. (1978). *Allomyces macrogynus*. *In* "Lower Fungi in the Laboratory" (Ed. M.S. Fuller). pp. 45-46. University of Georgia, Athens.

Murphy, M.N. and Lovett, J.S. (1966). RNA and protein synthesis during zoospore differentiation in synchronized cultures of *Blastocladiella*, *Developmental Biology* 14, 68-95.

Nielsen, T.A.B. and Olson, L.W. (1980). Mendelian mutants of *Allomyces*, *Transactions of the British Mycological Society* (submitted).

Nurse, P., Thuriaux, P. and Nasmyth, K. (1976). Genetic control of the cell division cycle in the fission yeast *Schizosaccharomyces pombe*, *Molecular and General Genetics* 146, 167-178.

Olson, L.W. (1974). Meiosis in the aquatic phycomycete *Allomyces macrogynus*, *Comptes Rendus des Travaux du Laboratoire Carlsberg* 40, 113-124.

Olson, L.W. (1980). *Allomyces neo-moniliformis* gametogenesis. The Cystogenes life cycle, *Protoplasma* (submitted).

Olson, L.W. and Rønne, M. (1975). Induction of abnormal male gametes and androgenesis in the aquatic phycomycete *Allomyces*, *Protoplasma* 84, 327-344.

Olson, L.W. and Borkhardt, B. (1978). Polyploidy and its control in *Allomyces macrogynus*, *Transactions of the British Mycological Society* 71, 65-76.

Olson, L.W. and Reichle, R. (1978). Synaptonemal complex formation and meiosis in the resting sporangium of *Blastocladiella emersonii*, *Protoplasma* 97, 261-273.

Olson, L.W. and Nielsen, T.A.B. (1980). Isolation of auxotrophic mutants of *Allomyces macrogynus*, *Mycologia* (in press).

Olson, L.W., Barr, C. and Nielsen, T.A.B. (1980). A simple device for replica plating aquatic Phycomycetes, *Mycologia* (in press).

Olson, L.W., Nielsen, T.A.B., Heldt-Hansen, H.P. and Grant, N.G. (1980). Maleness, its inheritance and control in the aquatic phycomycete *Allomyces macrogynus*, *Transactions of the British Mycological Society* (submitted).

Powell, M.J. (1978). Phylogenetic implications of the microbody-lipid globule complex in zoosporic fungi, *Biosystems* 10, 167-181.

Pringle, J.R. (1978). The use of conditional lethal cell cycle mutants for temporal and functional sequence mapping of cell cycle events, *Journal of Cellular Physiology* 95, 393-400.

Renaud, F.L. and Swift, A. (1964). The development of basal bodies and flagella in *Allomyces arbusculus*, *Journal of Cell Biology* 23, 339-354.

Silverman, P.M. and Epstein, P.M. (1975). Cyclic nucleotide metabolism coupled to cytodifferentiation of *Blastocladiella emersonii*, *Proceedings of the National Academy of Sciences U.S.A.* 72, 442-446.

Simchen, G. (1978). Cell cycle mutants, *Annual Review of Genetics* 12, 161-191.

Sparrow, F.K. (1969). "Aquatic Phycomycetes" (Ed. E.J. Brill). pp. 1187. The University of Michigan Press, Ann Arbor.

Stumm, C. and Croes, A.F. (1975). Polyribosomes in different stages of the life cycle of the water mold *Allomyces arbuscula*, *Archives of Microbiology* 102, 117-122.

Teter, H.E. (1944). Isogamous sexuality in a new strain of *Allomyces*, *Mycologia* 36, 194-210.

Thuriaux, P., Nurse, P. and Carter, B. (1978). Mutants altered in the control coordinating cell division with cell growth in the fission yeast *Schizosaccharomyces pombe*, *Molecular and General Genetics* 161, 215-220.

Timberlake, W.E., McDowell, L., Cheney, J. and Griffin, D.H. (1973). Protein synthesis during the differentiation of sporangia in the water mold *Achlya*, *Journal of Bacteriology* 116, 67-73.

Travland, L.B. and Whisler, H.C. (1971). The ultrastructure of *Harpochytrium hedinii*, *Mycologia* 63, 767-789.

Wheals, A.E., Grant, W.D. and Jockusch, B.M. (1976). Temperature-sensitive mutants of the slime mold *Physarum polycephalum*. I. Mutants of the amoebal phase, *Molecular and General Genetics* 149, 110-114.

Youatt, J. (1976). Sporangium formation in *Allomyces* throughout the growth cycle, *Transactions of the British Mycological Society* 67, 159-161.

ULTRASTRUCTURAL ASPECTS OF OOSPORE DIFFERENTIATION

G.W. Beakes

Department of Plant Biology, The University, Newcastle upon Tyne, NE1 7RU, U.K.

INTRODUCTION

The ultrastructural study of sexual resting spores in oomycete fungi has lagged behind that of the various asexual dispersive spores due to technical problems encountered in producing and, above all, satisfactorily fixing and embedding thick-walled metabolically inactive, spores. This, together with the reluctance of oospores to germinate in the laboratory has meant most ultrastructural accounts of oogenesis stop at, or before, the mature oospore stage. The ability to induce near synchronous oogonium development by the use of either correct mating types in heterothallic species (Hemmes and Bartnicki-Garcia, 1975) or by suitable manipulation of the environment in homothallic species (Haskins *et al.*, 1976; Beakes and Gay, 1977) means oogenesis is potentially an excellent system in which to correlate the structural, physiological and biochemical events associated with the progressive cessation and re-establishment of active metabolism. Already cell biologists are using sexual differentiation in *Achlya* as a model system in which to unravel the effects of steroid hormones at the cellular level (Horgen, 1977).

This present review is not intended to be comprehensive but rather assesses the contribution ultrastructural studies have made to the understanding of selected aspects of oospore differentiation which have particularly interested the author. In general, studies of differentiation in the two major groups of oomycete fungi, the Saprolegniales and Peronosporales has tended to occur independently. Therefore, it is intended to illustrate a possible developmental hierarchy ranging from the relatively unspecialized, largely saprophytic, Saprolegniaceae through the generally facultatively parasitic

Pythiaceae to the highly specialized, obligately parasitic, Peronosporaceae and Albuginaceae. Whether this represents a true evolutionary pathway or simply reflects changes as a result of an increasingly specialized life style is open to question.

The illustrations used in this account are mostly taken from studies of oospore differentiation in two homothallic *Saprolegnia* species, (*S. furcata*, Beakes and Gay, 1977, 1978a, 1978b and a largely emasculate isolate of *S. ferax*, Beakes, 1980a,b,c,d) and in *Pythium oligandrum* (Beakes, unpublished).

SPORE FORMATION

Sexual reproduction in oomycete fungi is generally characterized by the absence of motile gametes (Dick, 1969). However, the fact that in the marine parasite *Lagenisma coscinodisci* the male and female gametangia are simply modified cysts derived from zoomeiospores (Schnepf *et al.*, 1978a) points to the evolution of oogonia and antheridia from modified zoosporangia. In the majority of oomycete fungi, sexual oospores are differentiated within spherical or club shaped, often papillate, structures (oogonia) which usually develop as terminal swellings of the vegetative hyphae (Fig. 1). In most species the smaller male gametangia (antheridia) develop concomitantly with the oogonia (Fig. 1). Indeed, in *Achlya* gametangial differentiation has been shown to be regulated by

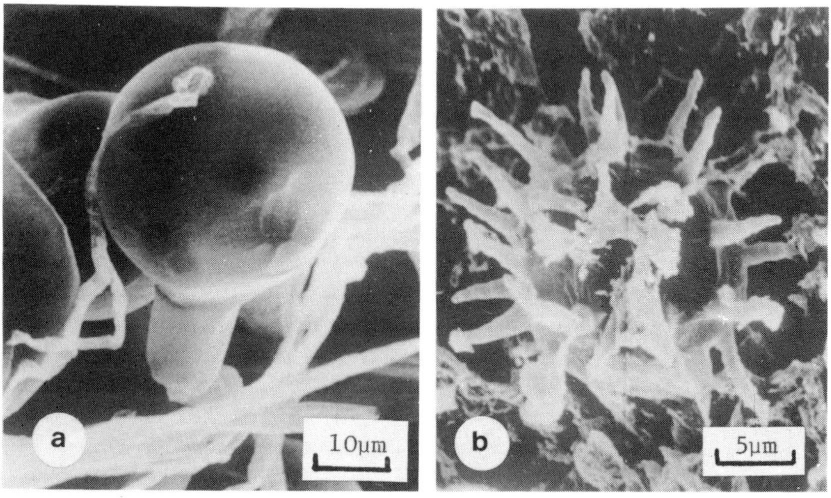

Fig. 1. Scanning electron micrographs of oogonia of Saprolegnia ferax *(1a) and* Pythium oligandrum *(1b)*

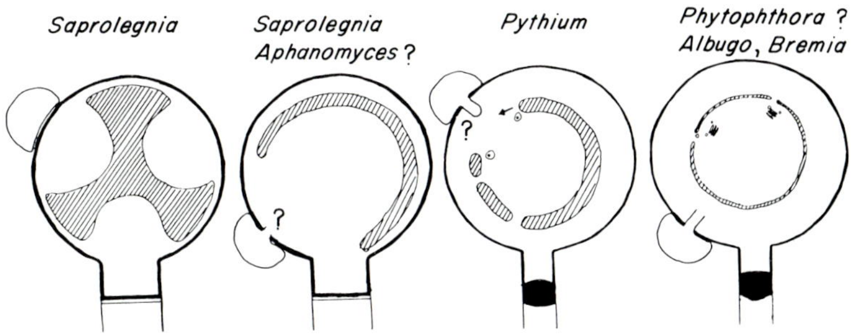

Fig. 2. Diagram summarizing the different patterns of oosphere cleavage in oomycete fungi

the sequential secretion of female (antheridiol) and male (oogonial) hormones (see Horgen, 1977). Apart from the demonstration that antheridiol induces the formation of clusters of wall vesicles preceeding antheridial branch initiation (Mullins and Ellis, 1974) little is known of the precise effects of these hormones at the ultrastructural level. Since, in the emasculate isolate of *S. ferax* normal differentiation occurs in the absence of antheridia, the oogonium clearly contains all the components necessary for oospore development (Beakes, 1980a).

In the Saprolegniales spore formation within multisporiate oogonia appears mechanistically similar to that described in zoosporangia (Gay *et al.*, 1971). Oosphere differentiation within delimited oogonia is a three phase process involving a) the development of a central vacuole followed by synchronous meiotic division of nuclei in the peripheral cytoplasm; b) the autolysis of supernumerary gametic nuclei and the redistribution of cytoplasm and vacuoles to form discrete, uninucleate, oosphere initials (Fig. 2a) and c) the fusion of adjacent tonoplast and plasma membranes around the hemispherical initials, with the exclusion and lysis of the small amounts of intervening cytoplasm (Fletcher, 1978; Beakes and Gay, 1978a). It has been suggested that the central vacuole is derived from the coalescence and expansion of dense body vesicles (DBV; Gay *et al.*, 1971; Fletcher, 1979; see Fig. 13). Calcium ions appear to be an essential requirement for normal cytoplasmic cleavage, since oosphere differentiation is severely reduced in both Ca^{2+} deficient oogonia of *S. diclina* (Fletcher, 1979) and in streptomycin (cation antagonist)-induced oogonia of *S. furcata* (Beakes and Gay, 1980). Fletcher (1979) has demonstrated that DBV fusion but not formation is reduced in Ca^{2+} deficient regimes. The microtubule inhibitor colchicine results in the development of irregularly sized spores (Slifkin, 1967; Schnepf and Heinzmann, 1980; Beakes,

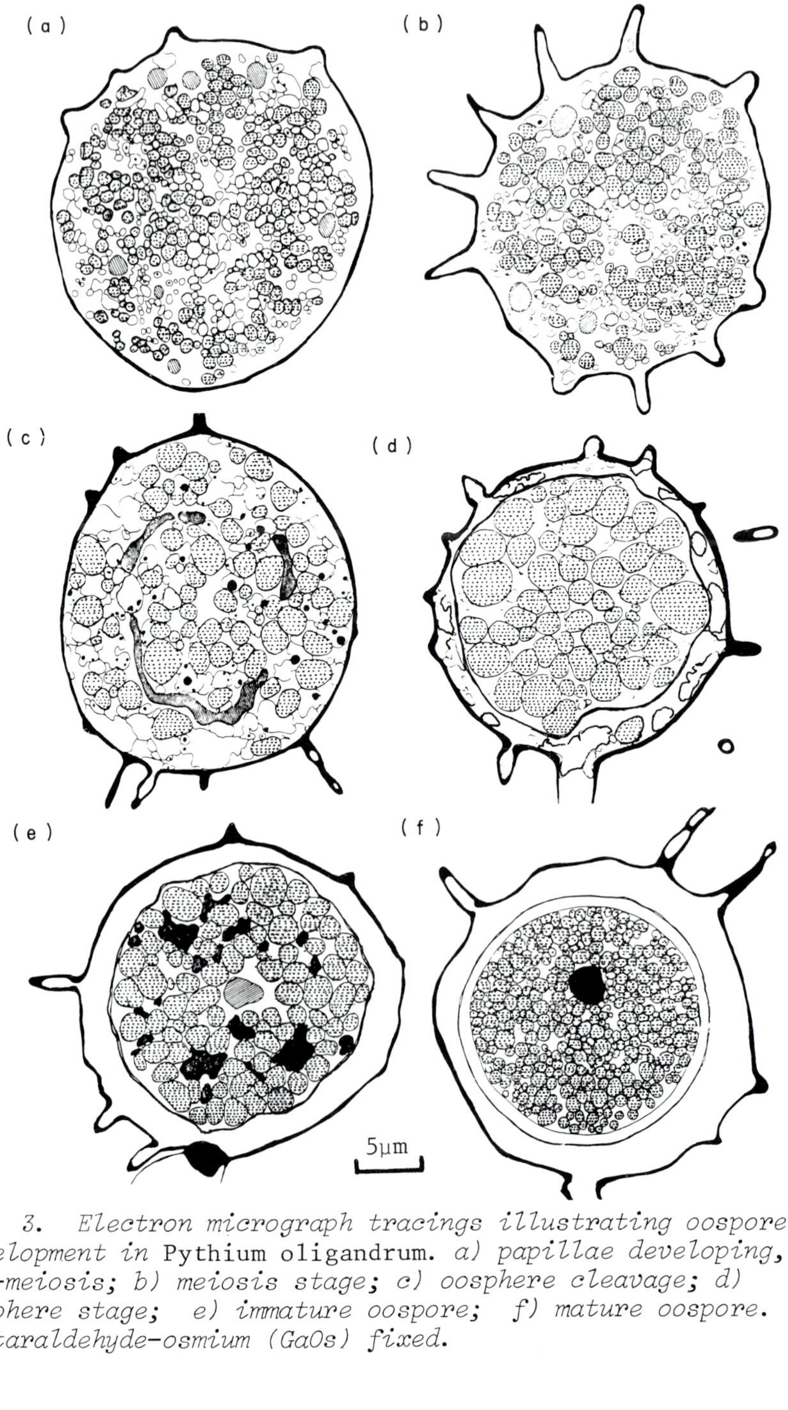

Fig. 3. Electron micrograph tracings illustrating oospore development in Pythium oligandrum. *a) papillae developing, pre-meiosis; b) meiosis stage; c) oosphere cleavage; d) oosphere stage; e) immature oospore; f) mature oospore. Glutaraldehyde-osmium (GaOs) fixed.*

unpublished observations). It is not known whether this is due to its effect on nuclear or cytoplasmic microtubules, although, so far, there has been no direct observation of cytoplasmic microtubules in oosphere initial development (Gay *et al.*, 1971; Beakes and Gay, 1978a).

Oosphere differentiation in unisporiate oogonia, characteristic of the Peronosporales and some members of the Saprolegniales, is much more difficult to follow microscopically. However, it seems likely that in the unisporiate Saprolegniales (such as *S. megasperma*, Beakes unpublished and *Aphanomyces*, Traquair and McKeen, 1980) cleavage is identical to that described above, since the oogonia first acquire a central vacuole, which presumably undergoes eccentric redistribution prior to cleavage (Fig. 2b). This does not occur in Peronosporalean oogonia, as shown in the micrograph tracings illustrating oospore development in *Pythium oligandrum* (Fig. 3). In the Peronosporales the oogonium cytoplasm becomes differentiated into an outer periplasmic and an inner oosphere phases (Fig. 3c,d) by the development of a specialized cleavage furrow (Fig. 4). In the Pythiaceae the periplasmic zone is usually narrow (0.5-1.5 μm; Fig. 2c, 4c; Hemmes and Bartnicki-Garcia, 1975; Haskins *et al.*, 1976; McKeen, 1975), whereas in the Peronosporaceae and Albuginaceae

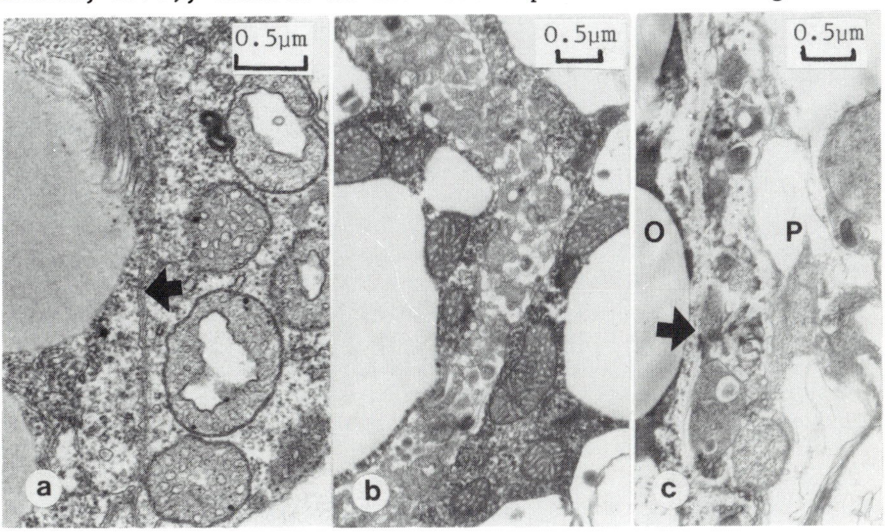

Fig. 4. Peronosporalean cleavage furrows: 4a) narrow cisternum in Bremia lactucae; *4b) broader vesicle system in* Pythium oligandrum; *note inclusions; 4c) differentiated oosphere (O) and periplasm (P) in* P. oligandrum. *Developing primary wall is arrowed GaOs fixed. (4a courtesy of Sargent* et al., *1977).*

it is much more extensive (5-15 μm; Fig. 2d; Sargent *et al.*, 1977; Tewari and Skoropad, 1977). Unfortunately, of all the accounts of Peronosporalean oosphere differentiation so far published only that for *Bremia lactucae* illustrates cleavage furrow development in any detail (Sargent *et al.*, 1977). In this species the oosphere is delimited by a narrow, 40 nm, cisternum (Fig. 4a) thought to be derived from Golgi-vesicles and showing connections with elaborate membranous whorls. The cleavage apparatus in *Pythium acanthicum* is also derived from a narrow, endoplasmic reticulum (ER)-like system (Haskins *et al.*, 1976). However, rather unexpectedly, the cleavage apparatus (Fig. 3c) in the closely related *P.oligandrum* forms a much wider, 0.8-1.0 μm diameter cleavage vesicle containing complex membranous and particulate inclusions, and electron dense granules (dense bodies?) (Fig. 4b). Following oosphere formation (Fig. 3d), the cleavage vesicle can still be discerned (Fig. 4c), although the inclusions are less uniform in appearance. A similar, 0.5-1.0 μm zone, also occurs around differentiated oospheres of *Albugo*, and shows similar staining properties to the oosphere dense body/finger print vesicles (Tewari and Skoropad, 1977). Therefore, although the cleavage apparatus is more specialized in Peronosporalean oomycetes, it also may be derived from the dense body apparatus or another component of the endomembrane system (Fig. 2).

NUCLEAR BEHAVIOUR AND FERTILIZATION

Perhaps the two aspects of oogenesis which have been the cause of most historical controversy have been the nature of gametangial nuclear divisions and the true occurrence of fertilization in the Saprolegniales (see review by Dick and Win-Tin, 1973). Ultrastructural studies of gametangial nuclear division in the Saprolegniales have supported the growing body of evidence demonstrating the occurrence of gametic meiosis in these fungi (Howard and Moore, 1970; Ellzey, 1974; Heath, 1974; Beakes and Gay, 1977; Beakes 1980d; Traquair and McKeen, 1980). Although the prophase I nuclei contain condensed axial elements, the Saprolegniales lack the well-defined intervening synaptonemal complexes (SC) reported in most other fungi (see Ellzey and Huizar, 1977). Similar structures have also been reported in prophase I nuclei of *Lagenisma* (Lagenidiales, Schnepf *et al.*, 1978b) and *Sapromyces* (Leptomitales, Gotelli, 1979). In the latter a well-defined central core and lateral filaments of typical SC dimensions are shown. To date, there have been no published accounts of meiosis in a Peronosporalean oomycete. In *P.oligandrum*, synchronous meiosis occurs after the papillae

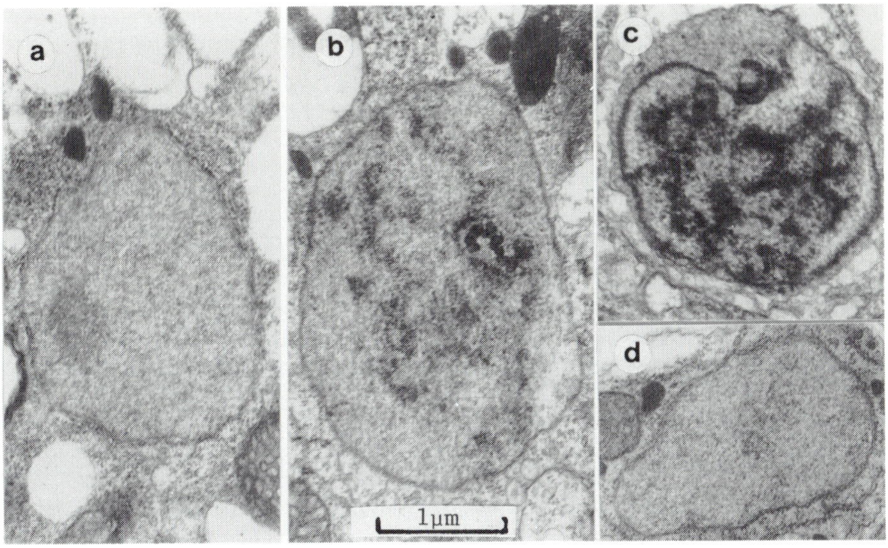

Fig. 5. Gametangial nuclei in P.oligandrum: *5a) pre-meiosis nucleus; 5b) prophase I (late?) nucleus; 5c) autolysing oogonial nucleus; 5d) gametic nucleus in antheridium. GaOs fixed.*

are formed (Fig. 3b). Prophase I nuclei are recognized by their larger size (compare Fig. 5a and b) and the presence of conspicuous, but ill-defined, electron dense axial elements (Fig. 5b).

One interesting feature of meiosis I in the Saprolegniales is that it is accompanied by characteristic changes in centriole structure which do not occur during mitosis (Howard and Moore, 1970; Beakes and Gay, 1977; Beakes, 1980d). The centrioles become embedded in a 'sheath' of amorphous electron dense material and surrounded by a 'cap' of endoplasmic reticulum. In some species the centrioles also double in length and separate slightly during prophase I; perhaps vestigial features associated with ancestral flagellum development (Beakes and Gay, 1977).

The second division of meiosis occurs within the same nuclear envelope as the first but is much more rapid and less synchronous. As a result ultrastructural accounts of meiosis II have only been given for *S.terrestris* (Howard and Moore, 1970) and *S.ferax* (Beakes, 1980d). The poles of the second divisional spindles are characterized by having single rather than paired centrioles.

Following meiosis the majority of the small gametic nuclei in the oogonia are autolysed (Fig. 5c), whereas those in the antheridium persist (Fig. 5d). In *Saprolegnia* nuclear autolysis occurs immediately after meiosis (Beakes and Gay, 1977; Beakes, 1980d) whereas in *Aphanomyces* (Traquair and McKeen, 1980) and the Peronosporales it is mostly delayed until after plasmogamy (Hemmes and Bartnicki-Garcia, 1975; Haskins *et al.*, 1976; McKeen, 1976; Traquair and McKeen, 1980). In these species the supernumery nuclei migrate to the oogonium periphery and in the Peronosporales become segregated into the periplasm at cleavage. One of the most remarkable and interesting features of oogonium nuclear behaviour is the precise regulation of nuclear degeneration so that only a single (or two in case of emasculate species) female gametic nucleus survives per oosphere. This may be under hormonal control since it is one of the few structural events clearly influenced by the presence of the antheridium (Beakes, 1980d).

In the *Saprolegnia*, the antheridium does not produce a fertilization tube until after oosphere differentiation (Fletcher, 1978; Beakes and Gay, 1978a). The fertilization tube ruptures both the oogonium and outer oosphere wall. In *Aphanomyces* (Traquair and McKeen, 1980) *Phytophthora* spp. (Hemmes and Bartnicki-Garcia, 1975; Hemmes and Ribeiro, 1977) fertilization tubes are initiated prior to oosphere differentiation, although it is uncertain whether plasmogamy preceeds cleavage (Fig. 2). In all cases fertilization tube development is associated with the accumulation of wall vesicles in both the adjacent male and female cytoplasm.

In all species so far examined ultrastructurally nuclear fusion (karyogamy) occurs before the onset of oospore maturation. Typically the fusion nuclei are larger than vegetative diploid nuclei, lack nucleoli and have fibrillar rather than particulate nucleoplasm (Hemmes and Bartnicki-Garcia, 1975; Beakes and Gay, 1977; Beakes, 1980d). This is consistent with the probable absence of RNA synthesis in mature oospores (Leary *et al.*, 1974).

MATURE OOSPORE CYTOPLASM

Despite the fact that the organization of mature oospores is a feature of major taxonomic importance (Dick, 1969) the true nature of the cytoplasmic components was poorly understood until the advent of electron microscopy (Howard, 1971). Although preservation of mature oospore structure is far from satisfactory, particularly with aldehyde fixatives (where shrinkage always occurs, Fig. 3f) sufficient detail can usually be obtained to identify most of the major components (Figs. 6,7). The conspicuous central reserve globule (described as granular cytoplasm by Dick, 1969), which is a prominent feature of both Saprolegnian and Peronosporalean oospores (Fig. 7), has been shown to be a vacuole (the ooplast) derived from the coalescence of dense body vesicles (Howard and Moore, 1970; Fukutomi *et al.*, 1971; Hemmes and Bartnicki-Garcia, 1975; McKeen, 1975; Sargent *et al.*, 1977; Beakes and Gay, 1978a). In *Saprolegnia* spp. the electron dense granules are retained throughout oospore differentiation (Fig. 7) although in some species (such as *S. ferax*, Beakes, 1980a) may be embedded in a moderately electron dense matrix (see Fig. 8a). In *Phytophthora capsici* (Hemmes and Bartnicki-Garcia, 1975) the ooplast is filled with a uniformly electron dense material, perforated by electron transparent holes (Fig. 7c) whereas in *P. macrospora* and *Bremia* it appears similar in contrast to that shown in Fig. 6 (Fukutomi *et al.*, 1971; Sargent *et al.*, 1977). In the rather well-fixed oospore of *Pythium* described by McKeen (1975) the ooplast has a uniform, moderately electron dense textured appearance (Fig. 7b). In *Pythium oligandrum* (Fig. 3d) and *Albugo* (Fig. 7d) the ooplast occupies a relatively small proportion of the oospore cytoplasm and is uniformly electron dense in appearance. The significance of these different ooplast structures is not known and an investigation of ooplast chemistry is long overdue.

Neutral lipid droplets are the major constituent of the peripheral cytoplasmic zone in all oospores (Figs. 3f, 6, 7).

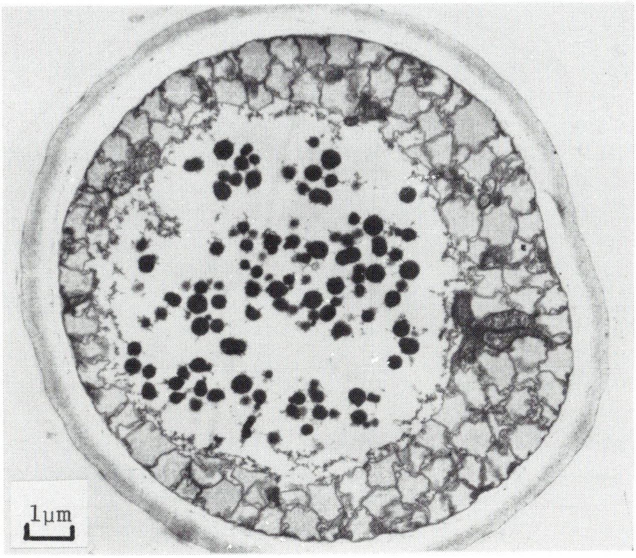

Fig. 6. Mature oospore of Saprolegnia furcata: *Permanganate (KM) fixed (from Beakes and Gay, 1978a).*

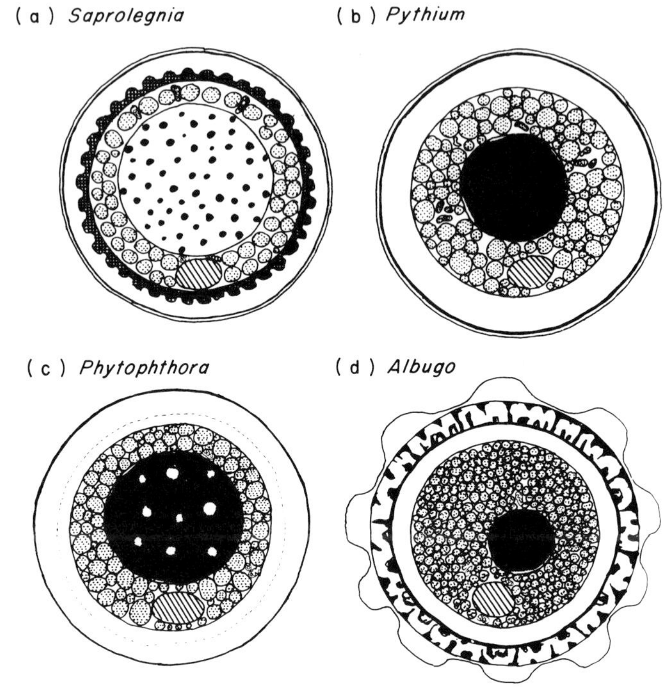

Fig. 7. Diagram summarizing structure of mature oospores.

In *Achlya* and some other members of the Saprolegniales the neutral lipid also coalesces into a single droplet during oospore maturation (Huizar, 1978) and may distort the ooplast vacuole (see Howard, 1971). Interspersed amongst the lipid droplets are the few surviving mitochondria and the fusion nucleus (Figs. 6, 7). Poor cytoplasmic preservation in mature oospores, however, prevents accurate assessment of ribosomes and membrane-bound organelles.

WALL DIFFERENTIATION

The complex changes in wall differentiation associated with oogenesis have been elegantly summarized by Hemmes and Bartnicki-Garcia (1975), who identified no less than eleven distinct phases of wall development. The oogonium wall appears to be secreted by Golgi-derived wall vesicles and is composed of an outer (primary) and inner (secondary) layers (layers 1a, 1b in Fig. 9). In *Phytophthora* spp. the inner layer is often made up of a series of lamellae, running parallel to the plasmamembrane (Hemmes and Bartnicki-Garcia, 1975; Hemmes and Ribeiro, 1977; Hegnauer and Hohl, 1978). In

Saprolegnia ferax window-like pits develop in the oogonium wall (Fig. 1a) as a result of the localized absence of secondary wall deposition. Pit development is associated with the accumulation of peripheral vesicles which suggests their differentiation may be regulated by localized modification of the plasma membrane (Beakes, unpublished observation).

Since cleavage in *Saprolegniales* involves the complete utilization of the existing plasmamembrane, the newly-differentiated oospheres undergo a brief (15-30 minutes) naked phase during which they show marked volume changes and intense surface activity (Fletcher, 1978; Beakes and Gay, 1978a). A thin (100 nm) outer wall is quickly secreted around the naked protoplasts to stabilise them. This primary oosphere wall (layer 2, Fig. 9) is equivalent to the epispore wall as defined by deBary (see Dick, 1969). Since oospheres in the Peronosporales are differentiated by an internal cleavage system, the oogonium plasmamembrane remains intact after oosphere formation. Thus the need for a stabilizing primary oosphere wall is obviated. Indeed such a wall layer (Fig. 9) is not shown in the *Pythium* sp. described by McKeen (1975), *Phytophthora capsici* (Hemmes and Bartnicki-Garcia, 1975), *Bremia lactucae* (Sargent *et al.*, 1977) and *Albugo* (Tewari and Skoropad, 1977). However, interestingly enough, it is formed in both *Pythium oligandrum* (Fig. 4b) and *P. acanthicum* (Haskins *et al.*, 1976) supporting the idea that the cleavage apparatus in the Peronosporales is derived from that of the Saprolegniales (see Fig. 2).

The development of the very electron dense, often multi-layered, wall layer (layer 3a, in Fig. 9) signifies the transformation from oosphere to oospore. This criterion is different from the usual one used (traditionally fertilization was used to distinguish the two, Dick, 1969) but makes sense physiologically since it is this layer which apparently governs the impermeability of mature oospores. Certainly, once it forms electron microscopy is no longer easy! This electron dense layer (3a, Fig. 9) is taken as the first of a series of secondary (endospore) layers which are deposited as the oospore matures. The composition of this layer is not known but because of its appearance and ability to withstand digestion during germination (Fig. 10d) it has been suggested it is composed of a resistant material, such as sporopollenin (Beakes and Gay, 1978b; Beakes, 1980b). Comparable layers have been observed in all oospores so far examined (Figs. 8, 9).

Permanganate fixation (Fig. 8) and freeze fracture studies (Hegnauer and Hohl, 1978) reveal the thick inner (endospore) oospore wall has a multilayered microfibrillar substrature (Layers 3b, c; Fig. 9). However, in glutaraldehyde fixation

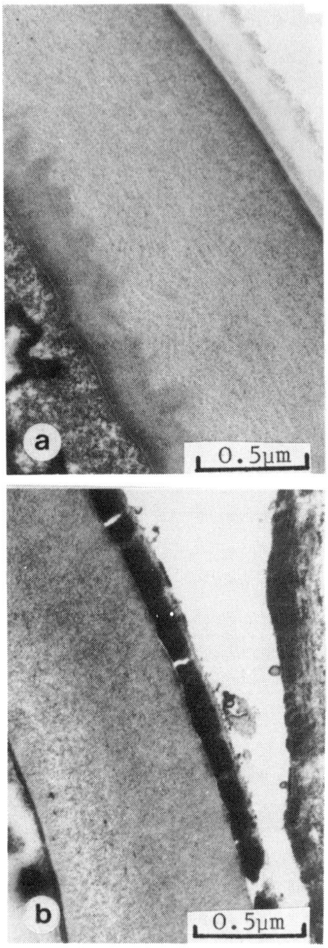

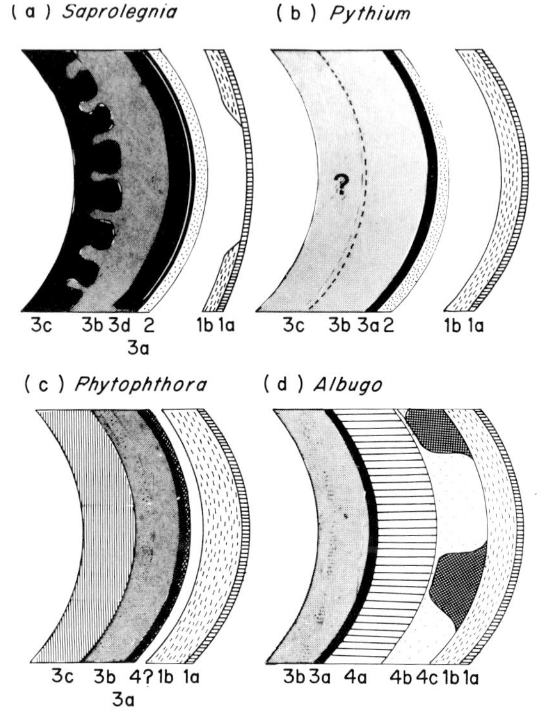

1a–b. Oogonium wall.
2. Epispore – first wall of oosphere.
3a–d. Endospore – secondary layers formed during oospore maturation.
4a–c. Exospore – wall layers formed within and derived from periplasm.

Fig. 8. Mature oospore walls of Saprolegnia furcata *(8a) and* Pythium oligandrum *(8b). KM-fixed.*

Fig. 9. Diagrams summarizing structure of mature oogonium and oospore walls. Layers are numbered/lettered in likely order of formation.

this layer usually looks uniformly electron transparent (Hemmes and Bartnicki-Garcia, 1975; McKeen, 1975; Sargent *et al.*, 1977). Biochemical analysis of mature oospore/oogonium walls in *Phytophthora* indicate the thick inner wall is largely composed of β1-3 glucans (Lippman *et al.*, 1974). In both *Saprolegnia* and *Phytophthora* the endospore wall is probably derived entirely from the oospore cytoplasm and is unusual in that its formation does not appear to be wall vesicle mediated (Hemmes and Bartnicki-Garcia, 1975; Beakes and Gay, 1978b; Beakes, 1980b). In other Peronosporales it has been suggested that endospore wall components may also be derived from the periplasm (McKeen, 1975). Perhaps, the most convincing evidence for this comes from *Bremia*, where curious tubule-like extensions of the inner periplasmic membrane appear during the period of inner oospore wall secretion (Sargent *et al.*, 1977). It has also been suggested that the condensation of the periplasmic cytoplasm may contribute additional oospore (exospore) wall layers (see Dick, 1969). Although such a layer was not observed in *Pythium* (Fig. 8b, 9b) it may be the origin of the thin outermost layer of the oospore wall reported in *P. cactorum* (layer 4, Fig. 9c, Hegnauer and Hohl, 1978). In *Albugo* a complex series of additional wall layers (4a,b,c; Fig. 9d) develop within the extensive periplasm, although the way in which they were differentiated was not described (Tewari and Skoropad, 1977).

OOSPORE GERMINATION

To date, ultrastructural aspects of oospore germination have only been described in *Pythium aphanidermatum* (Ruben and Stanghellini, 1978) and *Saprolegnia ferax* (Beakes, 1980a-d; Fig. 10). In both a very similar sequence of events is observed. The first detectable event of germination appears to be the digestion of the inner oospore wall, which is accompanied by a significant improvement in the quality of glutaraldehyde fixation (Fig. 10a). At this stage the central ooplast vacuole still appears uniformly electron dense, and there is no sign of embedded DB granules, some of which are present in the peripheral cytoplasm. The protoplasts of the activated oospores rapidly secrete around themselves a new germination wall layer (Fig. 10b,d) after which the spores once again fix well (Ruben and Stanghellini, 1978; Beakes, 1980a). The central vacuole expands and the electron opaque ooplast globule breaks down and is partially dispersed into the peripheral cytoplasm (Fig. 10b). During this period DBV again appear within the peripheral cytoplasm, and profiles of apparently disintegrating granules also are present in the central vacuole (Fig. 10b).

The activated spores swell and a germtube is produced whose wall is continuous with that of the germination wall layer (Fig. 10c). Oospore germtubes are often rather bulbous and seem to lack well-defined polarized growth, at least during the early stages of germination (Ruben and Stanghellini, 1978; Beakes, 1980b). The oospore body becomes increasingly vacuolate as the cytoplasm migrates into the germ tubes and

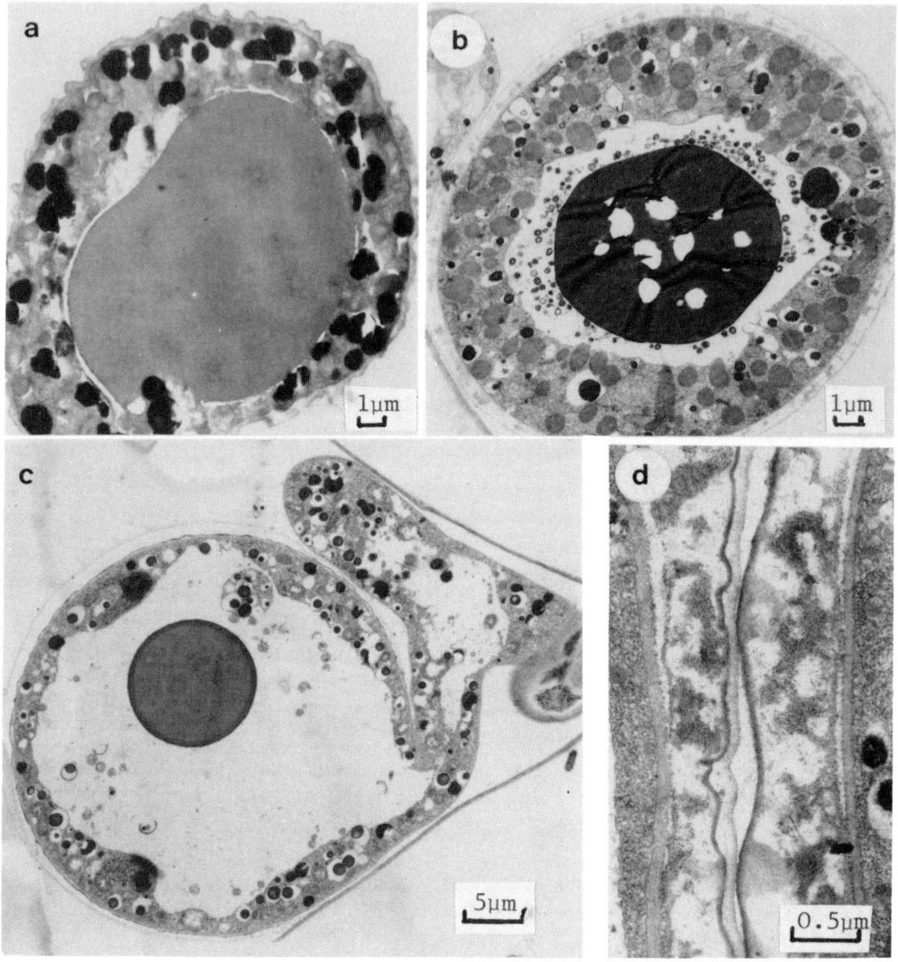

Fig. 10. Oospore germination in Saprolegnia ferax*: 10a) newly activated spore; 10b) activated spore after germination wall formation; 10c) young germling; 10d) detail of oospore wall after germination. GaOs fixed. (10b from Beakes, 1980a).*

the last vestiges of ooplast matrix are re-absorbed (Fig. 10c).

Nearly all of the fibrillar and amorphous components of the endospore wall are digested during germination, although the electron dense layer and the outer primary wall remain intact (Fig. 10d; Ruben and Stanghellini, 1978; Beakes, 1980b). In *Saprolegnia ferax*, most of the electron dense inner oospore layer also remains after germination and reveals a distinctive lamellar substructure (Fig. 10d).

GLUCAN AND LIPID RESERVES

The entire differentiation of oospores in *Saprolegnia* occurs in the absence of exogenous nutrients. Therefore, all of the complex structural changes which have just been outlined must be fueled and furnished by the organisms endogenous reserves, accumulated during vegetative growth. It also provides an excellent system in which to carry out comparative and quantitative assessments of the cellular organelle components which have been rarely undertaken in fungi (Beakes and Gay, 1978a; Fletcher, 1979; Beakes, 1980c).

Developing oogonia acquire two major reserve organelles; a vesicular fraction containing electron dense granules (Figs. 11,12) and droplets of neutral lipid (Fig. 3). Together, these may account for 60% of the organelle rich cytoplasm. The vesicular fraction has acquired a bewildering collection of synonyms (Beakes, 1980c) but, generally, have been referred to as dense body vesicles (DBV) in the Saprolegniales (Gay *et al.*, 1971) and finger print vesicles (FPV) in the Peronosporales (Hemmes and Bartnicki-Garcia, 1975). Although the appearance of this organelle fraction varies greatly during oogenesis (Figs. 12,13) the main structural features and likely chemical composition are summarized in Fig. 11. The latter is largely based on information derived from the isolation and analysis of comparable bodies from *Phytophthora* zoospores (Powell and Bracker, 1977) and the cytochemical localization of components at the ultrastructural level (Meyer *et al.*, 1976; Maxwell *et al.*, 1978; Beakes unpublished observations). The changes in the ultrastructural appearance

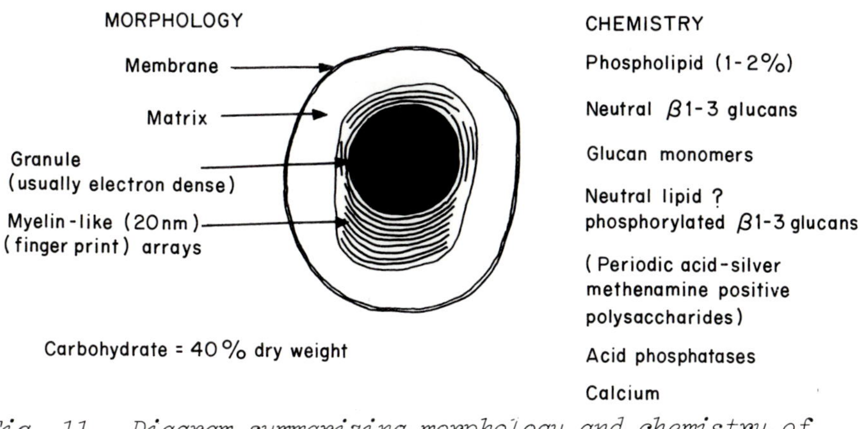

Fig. 11. Diagram summarizing morphology and chemistry of dense body/finger print vesicles

of the DBV's during oospore development are shown in Fig. 12 and summarized diagrammatically for *Saprolegnia* in Fig. 13, together with the cited roles of these vesicles during morphogenesis and their X-ray microanalysis spectra. The latter show the main chemical components of DB-granules after glutaraldehyde fixation analysed with a CORA analytical microscope. In young oogonia of both the Saprolegniales, (Fig. 12a, Gay *et al.*, 1971; Beakes, 1980c; Traquair and McKeen, 1980) and Peronosporales (Fig. 12d, Hemmes and Bartnicki-Garcia, 1975; McKeen, 1975; Sargent *et al.*, 1977) the dense body granules are condensed and lack the associated finger print matrix. In the Saprolegniales, finger print striations (Fig. 12b) develop around the granules during oosphere initial development and reach a maximum after oosphere formation (Beakes and Gay, 1978a; Traquair and McKeen, 1980). In many of the Peronosporales the entire vesicle becomes filled with 'finger print' material (Fig. 12e) which may develop either before (Fig. 12e and Haskins *et al.*, 1976) or after oosphere cleavage (Hemmes and Bartnicki-Garcia, 1975).

Initially the finger print arrays were thought to be composed of phospholipid (Gay *et al.*, 1971; Gay, 1972). However

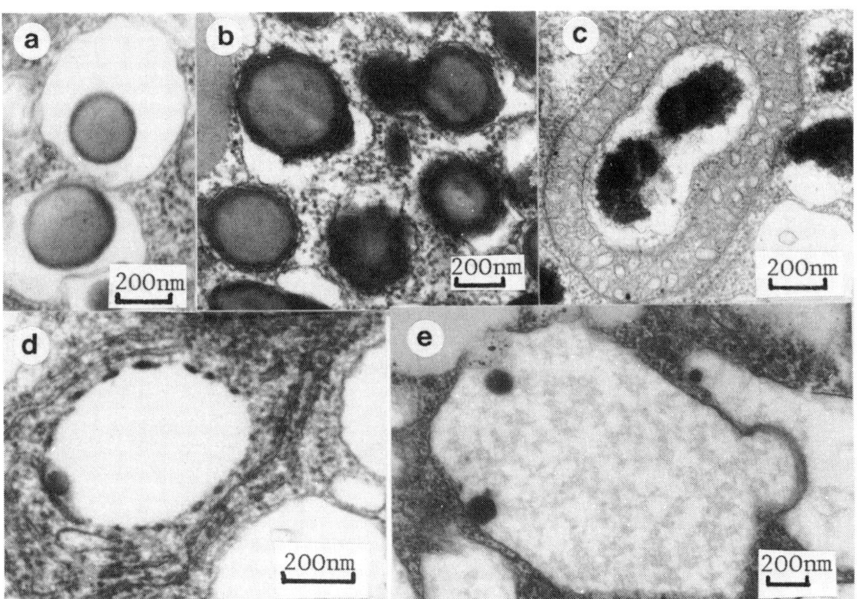

Fig. 12. Dense body/finger print vesicle structure in Saprolegnia ferax *(a–c) and* Pythium oligandrum *(d,e): 12a) in young oogonium; 12b) oosphere; 12c) germling; 12d) young oogonium; 12e) pre-cleavage oogonium. GaOs fixed.*

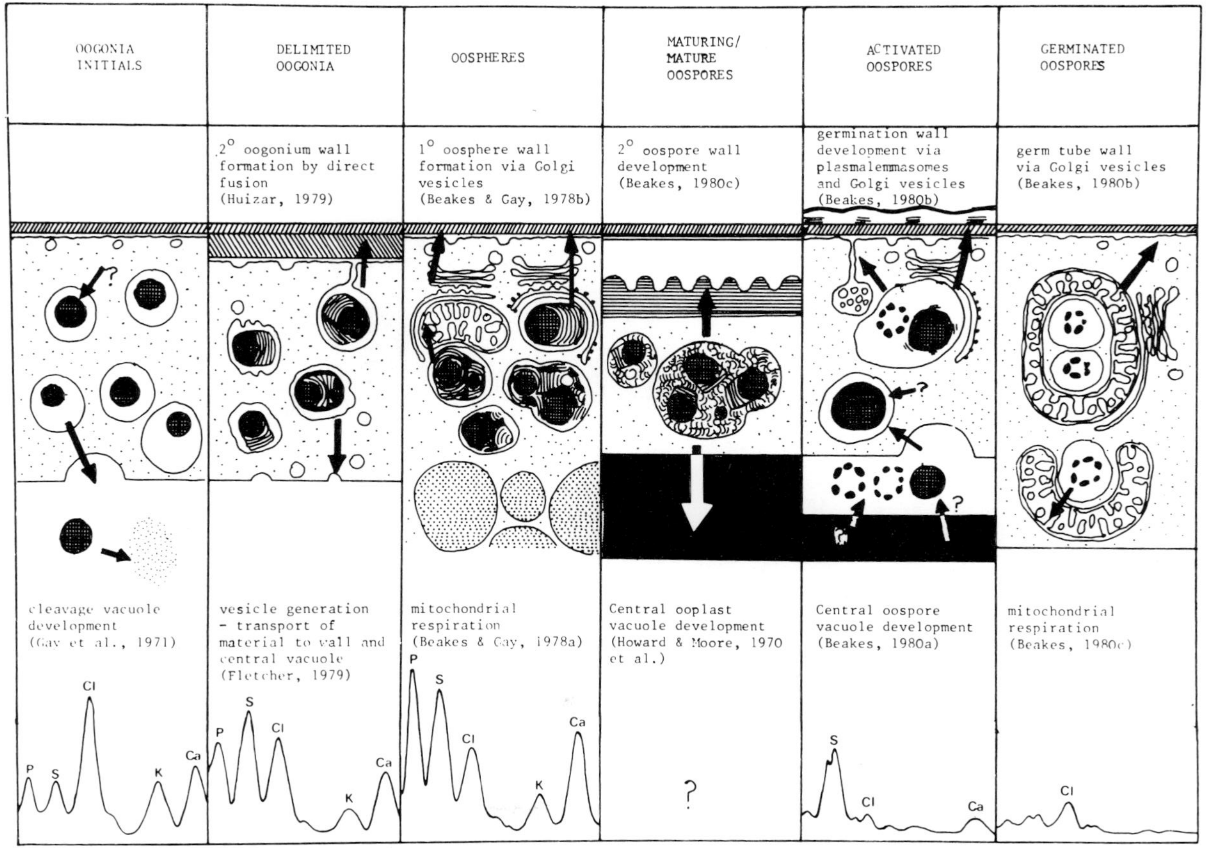

Fig. 13. Diagram illustrating changes in morphology, function and elemental composition of dense body vesicles during oospore differentiation in Saprolegnia. *Arrows represent possible flow of glucans*

the analysis of Powell and Bracker (1977) revealed carbohydrate as the major vesicle component, with phosphorylated β1-3 glucans as the most important constituent (Fig. 11). It is interesting that prominent phosphorous peaks in the X-ray spectra are mainly associated with DBV containing finger print arrays both of which are absent in germinating spores (Fig. 13). This suggests the finger print arrays rather than the electron dense granules are the likely subcellular site of the phosphorylated glucans. DBV's also contain calcium (Fig. 13) which may account for the deleterious effects Ca-deficiency has on oosphere differentiation (Fletcher, 1979). DBV's have been implicated in two major cellular roles; vacuole development and wall formation (Fig. 13). Both of these activities involve the transfer and incorporation of soluble glucans from the DBV's, the possible flow of which is indicated by arrows in Fig. 13. It should be added, however, that very few of the dense body functions indicated in Fig. 13 have been unequivocably confirmed. Since biochemical analysis of mature oospores of *Phytophthora* have indicated that most of the cellular glucan appears to be locked up in the oospore wall layers (Lippman *et al.*, 1974), it would be interesting to know if the mature ooplast still contains a glucan component.

In contrast to the DBV fraction, changes in the neutral lipid reserves during oogenesis are much easier to follow, since the globules represent the actual triacyl glyceride reserve product (Beakes, 1980c). In *Saprolegnia* and *Pythium* (Fig. 3) neutral lipid droplets account for 40-50% of the cytoplasmic volume. During oosphere differentiation there is a 20-30% drop in the neutral lipid reserves, probably due to its metabolism via the glyoxylate cycle to provide the carbohydrate required for wall synthesis (Beakes, 1980c). This is supported by the abundance of microbodies in the oosphere cytoplasm of *Saprolegnia*. During germination there is a steady decline in neutral lipid reserves, which again is associated with the presence of abundant microbodies in the germling cytoplasm in both *Pythium* (Ruben and Stanghellini, 1978) and *Saprolegnia* (Beakes, 1980c).

CONCLUSIONS

In spite of the technical problems involved, the ultrastructural study of oospore development has already helped resolve many of the long standing controversies associated with these spores. In particular, the much better understanding of the nature and origin of the cytoplasmic components in mature oospores has been of great value to oomycete taxonomists (Howard, 1971). Detailed knowledge of the mechanisms of oosphere cleavage and of the microstructure of the complex oospore walls may also be of value in indicating possible phylogenetic ralationships between the main groups of oomycete fungi.

Electron microscopy has made a useful contribution to resolving the nature of gametangial nuclear divisions and pinpointing the precise timing of plasmogamy and karyogamy. Nevertheless it would still be nice to have a full ultrastructural account of gametangial nuclear division in a Peronosporalean oomycete. These studies may also help unravel the mechanisms governing oospore dormancy. The inability to coax oospores to germinate in the laboratory has severely hampered both genetic and pathological studies of these often economically important fungi. Ultrastructural studies confirm that mobilization of the thick inner oospore wall is the first detectable event of germination and suggests that induction of endogenous glucanases may be the primary event triggering germination (Beakes, 1980b). We now have a sound structural understanding of the events associated with oospore formation and germination, and it is to be hoped that corroborative physiological and biochemical studies will follow soon.

ACKNOWLEDGEMENTS

I would like to thank Drs. J. Deacon, M. Dick and H. Fuller for supplying cultures; Bob Hewitt and Peter Webster for excellent technical assistance with electron microscopy, and Miss G. Williams and Mrs. S. Clothier for patiently preparing this manuscript. A special word of acknowledgement and thanks to Dr. John Gay who introduced me to the delights of sex in water moulds!

REFERENCES

Beakes, G.W. (1980a), Electron microscopic study of oospore maturation and germination in an emasculate isolate of *Saprolegnia ferax*. 1. Gross changes. *Canadian Journal of Botany* 58, 182-194.

Beakes, G.W. (1980b), *Ibid*. 2. Wall differentiation. *Canadian Journal of Botany* 58, 195-208.

Beakes, G.W. (1980c), *Ibid*. 3. Changes in organelle status and associations. *Canadian Journal of Botany* 58, 209-227.

Beakes, G.W. (1980d), *Ibid*. 4. Nuclear cytology. *Canadian Journal of Botany* 58, 228-240.

Beakes, G.W. and Gay, J.L. (1977), Gametangial nuclear division and fertilization in *Saprolegnia furcata* as observed by light and electron microscopy. *Transactions of British Mycological Society* 69, 459-471.

Beakes, G.W. and Gay, J.L. (1978a), Light and electron microscopy of oospore maturation in *Saprolegnia furcata*. 1. Cytoplasmic changes. *Transactions of British Mycological Society* 71, 11-24.

Beakes, G.W. and Gay, J.L. (1978b), *Ibid*. 2. Wall development. *Transactions of British Mycological Society* 71, 25-35.

Beakes, G.W. and Gay, J.L. (1980), Effects of streptomycin on the growth and sporulation of *Saprolegnia* spp. *Journal of General Microbiology* 119, 361-371.

Dick, M.W. (1969), Morphology and taxonomy of the oomycetes with special reference to Saprolegniaceae, Leptomitaceae and Pythiaceae. 1. Sexual Reproduction. *New Phytologist* 68, 751-775.

Dick, M.W. and Win-Tin (1973), The development of cytological theory in the oomycetes. *Biological Reviews* 48, 133-158.

Ellzey, J.T. (1974), Ultrastructural observations of meiosis within antheridia of *Achlya ambisexualis*. *Mycologia* 66, 32-47.

Ellzey, J.T. and Huizar, E. (1977), Synaptonemal complexes in antheridia of *Achlya ambisexualis* E87. *Archives for Microbiology* 112, 311-313.

Fletcher, J. (1978), Timing of events during oospore genesis in *Saprolegnia diclina*. *Transactions of British Mycological Society* 70, 417-422.

Fletcher, J. (1979), An ultrastructural investigation into the role of calcium in oosphere-initial development in *Saprolegnia diclina*. *Journal of General Microbiology* 113, 315-325.

Fukutomi, M., Akai, S. and Shiraishi, M. (1971), Fine structure of antheridia and oogonia of *Phytophthora macrospora*, the downy mildew fungus of rice plants. *Mycopathologia et Mycologia Applicata* 43, 249-258.

Gay, J.L. (1972), X-ray microanalysis in the development of oospores of the fungus *Saprolegnia*. *Micron* 3, 139-143.

Gay, J.L., Greenwood, A.D. and Heath, I.B. (1971). The formation of vacuoles (vesicles) during oosphere development and zoospore germination in *Saprolegnia*. *Journal of General Microbiology* 65, 233-241.

Gotelli, D. (1979), Synaptonemal complexes in the gametangia of *Sapromyces androgynus* (Leptomitales, Oomycetes). *Mycotaxon* 9, 90-92.

Haskins, R.H., Brushaber, J.A., Child, J.J. and Holtby, L.B. (1976), The ultrastructure of sexual reproduction in *Pythium acanthicum*. *Canadian Journal of Botany* 54, 2193-2203.

Hegnauer, H. and Hohl, H.R. (1978), Cell wall architecture of sporangia, chlamydospores, oogonia, and oospores in *Phytophthora*. *Experimental Mycology* 2, 216-233.

Heath, I.B. (1974), Mitosis in the fungus *Thraustotheca clavata*. *Journal of Cell Biology* 80, 204-220.

Hemmes, D.E. and Bartnicki-Garcia, S. (1975), Electron microscopy of gametangial interaction and oospore development in *Phytophthora capsici*. *Archives for Microbiology* 103, 91-112.

Hemmes, D.E. and Ribeiro, D.K. (1977), Electron microscopy of early gametangial interaction in *Phytophthora megasperma* var. *sojae*. *Canadian Journal of Botany* 55, 436-447.

Horgen, P.A. (1977), Steroid induction of differentiation: *Achlya* as a model system. *In* "Eucaryotic Microbes as Model Developmental Systems" (Ed. D. O'Day and P.A. Horgen). pp.272-294. Marcel Dekker, New York.

Howard, K.L. (1971), Oospore types in the Saprolegniaceae. *Mycologia* 63, 679-686.

Howard, K.L. and Moore, R.T. (1970), Ultrastructure of oogenesis in *Saprolegnia terrestris*. *Botanical Gazette* 131, 311-336.

Huizar, E.H. (1978), An ultrastructural study of the development of *Achlya recurva* oogonia. M.Sc. Thesis University of Texas, El Paso.

Leary, J.V., Roheim, J.R. and Zentmeyer, G.A. (1974). Ribosome content of various spore forms of *Phytophthora* spp. *Phytopathology* 64, 404-408.

Lippman, E., Erwin, D.C. and Bartnicki-Garcia, S. (1974), Isolation and composition of oospore-oogonium walls of *Phytophthora megasperma* var. *sojae*. *Journal of General Microbiology* 80, 131-141.

Maxwell, D.P., Hanssler, G. and Maxwell, M.D. (1978), Localization of acid phosphatase in *Pythium paroecandrum*. *Protoplasma* 94, 73-82.

McKeen, W.E. (1975), Electron microscopy studies of a developing *Pythium* oogonium. *Canadian Journal of Botany* 53, 2354-2360.

Meyer, R.,Parish, R.W. and Hohl, H.R. (1976), Hyphal tip growth in *Phytophthora*. *Archives for Microbiology* 110, 215-224.

Mullins, J.T. and Ellis, E.A. (1974), Sexual morphogenesis in *Achlya*: ultrastructural basis for hormonal induction of antheridial hyphae. *Proceedings of National Academy of Sciences U.S.A.* 71, 1347-1350.

Powell, M.J. and Bracker, C.E. (1977), Isolation of organelles from *Phytophthora palmivora*. *In* "Abstracts of Second International Mycological Congress" (Ed. H.E. Bigelow and E.G. Simmons). p.53 IMC-2 Inc., Tampa.

Ruben, D.M. and Stanghellini, M.E. (1978), Ultrastructure of oospore germination in *Pythium aphanidermatum*. *American Journal of Botany* 65, 491-501.

Sargent, J.A., Ingram, D.S. and Tommerup, I.C. (1977), Oospore development in *Bremia lactucae* Regel: an ultrastructural study. *Proceedings of Royal Society London* B 198, 129-138.
Schnepf, E., Deichgräber, G. and Drebes, G. (1978a), Development and ultrastructure of the marine parasitic oomycete, *Lagenisma conscinodisci* (Lagenidiales): sexual reproduction. *Canadian Journal of Botany* 56, 1315-1325.
Schnepf, E., Deichgräber, G. and Drebes, G. (1978b), Development and ultrastructure of the marine, parasitic oomycete, *Lagenisma conscinodisci* Drebes (Lagenidiales). Thallus, zoosporangium, mitosis and meiosis. *Archives for Microbiology* 116, 141-150.
Schnepf, E. and Heinzmann, J. (1980), Nuclear movement, tip growth and colchicine effects in *Lagenisma coscinodisci* (Oomycetes, Lagenidiales). *Biochem.Physiol.Pflanzen* 175, 67-76.
Slifkin, M. (1967), Nuclear division in *Saprolegnia* as revealed by colchicine. *Mycologia* 59, 431-445.
Tewari, J.P. and Skoropad, W.P. (1977), Ultrastructure of oospore development in *Albugo candida* on rapeseed. *Canadian Journal of Botany* 55, 2348-2357.
Traquair, J.A. and McKeen, W.E. (1980), Electron microscopy of developing *Aphanomyces* oogonia and oospores. *Mycologia* 72, 378-394.

II. ULTRASTRUCTURAL MORPHOGENESIS

HIGHER FUNGI

ULTRASTRUCTURAL MORPHOGENESIS IN HIGHER FUNGI: DISCUSSANT'S INTRODUCTION

M.F. Madelin

Department of Botany, University of Bristol, Bristol BS8 1UG, England

INTRODUCTION

The Higher Fungi are almost certainly descended from a common ancestor. This session is therefore concerned with morphogenesis in a group of organisms which share much genetical information and consequently many fundamental features. Because the structure and germination of spores were treated extensively in the Second International Fungal Spore Symposium, formation of spores will be emphasized here. In taking a synoptic view of sporogenesis, spores will be categorized as meiospores (ascospores and basidiospores) and mitospores (all others).

ASCOSPORES

Although the typical ascus is an eight-spored sac there is wide variation between, and sometimes within, species in the numbers of spores that form. Amongst Hemiascomycetes it ranges from one to about 100. Euascomycetes are known which regularly form 1,2,4,8,16,32,64,128,512 and *c.* 1000 spores per ascus, presumably reflecting synchronized nuclear divisions before any ascospores are delimited. A different process by which large numbers of ascospores are sometimes formed is that in which the first-formed spores bud secondary ones, a process of *in vivo* microcyclic sporulation.

The key feature of ascosporogenesis is that a diploid cell generates haploid spores by differentiation within a cytoplasmic mass of which some remains as the anucleate epiplasm. Differences exist in the ways that ascospores are initially delimited by membranes. These early steps lead to the complete envelopment of one or sometimes two haploid nuclei plus mitochondria and other cell constituents within a pair of closely apposed membranes, between which wall material will be laid down. In one process the double-membrane systems arise and develop singly about each prospective spore nucleus, in which case they are usually termed prospore walls or prosporal sacs; in the other they derive from a preceding double-membraned sac, the ascus vesicle, which partially surrounds all the prospective spore nuclei before it constricts about each one individually.

Individually assembled delimiting membranes have been suggested as proving characteristic of Hemiascomycetes (Ashton and Moens, 1979) while ascus vesicles are proving characteristic of Euascomycetes. However it seems that the complete picture is unlikely to remain as simple as this. Janex-Favre and Parguey-Leduc (1976) report that prosporal sacs (differing in a number of respects from prospore walls *sensu stricto:* see below) delimit individual spores in the Euascomycetes, *Tuber rufum* and *T. aestivum*. Also, not all Hemiascomycetes appear to produce wholly independent prosporal sacs or prospore walls, for in *Hansenula anomala* (Yoo *et al.*, 1973) and *Schizosaccharomyces octosporus* (Guth *et al.*, 1972) the delimiting membranes around each nucleus are initially connected.

The origins of double-membrane spore-delimiting systems are incompletely known, but probably there are several. The term 'prospore wall' was introduced by Moens (1971) for the membranous structure which lies on the cytoplasmic side of the spindle plaque of the young ascus in meiosis, and from which the individual ascospore wall is eventually derived. It was described first in *Saccharomyces cerevisiae*, and has also been reported in the yeast *Wickerhamia fluorescens* (Rooney and Moens, 1973b), four mycelial Hemiascomycetes (Ashton and Moens, 1979), and it probably also occurs in the fission yeast *Schizosaccharomyces pombe* (Yoo *et al.*, 1973) but not in *Schizosaccharomyces octosporus* (Rooney and Moens, 1973a). It first becomes visible in the cytoplasm adjacent to the outer of the two plaques of the spindle-pole body of Meiosis II, possibly arising from a small vesicle derived from the endoplasmic reticulum (ER) and attracted to the spindle-pole body. In *S. cerevisiae* it does not form in the absence of the spindle-pole body produced at the second meiotic division in the ascus, but it can form in the absence of normal budding of haploid nuclei from the single post-meiotic nucleus, with the result that anucleate spores may sometimes arise (Moens *et al.*, 1974). The prospore walls expand as pouches about the nuclei or, in those yeasts in which Meiosis I and II occur within a single nuclear envelope ('uninuclear meiosis'), about the four lobes of the nucleus which eventually cleaves.

In *Taphrina deformans* (Syrop and Beckett, 1972), *Tuber rufum* and *T. aestivum* (Janex-Favre and Parguey-Leduc, 1976) the ascospores are also delimited individually but by membranes which, because they appear to derive directly or indirectly from the plasmalemma, do not fit Moens' definition of 'prospore wall'.

An ascus vesicle has now been observed in many Euascomycetes. It characteristically consists of two layers of membrane which lie close to the ascus wall and constitute an open-ended cylinder with a number of fenestrations at the sides. Small flattened cisternae accumulate along the inside of the ascal plasmalemma, fuse and become continuous as the vesicle. Their origin is difficult to establish, but may include blebbing of the nuclear envelope, alone or in conjunction with elements of ER; ER itself; myelin figures and fungal 'mesosomes'; and invagination of the plasmalemma.

The precise mechanisms by which ascospore-delimiting membranes, once formed, actually delimit the young ascospores are obscure. Two processes are involved: growth and guidance of the membrane system. In several Hemiascomycetes the prospore wall forms a steadily expanding pouch (Moens, 1971) whose increase in size appears to result from actual growth rather than from the coalescence of existing cisternae (Rooney and Moens, 1973b). Like prospore walls, ascus vesicles also must increase in size during spore delimitation, for the combined surface of eight ascospores is much greater than that of the initial vesicle. Indeed in *Saccobolus kerverni*, nuclear blebbing, which is thought to be the process by which this species generates ascus vesicle, was seen to continue whilst the vesicle invaginated about the prospective ascospores (Carroll, 1969). The process by which it invaginates is however far from clear. The possibility that astral-ray microtubules shape the invaginating vesicle has been considered.

Although nuclei are not essential for the growth and shaping of ascospore-delimiting membranes, as the occasional anucleate ascospore reveals (Moens *et al.*, 1974; Simonet and Zickler, 1972), one or sometimes two are usually enclosed within the enveloping membranes. Subtle mechanisms must operate to optimise this occurrence. Even more precise mechanisms must operate in such a fungus as *Podospora anserina* which is secondarily homothallic because two *non-sister* haploid nuclei are enclosed in each of the four spores (Beckett and Wilson, 1968). Other important but obscure processes ensure the enclosure of mitochondria. In *Cephaloascus fragrans* the nearly complete prospore wall has an opening around a lobe of the ascal mitochondrion which detaches as the spore's mitochondrion (Ashton and Moens, 1979). In *Saccharomyces cerevisiae* a mitochondrion has been seen apparently in process of moving into the developing ascospore (Moens, 1971), and in *Schizosaccharomyces pombe* mitochondria appear to flow into the cytoplasm delimited by the prospore wall (Yoo *et al.*, 1973). These observations suggest that enclosure of mitochondria is not fortuitous.

The mature ascospore is enveloped in a firm, multi-layered, chemically complex wall which is laid down between the twin ascospore-delimiting membranes of which the interior becomes the plasmalemma of the ascospore and the outer becomes the spore-investing membrane. Walls form in several different ways. Three basic contributory events exist: centripetal addition of layers to the first-formed wall; centrifugal addition; and intercalation of layers by either alteration of existing material or introduction of new. The origins of wall material are however problematic. It is simplest to suppose that centrifugal growth depends on materials of epiplasmic origin and centripetal growth on sporoplasmic sources, but proof is not easily obtained. The existence of ascospore colour mutants in which wall colour is determined by the genome within the spore rather than by that of the parental ascus shows that information for the synthesis of these layers stem from the spore, but does not prove that the constituent materials do.

Many, perhaps most, ascospores enlarge before finally maturing within the ascus. For example, Delay's (1966) illustrations of *Ascobolus immersus* reveal an approximately thirty-fold volume increase of ascospores after their initial delimitation. Comparable illustrations reveal a nearly twenty-five-fold increase for *Podospora anserina* (Beckett and Wilson, 1968). Other spores increase much less. In view of the high final density of the sporoplasm it seems inconceivable that the increases result from uptake of only water. Equally, in view of the completeness of the delimiting membranes the increases cannot be the result of inward transfer of cytoplasm and organelles. Consequently one must invoke a trophic phase for young ascospores during which nutrients are transferred from epiplasm to sporoplasm in solution, and biosynthesis proceeds within the spore initials.

At late stages in the maturation of ascospores the volume of their cytoplasm may actually diminish. Centripetal wall growth in, e.g. *Sordaria brevicollis* (Hackett and Chen, 1975), *Poronia punctata* (Stiers, 1974) and *Saccharomyces cerevisiae* (Lynn and Magee, 1970) considerably reduces the cytoplasmic volume of the spore. This could lead to dehydration of the spore (Hawker and Madelin, 1976).

A number of special aspects of ascospore development take place after delimitation. There is sometimes multiplication of the spore nuclei so that even single-celled spores may contain two or more nuclei. Commonly, however, the multinucleate condition is associated with septation, which appears to represent a highly structured type of centripetal wall development.

BASIDIOSPORES

Almost all basidiospores are single-celled and simple in form. Sporogenesis on single-celled holobasidia has received most study so will be emphasised here. Although holobasidia usually bear four spores, the number may vary from one to eight. The basidiospores of Gasteromycetes are usually symmetrical, even globose, but those of Hymenomycetes are asymmetrical in very precise ways. The basic features of their production on the holobasidium are as follows. The basidium becomes filled with cytoplasm, expands to its full size, and then forms sterigmata, approximately coincident with the end of meiosis. Spherical spore initials then form at their tips and steadily grow. A large vacuole expands within the basidium while cytoplasm and nuclei pass through the sterigmata into the developing spores. The nucleus received by each spore commonly divides. In maturing, the spore sometimes becomes pigmented, develops ornamentation, or acquires a germ pore. Several of these steps in basidiosporogenesis merit particular comment.

Sterigmata arise as tapered outgrowths of the tips of the basidium. Vesicles are associated with their apical growth and appear to originate from a primitive Golgi apparatus (McLaughlin, 1973; Hoch and Setliff, 1976). In sterigmata of *Coprinus cinereus* the vesicles appear to fuse with the plasmalemma in the apical region which is essentially devoid of other organelles including ribosomes (McLaughlin, 1973). On basidia of certain mutants of *Schizophyllum commune* four or less abnormal sterigmata form which resemble vegetative hyphae. Control over the limitation of apical growth which characterizes normal sterigmata appears to have been lost (Schwalb, 1978). Cook (1972) concluded that formation of the sterigma-like conidium-bearing denticles on the conidiophore of *Oedocephalum roseum* initially involved only wall stretching, and observed that only small cytoplasmic vesicles (*c.* 75 nm diameter) were present. By contrast, when conidium initials began to form, wall synthesis commenced, and larger vesicles then appeared. There is some evidence that similar sequences of smaller and larger vesicles accompany sterigma and basidiospore development in *Poria latemarginata* and *Coprinus cinereus* (Hoch and Setliff, 1976; McLaughlin, 1977), but whether there is a similar transition from a phase of solely wall extension to one of synthesis is not known. That sterigmata are more than just diminutive branch hyphae is indicated also by their containing longitudinal microtubules (McLaughlin, 1973, 1977; Hoch and Setliff, 1976). Others are present in the basidium itself. A role in generating the precise architecture of the basidium and its sterigmata has been suggested (McLaughlin, 1973, 1977).

The basidiospore initial arises as a spherical enlargement of the sterigma tip. In species of *Coprinus* and *Boletus*, this initial when no more than 0.6-0.7 μm diameter already displays an internal asymmetry in the adaxial location of the hilar appendix body which marks the site of the future hilar appendix (Hugueney, 1975; McLaughlin, 1977; Yoon and McLaughlin, 1979). The initial asymmetry of the spore is compounded by the ensuing precisely regulated programme of wall expansion, thickening and setting. Wall growth itself is probably complex, for McLaughlin (1977) identifies five subcellular components associated with the process in *Coprinus cinereus*, namely, the hilar appendix body, small vesicles of undetermined origin, Golgi vesicles, peripheral ER and the plasma membrane, of which all but the first are elements of the endomembrane system.

The wall of the basidiospore in holobasidial Hymenomycetes and Gasteromycetes appears to consist of at least three layers, and in some Gasteromycetes its surface bears rodlets. The wall layers are not always formed in a centripetal sequence for intercalation may occur. In *Agaricus bisporus* it is the outer of the three layers which consolidates the shape of the spore, for even very small fragments of this layer retain their original curvature (Rast and Hollenstein, 1977).

Development of the basidiospore depends on an internal pressure which provides the force for its expansion. That the spore walls are highly elastic rather than merely plastic is revealed by an observation of Corner (1948). The basidiospores on a basidium usually enlarge synchronously. However, Corner saw a developing spore which began to swell rapidly to twice its diameter at which point it burst. The adjacent spore which was 10 μm wide concurrently shrank to become a sphere only 4 μm wide, an approximately six-fold elastic contraction of wall area. As Cook (1977) points out, synchronous expansion of spore initials connected to a common parent cell, and so under the same internal pressure, depends on their walls being of equal elasticity. A mechanism for achieving this equality must therefore exist in basidiospores developing on holobasidia. Perhaps the complex multi-layering of the walls of basidiospores on non-septate basidia helps to retain control of this critical elasticity during growth by providing multiple regulatory systems.

The widely cited hypothesis that cytoplasm is driven into the developing basidiospores by the expansion of vacuoles in the basidium is due to Buller (1933). This view appears to the author to be misleading for the following reasons. Vacuoles are surrounded by cytoplasm and can enlarge only by receiving water from the cytoplasm. Consequently, they can grow by only the same amount as the surrounding cytoplasm is diminished. Therefore, an expanding vacuole may replace the cytoplasm within a cell but cannot possibly displace it outwards. The partially dehydrated cytoplasm may nevertheless draw water through its plasmalemma, expand, and in so doing be compelled to flow elsewhere, but it would be its own imbibition of water that expelled it. Such a process would inject cytoplasm into developing basidiospores only if the cytoplasm could absorb water to replace that which it lost to the vacuoles. One should, therefore, distinguish two concurrent phenomena: (a) imbibition of water by the cytoplasm from contiguous sources, leading to increase in volume of the whole basidium-basidiospore system; and (b) concentration of solids in the basidial cytoplasm by transfer of water to the vacuoles, so leaving a partially dehydrated cytoplasm which, by virtue of its distal position, is that which enters the developing spores.

The provision of intracellular channels in the basidium and sterigmata through which cytoplasm flows and by which vesicles are directed to sites of wall synthesis may be a function of microtubules (McLaughlin, 1973), and microfilaments might play a role in movement of cell constituents (Gull, 1975). The movement of nuclei into the developing spores appears to be more than passive drift (Wells, 1964), and in *Lentinus edodes* is associated with cytoplasmic and intranuclear microtubules (Nakai and Ushiyama, 1978).

The mature basidiospore usually has dense cytoplasm with much lipid. Nuclear divisions within the delimited spore may exemplify polymerization by which the water potential of the spore is raised, so favouring its dehydration (Hawker and Madelin, 1976).

MITOSPORES

Mitospores include conidia, chlamydospores, spermatia, pycniospores, aeciospores, urediospores and teliospores. Although reproduction by mitospores is a highly developed feature of most Ascomycetes and of all Uredinales it has until recently been thought uncommon in other Basidiomycetes. However many do produce mitospores and do so by many methods (Kendrick and Watling, 1979).

The first Kananaskis Conference (Kendrick, 1971) emphasised the existence of two basic modes by which mitospores form in Higher Fungi; "blastic" denotes production by expansion of a new element which then becomes delimited by a septum; "thallic" denotes production by delimitation of part of an existing hypha by means of a septum, followed sometimes by its enlargement (i.e. by the conversion of a pre-existing element). That differences exist between spores formed by these two modes in respect of the origins of their walls and their contents is clear, but their implications are not. For example, in blastic ontogeny there is very largely *de novo* synthesis of the spore wall whereas in thallic ontogeny much of the wall is directly inherited from the parent conidiogenous cell, though there is often distension and elaboration of this wall. Likewise a blastic spore receives its cytoplasm and organelles by sequential transfer from the subtending sporogenous cell (very much as a basidiospore does) whilst the thallic spore starts with a full complement of contents (very much as an ascospore does). However conidia may linger in direct cytoplasmic communication with conidiophores from which materials may continue to be obtained. The relative contributions of transfer of cytoplasm from the conidiophore and *in situ* synthesis are little known, although studies which reveal the visible behaviour of nuclei, cytoplasm and its constituents provide some insight.

The Uredinales are commonly viewed as producing spores which are somehow distinct from conidia of other Higher Fungi, but the differences are probably not fundamental. Hughes (1970) showed that the mechanics of sporogenesis in rusts may be equated with processes well known in fungi of ascomycetous affinity. He interpreted the sporogenous cells in pycnia of *Gymnosporangium clavipes* as phialides, and surmised that this was the situation in all rusts. He equated catenulate aeciospores with meristem arthrospores. Urediospores are usually pedicellate, but Hughes pointed out that their initials in fact form in a sympodial succession on their sporogenous cells and then divide into two parts which become the urediospore and the pedicel. Teliospores are produced in three different ways. In the commonest, the pucciniaceous, the teliospore initials are again the equivalent of sympodioconidia, each dividing into teliospore and pedicel. In the melampsoraceous mode the spores are solitary terminal cells resembling chlamydospores while in the cronartiaceous mode they can, like aeciospores, be regarded as equivalent to meristem arthrospores. At the time Hughes wrote there was little relevant ultrastructural information available. However, most of his interpretations have been supported by subsequent fine-structural studies (for references, see Littlefield and Heath, 1979). Several such studies have revealed that pycniospore production is clearly phialidic (notwithstanding that two interpreted it as annellidic), spores forming blastically and repetitively at a fixed conidiogenous locus. The sympodioconidial type of blastic development of urediospores and pucciniaceous teliospores has been observed in several fine-structural studies. The area in which Hughes' views may require some revision concerns aeciospore production, for Rijkenberg and Truter (1974) and Codron (1978) both reported annellidic sporogenesis.

CONCLUSION

A problem in generalizing about the properties of spores concerns the homology or otherwise of the spores themselves. Is one comparing like with like? As remarked above, Higher Fungi are almost certainly descended from a common fungal ancestor. They also have almost all evolved in a terrestrial environment where they share many important selection pressures, so that convergent evolution is likely to have been common. Sporogenesis has undoubtedly evolved along many pathways of which four major routes - the ascosporic, basidiosporic, and blastic and thallic mitosporic - have been noted here. However within these four different systems one finds similar subcellular mechanisms operating. It is at this level that at least *partial* homologies may be sought, such as in the mechanisms which plasticize, displace, augment or alter walls; or move cytoplasm in bulk, or specific organelles in particular. One spore may not be homologous with another, but many of the processes which contribute to their formation and functioning may be. It is to this level of understanding that ultrastructural studies make a special contribution.

ACKNOWLEDGEMENTS

I thank the organisers of this symposium and the University of Bristol for grants which made my attendance possible, and Mrs. Mary Stodgell for patiently and skilfully typing this chapter.

REFERENCES

Ashton, M.L. and Moens, P.B. (1979). Ultrastructure of sporulation in the Hemiascomycetes *Ascoidea corymbosa*,

A. rubescens, *Cephaloascus fragrans*, and *Saccharomycopsis capsularis*. *Canadian Journal of Botany* 57, 1259-1284.

Beckett, A. and Wilson, I.M. (1968). Ascus cytology in *Podospora anserina*. *Journal of General Microbiology* 53, 81-87.

Buller, A.H.R. (1933). "Researches on Fungi" Vol. 5. Longmans, Green and Co., London.

Carroll, G.C. (1969). A study of the fine structure of ascosporogenesis in *Saccobolus kerverni*. *Archives of Microbiology* 66, 321-339.

Codron, D. (1978). Etude ultrastructurale de la formation des eciospores chez le *Puccinia urticae caricis* Kleb. *Revue de Mycologie* 42, 77-96.

Cook, B.E. (1972). The distribution and role of cytoplasmic membrane-bounded vesicles during the development of botryose solitary blastospores of two fungi. *New Phytologist* 71, 1135-1141.

Cook, B.E. (1977). The development of *Oedocephalum*-type conidia in the fungi *Cunninghamella echinulata* and *Fomes annosus*. *New Phytologist* 78, 637-642.

Corner, E.J.H. (1948). Studies in the basidium. I. The ampoule effect, with a note on nomenclature. *New Phytologist* 47, 22-51.

Delay, C. (1966). Etude de l'infrastructure de l'asque d'*Ascobolus immersus* Pers. pendant la maturation des spores. *Annales des Sciences Naturelles (Botanique et Biologie Végétale)*, *12e Série*, 7, 361-378.

Gull, K. (1975). Cytoplasmic microfilament organization in two basidiomycete fungi. *Journal of Ultrastructure Research* 50, 226-232.

Guth, E., Hashimoto, T. and Conti, S.F. (1972). Morphogenesis of ascospores in *Saccharomyces cerevisiae*. *Journal of Bacteriology* 109, 869-880.

Hackett, C.J. and Chen, K.-C. (1975). Ultrastructure of developing ascospores in *Sordaria brevicollis*. *Journal of Bacteriology* 126, 883-894.

Hawker, L.E. and Madelin, M.F. (1976). The dormant spore. *In* "The Fungal Spore: Form and Function" (Ed. D.J.Weber and W.M. Hess). pp. 1-70. John Wiley and Sons, Inc., New York.

Hoch, H.C. and Setliff, E.C. (1976). Sterigma and basidiospore development in *Poria latemarginata*. *Memoirs of the New York Botanical Garden* 28, 98-104.

Hughes, S.J. (1970). Ontogeny of spore forms in Uredinales. *Canadian Journal of Botany* 48, 2147-2157.

Hugueney, R. (1975). Morphologie, ultrastructure et développement de l'apicule des spores de quelques Coprinacées:

étude particulière du punctum lacrymans. *Bulletin de la Société Linnéenne de Lyon* 44, 249-256.

Janex-Favre, M.-C. and Parguey-Leduc, A. (1976). La formation des ascospores chez deux truffes: *Tuber rufum* Pico et *Tuber aestivum* Vitt. (Tubéracées). *Compte Rendu Hebdomadaire des Séances de L'Académie des Sciences, Série D*, 283, 1173-1175.

Kendrick, B. (Ed.) (1971). "Taxonomy of Fungi Imperfecti" University of Toronto Press, Toronto.

Kendrick, B. and Watling, R. (1979). Mitospores in Basidiomycetes. *In* "The Whole Fungus" (Ed. B. Kendrick). Vol. 2, pp. 473-545. National Museums of Canada, Ottawa.

Littlefield, L.J. and Heath, M.C. (1979). "Ultrastructure of Rust Fungi" Academic Press, Inc., New York.

Lynn, R.R. and Magee, P.T. (1970). Development of the spore wall during ascospore formation in *Saccharomyces cerevisiae*. *Journal of Cell Biology* 44, 688-692.

McLaughlin, D.J. (1973). Ultrastructure of sterigma growth and basidiospore formation in *Coprinus* and *Boletus*. *Canadian Journal of Botany* 51, 145-150.

McLaughlin, D.J. (1977). Basidiospore initiation and early development in *Coprinus cinereus*. *American Journal of Botany* 64, 1-16.

Moens, P.B. (1971). Fine structure of ascospore development in the yeast *Saccharomyces cerevisiae*. *Canadian Journal of Microbiology* 17, 507-510.

Moens, P.B., Esposito, R.E. and Esposito, M.S. (1974). Aberrant nuclear behaviour at meiosis and anucleate spore formation by sporulation-deficient *(spo)* mutants of *Saccharomyces cerevisiae*. *Experimental Cell Research* 83, 166-174.

Nakai, Y. and Ushiyama, R. (1978). Fine structure of shiitake, *Lentinus edodes*. VI. Cytoplasmic microtubules in relation to nuclear movement. *Canadian Journal of Botany* 56, 1206-1211.

Rast, D. and Hollenstein, G.O. (1977). Architecture of the *Agaricus bisporus* spore wall. *Canadian Journal of Botany* 55, 2251-2262.

Rijkenberg, F.H.J. and Truter, S.J. (1974). The ultrastructure of the *Puccinia sorghi* aecial stage. *Protoplasma* 81, 231-245.

Rooney, L. and Moens, P.B. (1973a). Nuclear division at meiosis in the ascomycetous yeast *Wickerhamia fluorescens*. *Canadian Journal of Microbiology* 19, 1383-1387.

Rooney, L. and Moens, P.B. (1973b). The fine structure of ascospore delimitation in the yeast *Wickerhamia fluorescens*. *Canadian Journal of Microbiology* 19, 1389-1392.

Schwalb, M.N. (1978). Regulation of fruiting. *In* "Genetics and Morphogenesis in the Basidiomycetes" (Ed. M.N. Schwalb and P.G. Miles). pp. 135-165. Academic Press, New York.

Simonet, J.M. and Zickler, D. (1972). Mutations affecting meiosis in *Podospora anserina* I. Cytological studies. *Chromosoma* 37, 327-351.

Stiers, D.L. (1974). Fine structure of ascospore formation in *Poronia punctata*. *Canadian Journal of Botany* 52, 999-1003.

Syrop, M.J. and Beckett, A. (1972). The origin of ascospore-delimiting membranes in *Taphrina deformans*. *Archives of Microbiology* 86, 185-191.

Wells, K. (1964). The basidia of *Exidia nucleata*. II. Development. *American Journal of Botany* 51, 360-370.

Yoo, B.Y., Calleja, G.B. and Johnson, B.F. (1973). Ultrastructural changes of the fission yeast (*Schizosaccharomyces pombe*) during ascospore formation. *Archives of Microbiology* 91, 1-10.

Yoon, K.S. and McLaughlin, D.J. (1979). Formation of the hilar appendix in basidiospores of *Boletus rubinellus*. *American Journal of Botany* 66, 870-873.

ASCOSPORE FORMATION

A. Beckett

*Department of Botany, University of Bristol,
Bristol BS8 IUG, England*

INTRODUCTION

Ascospore formation is of interest to the mycologist since it epitomises the sexual behaviour pattern diagnostic of the Ascomycotina. It is of interest to the geneticist and cytologist since it provides a means by which the products of meiosis may be isolated and analysed and it is of concern to the cell biologist since it exemplifies an unusual type of cytokinesis and cellular differentiation.

Ascospores are produced endogenously by compartmentalization of the ascus cytoplasm; originally termed "free cell formation" by Harper (1897). Despite differences reported in the literature for the mechanics of this process at the ultrastructural level, there is at present universal agreement on at least two aspects: *1)* Nucleate portions of cytoplasm are delimited by an envelope of two unit membranes. 2) Ascospore wall material is deposited between these two unit membranes which separate as the spore matures. The inner membrane forms the spore plasma membrane and the outer one forms the spore investing membrane. In thin sections, both membranes appear to be of plasma membrane type structure.

The differences referred to above have largely hinged upon: *1)* the origin and/or mode of formation of the delimiting membranes and 2) the type of membrane envelope formed and its configuration within the ascus. The mechanism(s) and controlling factors, by which these membranes, irrespective of their configuration, sequester nucleate portions of ascus cytoplasm remain largely unresolved. The physical and physiological requirements for the cleavage of the ascus protoplast into a variable but specific number of ascospores are probably similar for all ascomycete fungi. There is a need for the controlled synthesis of membrane material and its subsequent assembly at the morphological level. Portions of cytoplasm with normally one or sometimes two nuclei must be packaged within these membranes. Unpackaged cytoplasm, the epiplasm, remains outside the delimited spores and eventually aborts. Cell wall material for the developing spore must be synthesized and assembled between the membranes of the delimiting envelope. This may itself require that at least one and probably both of these membranes should be structurally and functionally specialized.

Here I shall attempt to show that these requirements are met in essentially the same way in all Ascomycetes so far studied. Differences in detail do exist however and these may prove useful for corroborating taxonomic groupings.

For convenience, ascospore formation may be considered as two phases; spore delimitation and spore maturation.

ASCOSPORE DELIMITATION

The origin and formation of spore delimiting membranes (SDMs)

Various suggestions for the origin of the membrane envelope have been published (Table *1*). In my view these variations represent different levels of organization of what is essentially one membrane-generating system. All of the proposed methods of membrane formation listed in Table *1* can be readily accommodated by the endomembrane concept of Morré, Mollenhauer and Bracker (1971). If, for example, these levels of organization are designated numerically from the nucleus outwards as shown in Figure *1*; then level *1* represents the formation of SDMs from the nuclear envelope, with or with out the participation of endoplasmic reticulum cisternae or myelin figures. It would also include the *de novo* synthesis of membrane-bounded vesicles in association with spindle pole bodies (SPBs). Level 2 involves Golgi apparatus equivalents, membrane-bounded vesicles and endoplasmic reticulum cisternae. Finally level 3 covers membrane synthesis and growth at the ascus plasma membrane and includes the so-called mesosome configurations. Since ascospore delimiting membranes are of plasma membrane type, then membranes generated within levels *1* and 2 must presumably undergo transformation prior to their fusion with the peripheral membrane cylinder (see later). In the case of *Saccharomyces cerevisiae* and most other *Hemiascomycetes* this transformation may occur as the membrane cisternae progressively envelope the nuclear lobes at the end of meiosis I and during meiosis II. An exception within this class is *Taphrina deformans* (Syrop and Beckett, 1972, 1976) where, following mitosis III, nuclei are delimited by direct invagination and growth of the ascus plasma membrane (Fig. *1*).

Membrane transformation is not presumably necessary. In *Saccobolus kerverni* (Carroll, 1967, 1969) and *Ascobolus stercorarius* (Wells, 1972) the evaginations of the nuclear envelope and the associated ER cisternae appear to be

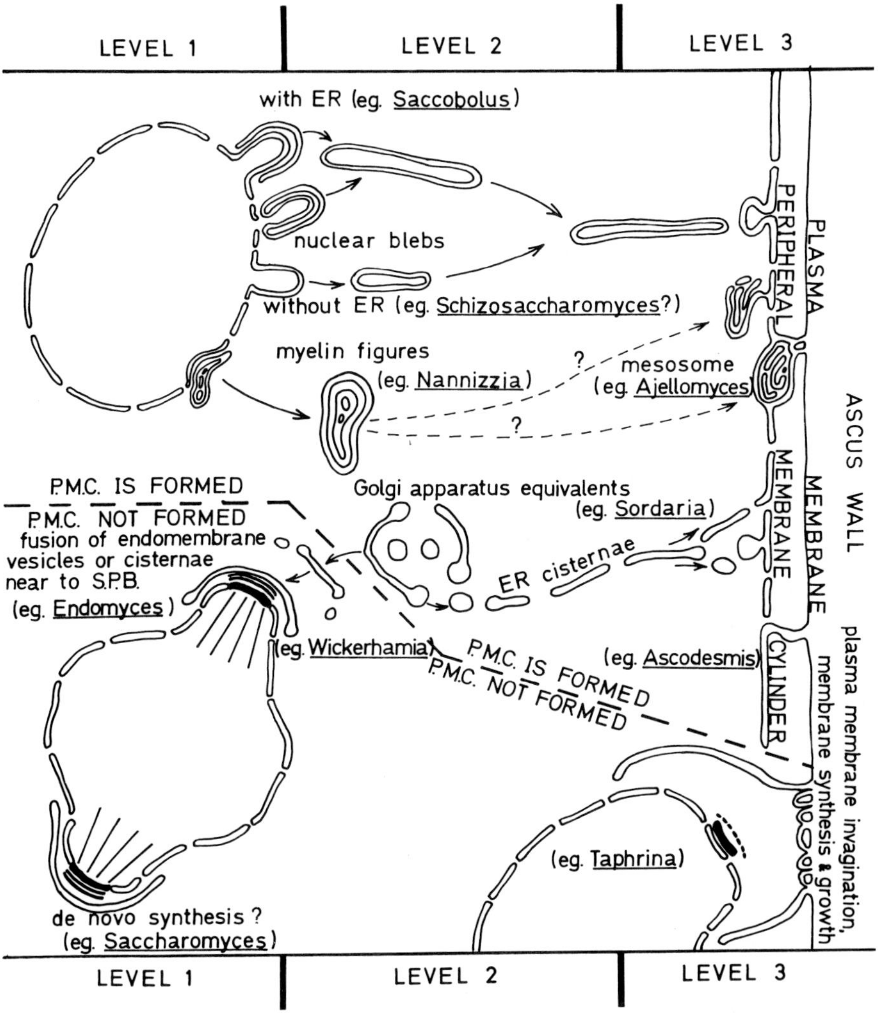

Fig. 1. A diagrammatic interpretation of the system which generates ascospore-delimiting membranes. PMC = peripheral membrane cylinder.

TABLE 1

Postulated Origins for Spore Delimiting Membranes (SDMs)

Classification[1] / Organism	Origin of SDMs	Reference
Hemiascomycetes		
Endomycetales		
Ascoidea corymbosa	*De novo* synthesis adjacent to SPB	Ashton & Moens, 1979
A. rubescens	"	Ashton & Moens, 1979
Cephaloascus fragrans	"	Ashton & Moens, 1979
Saccharomycopsis capsularis	"	Ashton & Moens, 1979
Saccharomyces cerevisiae	"	Moens, 1971; Guth *et al.*, 1972[2]; Moens & Rapport, 1973; Tingle *et al.*, 1973; Beckett *et al.*, 1973; Illingworth *et al.*, 1973; Zickler & Olson, 1975; Horesh *et al.*, 1979
Endomyces geotrichum	Endomembrane cisternae	Bracker *et al.*, 1974
Saccharomyces cerevisiae	Endoplasmic reticulum	Lynn & Magee, 1970[3]
Schizosaccharomyces pombe	Nuclear envelope	Yoo *et al.*, 1973[4]
Hansenula anomala	Plasma membrane	Bandoni *et al.*, 1967[5]
H. wingei	Endoplasmic reticulum	Black & Gorman, 1971
Wickerhamia fluorescens	"	Rooney & Moens, 1973[6]
Taphrinales		
Taphrina deformans	Plasma membrane	Syrop & Beckett, 1972, 1976

TABLE 1 continued

Plectomycetes		
Eurotiales		
Ajellomyces dermatitidis	Mesosomes (Plasma membrane)	Gil, 1973
Nannizzia gypsea	Myelin figures	Hill, 1975[7]
Penicillium vermiculatum	Endoplasmic reticulum	Wilsenach & Kessel, 1965
Pyrenomycetes		
Sphaeriales		
Cordyceps militaris	Endoplasmic reticulum	Moore, 1965
Sordaria fimicola	"	Mainwaring, 1967[8]; Furtado & Olive, 1970; Nogueira & Furtado, 1974
S. brevicollis	"	Hackett & Chen, 1976
Podospora anserina	Endomembrane vesicles	Zickler & Simonet, 1980
Xylosphaera polymorpha	Nuclear envelope & ER	Beckett & Crawford, 1970
Diatrype disciformis	Plasma membrane	Janex-Favre, 1978
Hypoxylon rubiginosum	"	Greenhalgh & Griffiths, 1970
Poronia punctata	"	Stiers, 1974
Rosellinia aquila	"	Beckett, unpubl. obs.
Sordaria humana	"	Beckett, unpubl. obs.
Xylosphaera polymorpha	"	Beckett & Crawford, 1970
Discomycetes		
Pezizales		
Ascobolus stercorarius	Nuclear envelope & ER	Wells, 1972
A. viridulus	"	Oso, 1969
Saccobolus kerverni	"	Carroll, 1967, 1969
Pustularia cupularis	Endoplasmic reticulum	Schrantz, 1967
Pyronema domesticum	"	Reeves, 1967

TABLE 1 continued

Phacidiales		
Lophodermella sulcigena	Plasma membrane	Campbell, 1973
Helotiales		
Sclerotinia sclerotiorum	Endomembrane vesicles	Codron, 1974
Tuberales		
Tuber aestivum	Plasma membrane	Janex-Favre & Parguey-Leduc, 1976;
T. rufum	"	Parguey-Leduc & Janex-Favre, 1977a & b; Janex-Favre, 1977
Loculoascomycetes		
Hemisphaeriales		
Placoasterella baileyi	Endoplasmic reticulum	Tyson & Griffiths, 1976

Notes. [1] Classification according to Ainsworth, Sparrow and Sussman (1973).

[2] Authors state origin of SDMs remains obscure. However, micrographs suggest *de novo* synthesis adjacent to SPBs.

[3] The micrographs published could also be interpreted as *de novo* synthesis adjacent to nuclei.

[4] Authors tentatively suggest nuclear envelope origin for SDMs but some micrographs would support *de novo* synthesis.

[5] Authors suggest invagination of plasma membrane as source of SDMs but during formation refer to them as endoplasmic reticulum cisternae.

[6] ER is only tentatively suggested as origin of SDMs - micrographs would also suggest *de novo* synthesis adjacent to SPBs.

[7] Author suggests myelin figures may form from convolutions of nuclear envelope.

[8] The nuclear envelope, ascus plasma membrane and *de novo* synthesis are also suggested as alternative sources of SDMs.

TABLE 2

List of Ascomycetes Showing the Presence (+) or absence (-) of a Peripheral Cylinder of SDMs (an "Ascus Vesicle")

Classification[1] / Organism	Peripheral Cylinder of SDMs ("Ascus Vesicle")	Reference
Hemiascomycetes		
Endomycetales		
Ascoidea corymbosa	-	)
A. rubescens	-	) Ashton & Moens, 1979
Cephaloascus fragrans	-	)
Endomyces geotrichum	-	Bracker *et al.*, 1974
Hansenula anomala	-	Bandoni *et al.*, 1967
H. wingei	-	Black & Gorman, 1971
Nematospora coryli	-	Plurad, 1972
Saccharomyces cerevisiae	-	Guth *et al.*, 1972; Lynn & Magee, 1970; Moens, 1971; Moens & Rapport, 1973; Beckett *et al.*, 1973; Illingworth *et al.*, 1973; Zickler & Olson, 1975; Horesh *et al.*, 1979
Saccharomycopsis capsularis	-	Ashton & Moens, 1979
Schizosaccharomyces pombe	-	Yoo *et al.*, 1973
Wickerhamia fluorescens	-	Rooney & Moens, 1973
Taphrinales		
Taphrina deformans	-	Syrop & Beckett, 1972, 1976

TABLE 2 continued

Pyrenomycetes		
Sphaeriales		
Ceratocystis fimbriata	–	Stiers, 1976[2]
Discomycetes		
Tuberales		
Tuber aestivum	–	) Janex-Favre, 1977; Janex-Favre &) Parguey-Leduc, 1976; Parguey-) Leduc & Janex-Favre, 1977a
T. rufum	–	
Plectomycetes		
Eurotiales		
Ajellomyces dermatitidus	+	Gil, 1973
Nannizzia gypsea	+	Hill, 1975
Pyrenomycetes		
Sphaeriales		
Ceratocystis fimbriata	+	Stiers, 1976
C. stenoceras	+	Garrison *et al.*, 1979
Diatrype disciformis	+	Janex-Favre, 1978
Hypoxylon fragiforme	+	Greenhalgh & Evans, 1968
H. rubiginosum	+	Greenhalgh & Griffiths, 1970
Neurospora lineolata	+	Hohl & Streit, 1975
Podospora anserina	+	Zickler, 1970
Poronia punctata	+	Stiers, 1974
Rosellinia aquila	+	Beckett, unpubl. obs.
Sordaria brevicollis	+	Hackett & Chen, 1976
S. fimicola	+	Mainwaring, 1967; Furtado & Olive, 1970; Nogueira & Furtado, 1974
S. humana	+	Sakai, 1974; Beckett, unpubl. obs.

TABLE 2 continued

Xylaria hypoxylon	+	) Beckett, unpubl. obs.
X. longipes	+	)
Xylosphaera polymorpha	+	Beckett & Crawford, 1970
Discomycetes		
Pezizales		
Ascobolus stercorarius	+	Wells, 1970, 1972; Zickler, 1970
A. viridulus	+	Oso, 1969
Pustularia cupularis	+	Schrantz, 1970
Pyronema domesticum	+	Reeves, 1967
Saccobolus kerverni	+	Carroll, 1967, 1969
Phacidiales		
Lophodermella sulcigena	+	Campbell, 1973
Helotiales		
Sclerotinia sclerotiorum	+	Codron, 1974
Loculoascomycetes		
Hemisphaeriales		
Placoasterella baileyi	+	Tyson & Griffiths, 1976
Laboulbeniomycetes		
Laboulbeniales		
Herpomyces	+	Hill, 1977

Notes. [1] Classification according to Ainsworth, Sparrow and Sussman (1973).
[2] Author suggests that delimitation of nuclei by SDMs may occur with or without the formation of a peripheral cylinder of membranes. The micrograph which is interpreted as illustrating delimitation in the absence of a peripheral cylinder could however represent a late stage during the invagination of such a cylinder of membranes.

particularly electron-opaque after potassium permanganate fixation in comparison with the non-evaginated portions of these membranes. This difference in fixation reaction and/or staining capacity may reflect a degree of differentiation towards plasma membrane type. Uranyl acetate and lead citrate produce a similar staining reaction with aldehyde-fixed specimens of *Xylosphaera polymorpha* (Beckett and Crawford, 1970).

Evidence for structural similarity between SDMs and the ascus plasma membrane is twofold: *1*) Clear structural continuity between the two has been demonstrated (Syrop and Beckett, 1972, 1976; Campbell, 1973; Bracker, pers. comm.). 2) Staining with a phosphotungstic acid-chromic acid mixture stains only the SDMs and the plasma membrane (Syrop and Beckett, 1972, 1976).

The form of the assembled SDMs within the ascus can be of two types (Ashton and Moens, 1979 and Table 2). One type is characterized by the direct envelopment of individual nuclei either by membranes formed in close association with SPBs as found for example in members of the *Endomycetales* (Table 2; Fig. *1*; Ashton and Moens, 1979), or by invagination and inward growth of the ascus plasma membrane (e.g. *Taphrina deformans*, Table 2; Fig. *1*; Syrop and Beckett, 1972, 1976). The second type is characterized by the development of a discontinuous cylinder of SDMs around the periphery of the ascus protoplast and which, immediately prior to delimitation, may more or less surround all of the interphase nuclei. This type includes most of the fungi which in some schemes are termed the "Euascomycetes" (Table 2). An exception to this division between the Hemiascomycetes and the "Euascomycetes" is the genus *Tuber* (Table 2) at least two species of which show delimitation by inward growth of the ascus plasma membrane in much the same way as described above for *Taphrina deformans* (Janex-Favre, 1977; Janex-Favre and Parguey-Leduc, 1976; Parguey-Leduc and Janex-Favre, 1977a). The situation in *Ceratocystis fimbriata* (Table 2 and Stiers, 1976) is inconclusive (see Note 2, Table 2 and Ashton and Moens, 1979).

On the basis of his study on *Pyronema domesticum*, Reeves (1967) likened the peripheral membrane cylinder to the so-called "ascus vesicle" described by Andrus and Harter (1933, 1937). Numerous workers have subsequently adopted this term. It should however be noted that the peripheral membrane cylinder as seen in electron micrographs cannot be resolved with the light microscope and, as Reeves (1967) suggested, is only partially analogous with the "ascus vesicle" and is not homologous with it.

Recent work with sporulation-deficient mutants of *Podospora anserina* by Zickler and Simonet (1980) has shown that initiation of the peripheral cylinder of SDMs is not correlated with nuclear division within the ascus since it can occur when nuclei have already aborted. There is however a correlation between normal formation and assembly of the membrane cylinder and the number of post meiotic mitoses in the ascus. Mutants in which the membrane cylinder is either not formed or remains incomplete and is not peripheral, exhibit numerous rapid post meiotic mitoses. The wild type with normal SDMs, shows only one such mitosis prior to spore delimitation. (Zickler and Simonet, 1980).

The mechanics of delimitation of the nuclei

The factors which control the envelopment of a specific number of nuclei by the SDMs are unknown. The observed association of cytoplasmic (astral) microtubules (MTs),microtubule organizing centres (MTOCs) and spindle pole bodies (SPBs) with the delimiting membranes as they enclose the interphase nuclei, has led to speculation as to their possible role(s) in this process (Carroll, 1966; Bracker, 1967; Beckett & Crawford, 1970; Zickler, 1970; Beckett, *et al.* 1974; Ashton and Moens, 1979).

Astral MTs radiate from the MTOC or SPB (for terminological distinction see Beckett, 1981) towards the SDMs (Bracker, 1967; Schrantz, 1967; Beckett and Crawford, 1970; Zickler, 1970; Wells, 1970, 1972). It has been suggested that these MTs may provide a physical barrier or template limiting invagination of the membrane envelope or they may act as contractile elements to physically pull in the membranes from the periphery of the ascus (Carroll, 1966; Bracker, 1967; Beckett and Crawford, 1970). Such speculations, however plausible, cannot be applied generally to ascomycete fungi because astral MTs have not been found associated with SDMs in any of the Hemiascomycetes so far studied (Syrop and Beckett, 1972, 1976; Zickler and Olson, 1975; Ashton and Moens, 1979). There is obviously a need for more examples to be investigated, but it may be significant that both the ontogeny of the SDMs and the method of delimitation of the nuclei in Hemiascomycetes Fig. *1*), differ from those in "Euascomycetes". In the Endomycetales, SDMs are initiated within a zone of exclusion close to the SPBs of the meiotic nucleus. Because of the configuration of the nuclear lobes in relation to the elongated spindles (Moens, 1971, Moens and Rapport, 1973; Zickler and Olson, 1975; Ashton and Moens, 1979) subsequent growth of the SDMs "automatically" envelopes these lobes. Any possible requirement for the displacement of the membranes towards or around the nuclei is therefore reduced.

If the membranes are formed by fusion of endomembrane vesicles (Fig. *1*) then the SPBs may be able to bind on to these vesicles and so retain them close to the nucleus. This was suggested by Rooney and Moens (1973) for *Wickerhamia fluorescens*.

In *Taphrina deformans* nuclei are located around the periphery of the ascus with their respective MTOCs located adjacent to the plasma membrane (Fig. *1*.) Inward growth of the plasma membrane at these specific points results in the immediate delimitation of each nucleus. As was stated by Syrop and Beckett (1972, 1976), but contrary to the recent interpretation by Alexopoulos and Mims (1979), this means that a peripheral cylinder of SDMs is not formed in *Taphrina*. Here also then, there may be less need for the mediation of microtubules.

In contrast to this, in those ascomycetes which have evolved a peripheral cylinder of SDMs there is a need for a mechanism by which this cylinder must be pulled inwards, broken and then reunited around a specific number of nuclei. For example, in the case of *Podospora anserina* (Beckett and Wilson, 1968; Zickler, 1970; Simonet and Zickler, 1972; Zickler and Simonet, 1980) this process must occur around each of four pairs of non-sister nuclei. The inward movement of the membranes probably involves their growth. Wells, (1972) cited the close ultrastructural similarities between the "sporoplasm" and epiplasm immediately after delimitation, as evidence to support such a concept. To this should be added the pertinent qualification that visually, the sporoplasm differs from the epiplasm only in the presence, in the former, of MTOCs and MTs associated with their respective nuclei. These microtuble complexes have never been demonstrated outside the areas bounded by the delimiting membranes and indeed in some cases MTs appear to make contact with these membranes (Beckett, unpubl. obs.). Are the MTOC, SPB/ microtubule complexes involved in the alignment of SDMs around nuclei? Information relevant to this question comes from work on sporulation-deficient (SPO) mutants (Moens, *et al.*, 1974; Simonet and Zickler, 1972; Lu and Galeazzi,1978; Zickler and Simonet, 1980). With temperature-sensitive mutants of *Saccharomyces cerevisiae*, Moens *et al.* (1974) showed that delimitation of nuclei did not occur in the absence of SPBs at meiosis II. Although microtubules are not involved, the SPB, which is probably a specialized MTOC (Fuller, 1976; Beckett, 1981), does seem to be essential. On the other hand studies on (SPO) mutants of *Podospora anserina* led Zickler and Simonet (1980) to conclude that although the delimitation of nuclei by SDMs was correlated with post meiotic mitotic spindle orientation, it was independent of both SPBs and nuclei. This was based on the observation of supernumerary anucleate spore initials that formed in some mutants. However, the abnormal delimitation of small, anucleate portions of cytoplasm in the absence of MTOCs, SPBs or MTs, does not necessarily preclude their involvement in the normal formation of nucleate spore initials. Heslot, (1958) found that MTOCs became disassociated from nuclei in mutants of *Sordaria* and Lu and Galeazzi (1978) have shown that isolated SPBs can delimit anucleate spores in the meiotic-1 mutant of *Neurospora crassa*. A role for MTOCs, SPBs and MTs in the normal compartmentalization of ascus cytoplasm by SDMs cannot therefore be ruled out even if their function is solely to stabilize specific regions of that cytoplasm.

Related to this may be the observation by Zickler and Simonet (1980) that the supernumerary nuclei from the several post meiotic mitoses in the *P. anserina* mutant strain *CS 12-1* are often arranged in four areas within the ascus. These regions may correspond to the normal location of the four pairs of nuclei in the wild type ascus. This suggests a regulatory mechanism which acts independently from those governing the formation of the peripheral membrane cylinder, spindle orientation etc., and which persists in the ascus after the *CS 12-1* mutation. Further ultrastructural studies on these and other mutants such as those described by Srb *et al.* (1973), and Raju and Newmeyer (1977), might elucidate many of these unresolved aspects of ascospore delimitation.

ASCOSPORE MATURATION

Maturation of the ascospore involves the acquisition of a characteristic shape and size, the formation of a typical wall structure and of a particular cytoplasmic configuration. Only shape, size and wall formation will be considered here.

Ascospore shape and size

Little is known of the ultrastructural features associated with ascospore shape and size. Cytological studies with the light microscope suggest a close correlation between the presence and number of nuclei within the young spore and the size, shape and wall-architecture of that spore (Raju and Newmeyer, 1977; Zickler and Simonet, 1980). In general ascospores containing supernumerary nuclei tend to be larger than wild type and are often abnormally shaped. This correlation however seems to depend upon the degree of variation from the normal. For example, ascospores of *Podospora anserina* which contain one or four nuclei are of the same shape as normal ones which contain two nuclei (Beckett and Wilson,

1968; Simonet and Zickler, 1972). It has also been suggested that the shape of the ascus at the time of spore delimitation may govern the shape of the spores produced within it (Srb *et al.*, 1973).

Ultrastructural studies of abnormal spores from mutant strains are needed but observations on some normal ascospores with characteristic shapes can provide interesting information. Spore shape is presumably acquired relatively soon after delimitation is completed and before the material of the wall sets and becomes inelastic. Such spores of *Xylosphaera polymorpha* (Beckett and Crawford, 1970), *Xylaria hypoxylon* and *Rosellinia aquila* (Beckett, unpubl. results) contain numerous MTs, aligned parallel with the long axis of the spore and which are associated with a MTOC situated in the basal tip.

These MTs may play a cytoskeletal role in the shaping of the spore (Beckett and Crawford, 1970).

Ascospore wall formation and structure

The multiplicity of specific terms used in the literature for ascospore wall layers, renders comparative analysis of developmental pattens difficult and often of little value (see Beckett, 1976a). The introduction of new terms does not necessarily solve this problem (e.g. Hackett and Chen, 1976). Simplification is desirable, and can in my view be achieved by adopting the system used for example by Carroll (1966, 1969) and Merkus (1973, 1974, 1975). Here, the initial deposit of undifferentiated wall material, between the two membranes of the delimiting envelope is termed the *primary wall*. All subsequent wall material that is formed, either by modification of the primary wall or by addition to it, can be termed *secondary wall* and may involve chemical changes that are not manifest at the ultrastructural level. The use of the term "tertiary" etc. (Beckett *et al.*, 1968; Mainwaring, 1972) is then unnecessary and layers of the secondary wall can be designated numerically from the outside inwards (e.g. Beckett, 1976a). Secondary wall material varies considerably in mode of development, structure and almost certainly composition. With few exceptions, it is secondary wall material that is involved in pigmentation and ornamentation of the spores.

Ascospore colour and ornament are useful characters for the classification of ascomycete species. It is not surprising therefore that there is no obvious common pattern of development at the ultrastructural level for secondary walls (see Tyson and Griffiths, 1976). The method of primary wall deposition is the only feature common to all.

Irrespective of the ultimate morphology of the wall and of the order in which the layers may be laid down; the origin of the material, either as precursor or as assembled units, is unclear. It is also not known at what point in time during the envelopment of the nuclei by SDMs, that primary wall deposition occurs. In yeasts for example, the use of such terms as "prospore wall" (Moens, 1971; Ashton and Moens, 1979) or "spore wall primordium" (Zickler and Olson, 1975) to describe the newly formed SDMs suggests that wall material is laid down between these membranes even before nuclear delimitation is completed. Indeed the spore wall primordium is described as consisting "of two membranes enclosing a lumen containing contrasted wall precursors" (Zickler and Olson, 1975). Obvious wall formation only becomes apparent however, once delimitation is complete and the two membranes begin to separate (Lynn and Magee, 1970; Moens, 1971; Illingworth *et al.*, 1973). The incompletely delimited spore illustrated by Beckett *et al.* (1973) for *S.cerevisiae* is probably atypical. In *Hansenula* (Bandoni *et al.*, 1967; Black and Gorman, 1971) the separation of the membranes and concomitant formation of wall material are initiated at the presumptive brims of the hat-shaped spores. The concentration and progressive deposition of polysaccharide are dramatically illustrated in material stained with silver methanamine (Black and Gorman, 1971), and suggest a precisely controlled process possibly involving the mediation of the relevant portions of the SDMs.

As mentioned previously, ascospore delimiting membranes are of plasma membrane type. Both have a protoplasmic and exoplasmic surface and it is reasonable to assume that in the very young spore initials at least, both the inner spore plasma membrane and the outer spore investing membrane are structurally and functionally similar. We may therefore speculate that both membranes could play a role in initiating and controlling wall formation. As the wall develops and secondary deposition occurs, it is likely that the spore investing membrane will become less involved. With the increase in wall thickness it will become spatially divorced from the nucleate spore protoplast and more importantly, its protoplasmic surface will be rendered non-functional with the abortion of the ascus epiplasm. In mature ascospores the investing membrane may be drastically restructured or may even break down. It is therefore to the spore initials that we should look for evidence of the regulation of wall formation by one or both of the SDMs.

Recent work on *Sordaria humana* using thiosemicarbazide (TSC) and thiocarbohydrazide (TCH) (Thiéry, 1967) to stain polysaccharide, shows that in this fungus, TCH positive material is detectable before nuclear delimitation is complete. After delimitation, prolonged staining for complex polysaccharides (Thiéry, 1969) with TSC suggests that both membranes may be involved in the export (and deposition) of these substances from the spore protoplast and the ascus epiplasm (Beckett, unpubl. obs.). Wall formation via the spore plasma membrane may also involve ER cisternae at the periphery of the protoplast as in *Pyronema domesticum* and *Saccharomyces cerevisiae* (Reeves, 1967; Lynn and Magee, 1970; see also Beckett *et al.*, 1974), the fusion of vesicles or lomasomes with the plasma membrane (Carroll, 1966; Beckett *et al.*, 1974; Stiers, 1974), or a banded, fibrillar apparatus (Bracker *et al.*, 1974). Whatever the means, it must be assumed that the spore nucleus provides the major regulating factor(s) for wall formation and architecture. This is confirmed by the absence of wall deposition in anucleate spores (Simonet and Zickler, 1972; Zickler and Simonet, 1980). In addition, the epiplasm may, while it is viable, play a role in wall deposition, ornamentation and pigmentation (Delay, 1966; Carroll, 1969; Lynn and Magee, 1970; Mainwaring, 1972; Illingworth *et al.*, 1973; Merkus, 1973, 1974, 1975). The incorporation of lipids from the epiplasm into the spore wall and/or investing membrane has been reported by Lynn and Magee (1970), Beckett *et al.* (1973) and Illingworth *et al.* (1973). The so-called "canals" of Marquardt (1963) probably represent collapsed lipid vesicles that have fused with the investing membrane. Epiplasmic lipids are also incorporated within the ascospore wall in *Endomyces geotrichum* (Bracker *et al.* 1974).

The ultrastructure of ascospore walls has so far been studied mainly by use of chemical fixation and thin sectioning procedures. Such techniques do not allow unequivocal differentiation between fibrillar and amorphous layers or structures within the wall (but see Greenhalgh and Evans, 1968). The use of freeze-etching (Byrne, 1975) and chemical extraction followed by metal shadowing (Beckett, 1976a, b), has clearly demonstrated the presence of fibrils within the innermost wall layer of ascospores of *Neurospora crassa* and *Daldinia concentrica* respectively. It is not known when during spore maturation, microfibrils are laid down but it is likely that the spore plasma membrane is involved in their synthesis. In *D. concentrica* (Beckett, 1976a, b) and *Sordaria humana* (Beckett, unpubl. obs.) microfibrils constitute an integral part of the germ slit or pore region.

Little is known of the chemical composition of ascospore wall layers in relation to their ultrastructure but Gooday *et al.* (1974) have shown that the ribbed layer in ascospores of

Neurospora crassa and *N. tetrasperma* ("perispore" Le Gal, 1947; Lowry and Sussman, 1968) is the site of sporopollenin. In *D. concentrica,* sporopollenin was thought to be present in two of the wall layers which overlie the germ slit (Beckett, 1976a).

CONCLUSIONS

1) All ascomycete genera have certain basic requirements for the process of ascospore formation. With few differences in detail, these are met in essentially similar ways.
2) The endomembrane system generates the spore delimiting membranes and different levels of organization can be recognized according to the state of transformation to plasma membrane type.
3) Two groups of fungi can be recognized depending on the presence or absence of a peripheral membrane cylinder. With the exception of *Tuber* these groups are exemplified by the "Euascomycetes" and the Hemiascomycetes.
4) Studies on sporulation-deficient mutants have elucidated some of the factors regulating the formation of ascospore delimiting membranes.
5) MTOC, SPB/microtubule complexes may be involved in the alignment of spore delimiting membranes around the nuclei.
6) Ultrastructural information relevant to spore shape is lacking but the presence and number of nuclei and possibly microtubules within spores may be important.
7) Both the spore plasma membrane and the investing membrane may play a role in regulating wall formation in young spore initials.
8) Differentiation of a primary wall gives rise to a variable secondary wall.
9) There is no common pattern of development for secondary wall formation.
10) The spore nucleus is presumed to play a major role in controlling spore wall deposition and architecture.
11) There is a need for further studies using enzyme extraction, chemical cleaning, shadowing and freeze-etching techniques for investigating the ultrastructure and composition of ascospore walls.

ACKNOWLEDGEMENTS

I wish to thank Mrs. Mary Stodgell and Miss Frances Wiggins for their patience and skill in typing this manuscript and Dr. Denise Zickler for providing me with a preprint of her paper on (SPO) mutants of *P. anserina.*

REFERENCES

Ainsworth, G.C., Sparrow, K.F. and Sussman, A.S. (1973). "The Fungi An Advanced Treatise" Vol. IVA, A Taxonomic Review with Keys: Ascomycetes and Fungi Imperfecti. Academic Press, New York:London.

Alexopoulos, C.J. and Mims, C.W. (1979). "Introductory Mycology" John Wiley & Sons, New York:Chichester:Brisbane: Toronto.

Andrus, C.F. and Harter, L.L. (1933). Morphology of reproduction in *Ceratostomella fimbriata*. *Journal of Agricultural Research* 46, 1059-1078.

Andrus, C.F. and Harter, L.L. (1937). Organization of the unwalled ascus in two species of *Ceratostomella*. *Journal of Agricultural Research* 54, 19-46.

Ashton, M.L. and Moens, P.B. (1979). Ultrastructure of Sporulation in the Hemiascomycetes *Ascoidea corymbosa*, *A. rubescens*, *Cephaloascus fragrans*, and *Saccharomycopsis capsularis*. *Canadian Journal of Botany* 57, 1259-1284.

Bandoni, R.J., Bisalputra, A.A. and Bisalputra, T. (1967). Ascopore development in *Hansenula anomala*. *Canadian Journal of Botany* 45, 361-366.

Beckett, A. (1976a). Ultrastructural studies on exogenously dormant ascospores of *Daldinia concentrica*. *Canadian Journal of Botany* 54, 689-697.

Beckett, A. (1976b). Ultrastructural studies on germinating ascospores of *Daldinia concentrica*. *Canadian Journal of Botany* 54, 698-705.

Beckett, A. (1981). The ultrastructure and behaviour of nuclei and associated structures within the meiotic cells of Euascomycetes. *In* "The Fungal Nucleus: Structure, Biochemistry and Genetics" (Ed. K. Gull and S.G. Oliver). Cambridge University Press, London. (*In Press*).

Beckett, A. and Wilson, I.M. (1968). Ascus cytology of *Podospora anserina*. *Journal of General Microbiology* 53, 81-87.

Beckett, A. and Crawford, R.M. (1970). Nuclear behaviour and ascospore delimitation in *Xylosphaera polymorpha*. *Journal of General Microbiology* 63, 269-280.

Beckett, A., Barton, R. and Wilson, I.M. (1968). Fine structure of the wall and appendage formation in ascopores of *Podospora anserina*. *Journal of General Microbiology* 53, 89-94.

Beckett, A., Illingworth, R.F. and Rose, A.H. (1973). Ascospore wall development in *Saccharomyces cerevisiae*. *Journal of Bacteriology* 113, 1054-1057.

Beckett, A., Heath, I.B. and McLaughlin, D.J. (1974). "An Atlas of Fungal Ultrastructure" Longman, London.

Black, S.H. and Gorman, C. (1971). The cytology of *Hansenula* III Nuclear segregation and envelopment during ascosporogenesis in *Hansenula wingei*. *Archiv für Mikrobiologie* 79, 231-248.

Bracker, C.E. (1967). Ultrastructure of fungi. *Annual Review of Phytopathology* 5, 343-374.

Bracker, C.E., Butler, E.E. and Montecillo, C.M. (1974). Subcellular morphogenesis of *Endomyces geotrichum*. *Eighth International Congress on Electron Microscopy, Canberra*, 2, 570-571.

Byrne, B. (1975). Freeze-etch and thin section studies of *Neurospora crassa* ascospores. *Protoplasma* 84, 357-377.

Campbell, R. (1973). Ultrastructure of asci, ascospores, and spore release in *Lophodermella sulcigena* (Rostr.) v. Hohn. *Protoplasma* 78, 69-80.

Carroll, G.C. (1966). A study of the fine structure of ascosporogenesis in *Saccobolus kerverni* and *Ascodesmis sphaerospora*. Ph.D. Thesis. University of Texas, Austin.

Carroll, G.C. (1967). The ultrastructure of ascospore delimitation in *Saccobolus kerverni*. *The Journal of Cell Biology* 33, 218-224.

Carroll, G.C. (1969). A study of the fine structure of ascoporogenesis in *Saccobolus kerverni*. *Achiv für Mikrobiologie* 66, 321-339.

Codron, D. (1974). Etude ultrastructurale de quelques points du développement des asques du *Sclerotinia sclerotiorum* (Lib.) de Bary. *Annales des Sciences Naturelles Botanique et Biologie Végétale* 15, 255-276.

Delay, C. (1966). Etude de l'infrastructure de l'asque d'*Ascobolus immersus* Pers. pendant la maturation des spores. *Annales des Sciences Naturelles Botanique et Biologie Végétale* 8, 361-420.

Fuller, M.S. (1976). Mitosis in fungi. *In* "International Review of Cytology" (Ed. G.H. Bourne and J.F. Danielli). Vol. 45, pp. 113-153. Academic Press, New York:London.

Furtado, J.S. and Olive, L.S. (1970). Ultrastructure of ascospore development in *Sordaria fimicola*. *Journal of the Elisha Mitchell Scientific Society* 86, 131-138.

Garrison, R.G., Mariat, F., Boyd, K.S. and Fromentin, H. (1979). Perithecial ultrastructure and formation of ascospores of *Ceratocystis stenoceras* (Robak) C. Moreau. *Annales de Microbiologie* 130A, 3-21.

Gil, F. (1973). Mesosomes: their role in the delimitation of the ascospore. *Mycophathologia et Mycologia applicata* 49, 243-247.

Gooday, G.W., Green, D., Fawcett, P. and Shaw, G. (1974). Sporopollenin formation in the ascospore wall of *Neurospora crassa*. *Archiv für Mikrobiologie* 101, 145-151.

Greenhalgh, G.N. and Evans, L.V. (1968). The developing ascospore wall of *Hypoxylon fragiforme*. *Journal of the Royal Microscopical Society* 88, 545-556.

Greenhalgh, G.N. and Griffiths, H.B. (1970). The ascus vesicle. *Transactions of the British Mycological Society* 54, 489-492.

Guth, E., Hashimoto, T. and Conti, S.F. (1972). Morphogenesis of ascospores in *Saccharomyces cerevisiae*. *Journal of Bacteriology* 109, 869-880.

Hackett, C.J. and Chen, K-C. (1976). Ultrastructure of developing ascospores in *Sordaria brevicollis*. *Journal of Bacteriology* 126, 883-894.

Harper, R.A. (1897). Kerntheilung und freie Zellbildung im Ascus. *Jahrbuch für wissenschaftliche Botanik* 30, 249-284.

Heslot, H. (1958). Contribution à l'étude cytogénétique et génétique des Sordariacées. *Revue de Cytologie et de Biologie Végétales* 19, 1-235.

Hill, T.W. (1975). Ultrastructure of ascosporogenesis in *Nannizzia gypsea*. *Journal of Bacteriology* 122, 743-748.

Hill, T.W. (1977). Ascocarp ultrastructure of *Herpomyces sp.* (Laboulbeniales) and its phylogenetic implications. *Canadian Journal of Botany* 55, 2015-2032.

Hohl, H.R. and Streit, W. (1975). Ultrastructure of ascus, ascospore and ascocarp in *Neurospora lineolata*. *Mycologia* 67, 367-381.

Horesh, O., Simchen, G. and Friedman, A. (1979). Morphogenesis of the synapton during yeast meiosis. *Chromosoma* (Berl.) 75, 101-115.

Illingworth, R.F., Rose, A.J. and Beckett, A. (1973). Changes in the lipid composition and fine structure of *Saccharomyces cerevisiae* during ascus formation. *Journal of Bacteriology* 113, 373-386.

Janex-Favre, M.C. (1977). La paroi des ascospores du *Tuber rufum* Pico. *Bulletin de la Société Mycologique de France* 93, 407-424.

Janex-Favre, M.C. (1978). Formation et évolution des ascospores chez le *Diatrype disciformis* (Hoffm.) Fries. étude ultrastructurale. *Revue de Mycologie* 42, 265-275.

Janex-Favre, M.C. and Parguey-Leduc, A. (1976). La formation des ascospores chez deux Truffes: *Tuber rufum* Pico et *Tuber aestivum* Vitt. (Tubéracées). Compte Rendu Hebdomadaire des Séances de l'Académie des Sciences Paris 283, 1173-1175.

Le Gal, (1947). Recherches sur les ornementations sporales des Discomycètes operculés. *Annales des Sciences Naturelles Botanique et Biologie Végétales* 8, 73-286.

Lowry, R.J. and Sussman, A.S. (1968). Ultrastructural changes during germination of ascospores of *Neurospora tetrasperma*. *Journal of General Microbiology* 51, 403-409.

Lu, B.C. and Galeazzi, D.R. (1978). Light and electron microscope observations of a meiotic mutant of *Neurospora crassa*. *Canadian Journal of Botany* 56, 2694-2706.

Lynn, R.R. and Magee, P.T. (1970). Development of the spore wall during ascospore formation in *Saccharomyces cerevisiae*. *The Journal of Cell Biology* 44, 688-692.

Mainwaring, H.R. (1967). The role of endoplasmic reticulum in ascomycete spore delimitation. *The Journal of Cell Biology* 35, 86-87A.

Mainwaring, H.R. (1972). The fine structure of ascospore wall formation in *Sordaria fimicola*. *Archiv für Mikrobiologie* 81, 126-135.

Marquardt, H. (1963). Elektronenoptische Untersuchungen über die Ascosporenbildung bei *Saccharomyces cerevisiae* unter cytologischem und cytogenetischem Aspekt. *Archiv für Mikrobiologie* 46, 308-320.

Merkus, E. (1973). Ultrastructure of the ascospore wall in Pezizales (Ascomycetes) I *Ascodesmis microscopica* (Crouan) Seaver and *A. nigricans* van Tiegh. *Persoonia* 7, 351-366.

Merkus, E. (1974). Ultrastructure of the ascospore wall in Pezizales (Ascomycetes) II Pyronemataceae sensu Eckblad. *Persoonia* 8, 1-22.

Merkus, E. (1975). Ultrastructure of the ascospore wall in Pezizales (Ascomycetes) III Otideaceae and Pezizaceae. *Persoonia* 8, 227-247.

Moens, P.B. (1971). Fine structure of ascospore development in the yeast *Saccharomyces cerevisiae*. *Canadian Journal of Botany* 17, 507-510.

Moens, P.B. and Rapport, E. (1973). Spindles, spindle plaques, and meiosis in the yeast *Saccharomyces cerevisiae* (Hansen). *The Journal of Cell Biology* 50, 344-361.

Moens, P.B., Esposito, R.E. and Esposito, M.S. (1974). Aberrant nuclear behaviour at meiosis and anucleate spore formation by sporulation-deficient (SPO) mutants of *Saccharomyces cerevisiae*. *Experimental Cell Research* 83, 166-174.

Moore, R.T. (1965). The ultrastructure of fungal cells. *In* "The Fungi An Advanced Treatise" (Ed. G.C. Ainsworth and A.S. Sussman). Vol. I, pp. 95-118. Academic Press, New York:London.

Morré D.J., Mollenhauer, H.H. and Bracker, C.E. (1971).

Origin and continuity of Golgi apparatus. *In* "Results and Problems in Cell Differentiation": Origin and Continuity of Cell Organelles" (Ed. J. Reinert and H. Ursprung). Vol. 2, pp. 82-126. Springer-Verlag, Berlin.

Nogueira, M.L.B. and Furtado, J.S. (1974). Origem da membrana de clivagem e delimitação de ascosporos no fungo pirenomiceto *Sordaria fimicola*. *Revista de Microbiologia* (S. Paulo) 5, 7-16.

Oso, B.A. (1969). Electron microscopy of ascus development in *Ascobolus*. *Annals of Botany* 33, 205-209.

Parguey-Leduc, A. and Janex-Favre, M.C. (1977a). L'organisation des asques de deux Truffes: *Tuber rufum* Pico et *Tuber aestivum* Vitt. *Revue de Mycologie* 41, 1-32.

Parguey-Leduc, A. and Janex-Favre, M.C. (1977b). L'organisation des ascospores chez le *Tuber aestivum* Vitt. *Travaux dédiés à G. Viennot-Bourgin* 307-323.

Plurad, S.B. (1972). Fine structure of ascosporogenesis in *Nematospora coryli* Peglion, a pathogenic yeast. *Journal of Bacteriology* 109, 927-929.

Raju, N.B. and Newmeyer, D. (1977). Giant ascospores and abnormal croziers in a mutant of *Neurospora crassa*. *Experimental Mycology* 1, 152-165.

Reeves, F. (1967). The fine structure of ascospore formation in *Pyronema domesticum*. *Mycologia* 59, 1018-1033.

Rooney, L. and Moens, P.B. (1973). The fine structure of ascospore delimitation in the yeast *Wickerhamia fluorescens*. *Canadian Journal of Microbiology* 19, 1389-1392.

Sakai, A. (1974). Centriolar plaque and spindle microtubules in the ascomycete *Sordaria humana*. *Botanical Magazine Tokyo* 87, 341-345.

Schrantz, J-P. (1967). Présence d'un aster au cores des mitoses de l'asque et de la formation des ascospores chez l'Ascomycète *Pustularia cupularis* (L) Fuck. *Compte Rendu Hebdomadaire des Séances de l'Académie des Sciences Paris* 264, 1274-1277.

Schrantz, J-P. (1970). Étude cytologique, en microscopie optique et électronique, de quelques Ascomycètes. I. Le noyau. *Revue de Cytologie et de Biologie Végétale* 33, 1-100.

Simonet, J.M. and Zickler, D. (1972). Mutations affecting meiosis in *Podospora anserina* I Cytological studies. *Chromosoma* (Berl.) 37, 327-351.

Srb, A.M., Basl, M., Bobst, M. and Leary, J.V. (1973). Mutations in *Neurospora crassa* affecting ascus and ascospore development. *The Journal of Heredity* 64, 242-246.

Stiers, D.L. (1974). Fine structure of ascospore formation in *Poronia punctata*. *Canadian Journal of Botany* 52, 999-1003.

Stiers, D.L. (1976). The fine structure of ascospore formation in *Ceratocystis fimbriata. Canadian Journal of Botany* 54, 1714-1723.

Syrop, M.J. and Beckett, A. (1972). The origin of ascospore-delimiting membranes in *Taphrina deformans. Archiv für Microbiologie* 86, 185-191.

Syrop, M.J. and Beckett, A. (1976). Leaf curl disease of almond caused by *Taphrina deformans* III Ultrastructural cytology of the pathogen. *Canadian Journal of Botany* 54, 293-305.

Thiéry, J.P. (1967). Mise en evidence des polysaccharides sur coupes fines en microscopie électronique. *Journal de Microscopie* 6, 987-1018.

Thiéry, J.P. (1969). Rôle de l'appareil de Golgi dans la synthèse des mucopolysaccharides, étude cytochimique I. *Journal de Microscopie* 8, 689-708.

Tingle, M., Singh Klar, A.J., Henry, S.A. and Halvorson, H.O. (1973). Ascospore formation in yeast. *In* "Microbial Differentiation" (Ed. J.M. Ashworth and J.E. Smith). Symposium of the Society for General Microbiology 23, 209-243.

Tyson, K. and Griffiths, D.A. (1976). Developmental morphology and fine structure of *Placoasterella baileyi.* II ascus and ascospore development. *Transactions of the British Mycological Society* 66, 263-279.

Wells, K. (1970). Light and electron microscope studies of *Ascobolus stercorarius* I nuclear divisions in the ascus. *Mycologia* 62, 761-790.

Wells, K. (1972). Light and electron microscopic studies of *Ascobolus stercorarius* II ascus and ascospore ontogeny. *University of California Publications in Botany* 62, 1-93.

Wilsenach, R. and Kessel, M. (1965). The role of lomasomes in wall formation in *Penicillium vermiculatum. Journal of General Microbiology* 40, 401-404.

Yoo, B.Y., Calleja, G.B. and Johnson, B.F. (1973). Ultrastructural changes of the fission yeast *Schizosaccharomyces pombe* during ascospore formation. *Archiv für Mikrobiologie* 91, 1-10.

Zickler, D. (1970). Division spindle and centrosomal plaques during mitosis and meiosis in some Ascomycetes. *Chromosoma* (Berl.) 30, 287-304.

Zickler, D. and Olson, L.W. (1975). The synaptonemal complex and the spindle plaque during meiosis in yeast. *Chromosoma* (Berl.) 50, 1-23.

Zickler, D. and Simonet, J-M. (1980). Identification of gene-controlled steps of ascospore development in *Podospora anserina. Experimental Mycology. (In Press).*

ONTOGENESIS AND ULTRASTRUCTURE OF SPORE WALLS IN HIGHER BASIDIOMYCETES

G.M. Olåh[1] and O.R. Reisinger[2]

[1]*Département de Phytologie, Faculté des Sciences de l'Agriculture et de l'Alimentation, Université Laval, Ste-Foy, Québec, Canada G1K 7P4*

[2]*Laboratoire de Botanique et Microbiologie Université de Nancy I, Case officielle n° 140, 54037 Nancy, France*

HOMMAGE POSTHUME A NOTRE MAITRE
MONSIEUR LE PROFESSEUR ROGER HEIM,
MEMBRE DE L'ACADEMIE DES SCIENCES.

INTRODUCTION

Fungal spores have always been important in the classification of these organisms. Spore colouring for example has been used by Fries to subdivide the Hymenomycetes. The form, size and ornamentation of some elements, as well as the succession of integuments which compose the spore wall, are being increasingly studied. Numerous investigations are being carried out to discover the relationship between the species, genera and families in order to establish natural systematic units and to better define the identity of the biological entities. Consequently, ultrastructural studies concerned with the organization of basidiospores and basides are numerous. This work is generally directed towards confirmation of the results provided by light microscopy. Special interest has been paid to the spore envelope, for example, and an extensive terminology has been created to describe the different layers of the spore walls. This nomenclature changes from one researcher to another and is based on results obtained by various techniques using spores of different physiological age. It is hardly valid to compare *chemically stressed*

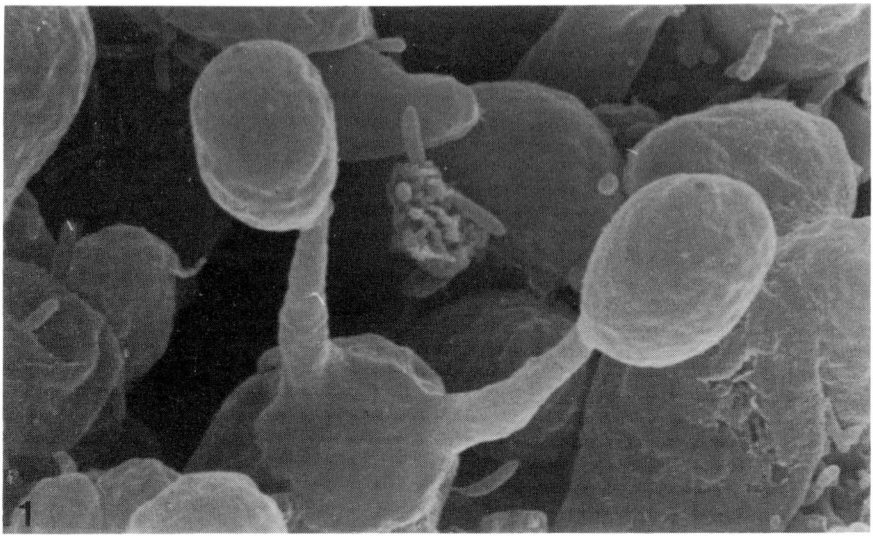

Fig. 1. Bisterigmatic basidium of Cantharellus cibarius Fr. Scanning E.M. 5 000 X

samples from a herbarium (e.g. as a result of autoclaving in an aqueous solution of KOH) with spores treated to maintain their ultrastructural integrity. The spore wall is, nevertheless, more resistant than the hyphal wall. During spore formation the wall structure undergoes a constant evolution. It is therefore vital to compare spores at the same stage of development.

These difficult problems have attracted the attention of several groups. Wells (1965); Sundberg (1971); Hugueney (1972, 1975b); McLaughlin (1973); Olåh and Reisinger (1974); Olåh *et al.* (1977); Weber and Hess (1974) have studied basidiospores and basides ultrastructure. Kühner (1973) and his co-workers and Kühner (1975) have studied the architecture of the spore wall but their findings do not agree with the dynamic approach of the study of the wall structure by Wells and others

These divergent interpretations can be attributed to the complexity of the subject, the variety of basidiospore types, the techniques applied and the heterogeneity of the materials employed. As a result there is a lack of universal agreement concerning nomenclature. Considering these difficulties we will concentrate on only a few examples of Basidiomycotonia such as Hymenomycetes and Agaricales. Simultaneously we shall seek parallels between the ontogeny of the wall of these *Fungi imperfecti* and that of the Basidiomycetes.

WHY ARE THE BASIDIOSPORES OF SUCH INTEREST ?

The word "spore" was used for the first time by Hedwig between 1778-1784. The principal role of the spore is to assure the survival and propagation of the species. The spores are commonly born at the tip of a sporogenous cell called basidium, a term proposed by Léveillée in 1837. The periods of activity and dormancy of sporogenesis represent two distinct phases both from the biochemical and karyological as well as the morphological point of view. The most recent studies center on the sporogenic cell, its ontogenesis and the engendered spore. This research should be useful for fungal classification.

MORPHOLOGICAL EVOLUTION OF BASIDIOSPORES AND CHANGES OF WALL ULTRASTRUCTURE ACCORDING TO SPORE TYPE AND PHYSIOLOGICAL AGE (Figs 1-3)

When ontogenetic studies were first undertaken, researchers were more interested in dormant spores and spore germination than in the morphogenetic evolution of the basidiospores themselves. Interest has gradually grown in the structure and chemical composition of the wall. Until maturation the spore is connected to the sporogenic basidial cell from which it evolves through the sterigmata. More recent dynamic studies have therefore covered the evolution of the wall structure in its entirety. According to Hugueney (1975a), McLaughlin (1974), Olåh *et al.* (1977), this approach is of great importance because, when accompanied by cytochemical investigations, it permits definition of the evolution of the spore wall structure as a function of age.

From the development of the sporogenous cell until spore liberation, Olåh *et al.* (1977) were able to establish six distinct phases. The first five phases are defined in relation to the meiotic divisions, while the last should be

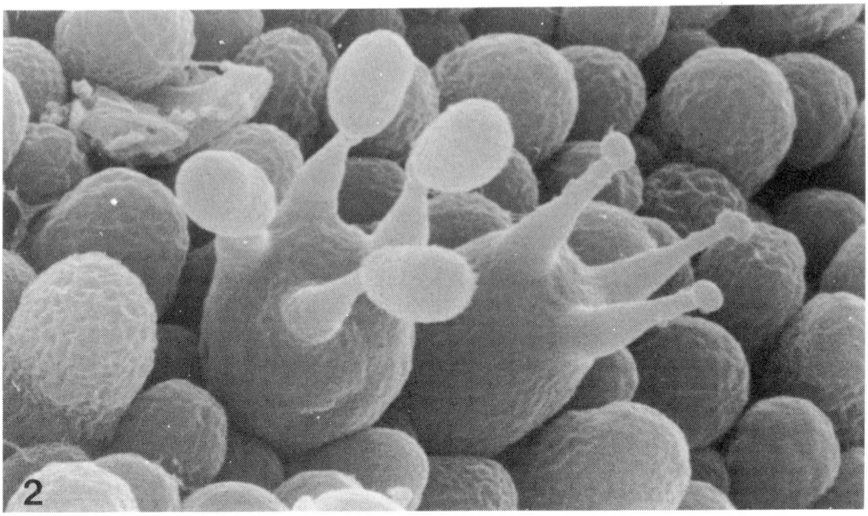

Fig. 2. Tri- and tetrasterigmatic basidium of Hygrophorus puniceus Fr. Scanning E.M. 5 000 X

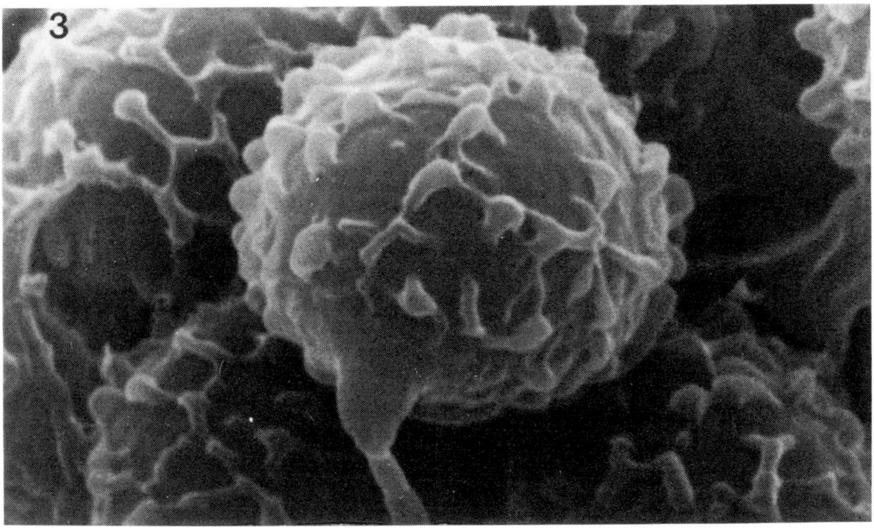

Fig. 3. Rough spore of Lactarius chrysorreus still attached to its sterigmata. Scanning E.M. 1 000 X

characterized with respect to the initiation and maturation of the spore.

These steps reflect the important changes undergone during development. Modifications in the endomembranous (vacuolization, vesiculation) and nuclear (karyogamy, meiosis) systems are characteristic for these steps (Figs 6 and 7).

Phase 1 or basidiole, first step in the evolution of the uni- or bi-nucleated basidium. At this phase there exist only two nuclei destined for pre-meiotic karyogamy. There are also small vacuoles and increased metabolic activity in the tip or apical zone.

Phase 2 or pre-meiotic, characterized in particular by apical growth and peripheral migration of the vesicles. Sterigmata formation commences; the young basidium increases in volume.

Phase 3 or meiotic I, depicts two meiotic nuclei; metabolic activity is intense. The vesicles situated along the periphery are retained in the membranous wall folds ensuring intercalary growth of the sporogenous cell (Piché *et al.*, 1976).

Phase 4 or meiotic II occurs in two steps : the second reductional division of meiosis and the development of sterigmata (Fig. 8). Meiosis II leads to duplication of the two preexistent nuclei; the spore becomes visible for the first time. The wall structure of the sporogenous cell gives rise in an *enteroblastic* way to the newly formed wall structure (Figs 4, 4a and 4b) while enveloping the young spore through a more or less visible initial wall.

Phase 5 or post-meiotic, the post-meiotic material migrates via the sterigmata into the young developing spore which is still in direct contact with the basidium (Figs 5 and 6).

Phase 6 or final formation of the spore. The elaboration of the spore wall (B-A) progresses through three physiological ages : *primary* (6-1), *secondary* (6-2) and *tertiary* (6-3). These different steps within phase 6 explain the stratification of the layers in the spore wall (Figs 9-13). The several layers identified by other workers reflect the evolution of the wall, implying increasing reorganization of the materials required for structure. These materials, which undergo physico-chemical stress during wall formation, are able to contract, creating a structural artifact appearing as a distinct structural entity. Three definitive layers constitute the wall structure. They can be designated

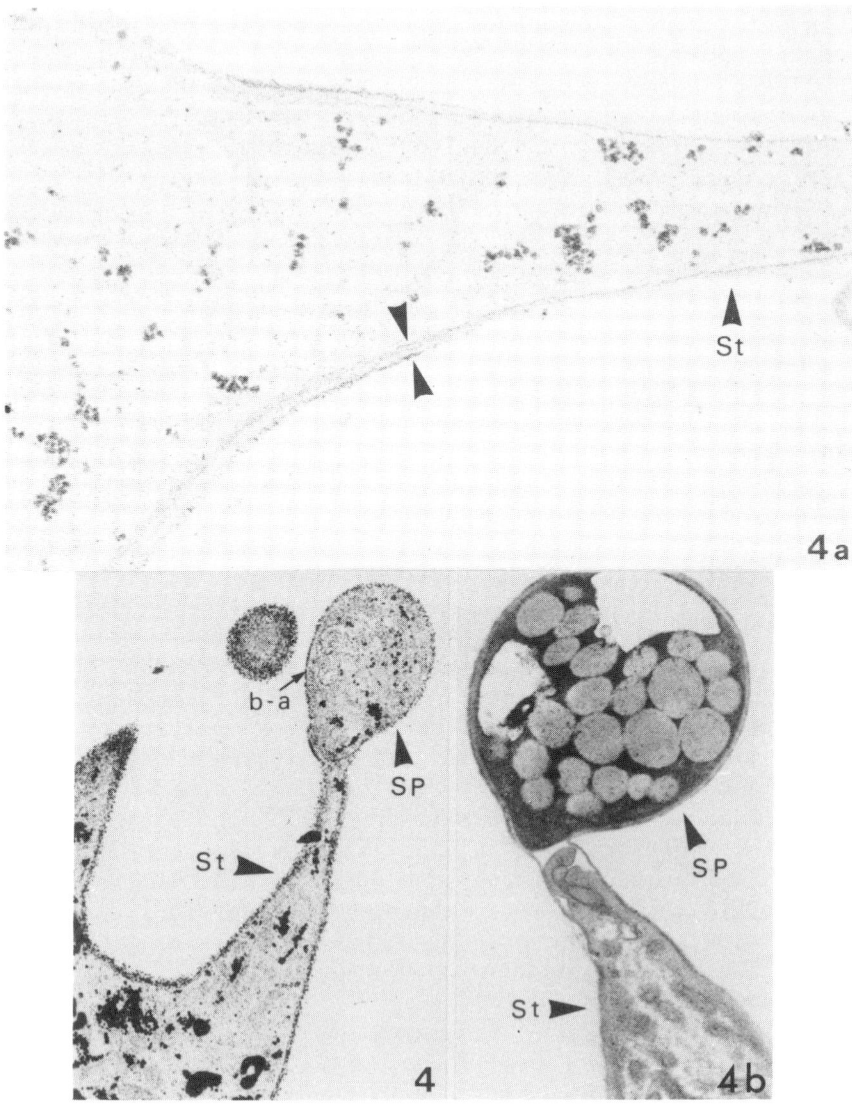

Figs 4 and 4a. Melanosporous Agaric showing the direct relationships between the spore initial and the sporogenic cell.

Fig. 4. Localization of the cystine.

Figs 4a and 4b. Localization of substances of polysaccharidic nature.

A, B and C, moving from the exterior towards the interior (Figs 11 and 13). The continuation of the wall (b-a) of the sterigmata surrounding the young spore may be permanent or may become fugacious, the unity no longer being functional. Between the wall of the sterigmata (b-a) (see Table 1), surrounding the basidiospore and the newly formed spore wall (CBA) there is a transitional space, lacking any apparent structure. This is designated a structural unit by many authors and considered part of the perispore or 5th layer (Perreau-Bertrand, 1967; Melendez-Howell, 1967; Heim, 1969; Olåh and Reisinger, 1974). This interstice can be more or less occupied depending on whether spores are smooth or rough. Furthermore, during the processes of degradation, the layer (b-a) is the first to disappear. Reisinger *et al.* (1977) have shown conclusively that the structure (B-A) is a fundamental layer of the spore.

EVALUATION OF RESULTS OBTAINED BY LIGHT, TRANSMISSION AND SCANNING ELECTRON MICROSCOPY

Studies of the spore wall architecture with dynamic approaches including cytochemistry, electron microscopy or microdissection have only been carried out during the last fifteen years, even though fundamental data concerning spores had already been provided by the end of the 18th century. Heim, Kühner and their collaborators have studied fungi with the light microscope since 1930. From 1929 to 1959 Malençon made numerous observations on sporogenesis as well as on spore ornamentation. Other schools such as Singer (1962), Smith (1964), Clemençon (1970), Buller (1933), Griffiths (1971), Heim (1962), Perreau-Bertrand (1967), Stocks and Hess (1970) and Hawker and Madelin (1974) have been particularly interested in classification. The results obtained using light and electron microscopy have not always been correctly related and interpreted. It follows that a too personalized and confused terminology has been created, complicating the data. However, cytochemical electron microscopy should permit description of well-defined phases of spore architecture and definition of its final structure (Olåh *et al.*, 1977) (see Table 1 and Figs 4 and 5). The elaboration of the spore wall (B-A) takes place progressively during phase 6 when a "stratification" or concentration of wall material occurs. The demarcation lines of such

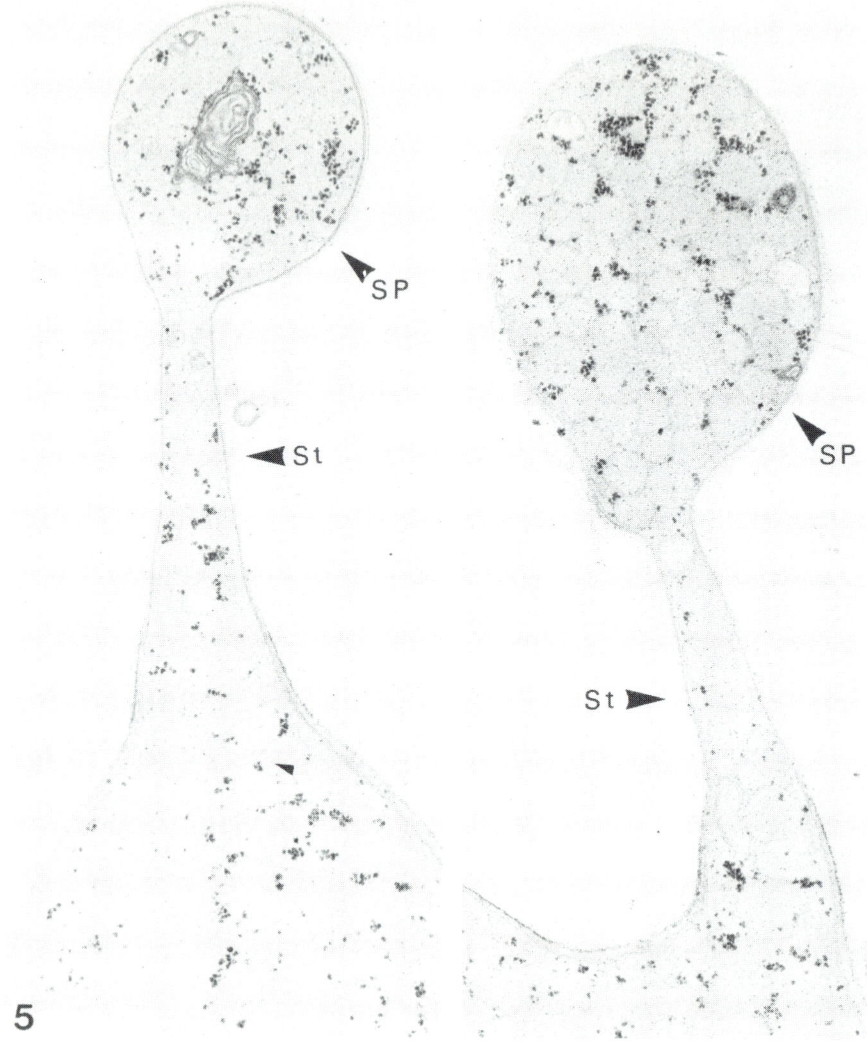

Fig. 5. Schematic view of the evolution of the spore initial in a melanosporous Agaric, Hygrophorus capreolaris Fr. showing the localization of substances of a polysaccharidic nature.

molecular rearrangements have no systematic value because at this stage it is a question of migration of the wall constitutive elements from the interior to the exterior. When the spores are mature and liberated from the sterigmata, layer "C" becomes more and more electron-transparent, reflecting migration of the precursor molecules of the wall structure into the "B-A" zone. Scanning electron microscopy indicates the external layer (b-a) is continuous on the sterigmata as well as on the spore (Fig. 14). These results agree with the observations of Stock and Hess (1977) and Griffiths (1971).

SIMILARITY OF STRUCTURE AND WALL ONTOGENY EXISTING BETWEEN "PROPAGULES" OF IMPERFECT FUNGI AND BASIDIOSPORES

A similarity exists between the mechanisms of basidiospore and "propagule" maturation in Deuteromycetes. Reisinger (1972), Mangenot and Reisinger (1976) have demonstrated a basic principle for the sequential appearance of the teguments of the conidial wall. They describe three distinct subunits (C-B-A), which correspond to the structure of the fundamental wall of the basidiospores (C, "B-A"). However, the external layer (b-a) recognized by several authors in Basidiomycotiana does not exist in the *Fungi imperfecti*. In a number of *Fungi imperfecti*, in addition to the Basidiomycetes, microvesicles are apparent which may be involved in transporting building materials to the wall (Olàh *et al.*, 1977).

If we consider other subdivisions of the Basidiomycetes (such as the Uredinales) and compare them from the point of view of spore ontogeny with certain *Fungi imperfecti*, we discover obvious similarities (Hughes, 1970).

CONCLUSIONS

Can we envisage a common denominator in the study of the wall structure ?

Based on the work of several laboratories, including our own, the following generalizations can be made. The evolution of the wall structure in Basidiomycetes is clearly related to their physiological age. Independently of the chemical composition of the wall, the materials required for its elaboration originate in the cytoplasm, migrate and reorganize themselves *in situ* at the wall level. They thus form a definite barrier or "parietal structure" (Bartnicki-Garcia, 1968 and 1970; Farkas, 1979; Olàh *et al.*, 1977). These definitive structures may be designated with the letter "C" for the most internal layer (Figs 11 and 13). In close contact with the plasma membrane, this zone becomes more and more electron-translucent with age as a result of the translocation of the elements utilized in the elaboration of the spore wall "B-A". In certain cases, the most external layer remains visible, surrounding the nascent and mature spores (Figs 14-16). It represents a wall structure originating from the sterigmata "b-a". In several Basidiomycetes we have observed a mechanism of wall elaboration very similar to that already described by Mangenot and Reisinger (in Weber and Hess, 1974) and Al-Rikabi (1978) in the *Fungi imperfecti*.

REMERCIEMENTS

Nous tenons à exprimer nos vifs remerciements à Mme O. Desbiens et M.A. Pusterla pour leur amicale et toujours bienveillante collaboration.

Figures 4 and 4a, 6-13 and Table 1 are reproduced by permission of the Société française de Phytopathologie (travaux dédiés à G. Viennot-Bourgin, 285-305, 1977).

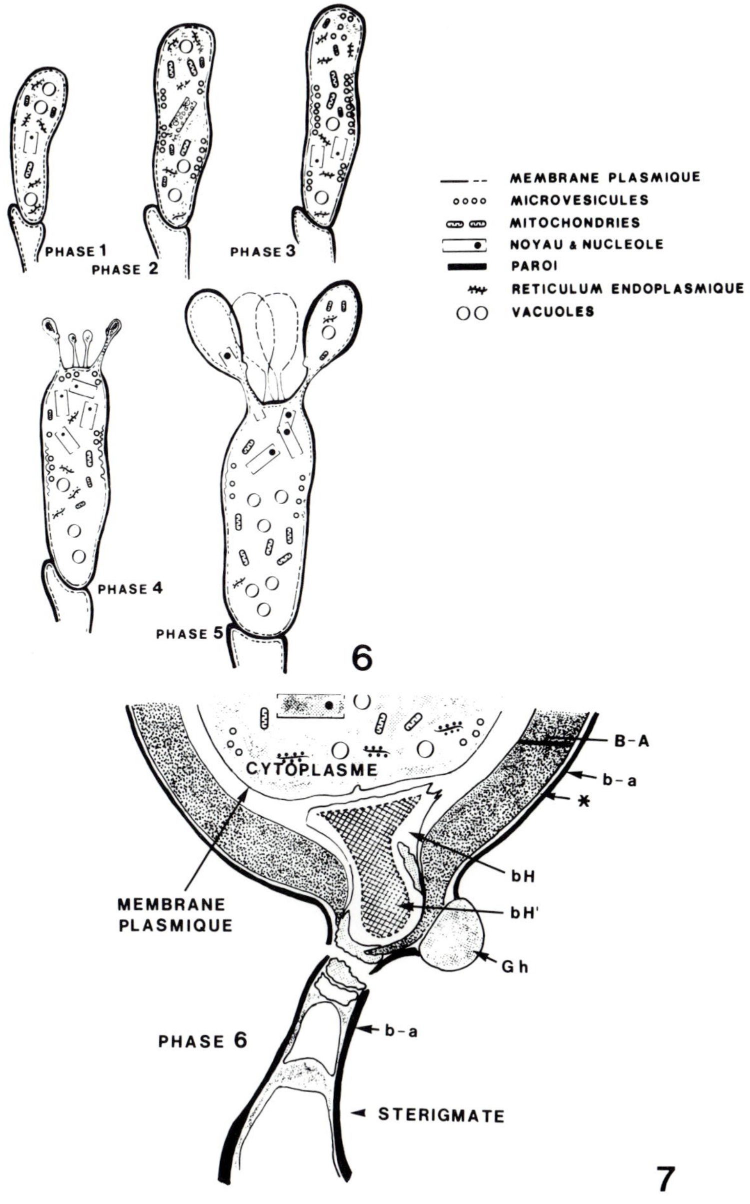

Fig. 6. Evolution of the sporogenic basidial cell and schematization of the ultrastructure illustrating the physiological age of this cell.

Fig. 7. Phase 6 of the ontogeny and the spore maturation demonstrating the correlation between the fundamental layers, namely, the true spore wall B-A and the layer b-a; originally formed from the sterigmatic wall.

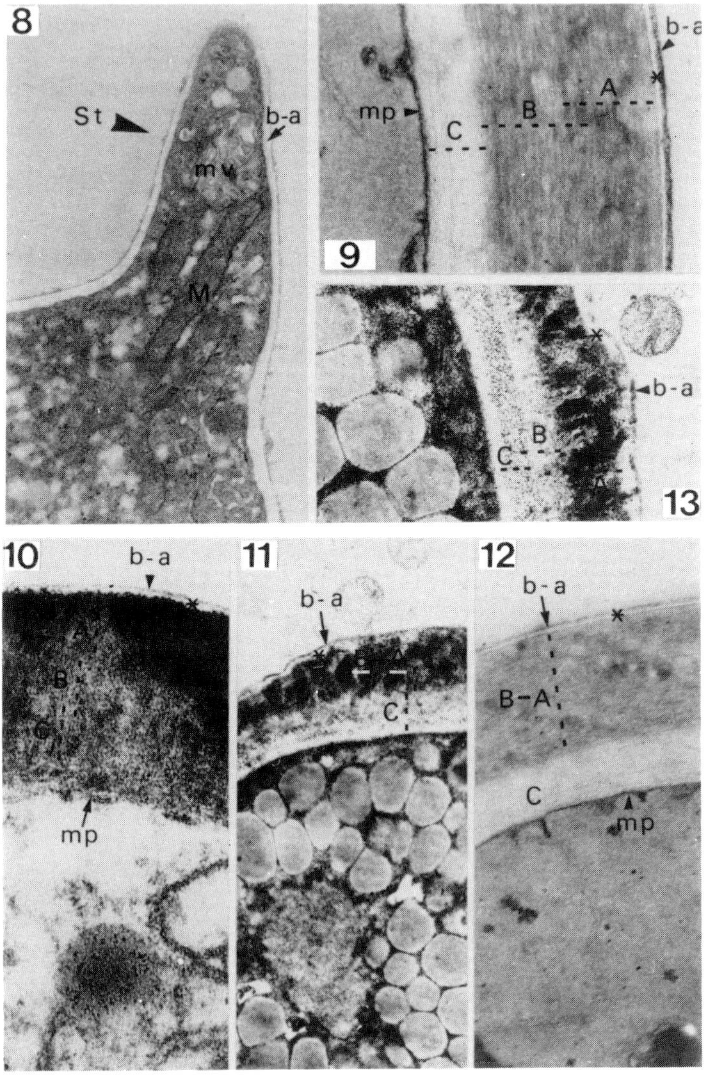

Fig. 8. Tip of a well-developed sterigmata of a leucosporous Agaric; at this stage the integrity of the cell wall is maintained (b-a). 8 000 X

Figs 9, 10 and 12. Evolution of the wall structure of a smooth spore of melanosporous Agaric.

Figs 11 and 13 idem, but observed in a melanosporous Agaric with rough-surfaced spore.

Fig. 9, 40 000 X; Fig. 10, 40 000 X; Fig. 11, 6 300 X; Fig. 12, 16 000 X; Fig. 13, 16 000 X.

Abbreviations :

-A : newly formed fundamental wall of the mother-cell;
-B idem; -b-a : fundamental wall of the mother-cell;
-C : newly formed fundamental structure; -M : mitochondria;
-mp : plasmic membrane; -mv : cytoplasmic microvesicle;
-St : sterigmata; -SP : spore.

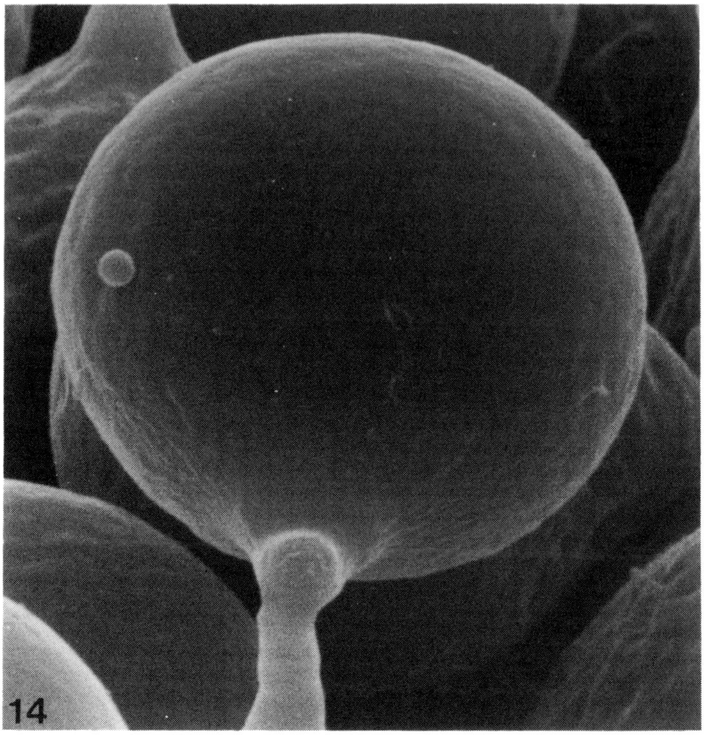

Fig. 14. The external layer of the spore wall is continuous with that of the sterigmata. Amanita vaginata Fr. Scanning E.M. 10 000 X.

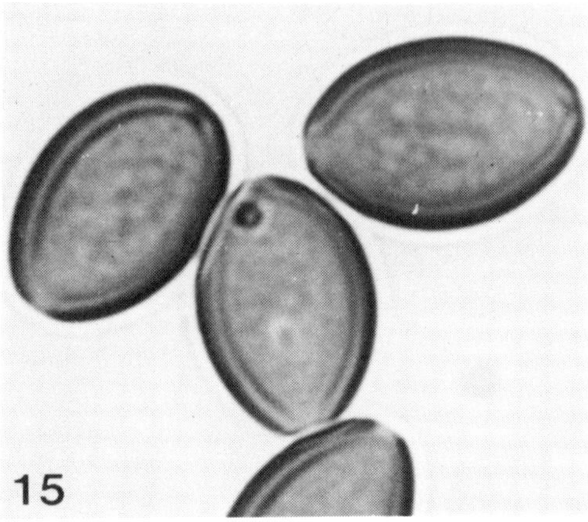

Fig. 15. Psilocybe quebecensis Olåh et Heim.

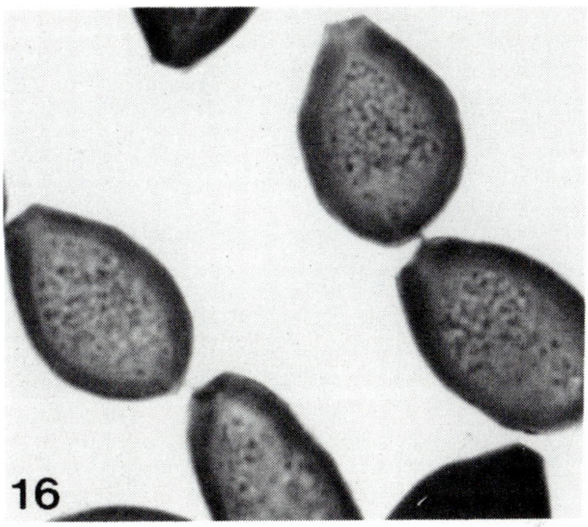

Fig. 16. Panaeolus foenisecii (Fr.) Kühn., in optical microscopy its spores, like those of Fig. 15, show a wall structure made of several layers. 1 200 X

TABLE 1

Comparison of the wall subunits (TEM)

1977 Olåh *et al.*	1974 Clåh Reisinger	1962 Heim 1967 Perreau-Bertrand	1973 Kühner		1970 Clemençon	1970 Stocks Hess	1974 Hawker Madelin
Cytoplasm	0 cytoplasm	cytoplasm					
Plasmic membrane	1st layer	plasmic membrane					
C	2nd layer	endospore		endosporium	corium leptotunica	W_2	W_3
BA	3rd layer	episporeum	eusporium	episporium	tunica coriotunica	W_1	W_2
					sclerosporium		
	4th layer	exospore			epitunica		
Interstice	5th layer	perispore	myxosporium		endosporothecium tectum 1 and 2		
b-a	6th layer	ectospore	ectosporium		ectosporothecium	F	W_1

W: wall; F: fibrous layer

REFERENCES

Al-Rikabi, K. (1978). Étude comparative des parois de la souche sauvage du champignon Drechslera sorokiniana et de son mutant albinos. Thèse de doctorat ès-sciences naturelles, Université de Nancy I.

Bartnicki-Garcia, S. (1968). Cell wall chemistry, morphogenesis, and taxonomy of Fungi. Ann. Rev. Microbiol., 22, 87-108.

Bartnicki-Garcia, S. (1970). Cell wall composition and other biochemical markers in fungi phylogeny. *In* J.B. Harborne Phytochemical Phylogeny, pp. 81-103. Academic Press, edit., London et New York.

Buller, A.H.R. (1933). Researches on fungi. V. Londres: 171-206.

Clemençon, H. (1970). Bau der Wände der Basidiosporen und ein Vorschlag zur Benenmung ihrer Schichten. Zeitschr. f. Pila: 113-133.

Farkas, V. (1979). Biosynthesis of cell walls of fungi. Microbiol. Rev. June 1979, p. 117-144.

Griffiths, D.A. (1971). The fine structure of Basidiospores of Paneolus campanulatus (L.) Fr. revealed by Freeze-Etching. *In* Arch. Mikrobiol. 76: 74-82.

Hawker, L.E. et Madelin, M.R. (1974). The dormant spore. *In* The Fungal Spore. Forms and Function, D.J. Weber et W.M. Hess Eds., Brigham Young University, John Wiley & Sons: 1-72.

Heim, R. (1962). L'organisation architecturale des spores de Ganodermes. Revue de Mycologie, 27: 199-212, 5 pl.

Heim, R. (1969). Champignons d'Europe, 2e édition, Paris, Boubée, 680 pp.

Hughes, S.J. (1970). Ontogeny of spore forms in Uredinales. Can. J. Bot. 48: 2147-2157

Hugueney, R. (1972). Ontogenèse des infrastructures de la paroi sporique de Coprinus cineratus Quel. var. nudisporus Kühner (Agaricales). C.R. Acad. Sci. Paris, D 275: 1495-1498.

Hugueney, R. (1975a). Morphologie ultrastructurale et développement de l'apicule des spores de quelques Coprinacées: étude particulière de Puncturus lacrymans. Bull. Soc. Linn. Lyon 8: 249-256.

Hugueney, R. (1975b). Ultrastructure de la paroi sporique de Coprinus congregatus Bull. ex. Fr. et ses variations liées à la symétrie bilatérale de la spore. Bull. Soc. Linn.

Kühner, R. (1973). Architecture de la paroi sporique des Hyménomycètes et de ses différenciations. Persoonia

7(2): 217-248.

Kühner, R. (1975). L'architecture de la paroi sporique des Rhodophyllacées (Basidiomycètes - Agaricales) comparée à celles des autres Hyménomycètes. C.R. Acad. Sci. Paris, D 280: 595-600.

Léveillée, J.H. (1837). Recherches sur l'hyménium des champignons. Annls. Sc. Nat., Bot., 2e série, 8: 321-338, 4 pl.

Malençon, G. (1929). Observations sur les ornements des spores chez les champignons. Arch. Bot., 3: 121-219

Malençon, G. (1942). La sporogenèse chez les Basidiomycètes. C.R. Acad. Sc., Paris, 215: 584-586

Malençon, G. (1945). L'ontogenèse des formes sporales complexes. Ibid., 221: 425-427.

Malençon, G. (1946). La formation des spores anguleuses chez les agarics rhodosporés. Ibid. 222: 912-914.

Malençon, G. (1931). Sur les spores des Russules et des Lactaires. Bull. Soc. Myc. Fr., 47: 72-86.

Malençon, G. (1950). A propos de la sporogenèse chez les Basidiomycètes angiocarpes. Revue de Mycologie, 15: 138-145, 5 fig.

Malençon, G. (1959). La sporogenèse des *Hymenogaster* et ses enseignements. Bull. Soc. Myc. Fr., 75: 99-131.

McLaughlin, D.J. (1973). Ultrastructure of sterigma growth and basidiospore formation in *Coprinus* and *Boletus*. Can. J. Bot., 51: 145-150.

McLaughlin, D.J. (1974). Ultrastructural localisation of carbohydrate in the hymenium and subhymenium of *Coprinus*. Evidence for the function of the golgi apparatus. Protoplasma, 82: 341-364.

Melendez-Howell, L.M. (1967). Recherches sur le pore germinatif des basidiospores. Annal. Sci. Nat. (Bot.) Paris, 12e série 8: 487-638.

Olåh, G.M. et Reisinger, O. (1974). L'ontogénie des téguments de la paroi sporale en relation avec le stérigmate et la gouttelette hilaire chez quelques agarics mélanosporés. C.R. Acad. Sci. Paris, D 278: 609-611.

Olåh, G.M., Cole, G.T. et Reisinger, O. (1977). Le rôle et la nature chimique des microvésicules secrétoires dans l'apex hyphal et dans les cellules sporogènes. Ann. Sci. Nat. Bot, 12e série 18(4): 301-318.

Olåh, G.M., Piché, Y et Reisinger, O. (1977). L'ontogénie pariétale des cellules sporogènes et des cellules sporales chez quelques Agarics. Trav. dédiés à G. Viennot-Bourgin. Soc. Fr. Phytopath., p. 275-305.

Piché, Y., Olåh, G.M. et Desbiens, O. (1976). La dynamique ultrastructurale du développement des cellules hyméniales de quelques Agarics. Troisième réunion annuelle. Société de microscopie du Canada. Univ. of Ottawa 3: 176-178.

Perreau-Bertrand, J. (1967). Recherches sur la différenciation et la structure de la paroi sporale chez les Homobasidiomycètes à spores ornées. Ann. Sci. Nat. 12(8): 487-638.

Reisinger, O. (1972). Contribution à l'étude ultrastructurale de l'appareil sporifère chez quelques Hyphomycètes à paroi mélanisée (Genèse, modification et décomposition). Thèse d'Etat, Université de Nancy.

Reisinger, O., Kiffer, E., Mangenot, F. et Olàh, G.M. (1977). Ultrastructure, cytochimie et microdissection de la paroi des hyphes et des propagules exogènes des Ascomycètes et des Basidiomycètes. Rev. Mycol., 41(1) : 91-117.

Singer, R. (1962). The Agaricales in modern taxonomy. Weinheim.

Smith, A.H. et Singer, R. (1964). A monograph on the genus *Galerina* Earle. New York and London.

Stocks, D.L. et Hess, W.M. (1970). Ultrastructure of dormant and germinated basidiospores of a species of *Psilocybe*. Mycologia 62(1): 176-191.

Sundberg, W. (1971). A study of basidial ontogeny and meiosis in *Schizophyllum*. Thèse Ph.D., University of California, Davis Botany, 156 p.

Weber, D.J. et Hess, W.M. (1974). Fungal spore, form and function. A Wiley-interscience publication, John Wiley & Sons, New York, 895 p.

Wells, K. (1965). Ultrastructural features of developing and mature basidia and basidiospores of *Schizophyllum commune*. Mycologia 57: 236-261.

CONIDIOGENESIS

T. M. Hammill

Department of Biology
SUNY College of Arts and Science
Oswego, New York 13126, U.S.A.

1. INTRODUCTION

One can read many definitions of taxonomy. They range from being imprecise and not very useful to the opposite. In the *Dictionary of Fungi* (Ainsworth et al., 1971), taxonomy is simply defined as "systematic classification," and no mention is made of evolution. Talbot (1971) and Snell and Dick (1971) defined taxonomy with scarcely more precision. Simmons (1966) noted that, while classification is used to designate any systematic arrangement of organisms in groups or categories based upon some definite scheme, taxonomy is a special sort of classification of organisms according to their natural relationships. Ainsworth (1973) stated that there is general agreement that taxonomy should endeavor to reflect phylogeny, and that in the virtual absence of a fossil record, much of the detailed phylogeny of microorganisms is speculative. Perhaps the most useful short definition of taxonomy appears in Gove & Staff (1976) : "orderly classification of plants and animals (sic) according to their presumed natural relationships forming a basic biological discipline involving during its Linnaean period the firm establishment of binomial nomenclature and acceptance of the static concept of fixity of the species, during its Darwinian period the dynamic concept of speciation by natural selection, and during its modern Mendelian epoch an expansion to include study of the mechanisms underlying speciation and related processes." Thus, despite the lack of precision in some definitions, we arrive at the notion that taxonomy fundamentally is the study of evolution.

It is interesting to note that the starting points for fungal nomenclature are (Ainsworth et al., 1971): May 1,

1753 (Linnaeus, *Species Plantarum*, ed. 1), Dec. 31, 1801 (Persoon, *Synopsis methodica fungorum)*, Jan. 1, 1821 to Jan. 1, 1832 (Fries, *Systema mycologicum*, Vol. 1-3), and Dec. 31, 1820 (Sternberg, *Flora der Vorwelt*, versuch 1). These starting points predate by several decades, and, in one case, more than a century, Darwin's 1859 publication, *On the Origin of Species by National Selection* ....

Classification of organisms as an intellectual activity largely began with the Greeks, most notably Aristotle (384-322 B.C.). Before Darwin's work, and in many--perhaps most--quarters for many years after Darwin, *classification* of organisms was a very different type of activity from the *taxonomy* of today. Modern taxonomy was made possible by Charles Darwin and a handful of his contemporaries who conceived of a revolutionary new way of looking at organisms.

Organic evolution has become the most significant paradigm in the history of biology. Indeed, Dobzhansky (1973) said "nothing in biology makes sense except in the light of evolution." Accordingly, biologists (particularly in recent decades) have sought to integrate data into taxonomic models which are hypothesized to reflect evolutionary relationships.

With the foregoing in mind, let us examine the group of microfungi called Hyphomycetes. Until relatively recently, classification of Hyphomycetes has been based largely upon morphology of the sporulation apparatus as studied by light microscopy which had become greatly refined by 1900. By 1880, tens of thousands of microfungal names existed, but for lack of any sort of comprehensive classification scheme, mycological literature lay in chaos. Saccardo developed a classification system designed to organize and categorize all fungal descriptions. It was based upon characters such as spore size, shape, color, ornamentation, and septation, and the color and arrangement of the sporulation apparatus (Saccardo et al., 1882-1972). While the Saccardoan system allowed for storage and retrieval of data about Hyphomycetes, it is taxonomically useless, being wholly artificial and completely without phylogenetic significance (Wolf and Wolf, 1947).

Following the lead of several earlier mycologists (notably: Costantin, Vuillemin, and Mason), Hughes (1953) proposed a comprehensive classification system for Hyphomycetes based upon developmental patterns of conidiogenesis. Hughes believed that a more rational classification of Fungi Imperfecti would be possible only when the precise method of conidial origin was understood and emphasized. While Hughes did not state directly that his proposal was a quest for phylogenetic significance, there is little doubt that this is how the Hughesian proposal generally was interpreted, and much conidiogenic research was stimulated. Electron microscopic data about conidiogenesis began to be published in 1963 (Tanaka and Yanagita, 1963), but most such data have appeared only during the past decade, and their full impact is yet to be known.

Phialides vs. Annellides

The Hughesian proposal consisted of eight developmental sections (plus two subsections), each characterized by a particular conidiogenic pattern. Hughes included in his Section III (in part) those Hyphomycetes whose conidiogenous cells are annellides and in Section IV those whose conidiogenous cells are phialides. These two types of conidiogenous cells have been considered during most of the past quarter century to be fundamentally different, and considerable controversy arose in discussions about them.

Phialides were first described as a type of conidiogenous cell by Vuillemin (1910); the first published electron micrograph of a phialide appeared in 1963 (Tanaka and Yanagita, 1963), but most of our understanding about phialoconidiogenesis required several more years cumulation of data. For example, as late as 1971, there still was speculation about whether secondary and subsequent conidial initials were naked masses of protoplasm first and then surrounded by a new cell wall formed *de novo* in each case (Subramanian, 1971); such a speculation was dismissed as improbable by Hammill (1974).

Although an annellide was first figured by Pinkerton (1936), annellides were described in detail by Hughes (1951, 1953); the first published electron micrographs of an annellide appeared in 1971 (Cole and Aldrich, 1971; Hammill, 1971a). At that time, there was little doubt that annellides and phialides differed fundamentally. For example, the following quote from Kendrick and Chang (1971) is illustrative: "The annellide superficially resembles the phialide. It has a deceptive air of possessing a definite length, and its contours frequently match those of phialides: an elongated, often somewhat widened body, and a narrower neck.... The fixed endogenous meristem of the characteristic phialide and its disinclination to increase in length with conidium production clearly delineate it from the average annellide, with its repeated, short percurrent proliferations and its consequent multiplicity of wall lamellae ...."

It is also significant to note that at the first international workshop on taxonomy of Fungi Imperfecti (Kananaskis I, held in 1969), the proceedings of which were edited by Kendrick (1971), annellidic and phialidic fungi were separated into two basically different developmental categories termed holoblastic and enteroblastic, respectively. Further, Hughes (1971), and Cain (1972) speculated that annellides evolved from phialides.

Data from electron microscopy, published during the years following Kananaskis I, made a rigid distinction between annellides and phialides increasingly difficult to maintain, prompting Hammill (1971b, 1974), Morgan-Jones et al. (1972), and Cole and Samson (1979) to envision a developmental continuum between annellidic and phialidic forms. The data which led to that reconciling hypothesis were reviewed in the definitive summary of conidiogenic research through 1977, *Patterns of Development in Conidial Fungi* (Cole and Samson, 1979).

In the proceedings of Kananaskis II (held in September, 1977), *The Whole Fungus* (Kendrick, 1979b), several chapters were devoted to anamorphic development and to anamorphic-teleomorphic correlations. Especially notable vis-a-vis the discussion which follows below were chapters by Kendrick and Nag Raj (1979), Luttrell (1979), Madelin (1979), Subramanian (1979), and Wang (1979).

Below are presented and briefly discussed some electron micrographic data about conidiogenesis in several species of Hyphomycetes. Most of the data refer to phialidic and annellidic conidiogenesis, and were previously unpublished.

2. MATERIALS AND METHODS

Species of Fungi Imperfecti figured here include *Dendrostilbella chlorina* Malençon (isolated from forest soil in Clarke County, Georgia, during the summer, 1968), *Metarrhizium anisopliae* (Metsch.) Sorokin (Hammill, 1970, 1972a), *Trichurus spiralis* Hasselbring (Hammill, 1977b), *Scopulariopsis koningii* (Oud.) Vuillemin (Hammill, 1971a), *Doratomyces nanus* (Ehrenb. ex Link) Morton & Smith, (Hammill, 1972b), *Trichoderma saturnisporum* Hammill (1970, 1974), *Tritirachium roseum* van Beyma (Hammill, 1973), and *Pseudobotrytis terrestris* (Timonin) Subramanian. Original publications can be consulted for the methods used to prepare previously published micrographs (Fig. 9, 14, 16, 18, 20). For previously unpublished micrographs (Fig. 1-8, 10-13, 15, 17, 19, 21), the following preparatory procedure was used (a similar procedure was used for *Mucor* # 218 shown in the paper about mucoralean sporangiosporogenesis, this volume). Cultures were flooded with 0.8% glutaraldehyde buffered with 0.05M sodium cacodylate (pH 7.2) to which had been added a small drop of Tween-20 to facilitate wetting of cell walls. After 10-30 minutes of fixation, material was rinsed in buffer for 1-3 hours (several changes), and fixed 3-11 hours in 1-2% OsO_4 (same buffer). After rinsing in buffer, with a transition to distilled water (30-60 minutes), material was soaked overnight in unbuffered 0.5% uranyl acetate, rinsed 30 minutes in several changes of distilled water, dehydrated in an ethanol series, and embedded in Araldite 502 or Spurr's medium. Sections were stained 25 minutes in 0.5-1% uranyl acetate, rinsed, and stained 10 minutes in lead citrate. Electron microscopy was done with an RCA EMU-4 electron microscope at 100 kV.

3. RESULTS AND DISCUSSION

Figures 1-6 are of conidiogenesis in the phialidic *Dendrostilbella chlorina*, Fig. 7-8 are of the phialidic *Metarrhizium anisopliae*, Fig. 9-13 are of the annellidic *Trichurus spiralis*, and Fig. 14-21 are of conidial bases of those species plus an additional five species.

Dendrostilbella chlorina (Hammill culture # 22-68)

Dendrostilbella is a synnematal genus in which the conidiophores support a dendritic, verticillate arrangement of conidiogenous cells (Fig. 1) from which are produced numerous, small, hyaline amerospores in mucus (Barron, 1968). The genus has been considered moniliaceous (Barron, 1968), but that concept is under revision (Hammill and Shipman, in prep.). Barron (1968) also indicated that conidiogenous cells of *Dendrostilbella* appear to be phialides, but that there is still some doubt because the type species (*D. prasinula* Höhn.) has not been re-examined.

The conidiogenous cells of this isolate of *D. chlorina* could be shown unequivocally to be phialides only with electron microscopy. With light microscopy, phialides measured ca. 15-25 X 1-1.5 µm (in Fig. 1, phialides a, b, and c measured 18 µm, 16 µm, and 17 µm, respectively). The conidia of *D. chlorina* are minute (Fig. 1, 4-6), with mean light microscopic measurements of 3.21 X 1.92 µm.

Conidiogenesis in *D. chlorina* exhibited the essentials of phialoconidiogenesis, viz., (A) multivesiculate conidial initials (Fig. 2) "blew out" from collarette-encircled conidiogenous cell apices and continued to enlarge (Fig. 3) until the approximate size of mature conidia was achieved, (B) conidia then were delimited by centripetally developing transverse septa which became functionally bilayered (Fig. 4), and (C) the conidiogenous cycle (Madelin, 1979) was repeated many times. Figures 5-6 illustrate early and later stages in the repetition of the conidiogenous cycle. A distinct conidial initial began to round up, initiating

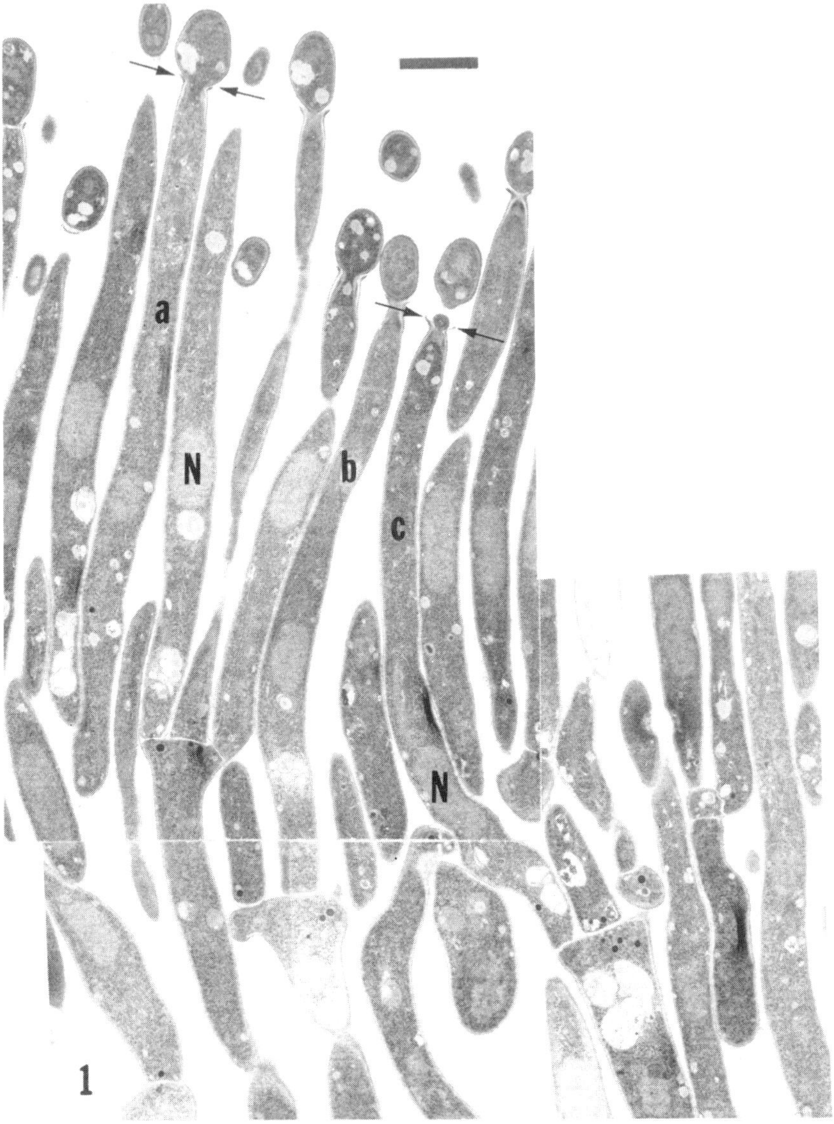

Fig. 1. Dendrostilbella chlorina. *Section from synnematal head. Arrows point to collarettes of phialides. Lengths of phialides a, b and c measured ca. 18 µm, 16 µm and 17 µm, respectively. N = nucleus. Magnification bar = 2 µm.*

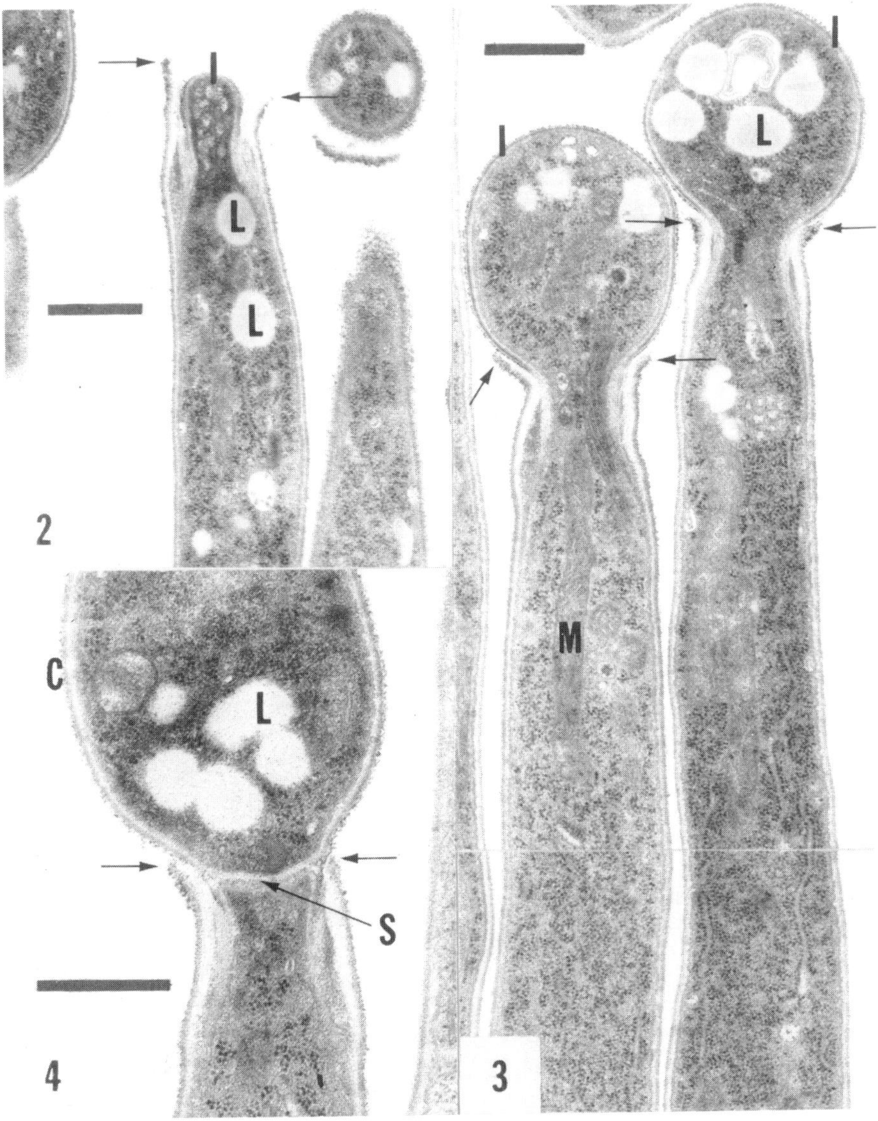

Fig. 2-4. **Dendrostilbella chlorina.** *Fig. 2--Multivesiculate conidial initials (I) emerging from within collarette (arrows). Fig. 3--Conidial initials (I) at a later stage of development. Arrows point to collarettes. Fig. 4--Conidium (C) delimited by functionally bilayered septum (S) within the collarette (arrows). L = lipid globules, M = mitochondrion. All magnification bars = 0.5 μm.*

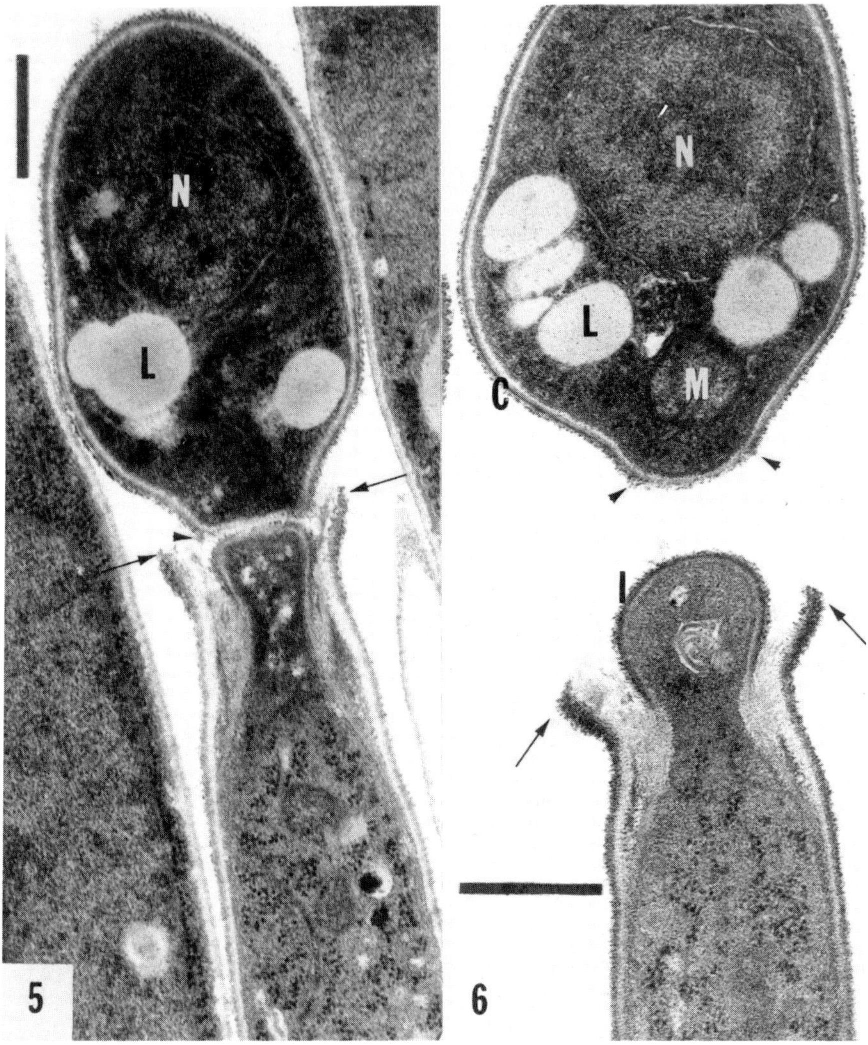

Fig. 5-6. Dendrostilbella chlorina. *Fig. 5--Conidial initial beginning to round up. Previously formed conidium beginning to secede and to show a basal frill (arrow head). Fig. 6--Later stage. Conidial initial (I) emerged beyond collarette; conidium (C) seceded; basal frill (arrow heads) detectable. Arrows = collarette, N = nucleus, L = lipid globules, and M = mitochondrion. Magnification bars = 0.5 μm. (Fig. 6 also was published in Alexopoulos and Mims, 1979.)*

secession of the previously formed conidium (Fig. 5), the contents of which had become increasingly dense. Conidial secession accompanied continued enlargement of conidial initials (Fig. 6). Collarettes of *D. chlorina* were not able to be resolved with light microscopy because of the extremely small dimensions of phialidic apices. With electron microscopy, collarettes were quite conspicuous (Fig. 1-6). Wall material between the electron-opaque collarette and the conidial initial (Fig. 6) is presumed to result from secession of several conidia. Basal frills on seceded conidia (Fig. 5-6) resulted from secession, and are the scars of former attachments of conidia and phialides.

From the examples given by Subramanian (1979), examples which included *Trichoderma saturnisporum*, the phialides of *D. chlorina* would appear to fit into his type 3 category, i.e., those phialides which he defined as pseudosynechidic, tunicogenous (penititunicogenous at least after the primary conidium), porrectic, percurrent, solitary, glomerate, and slimy.

Metarrhizium anisopliae (Hammill Culture # 39-68).

Phialoconidiogenesis in *M. anisopliae*, which produces dark green columns of cylindrical conidia from closely packed and parallelly oriented phialides borne upon a sporodochium-like mass of hyphae, was the subject of previous reports (Hammill, 1972a, 1977a). Figures 7-8 illustrate two stages of conidiogenesis in this species. A conidial initial is shown "blowing out" of the phialide and assuming the cylindrical shape (Fig. 7). The base of the previously formed conidium is shown above the conidial initial, and the phialide possesses an extremely minute collarette. Conidial delimitation is by centripetal septation across the protoplasmic continuity between the phialide and the nucleated conidial initial (Fig. 8). Subsequently, conidial maturation occurs while the conidiogenous cycle is repeated (Hammill, 1972a).

Subramanian (1979) included *M. anisopliae* among his type 2 phialides, i.e., those which he defined as being synechidic, novi-cum-penititunicogenous, porrectic, catenate, and dry.

Trichurus spiralis (Hammill culture # 210-72).

Trichurus spiralis is a synnematal species with tortuous, sterile setae in the conidiogenous region of synnemata. Electron microscopy of conidiogenesis in *T. spiralis* was the subject of a previous report (Hammill, 1977b) in which annelligenesis was clearly demonstrated. Figure 9 is from that paper (Fig. 9 here also was Fig. 9 there); it

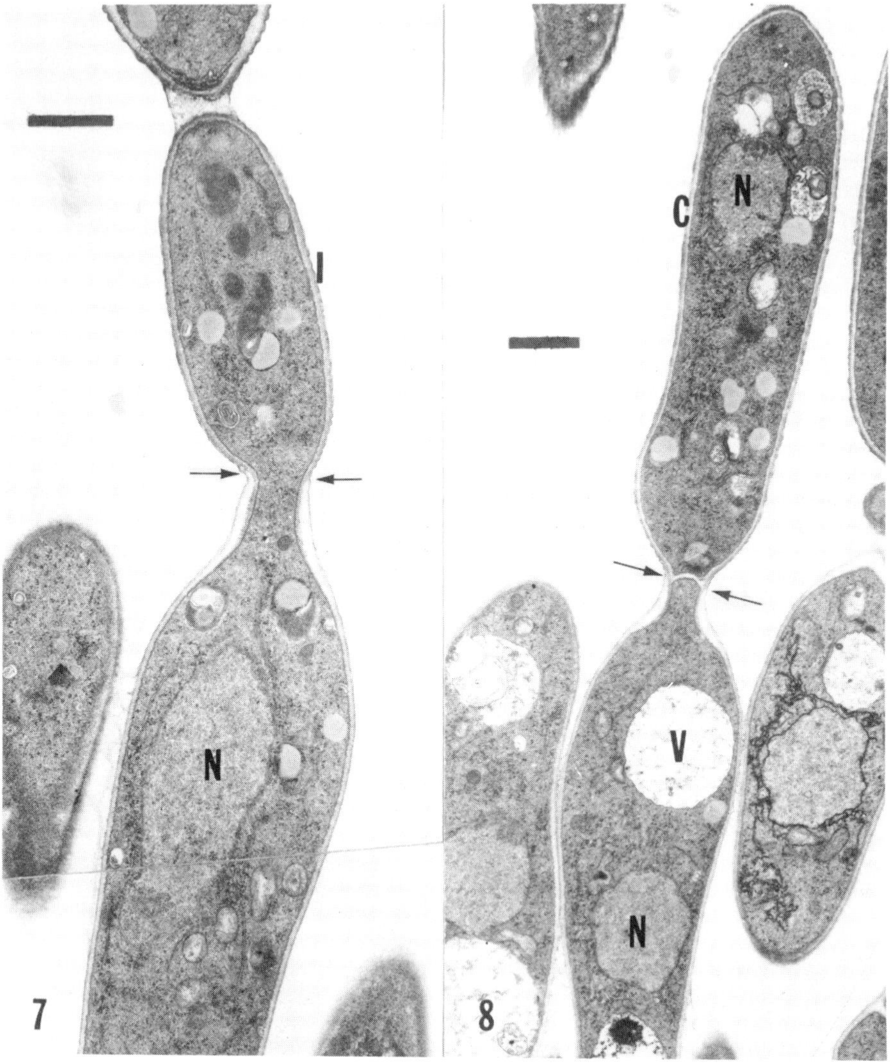

Fig. 7-8. Metarrhizium anisopliae. *Fig. 7--Conidial initial (I) emerging from tip of phialide. Arrows point to minute collarette. Base of previously formed conidium at top of micrograph. N = nucleus. Fig. 8--Conidium (C) delimited by septum. Arrows point to septum and minute collarette. N = daughter nuclei, V = vacuole. Magnification bars = 1 μm.*

demonstrates the essential features of an annellide, viz., successively more distal planes of septation produce, as a result of conidial secession, annellate neck regions on conidiogenous cells. The oldest annellations (secession scars resulting from circumscissile ruptures of the lateral wall connecting conidia to the conidiogenous cell) are at the base of the neck region (a, in Fig. 9), while successively younger annellations are more distal to that point (the youngest is at point b in Fig. 9). Figure 9 also shows a perforate septum at the conidial base, and a Woronin body close to the septal pore. Woronin bodies function as septal pore plugs when conidia of *T. spiralis* secede (see Fig. 10 in Hammill, 1977b). The conidial delimiting septum in Fig. 9 had increased in thickness from deposition of wall material on the conidial side of the secessional plane and had developed a layer of electron-opacity immediately distal to the secessional plane. Conidial secession resulted in basal frills on conidia and new annellations on conidiogenous cells.

Hammill (1977b) reported that a developmental sequence in formation of the first conidium by an annellide of *T. spiralis* had not been observed during that study, but that attempts were being made to rectify that lack of information. Figures 10-13 demonstrate four stages in development of a primary conidium in *T. spiralis*.

In the differentiation of a conidiogenous cell (Fig. 10), minute vesicles appeared around its elongating tip. Such vesicles are generally hypothesized to be integrally involved in growth and cell wall formation (Cole and Samson 1979). Formation of a conidiogenous cell "beak" is followed by its enlargement (Fig. 11), nucleation, septal delimitation (Fig. 12), septal and conidial maturation (Fig. 13), and conidial secession (Fig. 13). In Fig. 12, an electron-translucent layer can be seen; that layer represents the secession plan immediately distal to which the septum develops an electron-opaque layer and increases in thickness with the concurrent increase in conidial wall thickness. In Fig. 13, the initiation of secession can be seen at the point labelled R, the beginning of the circumscissile rupture, characteristic of conidial secession in *T. spiralis* in particular and in conidial fungi in general. Until that stage illustrated by Fig. 13, there was no discontinuity in the cell wall of the conidiogenous cell--conidium arrangement.

Formation of the primary conidium in *T. spiralis* is therefore clearly holoblastic (Cole and Samson, 1979; Kendrick, 1971) and totitunicogenous (Subramanian, 1972, 1979).

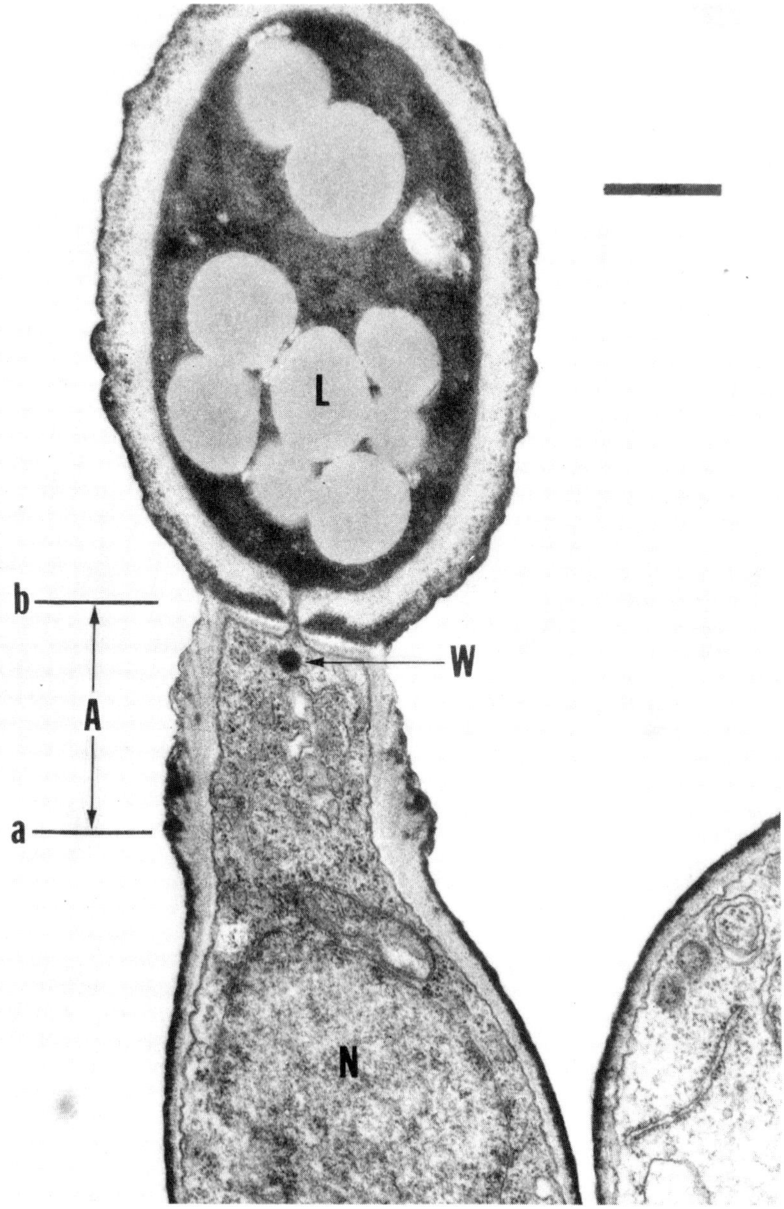

9

Fig. 9. Trichurus spiralis. *A = annellate neck region between oldest annellation (a) and youngest (b). W = woronin body, N = nucleus, L = lipid globule. Magnification bar = 0.5 μm. (From Hammill, 1977.)*

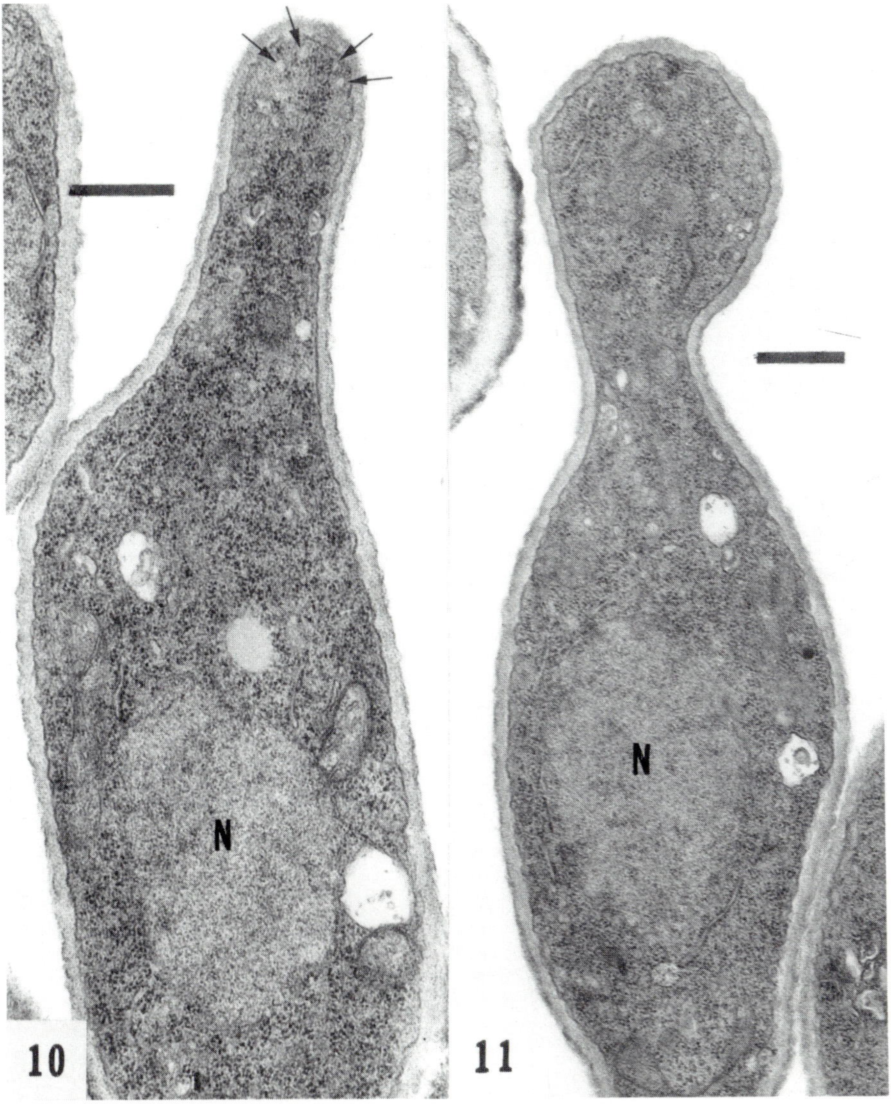

Fig. 10-11. Trichurus spiralis. *Two stages in development of a primary conidium. Fig. 10--Annellide differentiating; vesicles (arrows) underlying plasmalemma, and presumably contributing to wall deposition. Fig. 11--Conidial initial "blowing out." Wall is continuous around conidial initial and conidiogenous cell. N = nucleus. Magnification bars = 0.5 μm.*

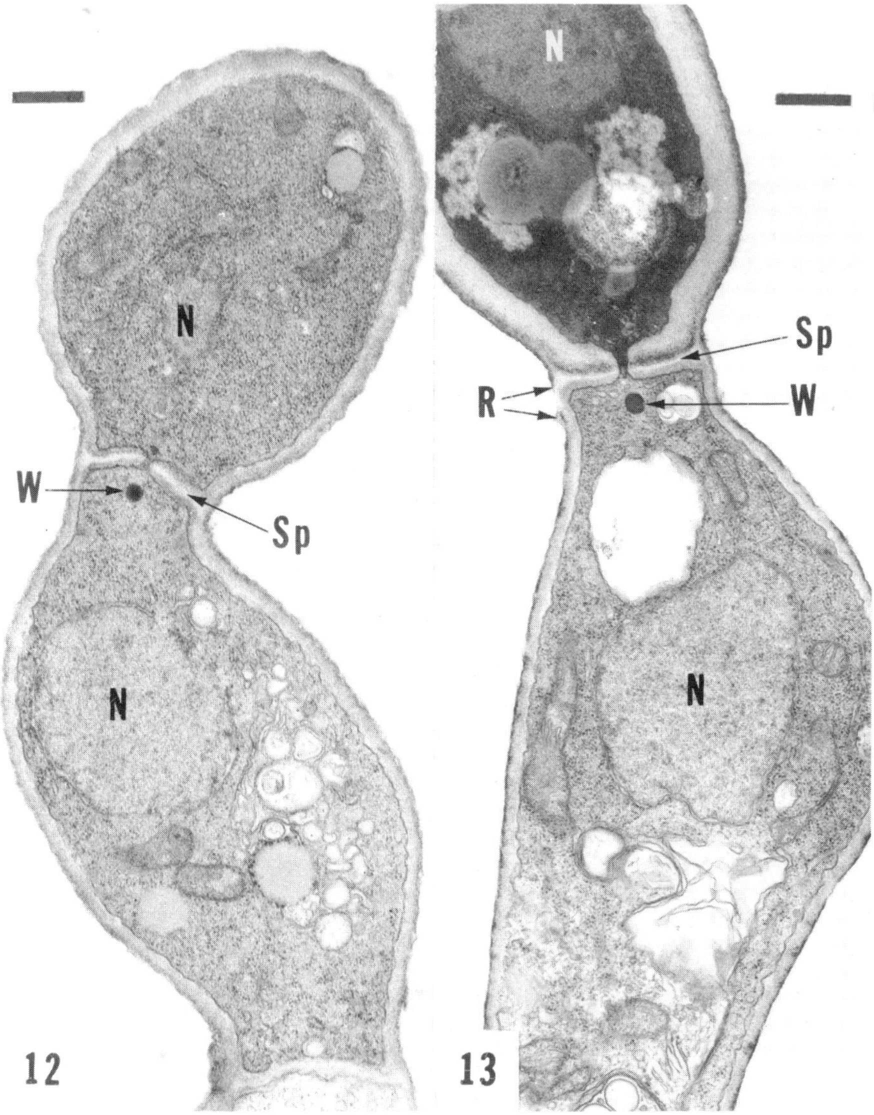

Fig. 12-13. **Trichurus spiralis.** *Two later stages in development of a primary conidium. Fig. 12--Conidium delimited by functionally bilayered septum. Secession plane (Sp) apparent. Fig. 13--Primary conidium beginning to secede at secession plane (Sp). R = beginning of circumscissile rupture which produces basal frill on conidium and annellation on conidiogenous cell. W = Woronin bodies, N = nuclei. Magnification bars = 0.5 μm.*

Conidial Bases

During the period following Hughes' (1953) proposal, considerable attention was paid to clarifying the nature of conidial attachment and secession, and to whether the width of the attachment point of conidia to conidiogenous cells was of classificatory/taxonomic significance (e.g., Luttrell, 1963; Madelin, 1966). For example, Madelin (1966) discussed variations in phialides partially on the basis of whether conidia were delimited by "budding" or by "transverse septation." Hammill (1972a) opined that the difference is simply a matter of degree, and that probably all conidia are delimited by centripetally developing transverse septa, or modifications thereof (Hammill, 1974). That view has not been invalidated.

Cole and Samson (1979) discussed the two mechanisms of conidial secession found among Fungi Imperfecti, viz., rhexolytic and schizolytic. In the former, conidial secession does *not* involve splitting of a functionally bilayered septum, but instead involves breakage or lysis of a cell below the seceding conidium. Schizolysis, as the more common alternative of conidial secession, involves splitting of a functionally bilayered septum, with the distal "half septum" forming the conidial base and the proximal "half septum" forming the conidiogenous cell apex. (The term "half septum" is put into quotation marks because septa may not split into absolute halves.)

Figures 14-19 are of six putatively schizolytic species. Three of the species are annellidic (*Scopulariopsis koningii*, Fig. 14; *Doratomyces nanus*, Fig. 15; *Trichurus spiralis*, Fig. 16), and three of the species are phialidic (*Metarrhizium anisopliae*, Fig. 17; *Trichoderma saturnisporum*, Fig. 18; *Dendrostilbella chlorina*, Fig. 19). The term schizolysis suggests enzymic mediation of septal splitting, but whether that always is the case in reputedly schizolytic species is unknown.

Conidia which secede by septal splitting exhibit truncate bases ringed by basal frills, however minute. Such basal structures are shown in Fig. 14-19. Figures 20-21 are of the holoblastic species, *Tritirachium roseum* (Fig. 20) and *Pseudobotrytis terrestris* (Fig. 21). Both species have very thickened septa between conidia and conidiogenous cells, and conidial secession results from splitting of these septa. Truncate bases with basal frills are evident, but whether the secessional process is schizolytic or schizomechanic (i.e., splitting as a result of mechanical processes), or a combination, is similarly unknown.

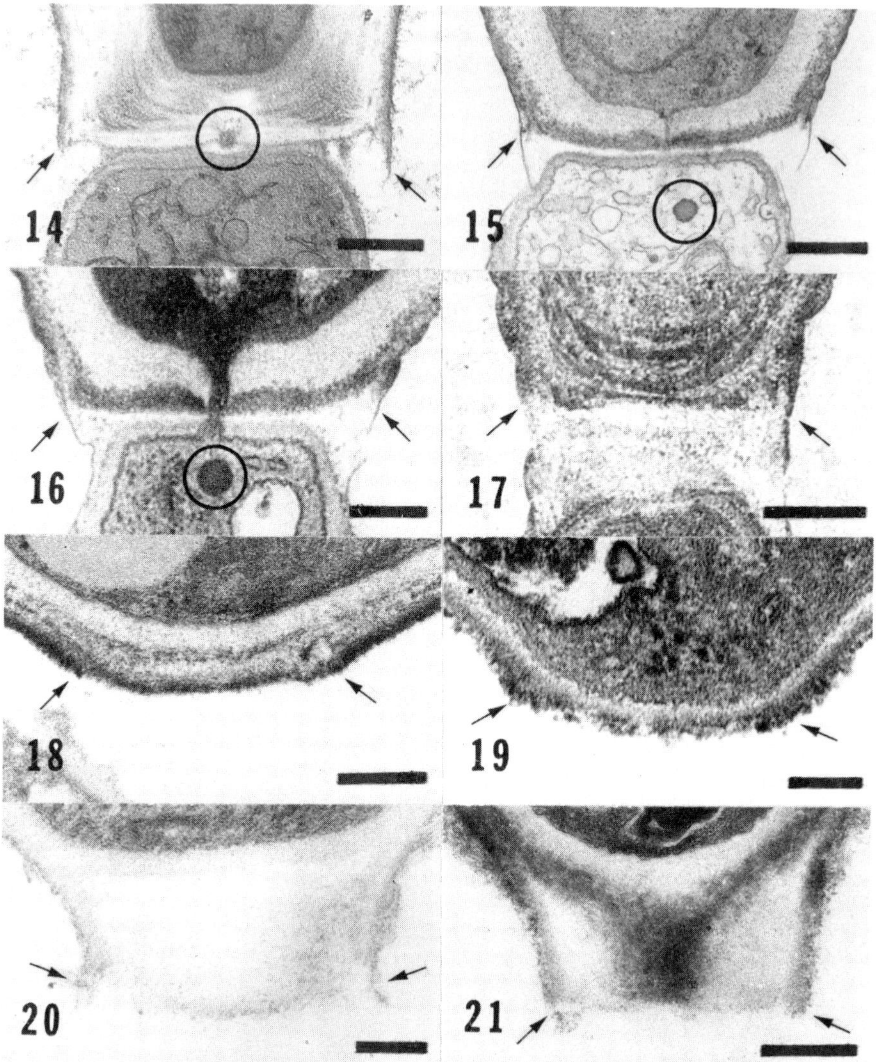

Fig. 14-21. Conidial bases showing basal frills (arrows). Fig. 14--Scopulariopsis koningii *(from Hammill, 1971). Fig. 15*--Doratomyces nanus. *Fig. 16*--Trichurus spiralis *(from Hammill, 1977). Fig. 17*--Metarrhizium anisopliae. *Fig. 18*--Trichoderma saturnisporum *(from Hammill, 1974). Fig. 19*--Dendrostilbella chlorina. *Fig. 20*--Tritirachium roseum. *Fig. 21*--Pseudobotrytis terrestris. *Woronin bodies and septal pore plugs are circled in Fig. 14-16. Magnification bars = 0.5µm in Fig. 14-15; 0.25 µm in Fig. 16-18 & 21; 0.1 µm in Fig. 19-20.*

4. CONCLUSIONS

The major differences between phialidic and annellidic development appear to be as follows (Cole and Samson, 1979):

A. The outer wall layer of annellides usually rupture at the level of the conidial-delimiting septum, leaving an annellation; the outer wall layer of phialides may rupture at any point from the level of the septum to a more distal level, leaving a collarette.

B. Annellides undergo a series of percurrent proliferations of an inner wall layer, usually with sequentially more distal septal delimitations of conidia and the consequent annellate secession scars; phialides do not elongate detectably, but secession scars may be left within the collarette of some phialides (e.g., in *Dendrostilbella chlorina* and *Trichoderma saturnisporum*).

C. Septal pores between conidia and annellides are sealed with Woronin bodies (see Fig. 14-16); Woronin bodies do not appear to plug septa in most phialides (possible exceptions were reported by Hawes and Beckett, 1977, and Khan and Aldrich, 1973).

On the basis of present data, it is difficult to formulate distinct concepts of phialidic and annellidic development (Cole and Samson, 1979). Indeed, there appear more similarities than differences, and it is unlikely that the similarities are merely analogous rather than homologous. As Cole and Samson (1979) said, "We have come full circle in our consideration of distinguishing characters of annellidic and phialidic development ... annellides have annellations, phialides have collarettes. Both kinds of conidiogenous cells give rise to a basipetal succession of conidia."

Though we may have devised a "new classification" based upon conidiogenesis, one which overcomes some of the obvious Saccardoan limitations, we have not yet done anything significantly *taxonomic* while using that data *exclusively*. In fact, comments by Luttrell (1963) presaged a general conclusion reached at Kananaskis II; he stated,"The increasing volume of research stimulated by Hughes's concepts has been devoted, for the most part, to taxonomic revisions utilizing these new criteria in advance of critical evaluation of the criteria themselves." The intervening years, particularly with the aid of electron microscopy, witnessed critical evaluation of conidiogenic concepts, and the following comments by Kendrick at Kananaskis II illustrate what is probably the general sentiment at present. He said (Kendrick, 1979a) that the chapters by Madelin (1979) and

Wang (1979) "firmly establish that conidium ontogeny cannot, after all, form the base on which a stable classification of Fungi Imperfecti might be erected." Kendrick also said, in the discussion following the chapter by Luttrell (1979), "...that we must now surely realize that the dream of a 'natural' classification of Fungi Imperfecti based exclusively on conidiogenesis has evaporated." But, while conidiogenesis may not have had, *by itself*, the taxonomic impact for which there had been hope, the conidiogenic data generated by that hope can only improve upon our understanding of and perhaps our taxonomy of *The Whole Fungus*.

5. ACKNOWLEDGEMENTS

This chapter is based in part upon my unpublished contribution to the *Symposium on Trends in Fungus Taxonomy*, held at the XII International Congress of Microbiology, Munich, West Germany (Sept., 1978). A portion of the research described herein was supported by NSF Grant No. B042219. Electron Microscopy was done at the Nelson C. Brown Laboratory for Ultrastructural Studies, SUNY College of Environmental Science and Forestry, Syracuse, New York. Dr. W. A. Côté, Jr., and Dr. Robert Hanna are thanked for permission to use that facility. Organizers of 3IFSS and NSF Grant No. DEB-8010613 are both acknowledged for travel support. Ms. Mary Garcia and Mr. Robert Sheitz, former undergraduate students, are thanked for their research assistance, and Ms. Lorretta Beckwith is thanked for typing the manuscript.

6. REFERENCES

Ainsworth, G. C. (1973). Introduction and keys to higher taxa. *In* "The Fungi, An Advanced Treatise. A Taxonomic Review with Keys: Ascomycetes and Fungi Imperfecti" (Ed. G. C. Ainsworth, F.K. Sparrow and A. S. Sussman). Vol. IVA, pp. 1-7. Academic Press, New York.

Ainsworth, G. C., James, P. W., and Hawksworth, D. L. (1971). "Ainsworth & Bisby's Dictionary of the Fungi, 6th Edition, Including the Lichens" Commonwealth Mycological Institute, Kew, England.

Alexopoulos, C. J., and Mims, C. W. (1979). "Introductory Mycology, 3rd Ed." John Wiley & Sons, New York.

Barron, G. L. (1968). "The Genera of Hyphomycetes from Soil" The Williams and Wilkins Company, Baltimore, Maryland.

Cain, R. F. (1972). Evolution of the fungi. *Mycologia* 64, 1-14.

Cole, G. T., and Aldrich, H. C. (1971). Ultrastructure of conidiogenesis in *Scopulariopsis brevicaulis*. *Canadian*

Journal of Botany 49, 745-755.

Cole, G. T., and Samson, R.A. (1979). "Patterns of Development in Conidial Fungi" Pitman Publishing Ltd., London, England.

Dobzhansky, T. (1973). Nothing in biology makes sense except in the light of evolution. *The American Biology Teacher* 35, 125-129.

Gove, P. B., and Staff. (1976). "Webster's Third New International Dictionary of the English Language, Unabridged" G. & C. Merriam Co., Springfield, Mass.

Hammill, T. M. (1970). *Paecilomyces clavisporis* sp. nov., *Trichoderma saturnisporum* sp. nov., and other noteworthy soil fungi from Georgia. *Mycologia* 62, 107-122.

Hammill, T. M. (1971a). Fine structure of annellophores. I. *Scopulariopsis brevicaulis* and *S. koningii*. *American Journal of Botany* 58, 88-97.

Hammill, T. M. (1971b). Fine structure of conidiogenesis in selected species of Fungi Imperfecti. Ph.D. Thesis, SUNY College of Forestry at Syracuse University, Syracuse, New York.

Hammill, T. M. (1972a). Electron microscopy of phialoconidiogenesis in *Metarrhizium anisopliae*. *American Journal of Botany* 59, 317-326.

Hammill, T. M. (1972b). Fine structure of annellophores. II. *Doratomyces nanus*. *Transactions of the British Mycological Society* 59, 249-253.

Hammill, T. M. (1973). Fine structure of conidiogenesis in the holoblastic, sympodial *Tritirachium roseum*. *Canadian Journal of Botany* 51, 2033-2036.

Hammill, T. M. (1974). Electron microscopy of phialides and conidiogenesis in *Trichoderma saturnisporum*. *American Journal of Botany* 61, 15-24.

Hammill, T. M. (1977a). Additional electron microscopy of phialoconidiogenesis in *Metarrhizium anisopliae*: Microtubules in phialidic necks. *Mycologia* 69, 1058-1061.

Hammill, T. M. (1977b). Transmission electron microscopy of annellides and conidiogenesis in the synnematal hyphomycete *Trichurus spiralis*. *Canadian Journal of Botany* 55, 233-244.

Hammill, T. M., and Shipman, J. S. (in prep.) On the synnematal genus *Dendrostilbella*, with notes about *D. chlorina*.

Hawes, C. R., and Beckett, A. (1977). Conidium ontogeny in the *Chalara* state of *Ceratocystis adiposa*. II. Electron microscopy. *Transactions of the British Mycological Society* 68, 267-276.

Hughes, S. J. (1951). *Annellophora* nom. nov. (≡ *Chaetotrichum* Syd. non Rabenh.). *Transactions of the British Mycological Society* 34, 544-550.
Hughes, S. J. (1953). Conidiophores, conidia, and classification. *Canadian Journal of Botany* 31, 577-659.
Hughes, S. J. (1971). Annellophores. *In* "Taxonomy of Fungi Imperfecti" (Ed. B. Kendrick). pp. 132-139. University of Toronto Press, Toronto, Canada.
Kendrick, B., Ed. (1971). "Taxonomy of Fungi Imperfecti" University of Toronto Press, Toronto, Canada.
Kendrick, B. (1979a). Introduction. *In* "The Whole Fungus" (Ed. B. Kendrick). Vol. 1, pp. 11-15. National Museum of Natural Sciences, National Museums of Canada, Ottawa.
Kendrick, B., Ed. (1979b). "The Whole Fungus. Vol. 1 & 2." National Museum of Natural Sciences, National Museums of Canada, Ottawa.
Kendrick, B., and Chang, M. G. (1971). Karyology of conidiogenesis in some hyphomycetes. *In* "Taxonomy of Fungi Imperfecti" (Ed. B. Kendrick). pp. 279-291. University of Toronto Press, Toronto, Canada.
Kendrick, B., and Nag Raj, T. R. (1979). Morphological terms in Fungi Imperfecti. *In* "The Whole Fungus" (Ed. B. Kendrick). Vol. 1, pp. 43-61. National Museum of Natural Sciences, National Museums of Canada, Ottawa.
Khan, S. R., and Aldrich, H. C. (1973). Conidiogenesis in *Termitaria snyderi* (Fungi Imperfecti). *Canadian Journal of Botany* 51, 2307-2314.
Luttrell, E. S. (1963). Taxonomic criteria in *Helminthosporium*. *Mycologia* 55, 643-674.
Luttrell, E. S. (1979). Deuteromycetes and their relationships. *In* "The Whole Fungus" (Ed. B. Kendrick). Vol. 1, 241-264. National Museum of Natural Sciences, National Museums of Canada, Ottawa.
Madelin, M. F. (1966). The genesis of spores in Higher Fungi. *In* "The Fungus Spore" (Ed. M. F. Madelin). pp. 15-37. Butterworths, London, England.
Madelin, M. F. (1979). An appraisal of the taxonomic significance of some different modes of producing blastic conidia. *In* "The Whole Fungus" (Ed. B. Kendrick). Vol. 1, pp. 63-79. National Museum of Natural Sciences, National Museums of Canada, Ottawa.
Morgan-Jones, G., Nag Raj, T. R., and Kendrick, B. (1972). Conidium ontogeny in coelomycetes. IV. Percurrently proliferating phialides. *Canadian Journal of Botany* 50, 2009-2014.
Pinkerton, M. E. (1936). A comparative study of conidial formation in *Cephalosporium* and some related hyphomycetes.

Annals of the Missouri Botanical Garden 23, 1-69.

Saccardo, P. A. et al. (1882-1972). "*Sylloge Fungorum Omnium Hucusque Cognitorum*, Vols. 1-26" Johnson Reprint Corp., New York.

Simmons, E. G. (1966). Symposium on stability in dynamic microbial systems. V. The theoretical bases for classification of the Fungi Imperfecti. *The Quarterly Review of Biology*, 41: 113-123.

Snell, W. H., and Dick, E. A. (1971). "A Glossary of Mycology, Revised Edition" Harvard University Press, Cambridge, Mass.

Subramanian, C. V. (1971). The phialide. *In* "Taxonomy of Fungi Imperfecti" (Ed. B. Kendrick). pp. 92-114. University of Toronto Press, Toronto, Canada.

Subramanian, C. V. (1972). Conidium ontogeny. *Current Science* 41, 619-624.

Subramanian, C. V. (1979). Phialidic Hyphomycetes and their teleomorphs-an analysis. *In* "The Whole Fungus" (Ed. B. Kendrick). Vol. 1, pp. 125-151. National Museum of Natural Sciences, National Museums of Canada, Ottawa.

Talbot, P. H. B. (1971). "Principles of Fungal Taxonomy" St. Martin's Press, New York.

Tanaka, K., and Yanagita, T. (1963). Electron microscopy on ultrathin sections of *Aspergillus niger*. II. Fine-structure of conidia-bearing apparatus. *Journal of General and Applied Microbiology* 9, 189-203.

Vuillemin, P. (1910). Les conidiospores. *Bulletin Société des Sciences de Nancy* 12, 151-175.

Wang, C. J. K. (1979). Pleomorphic Fungi Imperfecti. *In* "The Whole Fungus" (Ed. B. Kendrick). Vol. 1, 81-90. National Museum of Natural Sciences, National Museums of Canada, Ottawa.

Wolf, F. A., and Wolf, F. T. (1947). "The Fungi, Vol. 1" John Wiley & Sons, New York.

MUCORALEAN SPORANGIOSPOROGENESIS

T. M. Hammill

Department of Biology
SUNY College of Arts and Science
Oswego, New York 13126, U.S.A.

1. INTRODUCTION

Zygomycetes constitute a major class of eumycotic fungi. Their general characteristics can be summarized as follows (Hesseltine and Ellis, 1973; Ingold, 1978; Talbot, 1971):
(1) The growth form usually is filamentous, particularly under aerobic, saprobic conditions. Modifications of hyphae include chlamydospores, gametangia, zygospores (also azygospores) plus appendages, and a variety of sporangia, etc. Some Zygomycetes assume a yeast-like growth form when grown anaerobically (Webster, 1970) or when parasitizing animals (Stewart and Rogers, 1978).
(2) Motile cells are lacking completely from all stages of growth, i.e., Zygomycetes produce neither zoospores nor gametes.
(3) Sexual reproduction, when known, involves gametangial conjugation. After gametangia plasmogamize, zygospores form. Azygospores, when formed, result from parthenogenesis.
(4) Asexual reproduction (other than azygosporogenesis) includes the formation of (a) multispored sporangia, (b) sporangiola or merosporangia with fewer spores, (c) modified sporangial units which function as conidia, and (d) monospored sporangia, some of which may be true conidia.

The class name, Zygomycetes, is derived from the nature of sexual reproduction in these fungi, i.e., Zygomycetes form zygospores following gametangial conjugation. However, many species classified as Zygomycetes are not known to reproduce sexually. Nevertheless, those are confidently classified as Zygomycetes because of other characteristics which correlate with such a relationship (Alexopoulos, 1962), e.g., common methods of asexual reproduction, hyphal appearance, ecology, physiology, etc. (Alexopoulos and Mims, 1979).

Modes of sporangial development in the Mucorales and Kickxellales, whose taxa were included in the Mucorales by Hesseltine and Ellis (1973), provided a basis for Cole and Samson (1979) to categorize such fungi into five groups, viz., those which produce (1) multispored sporangia, (2) few-spored sporangiola, (3) monosporous sporangiola on vesicles, (4) few-spored merosporangia, and (5) monosporous merosporangia.

Among these, groups two through five have been studied by electron microscopists more than has group one (Baker *et al.*, 1977; Dykstra, 1974; Fletcher, 1972, 1973a,b; Jeffries and Young, 1975; Khan and Talbot, 1975). In addition to the ten species covered by those seven research publications, Cole and Samson (1979) showed electron micrographs of another eleven species which produce other than multispored sporangia. O'Donnell (1979) published scanning electron micrographs exhibiting the various developmental modes of sporangia in 78 species of Zygomycetes. Of these, 44 species were of modes other than multispored sporangia. Although the 78 species treated by O'Donnell (1979) were narrowly arranged according to the taxonomy of Benjamin (1979) and were distributed among the Mucorales, Kickxellales, Dimargaritales, and Zoopagales, all 78 species formerly were included in the Mucorales by Hesseltine and Ellis (1973).

Mucoralean species in group one, those which produce multispored sporangia, have been ignored, in a developmental sense, almost completely by electron microscopists. In addition to O'Donnell's (1979) brief treatment of several mucoralean species with multispored sporangia, Hirahare and Seshita (1976) conducted a scanning electron microscopic examination of the sporangial surface of several species of *Mucor*, and Tu and Malhotra (1977) examined sequential changes in the plasma membrane of the sporangial apparatus during sporangiosporogenesis in *Phycomyces blakesleeanus* Burgeff. However, the only comprehensive developmental study of sporangiosporogenesis in multispored sporangia is that of Bracker (1966, 1968) in *Gilbertella persicaria* (Eddy) Hesseltine. That study provided the empirical basis for the most recent model of development in multispored sporangia, a model included in the introductory textbook by Moore-Landecker (1972) and the advanced works of Cole and Samson (1979) and O'Donnell (1979). Bracker's model is reproduced and discussed below (Fig. 1).

Figure 1 is modified from Fig. 88 of Bracker's (1968) paper. It includes the complete developmental sequence in one diagram which is subdivided into segments A (precleavage), B (cleavage), and C (postcleavage). Those segments are further subdivided into early, mid-, and late stages.

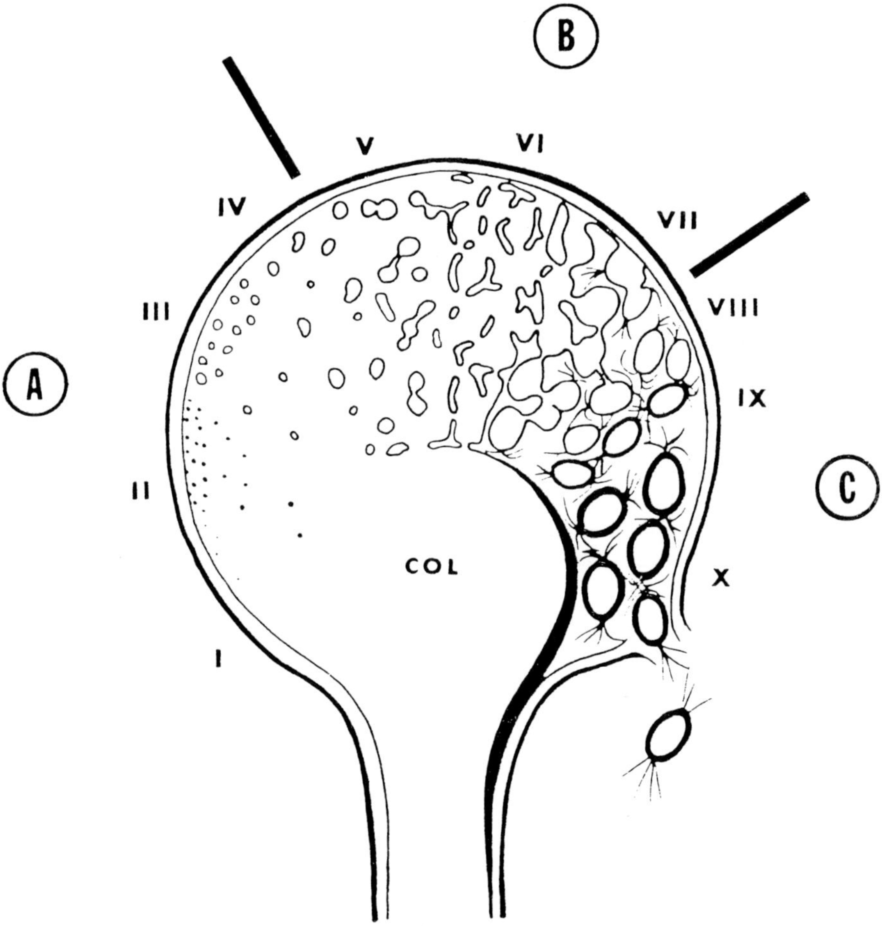

Fig. 1. Diagram (modified from Bracker, 1968) summarizing the process of sporangiospore formation. Only the cleavage apparatus and walls are indicated. Spatial distribution of developmental stages if for convenience in illustration only, and does not represent spatial progression within a single sporangium. STAGE A: PRECLEAVAGE. Early precleavage (I-II), mid-precleavage (III), late precleavage (IV). STAGE B: CLEAVAGE. Early cleavage (V), mid-cleavage (VI), late cleavage (VII). STAGE C: POSTCLEAVAGE. Early postcleavage (VIII), mid-postcleavage (IX), late postcleavage or spore maturity (X). The discontinuity in the sporangial wall near X represents rupturing of the sporangium as it splits into halves at maturity.

The Brackerian Model

During precleavage in *G. persicaria* (Fig. 1, A I-IV), dividing nuclei were not observed, but there appeared multiple continuities between nuclei and endoplasmic reticulum (ER). Lipid globules and glycogen granules were present, short spines on the outer sporangial wall were observed, and vacuoles could be found in the region of subsequent columellar development. Two types of early precleavage sporangia appeared (Fig. 1, A I-II), one with few peripheral vesicles, the other (interpreted by Bracker as a later stage) with many cytoplasmic vesicles. Such vesicles were not seen beyond early precleavage, and were not similar in appearance or size to microbodies (which also were present). The cytoplasmic vesicles appeared to bud from ring-like cisternae, considered by Bracker to represent the functional equivalents of Golgi bodies. At mid-precleavage (Fig. 1, A III), small vesicles were replaced by larger vesicles, which often were associated with ER. Such vesicles, termed cleavage vesicles by Bracker, were lined with electron-opaque granules which measured from less than 80 Å to more than 200 Å in diameter. These granules were situated about 50 Å from the inner surface of the cleavage vesicle membrane, and provided a marker by which cleavage vesicles could be identified in section. Cleavage vesicles were absent from the columellar zone of precleavage sporangia, and, at late precleavage, measured ca. 1.5 μm in diameter (Fig. 1, A IV).

Cleavage in *G. persicaria* (Fig. 1, B V-VII) was initiated by the coalescence of cleavage vesicles (Fig. 1, B V). The presence of many cisternal rings suggested that new cleavage vesicles were being formed. As cleavage vesicles coalesced, a coarse tubular network of membranes appeared, marking the formation of a rudimentary cleavage apparatus at mid-cleavage (Fig. 1, B VI). Eventually, the cleavage apparatus ramified throughout the sporogenous region of developing sporangia, and flattened into a branched cisternal system (Fig. 1, B VII). Bracker (1968) indicated that ramification of the cleavage apparatus appeared to result mainly from extension rather than addition of new vesicular material. But, whatever the mechanism, the result was delimitation of wall-less spore protoplasts. Cisternal rings, common at mid-cleavage, were less common at late cleavage, and were not observed in postcleavage protoplasts of spore initials. Columellae were formed by a similar process of membrane delimitation, with the columellar region membrane-bounded by mid-cleavage (Fig. 1, B VI). Columellae, completely separated from the sporogenous region of sporangia before late cleavage (between stages B VI and B VII of Fig. 1), contained organelles, plus degenerated mitochondria and crystals in vacuoles. Cleavage vesicles failed to fuse with the sporangial plasma membrane until late cleavage, at which time most spore protoplasts were independent units. Membranes of the cleavage apparatus became plasma membranes of the spore protoplasts. The electron-opaque granules which appeared during precleavage (Fig. 1, A III) inside of cleavage vesicles, now were positioned external to the spore protoplasts as a result of vesicular fusion during development of the cleavage apparatus.

Once the sporogenous region of a sporangium was filled with spore initials (wall-less protoplasts), events of postcleavage (Fig. 1, C VIII-X) commenced. The electron-opaque granules (outside spore initials) associated into small groups, then appeared to fuse into rods which formed a network and eventually a porous envelope around spore initials. Porosity of the envelope decreased until each spore initial was surrounded by a continuous, electron-opaque layer situated at ca. 80 Å from its plasma membrane. That layer marked the outer boundary of each spore. A second electron-opaque layer appeared inside the outer boundary, and increased somewhat in thickness while a secondary wall layer was being laid down inside of it. Walls of nearly mature spores measured ca. 0.4 μm thick. Columellae at this time varied from being densely protoplasmic (mostly found in smaller sporangia with fewer spores) to highly vacuolate. Columellar walls appeared fibrillar and structurally distinct from sporangial walls and walls of sporangiophores, inside of which columellar walls tapered to an end below the base of sporangia.

Appendages of sporangiospores were non-motile, and developed within cleavage vesicles between late precleavage and mid-cleavage. As sporangiospores formed, appendages attached to the outermost wall layer. Also, as sporangiospores matured, their ground plasm became very dense, with membranous structures difficult to resolve. Quiescent cytoplasm of mature sporangiospores was very electron-opaque. Mature sporangiospores were released through rupturing of sporangial walls (Fig. 1, C X).

While Bracker's study (1966, 1968) provided the fine-structural foundation for an understanding of development in multispored sporangia, almost nothing has been added to the Brackerian model since then. This is especially puzzling because the dozen or so years since Bracker's work on *G. persicaria* have been a period of rapid growth in the study of fungi by electron microscopy.

Among the Mucorales, multispored sporangia are produced chiefly by taxa in the families Mucoraceae and Mortierellaceae, with the former producing columellate sporangia and the latter producing non-columellate sporangia (Cole and Samson, 1979; Hesseltine and Ellis, 1973; Zycha, Siepmann, and Linnemann, 1969). In addition, members of the Choanephoraceae, Pilobolaceae, Saksenaeaceae, and Thamnidiaceae produce multispored sporangia, but are distinguished from the Mucoraceae and Mortierellaceae on the basis of other differentiating characters (Cole and Samson, 1979; Ellis and Hesseltine, 1973). *Gilbertella persicaria* was included in the Choanephoraceae by Ellis and Hesseltine (1973), but in the Mucoraceae by Benjamin (1979) and O'Donnell (1979). It resembles the genus *Mucor* except for the presence of bipolar appendages on sporangiospores, and a sporangial wall that splits into halves at maturity (Bracker, 1968), a characteristic which is well shown on page 107 of O'Donnell (1979).

The cleavage pattern and associated phenomena reported by Bracker (1966, 1968) were not claimed to occur in all members of the Mucorales. Bracker said in 1966, "Before a detailed concept of cleavage can be formulated within the Mucorales it will be necessary to study intensively the cleavage process in other representative species." In 1968, Bracker added, "Because of the apparent diversity in the cleavage mechanisms among the Mucorales..., it is not yet possible to equate structures found in *G. persicaria* using the electron microscope with those in other organisms studied by light microscopy." Despite those assertions, no further developmental studies of sporangiosporogenesis in multispored sporangia have appeared since Bracker's work.

2. SPORANGIOSPOROGENESIS IN *MUCOR:* PRELIMINARY RESULTS

The following reports preliminary results of an electron microscopic study of sporangiosporogenesis in an isolate of *Mucor* (Hammill culture # 218) which shows many of the diagnostic characteristics of *M. mucedo* (L.) Fresenius. While the isolate tentatively is being called *M. mucedo*, it varied from some of the "usual" physiological responses of *M. mucedo* to the cultural conditions described by Schipper (1975; personal communication). Isolate # 218 also failed to exhibit any mating response to tester strains of *M. mucedo* in the laboratory of Dr. M. A. A. Schipper, or to strains of *M. mucedo* or several other *Mucor* spp. from the Section Mucedo in the laboratory of Dr. J. J. Ellis (personal communications). In addition, sporangiospores of isolate # 218 are multinucleate (rough estimate = 15-20 nuclei per spore), while Storck and Morrill (1977) reported that sporangiospores of *M. mucedo*

(ATCC 7941) were predominately uninucleate (96.8%).

Preparatory procedures for electron microscopy were essentially the same as those described in the chapter on conidiogenesis in this volume except that, for Fig. 4, the procedure of Bracker (1968) was used.

Figures 2-13 are arranged in a putative developmental sequence including sporangial initiation (Fig. 2, 3), early precleavage (Fig. 4, 5), mid-cleavage (Fig. 6), late cleavage (Fig. 7-10), and mid-to-late postcleavage (Fig. 11-13).

Sporangial initials (Fig. 2) developed by a blastic process. The tip of a sporangiophore "blew out," as multinucleate protoplasm translocated into the developing sporangial initial from the sporangiophore. The sporangial initial was surrounded by a cell wall which was only ca. 40 % as thick as the wall of the sporangiophore. Numerous "typical" eukaryotic cellular organelles were present, including nuclei, mitochondria, ribosomes, vesicles, and cisternal rings, considered the functional equivalents of Golgi bodies in fungi. Microbodies were numerous within only a few micrometers below the level of the sporangiophore shown in Fig. 2. Microtubules also were present in the neck of the sporangial initial shown in Fig. 3, and perhaps play a role in organellar translocation into developing initials. Although not figured, mitotic nuclei frequently were observed at this stage and later.

Sporangial initiation continued until precleavage sporangia were formed. Figure 4 shows an early precleavage sporangium filled with "typical" organelles plus protein crystals. The hemispherical arrangement of nuclei and mitochondria suggests that protoplasm still was being translocated into the specimen when it was killed and fixed.

Early precleavage sporangia (Fig. 5) contained numerous peripheral vesicles which measured ca. 0.1-0.3 μm in diameter. Although not figured here, these vesicles eventually were replaced by larger vesicles which measured ca. 1 μm. The larger vesicles appeared to associate into rudimentary planes which may have defined cleavage furrow initials. During the transition from sporangial initiation to precleavage sporangia, spines appeared on the sporangial surface (Fig. 5). In the Mucorales, such spines are thought to be calcium oxalate (Jones, *et al.*, 1976).

Figure 6 shows a mid-cleavage sporangium. A small portion of the columellar region is at the lower left. Cleavage furrows are well defined in the extracolumellar region of the sporangium. The columellar boundary appears to form concurrently with and as a part of the process of cleavage. The cleaving sporangial protoplasm contains abundant organelles which apparently are randomly segregated into individual spore protoplasts during cleavage.

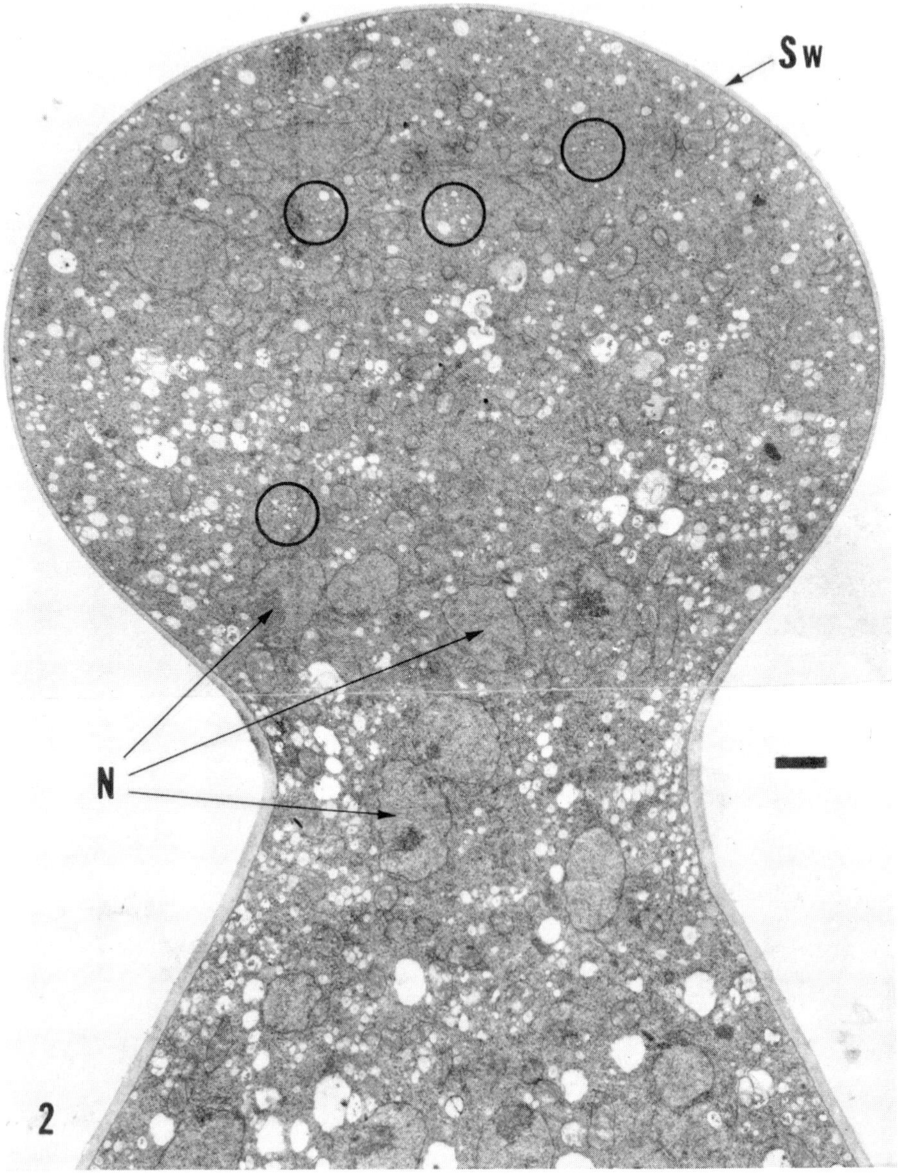

Fig. 2. Sporangial initial developing atop sporangiophore; completely surrounded by cellular wall (Sw); multinucleate (N); with numerous "typical" organelles (e.g., mitochondria, vesicles, ribosomes) and ring cisternae (circled). Magnification bar = 1 μm.

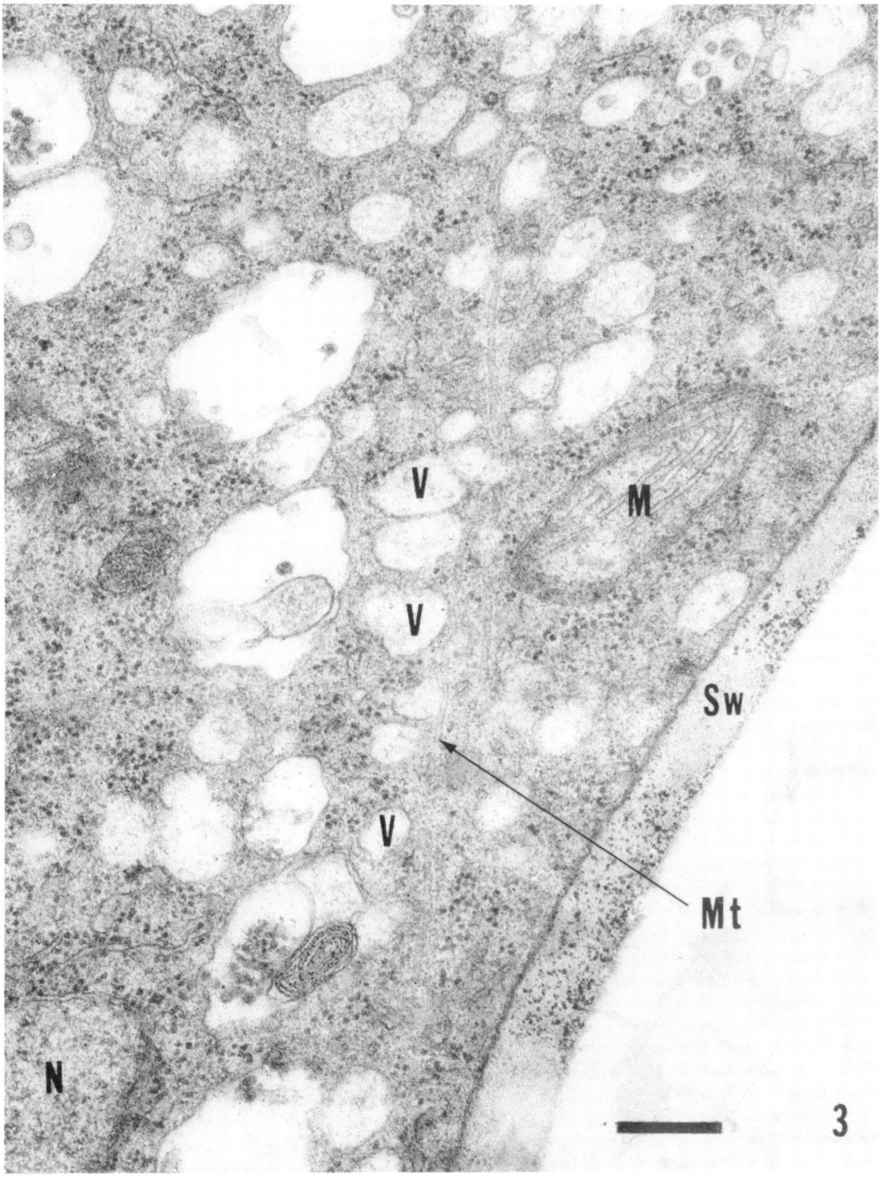

Fig. 3. Serial section of sporangial initial shown in Fig. 2. Microtubules (Mt) with associated vesicles (V) and a mitochondrion (M) shown in neck of sporangial initial. Flecks on cellular wall (Sw) are an artifact of staining. Magnification bar = 0.25 µm.

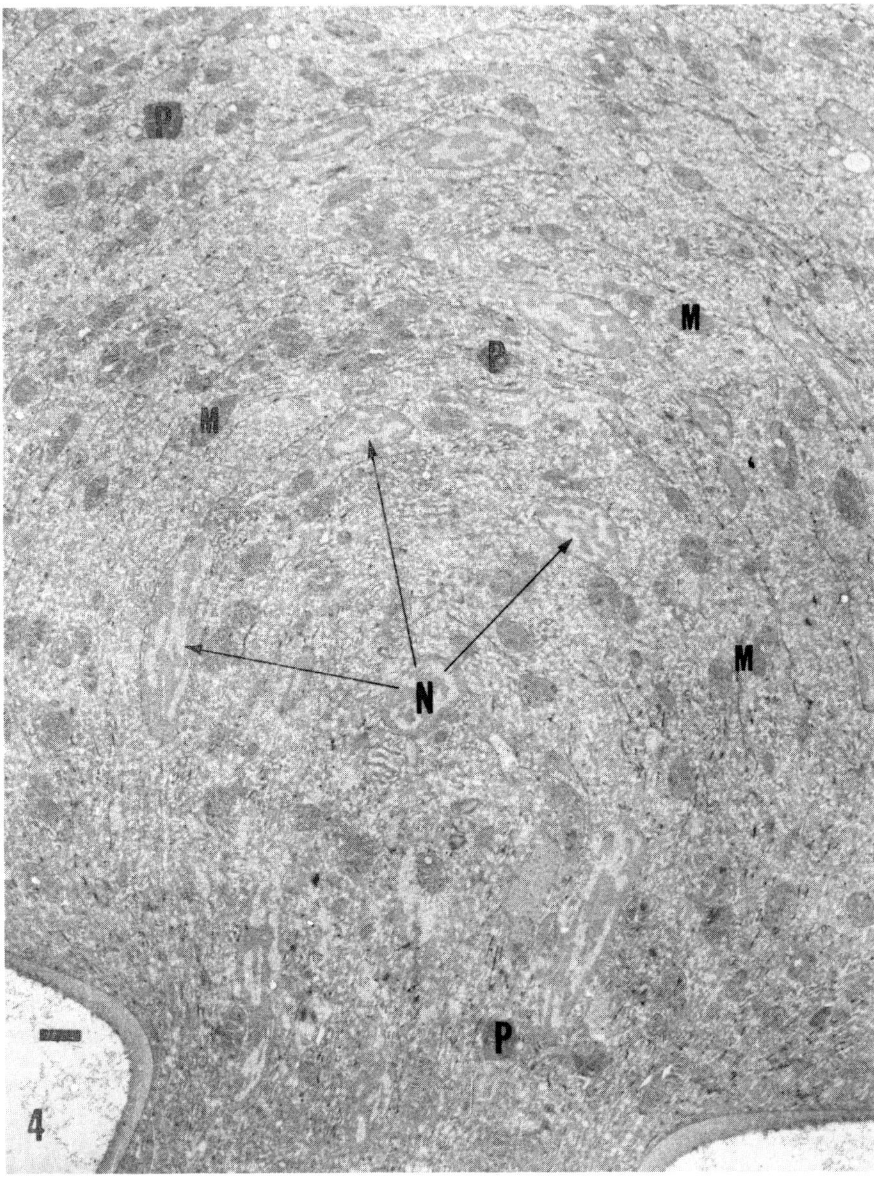

Fig. 4. Early precleavage sporangium, $KMnO_4$ fixation. Nuclei (N) and mitochondria (M) are numerous. Hemispherical arrangement of organelles suggests continuing "mushrooming" of developing sporangium. P = protein crystal. Magnification bar = 1 μm.

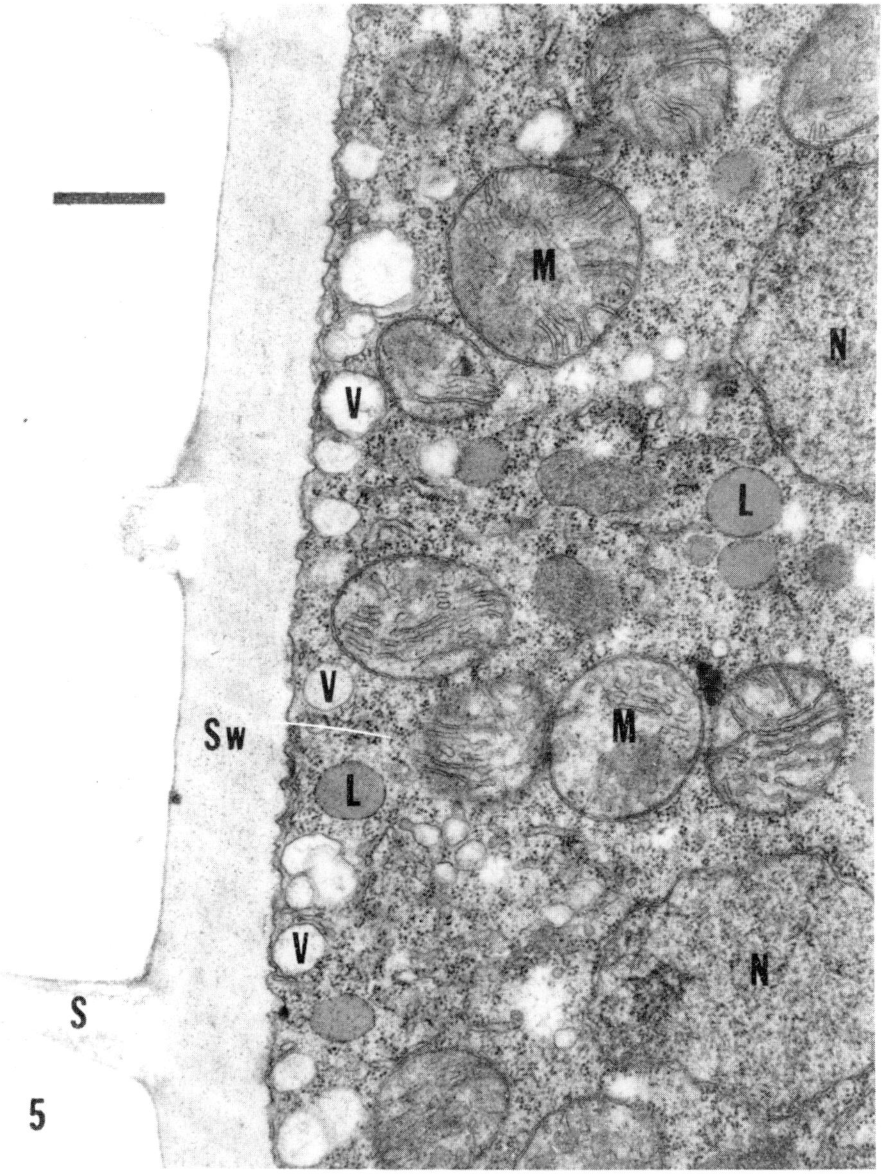

Fig. 5. Periphery of early precleavage sporangium. Numerous vesicles (V) measuring ca. 0.1-0.3 μm in diameter are present. Note nuclei (N), mitochondria (M), lipids (L), and spines (S) attached to sporangial wall (Sw). Magnification bar = 0.5 μm.

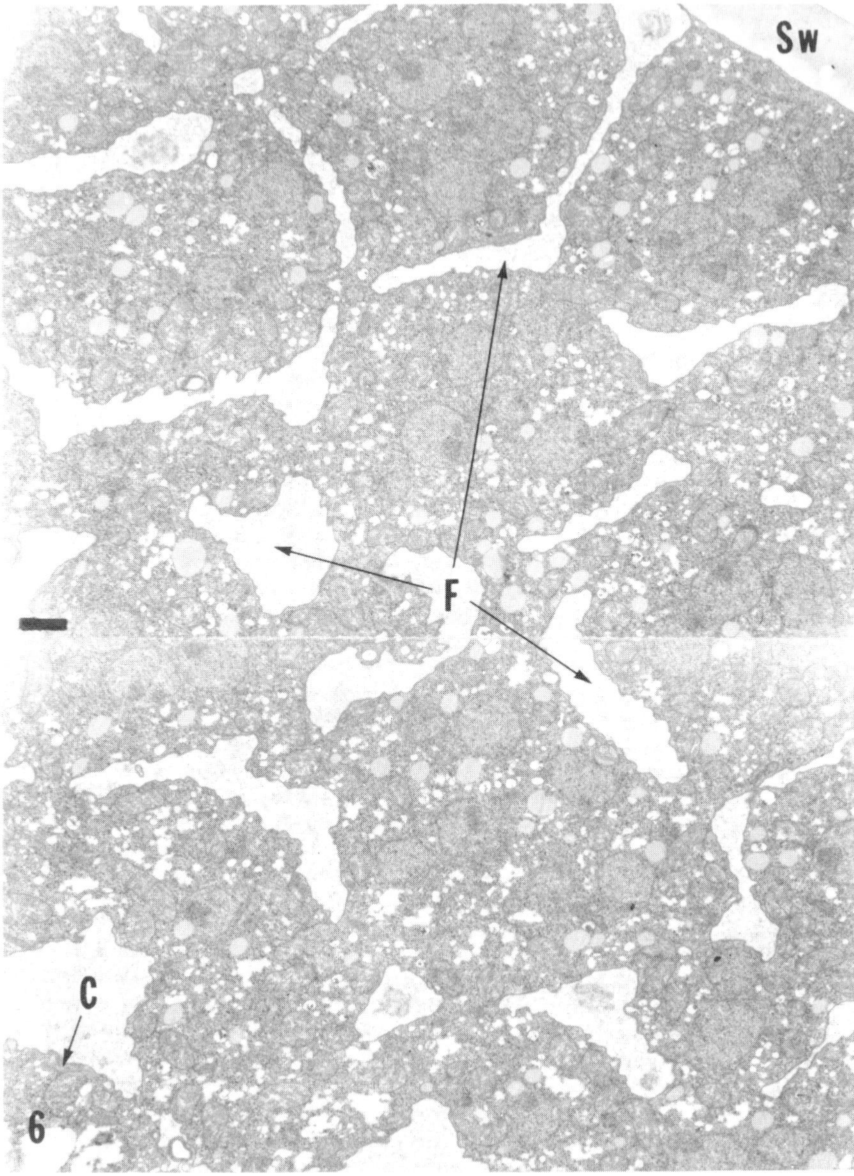

Fig. 6. Section of mid-cleavage sporangium. Note sporangial wall (Sw) at upper right of micrograph, and periphery of columellar region (C) at lower left. Cleavage furrows (F) were delimiting multinucleate spore protoplasts. Magnification bar = 1 μm.

The late cleavage sporangium shown in Fig. 7 is characterized by the presence of spore protoplasts which were nearly completely separated from each other. The micrograph is positioned as in Fig. 6, and each spore protoplast shows several nuclei. Figures 8-10 are enlargements of the sporangium illustrated in Fig. 7 or of other sporangia from an equivalent stage of development. The differentiating spore protoplasts (Fig. 8) contained numerous cisternal rings and vesicles. The vesicles contained a material which fixed and stained as an amorphous, electron-opaque substance. The appearance of these vesicles in micrographs suggests that they contributed to the advancing front of cleavage furrows. A photographic enlargement of a cleavage furrow boundary (Fig. 9) clearly shows that cleavage furrows in *Mucor* # 218 are lined with an electron-opaque layer situated ca. 15 μm from the plasmalemma of the differentiating spore protoplasts. It is hypothesized that the vesicular membrane contributes to the plasmalemma of the spore protoplasts, while the amorphous, electron-opaque, vesicular contents organize into the electron-opaque lining of cleavage furrows. During postcleavage, spore wall material is deposited between the electron-opaque layer and the spore plasmalemma.

Figure 10 is taken from the lower left of the columellar region of a sporangium in late cleavage. The portion of the spore protoplast shown in the upper left of the micrograph illustrates that the developmental stage is equivalent to Fig. 7. While the spore protoplast lacks a cell wall, the columellar wall, which formed beneath the sporangial wall, is present and nearly at its final thickness of ca. 0.2 μm. Clearly, the columellar wall is deposited before spore walls form. Also, the columellar wall, which tapers and merges into the sporangial wall as it gets closer to the central axis of the sporangium, appears structurally distinct from the sporangial wall.

Figures 11-13 are from postcleavage sporangia. In Fig. 11, the columella appears highly vacuolate, but with peripheral protoplasm containing numerous "typical" organelles. Sporangiospores exhibit a thick cell wall with electron-opaque and electron-transparent layers. Contents of the sporangiospores are very dense, with cellular details difficult to resolve. The structurally distinct columellar wall maintains its final thickness, while the sporangial wall appears to be thinner than at earlier stages, perhaps as a result of partial enzymatic dissolution. Higher magnification micrographs of the columellar periphery illustrate microtubules (Fig. 12) as well as other organelles, including nuclei, mitochondria, ribosomes, etc. It is noteworthy that autophagic vacuoles and microbodies are present in columellar protoplasm (Fig. 13),

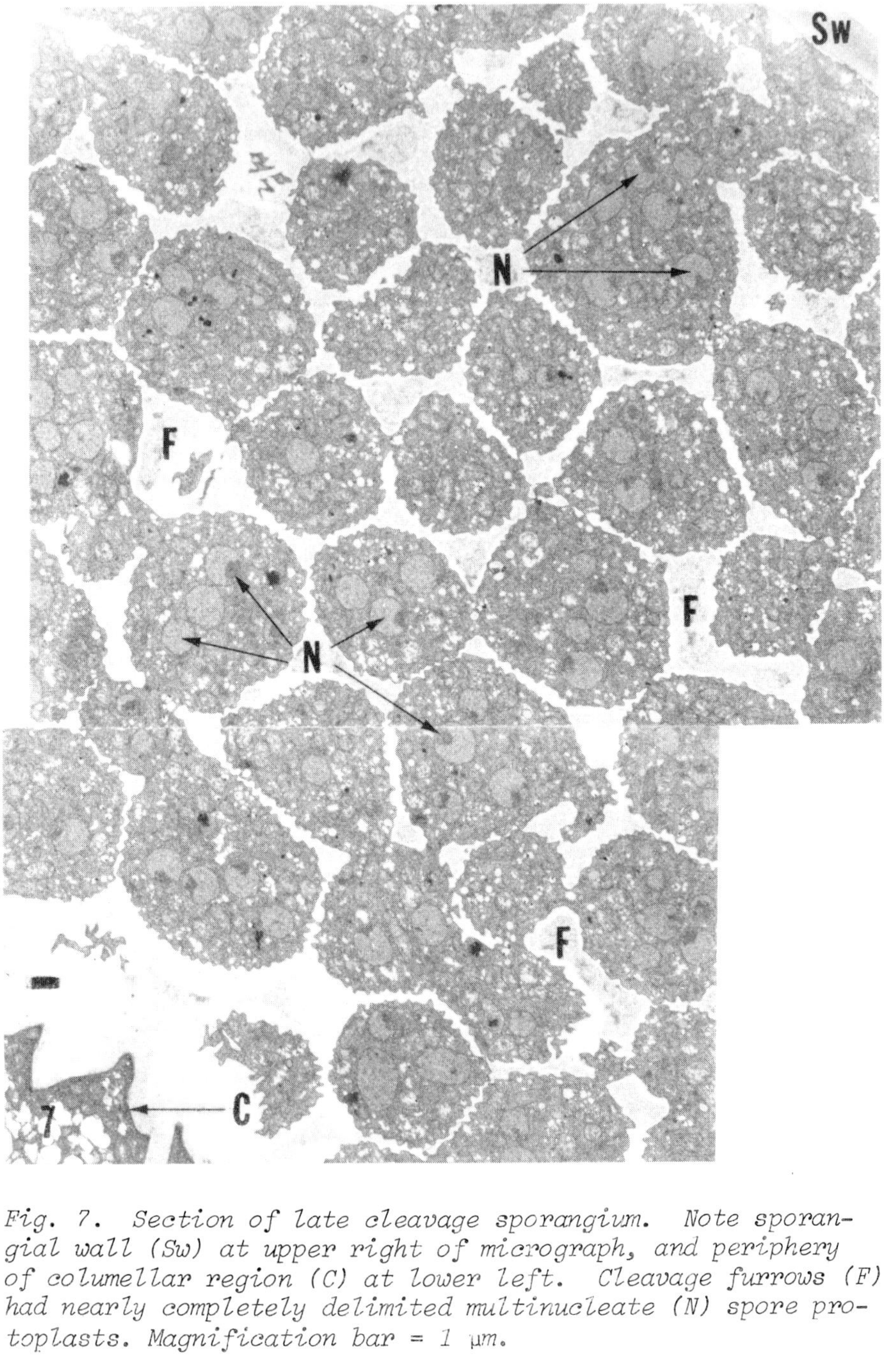

Fig. 7. Section of late cleavage sporangium. Note sporangial wall (Sw) at upper right of micrograph, and periphery of columellar region (C) at lower left. Cleavage furrows (F) had nearly completely delimited multinucleate (N) spore protoplasts. Magnification bar = 1 μm.

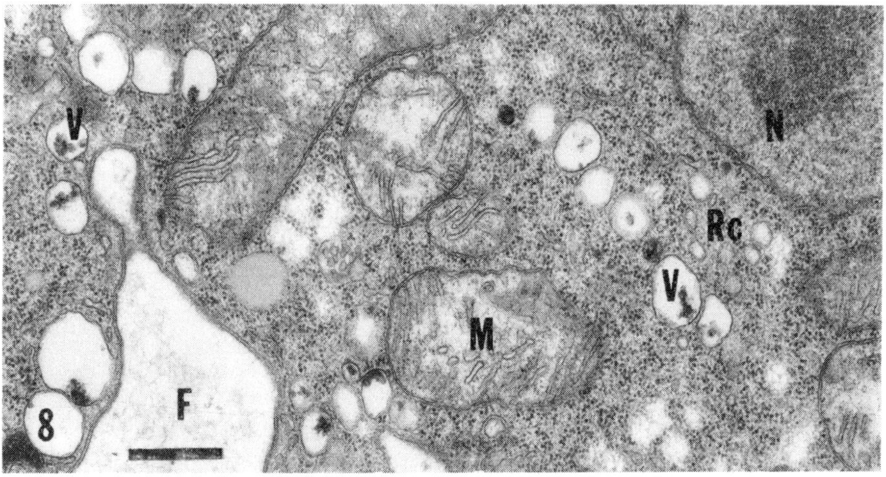

Fig. 8. Portion of spore protoplast from late cleavage sporangium in Fig. 7. Cleavage furrows (F) appear to develop by addition of vesicles (V) to the advancing front. Vesicles contain amorphous, electron-opaque contents and presumably develop from ring cisternae (Rc) which are common. Magnification bar = 0.5 µm.

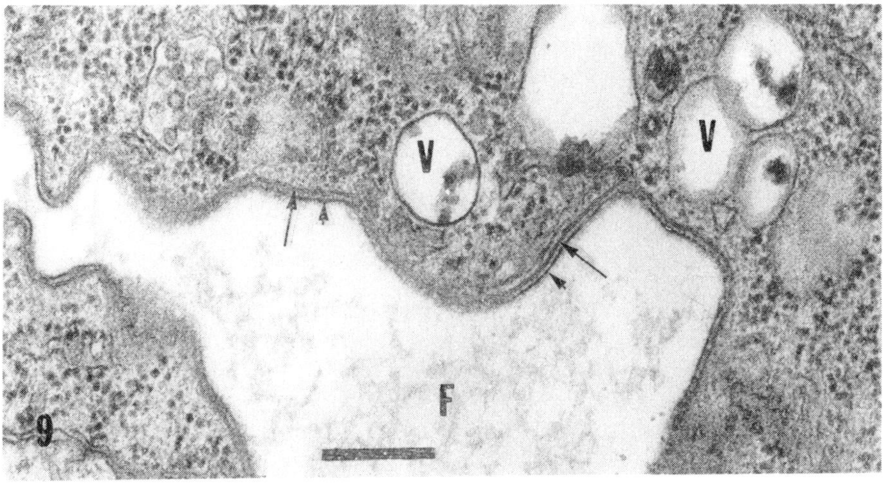

Fig. 9. Portion of similar spore protoplast as in Fig. 8. Cleavage furrow (F) lined with an electron-opaque layer (arrow heads) believed to be developed from electron-opaque contents of vesicles (V). Plasmalemma of spore protoplasts (arrows) underlies electron-opaque layer. Magnification bar = 0.25 µm.

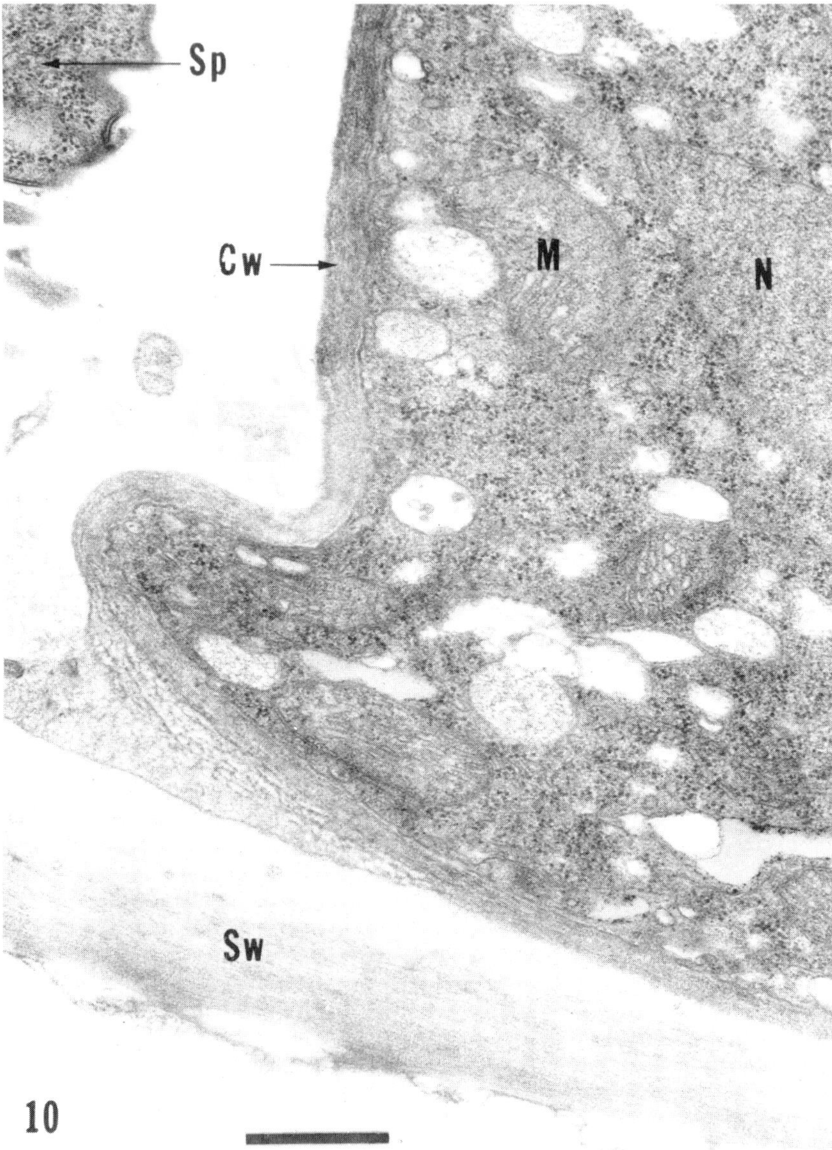

Fig. 10. Base of columella in late cleavage sporangium (same stage as Fig. 7). Spore protoplast (Sp) without cell wall. Columellar wall (Cw) present, having been laid down inside of sporangial wall (Sw). Fold in columellar wall is an artifact of preparation. M = mitochondrion; N = nucleus. Magnification bar = 0.5 μm.

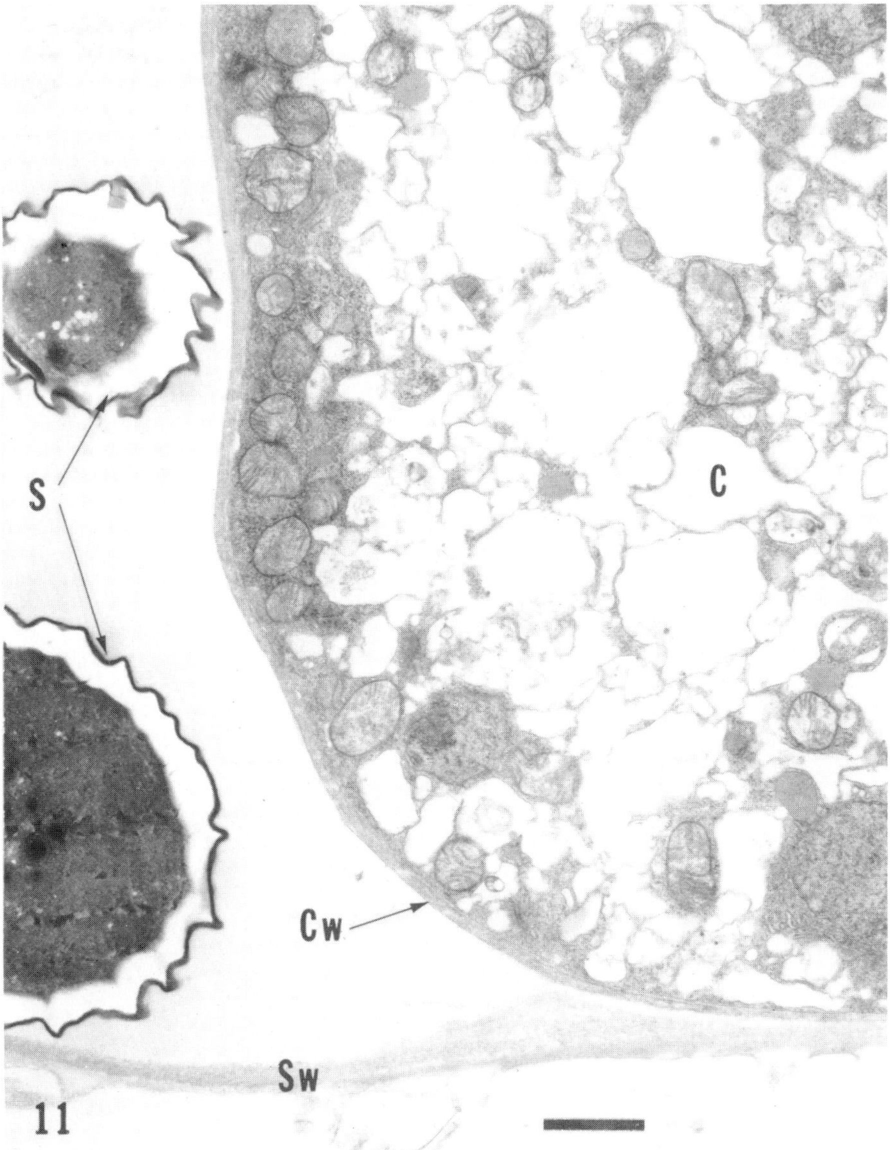

Fig. 11. Base of columella in mid-to-late postcleavage sporangium. Quiescent spores (S) very dense. Columellar wall (Cw) appears structurally distinct from thinning sporangial wall (Sw). Columella (C) highly vacuolate, but with peripheral protoplasm containing "typical" organelles. Magnification bar = 1 μm.

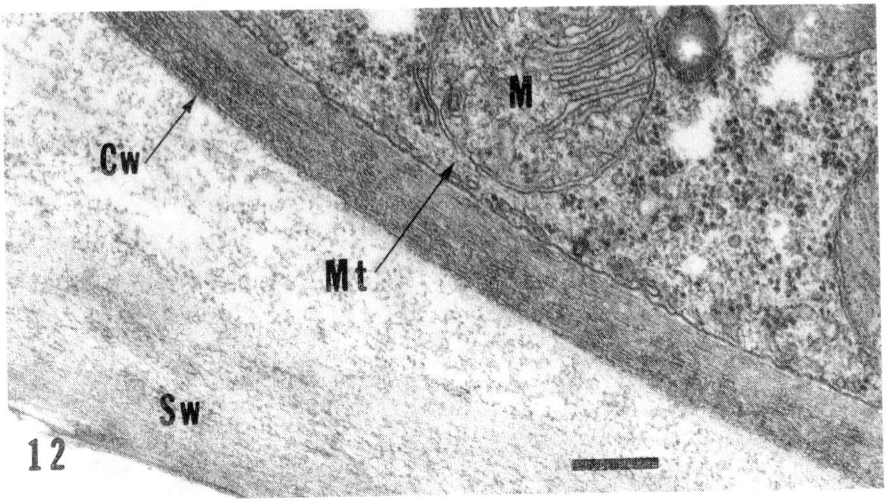

Fig. 12. Portion of columellar base in mid-to-late post-cleavage sporangium. Columellar wall (Cw) appears structurally distinct from sporangial wall (Sw). Note mitochondrion (M) and microtubule (Mt) in peripheral protoplasm of columella. Magnification bar = 0.25 μm.

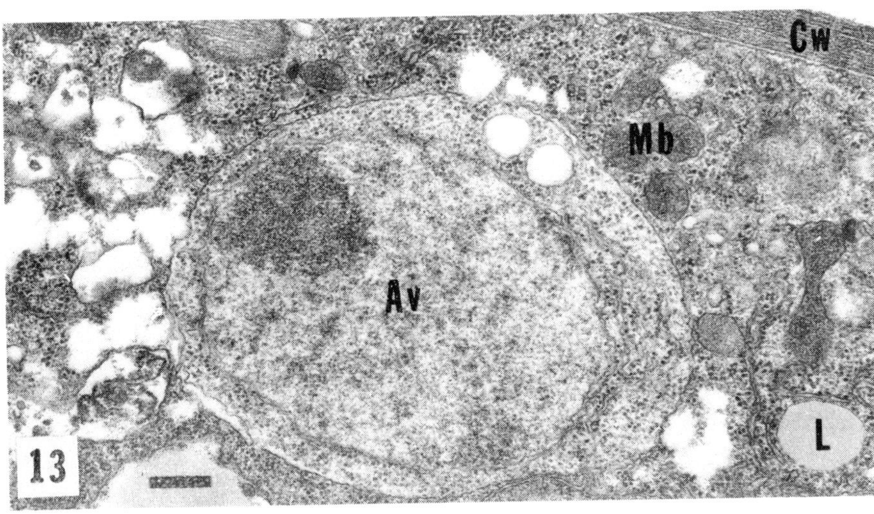

Fig. 13. Portion of columellar periphery in late postcleavage sporangium. Note conspicuous autophagic vacuole (Av) containing nucleus, ribosomes, and some membranous materials. Also note microbodies (Mb), columellar wall (Cw), and lipid globule (L). Magnification bar = 0.25 μm.

perhaps being involved in turnover of columellar contents.

3. SUMMARY

The sporangiosporogenic process in *Mucor* # 218 appears to be essentially similar to that described by Bracker (1966, 1968) in *G. persicaria*. However, additional data, derived from the preliminary study of *Mucor* # 218, are listed below.

(A) Sporangial cleavage in *Mucor* # 218, as in *G. persicaria*, appears to be a function of membranous activity. Numerous vesicles of various types and ring cisternae leave little doubt as to the dynamism of membranous interactions during sporangial development.

(B) Whereas Bracker reported the decreased number of cisternal rings in *G. persicaria* from mid-cleavage onward, and their complete absence during postcleavage, such rings were numerous and conspicuous during all stages in *Mucor* # 218 from sporangial initiation through cleavage. Cisternal rings were observed during postcleavage, and may be involved in the development of spore walls.

(C) Electron-opaque granules of the type which Bracker described as lining cleavage vesicles, and which later played a role in the wall development of sporangiospores in *G. persicaria*, did not appear in *Mucor* # 218. Instead, cleavage vesicles in the latter specimen contained an amorphous, electron-opaque substance which also appeared to play a role in wall development of sporangiospores.

(D) Numerous, small, peripheral vesicles in precleavage sporangia of *Mucor* # 218 were similar to those reported by Bracker in *G. persicaria*. Such vesicles may be involved in sporangial wall formation from sporangial initiation to the point of organization of the cleavage apparatus. Sporangia, despite a great increase in surface area prior to cleavage, never are devoid of a cell wall, which appears to be of relatively constant thickness during sporangial enlargement. One must assume that sporangial wall synthesis is concurrent with sporangial enlargement.

(E) Bracker reported that ramification of the cleavage apparatus in *G. persicaria* appeared to result from extension of a tubular network rather than from addition of new vesicular material to the cleavage apparatus. In *Mucor* # 218, however, addition of new vesicular material to the advancing fronts of cleavage furrows appears to be the functional system.

(F) Bracker did not observe dividing nuclei in *G. persicaria*. Such nuclei frequently were seen during sporangial initiation in *Mucor* # 218 and, in one case, there appeared a dividing nucleus in a spore protoplast during late cleavage.

(B) Bracker reported microbodies during precleavage and in

columellar protoplasm of *G. persicaria*. He also reported that some vacuoles in columellae contained degenerate mitochondria or crystals. In *Mucor* # 218, microbodies were numerous in the upper sporangiophore during sporangial initiation (mostly below the level shown in Fig. 2) and in columellar protoplasm. The autophagic vacuoles in columellae of *Mucor* # 218 may be similar to the vacuoles mentioned by Bracker. Turnover of columellar contents may be worth investigating further, especially as it may relate to spore release from sporangia.

(H) In *Mucor* # 218, columellar walls appear structurally distinct from either sporangial or sporangiophore walls. Columellar wall material is deposited prior to the deposition of spore wall material, and inside of the base of the sporangial wall.

(I) Microtubules were observed in sporangia, sporangiophores and columellar protoplasm of *Mucor* # 218. It is assumed that their functions at those locations are as commonly conceived, i.e., bulk movement of cellular components and/or a structural framework.

4. ACKNOWLEDGEMENTS

Dr. Wilfred A. Côté and Dr. Robert B. Hanna of the Nelson C. Brown Laboratory for Ultrastructural Studies, SUNY College of Environmental Science and Forestry, Syracuse, N. Y., are thanked for their permission to use those facilities. Dr. M. A. A. Schipper and Dr. J. J. Ellis are thanked for their examinations of *Mucor* # 218. Dr. Charles Bracker is thanked for permission to use his diagramatic model of sporangiosporogenesis in *G. persicaria*. Ms. Andrea Sant and Mr. Paul Roubal, two former undergraduate students, are thanked for their research assistance, and Ms. Lorretta Beckwith is thanked for typing the manuscript.

5. REFERENCES

Alexopoulos, C. J. (1962). "Introductory Mycology, 2nd Ed." John Wiley & Sons, New York.

Alexopoulos, C. J. and Mims, C. W. (1979). "Introductory Mycology, 3rd Ed." John Wiley & Sons, New York.

Baker, K.L., Hooper, G.R. and Beneke, E.S. (1977). Ultrastructural development of merosporangia in the mycoparasite *Syncephalis sphaerica* (Mucorales). *Canadian Journal of Botany* 55, 2207-2215.

Benjamin, R.K. (1979). Zygomycetes and their spores. *In* "The Whole Fungus" (Ed. B. Kendrick). Vol. 2, pp. 573-616. National Museum of Natural Sciences, National Museums of Canada, Ottawa.

Bracker, C.E. (1966). Ultrastructural aspects of sporangiospore formation in *Gilbertella persicaria*. *In* "The Fungus Spore" (Ed. M.F. Madelin). pp. 39-60. Butterworths, London, England.
Bracker, C.E. (1968). The ultrastructure and development of sporangia in *Gilbertella persicaria*. *Mycologia* 60, 1016-1067.
Cole, G.T. and Samson, R.A. (1979). "Patterns of Development in Conidial Fungi" Pitman Publishing Ltd., London, England.
Dykstra, J.J. (1974). An ultrastructural examination of the structure and germination of asexual propagules of four mucoralean fungi. *Mycologia* 66, 477-489.
Fletcher, J. (1972). Fine-structure of developing merosporangia and sporangiospores of *Syncephalastrum racemosum*. *Archiv für Mikrobiologie* 87, 269-284.
Fletcher, J. (1973a). Ultrastructural changes associated with spore formation in sporangia and sporangiola of *Thamnidium elegans* Link. *Annals of Botany* 37, 963-971.
Fletcher, J. (1973b). Distribution of cytoplasmic vesicles, multivesicular bodies, and paramural bodies in elongating sporangiophores and swelling sporangia of *Thamnidium elegans* Link. *Annals of Botany* 37, 955-962.
Hesseltine, C.W. and Ellis, J.J. (1973). Mucorales. *In* "The Fungi. A Taxonomic Review with Keys: Basidiomycetes and Lower Fungi" (Ed. G.C. Ainsworth, F.K. Sparrow and A.S. Sussman). Vol. IV B, pp. 187-217. Academic Press, New York.
Hirahare, S. and Seshita, Y. (1976). Scanning electron microscopic studies on *Mucors*. *Bulletin of the Hiroshima Jogakuin College* 26, 119-126.
Ingold, C.T. (1978). "The Biology of *Mucor* and its Allies" Edward Arnold Publishers, Ltd., London, England.
Jeffries, P. and Young, T.W.K. (1975). Ultrastructure of the sporangiospore of *Piptocephalis unispora* (Mucorales). *Archiv für Mikrobiologie* 105, 329-333.
Jones, D., McHardy, W.J. and Wilson, J.J. (1976). Ultrastructure and chemical composition of spines in Mucorales. *Transactions of the British Mycological Society* 66, 153-157.
Khan, S.R. and Talbot, P.H.B. (1975). Monosporous sporangiola in *Mycotypha and Cunninghamella*. *Transactions of the British Mycological Society* 65, 29-39.
Moore-Landecker, E. (1972). "Fundamentals of the Fungi" Prentice-Hall, Inc., Englewood Cliffs, New Jersey.
O'Donnell, K.L. (1979). "Zygomycetes in Culture" Department of Botany, University of Georgia, Athens.

Schipper, M.A.A. (1975). On *Mucor mucedo, Mucor flavus* and related species. *Studies in Mycology No. 10,* Centraalbureau voor Schimmelcultures, Baarn, The Netherlands.

Stewart, P.R. and Rogers, P.J. (1978). Fungal dimorphism: A particular expression of cell wall morphogenesis. *In* "Developmental Mycology. The Filamentous Fungi" (Ed. J.E. Smith and D.R. Berry). Vol. 3, pp. 165-196. John Wiley & Sons, New York.

Storck, R. and Morrill, R.C. (1977). Nuclei, nucleic acids and protein in sporangiospores of *Mucor bacilliformis* and other *Mucor* species. *Mycologia* 69, 1031-1041.

Talbot, P.H.B. (1971). "Principles of Fungal Taxonomy" St. Martin's Press, New York.

Tu, J.C. and Malhotra, S.K. (1977). Plasma membrane of *Phycomyces:* Sequential changes in the structure during spore formation. *Cytobios* 20, 121-132.

Webster, J. (1970). "Introduction to Fungi" Cambridge University Press, London, England.

Zycha, H., Siepmann, R. and Linnemann, G. (1969). "Mucorales. Eine Beschreibung aller Gattungen und Arten dieser Pilzgruppe, mit einem Beitrag zur Gattung *Mortierella*" J. Cramer, Lehre, West Germany.

SURFACE WALL COMPONENTS OF ASPERGILLUS NIGER CONIDIA

G. T. Cole* and L. M. Pope#

Departments of Botany and Microbiology#*
University of Texas, Austin, Texas 78712

INTRODUCTION

Although the importance of microbial surfaces in cell recognition, adherence, colonization and pathogenicity is well appreciated (Smith, 1977; Frazier and Glaser, 1979) few studies have dealt with ultrastructural and biochemical aspects of fungal cell surfaces. Since the conidial fungi are causative agents of the majority of plant and animal mycoses (Cole and Kendrick, 1981), it is fitting that we direct our attention to the surface of the conidium. In an earlier investigation (Cole *et al.*, 1979) we examined the fine structure of conidial walls of several imperfect fungi and demonstrated the presence of a common fibrillar surface component referred to as the "rodlet" layer (Hess *et al.*, 1968). Surface wall constituents which are similar in morphology to the rodlets of conidia have also been demonstrated on spores produced by other groups of fungi (Bronchart and Demoulin, 1971; Hess *et al.*, 1972; Wessels *et al.*, 1972) as well as bacterial spores of *Bacillus* sp., *Streptomyces* sp., and *Micropolyspora* sp. (Bradley and Ritzi, 1968; Holt and Leadbetter, 1969; Gould *et al.*, 1970; Williams *et al.*, 1972; Williams *et al.*, 1974; Aronson and Fitz-James, 1976; Takeo, 1976; Smucker and Pfister, 1978; Wehrli *et al.*, 1980). The chemical composition of the rodlet layer of conidia is considered to be mainly proteinaceous (Hashimoto *et al.*, 1976; Wu-Yuan and Hashimoto, 1977; Beever *et al.*, 1979; Cole *et al.*, 1979). However, the morphologically equivalent layer on the hyphal wall of the basidiomycete *Schizophyllum commune* has been identified as a crystallized S-glucan (Wessels *et al.*, 1972), and on the aerial spore wall of *Streptomyces coelicolor* as chitin (Smucker and Pfister, 1978).

Using a Ribi Cell Fractionator, Cole *et al.* (1979) separated the conidial wall of *Aspergillus niger* (Strain IFO 6341) into an outer and inner layer. The outer layer, which included the rodlets, was composed mainly of protein (63.4%) and lipid (24.8%). However, the rodlets were not purified during this investigation. Because of the hydrophobic nature of *A. niger* conidia, which may be partly due to the presence of the rodlet layer (Beever and Dempsey, 1978), we decided in this study to initially harvest conidia from sporulating plate cultures using a nonionic detergent followed by either sonication or sodium hydroxide treatment of the washed conidial pellet. Employing these isolation procedures it was possible to obtain homogeneous rodlet preparations and more accurately define the morphology of the rodlet layer and other surface components of the conidial wall. The chemical data on cell wall composition was difficult to obtain primarily because of the presence of large amounts of pigment (melanin?) which interfered with certain spectrophotometric procedures (e.g., Lowry and Anthrone methods). No investigations to the authors' knowledge of fungal cell wall composition have involved detergent treated dematiaceous forms and it was necessary, therefore, to modify existing analytical procedures. The chemical composition of conidial wall fractions of *A. niger* are reported.

MATERIALS AND METHODS

Culturing and harvesting

The isolate of *A. niger* (UT 639) used in this study was an air contaminant. The mycelium from which conidia were harvested was grown on potato sucrose agar (PSA) medium (Cole *et al.*, 1979) in 15 cm plastic Petri dishes and incubated in continuous darkness at 28°C for 4 to 6 weeks.

Conidia were harvested using a 0.01% solution of one of the detergents listed below. The open Petri dish was tilted at about 45° while drops of the detergent solution released from a pipette were allowed to roll down and over the surface of the mycelium. The suspension collected from the bottom of the dish contained only conidia. The concentration of conidia used in each experiment was approximately the same and was determined by hemacytometer counts. The following detergents were used: Nonidet P-40 (Sigma), SDS, Triton X-100 (Sigma), deoxycholic acid (Sigma), 1-O-n-octyl-β-D-glucopyranoside (Sigma), Zwittergent 3-10 (Calbiochem) and Zwittergent 3-12 (Calbiochem). The conidial suspension was transferred to an Erlenmeyer flask and the concentration of the detergent was increased to 1.0%. The flask was then placed in a New Brunswick shaker-incubator at 25°C and agitated at 200 rpm for 12-16 h.

The viability of conidia after incubation in each detergent was examined. A droplet of the conidial suspension was placed on a Nucleopore filter (1.0 μm pore size) overlying the surface of PSA medium in a Petri dish. The culture was incubated at 28°C for 12 h. The Nucleopore filter was then removed from the surface of the agar, mounted on a microscope slide and examined with a light microscope for germinating conidia.

The conidial suspension was centrifuged at 4,100*g* for 15 min and the supernatant, which contained detergent-extracted wall material from the conidia, was prepared for chemical analysis. The supernatant analysed was derived from Nonidet P-40 (NP-40) treated conidia (Table 1).

Preparation of the detergent (NP-40) soluble conidial wall fraction for chemical analysis

The detergent was removed from the sample by the following procedure. The supernatant was first treated with cleaned Amberlite XAD-2 (Rohm and Haas) beads (1 g/ml) in an Erlenmeyer flask which was placed on a wrist-action shaker at room temp. for 2 h. The supernatant was then carefully removed and treated with ammonium sulfate (3.52 g $(NH_4)_2SO_4$/ 20 ml supernatant). The solution was subsequently centrifuged at 5,900*g* for 20 min and the pellet was resuspended in dH_2O and dialysed against dH_2O at 4°C for 48 h. The retentate was then treated with Bio-Beads SM-2 (Bio-Rad Lab., Richmond, Cal.) at a concentration of 0.3 g/ml of retentate. The sample was removed from the Bio-Beads and lyophilized in preparation for chemical analysis (Table 1).

Isolation and purification of conidial wall layers

The conidial pellet was washed in dH_2O and centrifuged (4,100*g*) four times. The weight of the wet pellet was estimated and an equal weight of Bio-Beads SM-2 was added. The mixture was resuspended in 50 ml dH_2O in a 250 ml Erlenmeyer flask and placed on a wrist-action shaker at room temp. for 4 h. The conidial suspension was then carefully removed, washed in dH_2O and centrifuged (4,100*g*) two times. The conidia were subsequently subjected to either sonication or treatment with 0.1 N NaOH.

Sonication

The conidia were resuspended in 50 ml dH_2O and sonicated in an ice bath for a total of 30 min (six 5 min bursts, 50 Watts, setting 4, Branson Sonifier, Branson Sonic Power Co., Danbury, Conn.). The sonicated suspension was then centrifuged at 365*g* for 15 min. The supernatant contained rodlets as well as some underlying wall components while the pellet was composed of intact conidia lacking the outer wall layer. The supernatant was centrifuged at 125,000*g* for 9 h and the resulting pellet (Fig. 5) was lyophilized in preparation for chemical analyses (Table 1). The 125,000*g* pellet was also treated with 0.1 N NaOH and 0.1 N HCl in an attempt to further purify the rodlets. The outer wall components were resuspended in 10 ml of 0.1 N NaOH and incubated at 25°C in the shaker incubator (200 rpm) for 20 h. The suspension was then centrifuged at 125,000*g* for 9 h and the pellet was replicated in preparation for electron microscopy (Fig. 6). These same conditions were used for treatment of the outer wall sample with 0.1 N HCl (Fig. 7).

The 365*g* pellet containing intact conidia with the outer wall layer removed was resuspended in dH_2O and homogenized in a Braun Model MSK cell homogenizer (Bronwill Scientific, Inc., Rochester, New York) at 2,000 c.p.m. (three 15 sec bursts) using 0.45-0.50 mm glass beads which ruptured the cells (approx. 90%) and permitted isolation of the inner conidial wall layer (Fig. 8). The sample was cooled with liquid CO_2 during the homogenization process. The inner wall suspension was centrifuged at 5,900*g* for 15 min. The pellet was then washed in dH_2O, centrifuged (5,900*g*) five times and lyophilized in preparation for chemical analyses (Table 1).

Treatment with 0.1 N NaOH

The washed and Bio-Bead SM-2 treated conidial pellet was weighed (wet wt.) and resuspended in 0.1 N NaOH in a ratio of 5 ml NaOH/g conidia. The sample was then placed in the incubator shaker at 25°C and 200 rpm for 12 h. The suspension was subsequently centrifuged 365*g* for 15 min. The supernatant primarily contained rodlets (Fig. 10) while the intact conidia showed patches of smooth wall (Fig. 9). The effectiveness of 0.1 N NaOH in removing rodlets during incubation periods of 4-22 h was monitored by replica preparation and electron-microscopic examination of the treated conidial pellet and supernatant. In preparation for chemical analysis, the supernatant was first dialyzed against

dH_2O (24 h at 4°C) and then centrifuged at 125,000*g* for 2 h. The pellet contained some rodlets as well as amorphous wall material removed by the 0.1 N NaOH treatment and was discarded. The resulting supernatant was centrifuged at 125, 000*g* for 9 h and the pellet (Fig. 10) was lyophilized and later used for chemical analysis (Table 2).

Electron microscopy

Single drops of intact conidia or conidial wall preparations dispersed in dH_2O were placed on small pieces of split mica and air dried. The samples were then shadowed with carbon-platinum at 45° in a Balzers vacuum evaporator. The mica and replicated samples were then placed in 50% chromic acid and warmed to 50°C which cleaned the replicas and dislodged them from the mica. The replicas were then transferred several times to fresh distilled water using a platinum loop, mounted on Formvar-coated grids and examined in a Hitachi HU-11E transmission electron microscope (Figs. 2, 4-10).

Chemical analysis

The methodology used for carbohydrate, amino sugar and lipid analyses was described in an earlier publication (Cole *et al.*, 1979). Total protein was determined by the Lowry method (Lowry *et al.*, 1951) using bovine serum albumin as a standard. The spectrophotometric effects of detergents on these protein determinations was taken into account by adding detergent to the standard at a concentration equal to that in the sample (Fig. 1). Interference by detergents in the Lowry reaction was eliminated by the addition of sodium dodecyl sulfate (SDS) to samples and standards (Dulley and Grieve, 1975). Protein was also estimated by the Yemm and Cocking (1955) ninhydrin method and the Kjeldahl-Nessler method (Kabat and Mayer, 1961; Campbell *et al.*, 1970) based on total nitrogen content. The assumption for the Kjeldahl-Nessler method is that nitrogen primarily occurs in amino acids and the nitrogen content of phospholipids is comparatively small (Beever *et al.*, 1979). For amino acid analysis, 1.5-2.0 mg of wall material was hydrolysed in 6 N HCl (1 mg/ml) in sealed ampoules at 110°C for 16 h and the hydrolysate was analysed in a Beckman Model 121 Amino Acid Analyser. Total protein was also estimated by summation of the amounts of each amino acid (Tables 1, 2).

For determination of ash, wall samples were heated in a muffle furnace at 1000°C for 3 h and the residue was weighed (Table 1).

OBSERVATIONS

Effects of detergent treatment of conidia

Treatment of *A. niger* conidia with all detergents reduced the hydrophobicity of the cells. After suspension and agitation in 1% detergent for 12-16 h, the conidia were still viable and germinated on the Nucleopore filters. However, conidia suspended in 1% SDS showed some inhibition of growth while conidial samples derived from all other detergent suspensions showed the same amount of growth.

A comparison of the effect of varying concentrations of the nonionic detergents NP-40 and Triton X-100 with equal concentrations of conidia (approx. 1.5×10^7 cells/ml) over a 12 h period is presented in Fig. 1. In addition to black pigment (melanin?), protein was released from conidia into the detergent solution. The amount of protein present was determined by the modified Lowry procedure. In both cases,

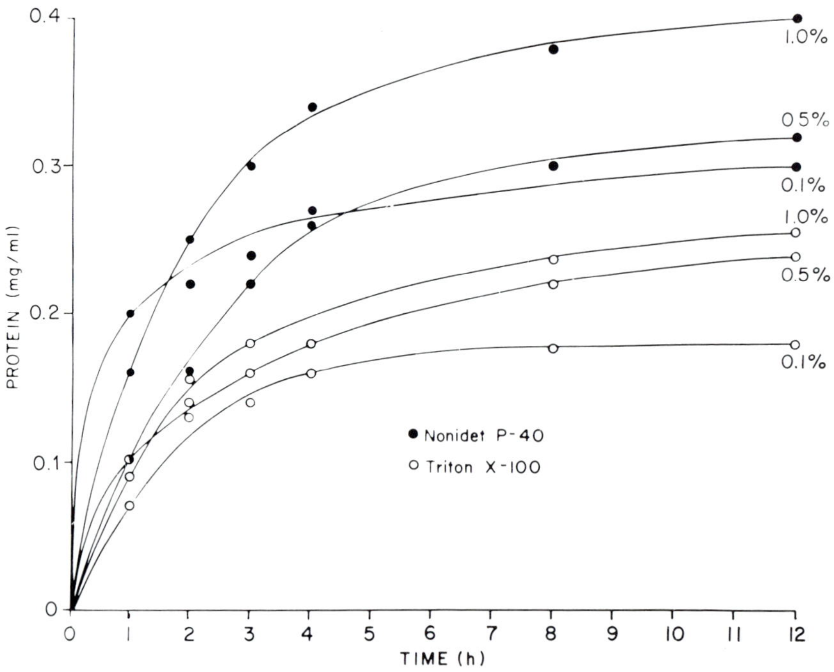

Fig. 1. Protein released from conidia during incubation in varying concentrations (0.1%, 0.5%, 1.0%) of nonionic detergents (Nonidet P-40 and Triton X-100) over a 12 h period. Protein determinations were performed using a modified Lowry method.

1.0% detergent solution showed the greatest effects and most protein was released during the first 2-3 h of incubation. Nonidet P-40 was generally more efficient in wetting the conidia and releasing protein than Triton X-100 and the former was, therefore, used in all later experiments.

Electron microscopy

Surface replicas of untreated, dry conidia of *A. niger* (UT 639) taken directly from plate culture revealed distinct fascicles of rodlets (Fig. 2) and the absence of an outermost amorphous layer (c.f. Figs. 8 and 11 in Cole *et al.*, 1979). After treatment with NP-40 for 12-16 h, most conidia had swelled (Fig. 3) but rodlets were still present on the outer conidial wall surface (Fig. 4). A slight change in the texture of the rodlet layer is evident from a comparison of Figs. 2 and 4. After the conidia were removed from the detergent suspension by slow-speed centrifugation, the

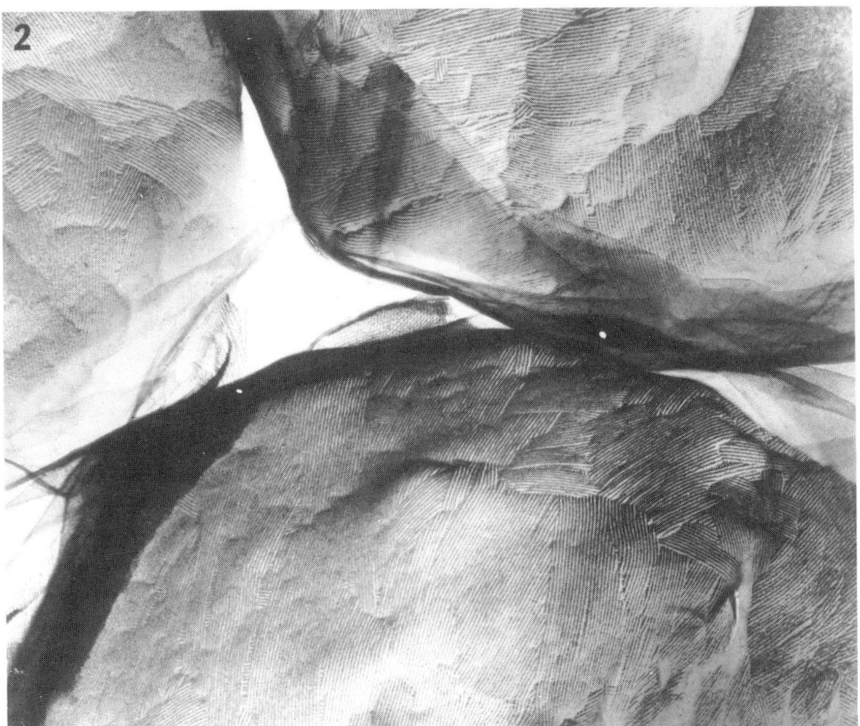

Fig. 2. Replica of untreated conidia of A. niger taken directly from a 5 week PSA plate culture. X 41,000.

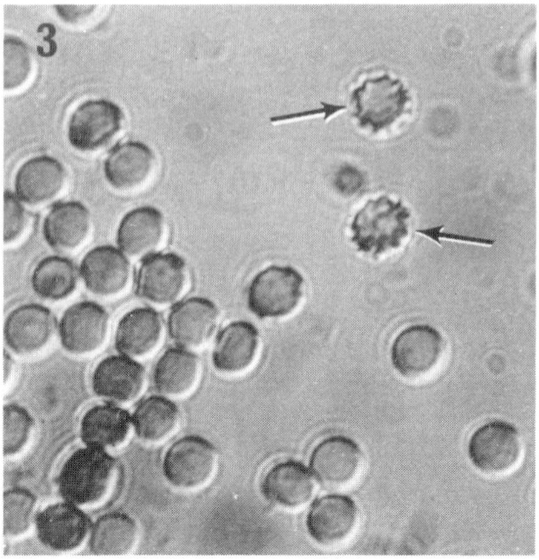

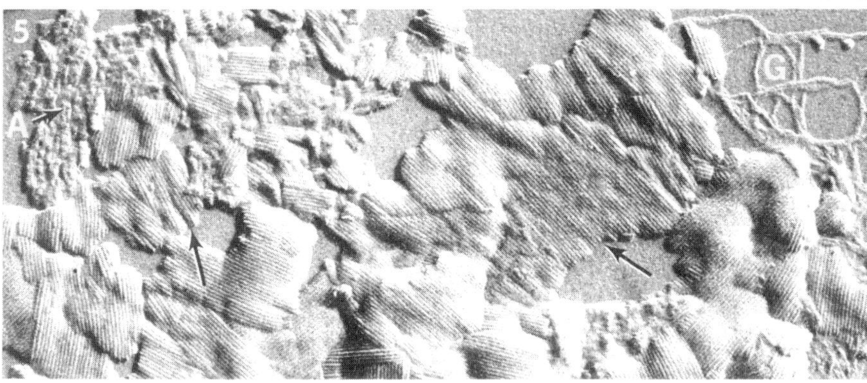

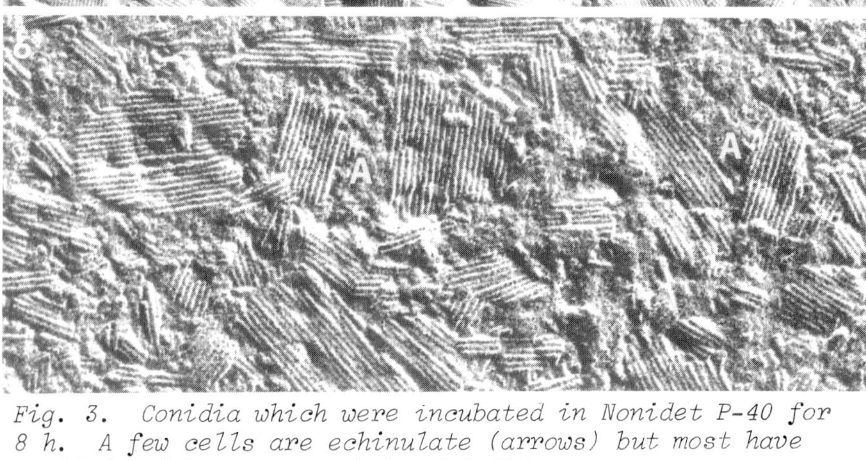

Fig. 3. Conidia which were incubated in Nonidet P-40 for 8 h. A few cells are echinulate (arrows) but most have swelled and become rounded. X 2,400.

Fig. 4. Surface replica of Nonidet P-40 treated conidium showing distinct rodlet fascicles. X 65,000.

supernatant (detergent soluble wall fraction) was centrifuged at 125,000*g* for 9 h and replicas of the pellet were prepared for electron microscopy. Only amorphous material was visible - no rodlets were found.

Figure 5 is a replica of the outer wall layer removed from the detergent treated conidia as a result of sonication. The pellet in Fig. 5 was obtained by removing the intact conidia from the sonicated suspension by centrifugation (365*g*) and then centrifuging the supernatant at 125,000*g* for 9 h. The rodlets show transverse fractures and some longitudinal separations (arrows) but otherwise appear intact. The amorphous material (A) lies below the rodlet layer and is shown attached to the undersurface of some rodlet fascicles. Below the amorphous layer of the conidial wall is the distinct fibrous material (G) shown in Fig. 5. When this same outer conidial wall preparation was treated with 0.1 N NaOH in a shaker incubator (25°C, 200 rpm, 20 h), the fibrous material disappeared but the amorphous component associated with the rodlet fascicles persisted (Fig. 6). On the other hand, if the outer wall sample derived from the sonication procedure was first treated with 0.1 N NaOH as above, washed in dH_2O and then treated with 0.1 N HCl (same conditions as 0.1 N NaOH), both the fibrous component and the rodlets were lost from the 125,000*g* pellet (Fig. 7). The remaining wall material tended to become clumped in the acid.

The sonicated, intact conidia revealed a rough outer surface which was almost totally free of rodlets. After homogenization, the smooth inner wall (I, Fig. 8A) of the fractured conidia was visible. Fibrous material comparable to that shown in Fig. 5 appeared along transverse fractures of the conidial wall (G, Fig. 8B).

When detergent treated conidia were incubated in 0.1 N NaOH at 25°C for 12 h (continuous agitation), intact rodlet fascicles were removed along with some underlying amorphous

---

Fig. 5. Replica of outer conidial wall layer obtained by sonication. Arrows indicate transverse fractures and longitudinal separations of rodlet fascicles. A, amorphous wall material; G, fibrous glucan-like material. X 39,000.
Fig. 6. Replica of outer conidial wall layer obtained by sonication and subsequently treated with 0.1 N NaOH (c.f. Fig. 5). A, amorphous wall material. X 49,000.
Fig. 7. Replica of outer conidial wall layer obtained by sonication and then treated with 0.1 N NaOH followed by 0.1 N HCl (c.f. Figs. 5, 6). Most rodlet fascicles are gone and the amorphous wall material has clumped. X 49,000.

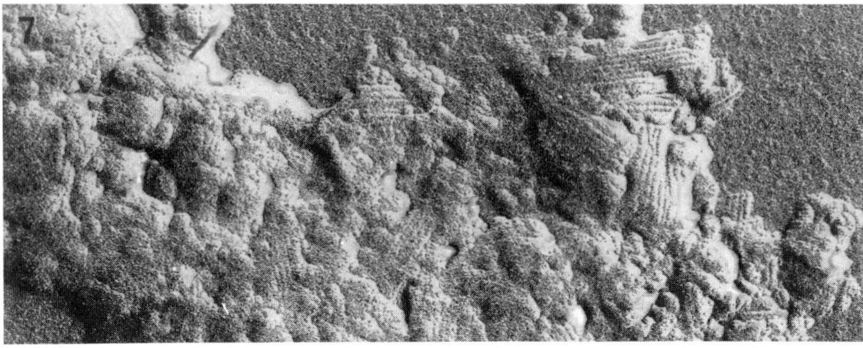
7

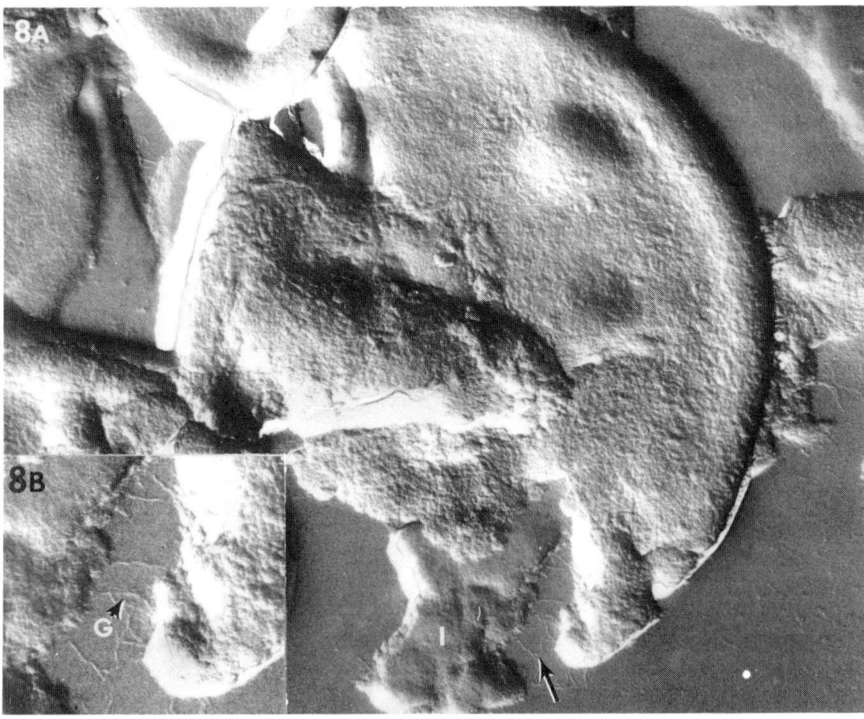
8A
8B
G
I

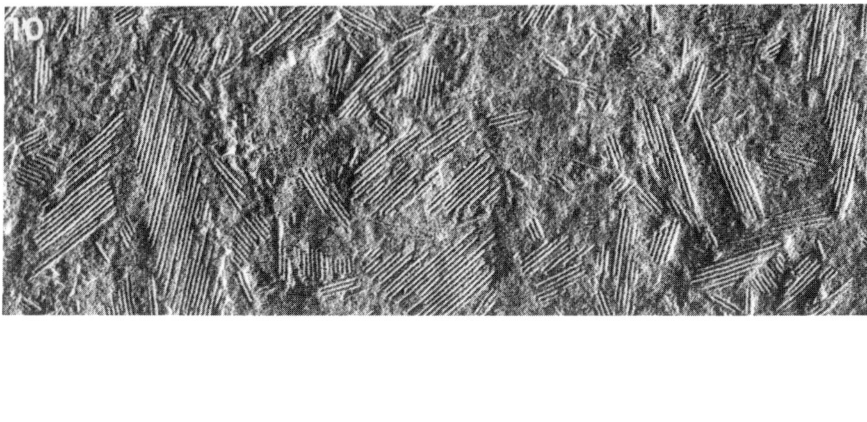
10

Fig. 8. Replica of homogenized inner conidial wall layer (A). Note rough outer and smooth inner (I) surface of this wall layer. Fibrous glucan-like material (G, Fig. 8B) is shown at regions where the wall is fractured (arrow, Fig. 8A). A, X 18,350; B, X 58,000.
Fig. 9. Replica of 0.1 N NaOH treated conidia (12 h, 25°C). Some rodlet fascicles (arrows) are still attached to the wall surface. X 20,500.
Fig. 10. Replica of rodlets isolated from conidia treated with 0.1 N NaOH (c.f. Fig. 9). X 47,500.

---

wall material. The suspension was first centrifuged (365*g*) to remove intact conidia (Fig. 9) which revealed that most rodlets were removed from the wall surface. Some separation of adjacent rodlets within fascicles was apparent. If the intact conidia were then resuspended in 0.1 N NaOH and incubated under the same conditions for another 12 h (total of 24 h), almost all rodlets were removed. However, additional amorphous material was also extracted from the wall. When the conidial free supernatant from the first alkali treatment was centrifuged at 125,000*g* for 2 h and the resulting supernatant then centrifuged at 125,000*g* for 9 h, a pellet was obtained which is shown in Fig. 10. Although some amorphous wall material remains, this pellet contains a much higher percentage of rodlets than Figs. 5 or 6 and was used for chemical analysis (Table 2). The supernatant was concentrated and a replica was prepared from the concentrate. No structural material was visible under the electron microscope.

Wall chemistry

A summary of the differences in chemical composition of the NP-40 extracted conidial wall fraction, sonicated outer wall layer (Fig. 5) and homogenized inner wall layer is presented in Table 1. The values represent averages of data from at least three repetitions of each analytical procedure. The percentage recovery of each cell wall fraction is low which cannot be explained.

TABLE 1

Summary of Chemical Composition of NP-40 Detergent Soluble Wall Fraction, Sonicated Outer Wall Layer and Homogenized Inner Wall Layer of Aspergillus niger Conidia

	Detergent Soluble Fraction (%)	Outer Layer (%)	Inner Layer (%)
Total Carbohydrate	3.5	6.9	28.2
Mannose	(15.6)[a]	(30.8)	(37.4)
Galactose	(48.7)	(44.2)	(27.3)
Glucose	(35.7)	(25.0)	(35.3)
Hexosamine	n.d.[b]	3.3	3.0
Peptides	33.6[c]	34.9[c]	6.9[e]
	(38.0)[d]	(35.6)[d]	
		(38.4)[e]	
Lipids			
Readily extractable lipids	17.0	20.0	8.4
Bound lipids	15.5	10.0	22.0
Ash	1.5	n.d.	n.d.
Recovery	71.1	75.1	68.5

[a]Monosaccharide content expressed as a percentage of total carbohydrate

[b]Not determined

[c]Determined by the Yemm and Cocking (1955) ninhydrin method

[d]Determined by the Kjeldahl-Nessler method (Kabat and Mayer, 1961; Campbell *et al.*, 1970)

[e]Determined by summation of amino acids.

Carbohydrate

The major component of the inner conidial wall layer was carbohydrate which sharply contrasted with the other two wall components analysed by the anthrone procedure in Table 1. Based on glc data, galactose was the principal monosaccharide in the detergent soluble and outer wall fractions but showed the lowest percentage in the inner wall layer. Of interest also is the low percentage of mannose in the detergent soluble fraction.

Hexosamine

As expected, the hexosamine content of the conidial wall layers is quite low (Cole *et al.*, 1979).

Peptides and amino acids

The detergent treated conidia released a large amount of black pigment from all wall fractions examined which was particularly troublesome for the spectrophotometric analyses of the Lowry procedure (Lowry *et al.*, 1951). On the other hand, fewer problems and more consistent data resulted when total protein was determined by the ninhydrin, total nitrogen and amino acid summation methods (Table 1). A surprisingly low protein content based on amino acid analysis was found in the inner wall layer. This is presently under reexamination using other methods of protein determination. The protein composition of the detergent soluble and outer wall fractions were confirmed by several procedures and demonstrated a close similarity (Table 1). The amino acid composition of the 0.1 N NaOH extracted rodlet fraction (Fig. 10) is presented in Table 2. The principal amino acids were aspartic acid and histidine while relatively large amounts of serine, proline, glutamic acid and leucine were also detected. The total protein content of the rodlet fraction, determined by summation of amino acids present, was 8.1%. Different rodlet isolations prepared by this same procedure had similar values for total protein.

Lipids

The lipid compositions of the conidial wall fractions were compared using the Bligh and Dyer (1959) method for readily extractable lipids and the procedure of Bartnicki-Garcia and Nickerson (1962) for bound cell wall lipids. The highest percentage of readily extractable lipids occur-

TABLE 2

Amino Acid and Total Protein Composition of Rodlets Isolated by 0.1N NaOH Extraction Procedure

Amino Acid	Rodlets	
Aspartic acid	87.43[a]	0.012[b]
Threonine	45.43	0.005
Serine	67.56	0.007
Proline	43.80	0.005
Glutamic acid	58.54	0.009
Glycine	81.51	0.006
Alanine	50.22	0.005
Cystine	n.d.[c]	
Valine	34.99	0.004
Methionine	n.d.	
Isoleucine	28.76	0.004
Leucine	49.17	0.007
Tyrosine	10.39	0.002
Phenylalanine	13.17	0.002
Histidine	57.67	0.009
Lysine	19.53	0.003
Arginine	6.54	0.001
Percent Protein/mg Dry Weight[d]		8.1

[a]Values expressed as nmol amino acid per mg dry weight

[b]Values expressed as percent amino acid per mg dry weight

[c]Not detected

[d]Determined by summation of amino acids

red in the detergent soluble and outer wall fractions, while most bound lipid was associated with the inner wall layer (Table 1).

Ash

The ash content of only the detergent soluble fraction was determined. However, based on an earlier investigation of *A. niger* (Cole *et al.*, 1979), the ash of the outer and inner conidial wall layers would be expected low (less than 2%).

DISCUSSION

Although the applications of nonionic, ionic and zwitterionic detergents in isolation of proteins are well documented (Tanner and Boxer, 1972; Tanford *et al.*, 1974; Baron and Thompson, 1975; Gonenne and Ernst, 1978; Helenius *et al.*, 1979) use of detergents in isolation of surface wall components from hydrophobic fungal cells has not been extensively explored. The detergents employed in this study showed different effects on *A. niger* conidia. Sodium dodecyl sulfate and 1-O-n-octyl-β-D-glucopyranoside were the least efficient in wetting the cells. After incubation of conidia in these detergents, the supernatant was yellow to tan which contrasted with the black supernatants obtained after incubation in other detergents used in this investigation. In addition, SDS and 1-O-n-octyl-β-D-glucopyranoside were atypical because no detectable protein was removed from the conidia (determined by the modified Lowry method) after incubation in 1% detergent for 12 h (data not shown). On the other hand, both nonionic detergents (NP-40 and Triton X-100) solubilized significant amounts of protein (Fig. 1) without showing any obvious structural changes to the rodlet layer. More than 30% of the NP-40 soluble fraction was protein while an additional 32.5% was lipid (Table 1). In spite of removal of these components, the conidia remained viable. This suggests that most detergents tested solubilized wall components but do not penetrate to or affect the plasmalemma. The conidial swelling which occurs during NP-40 treatment (Fig. 3) may be due simply to hydration of the dry cell wall. Treatment with 1% SDS, however, showed some inhibition of conidial germination, and may have metabolic effects on *A. niger*.

Freeze-etch replicas of deep-etched *Cladosporium macrocarpum* conidia have revealed that an outermost amorphous layer covers the rodlet fascicles. A morphologically similar layer on the surface of microconidia produced by *Trichophyton mentagrophytes* was estimated to be 15 to 20 nm thick, was extracted by treating the cells with a mixture of 4 or 8 M urea, 1% mercaptoethanol, and 1% SDS, and was identified as a glycoprotein-lipid complex (Wu-Yuan and Hashimoto, 1977). Although an amorphous layer was also observed on the outermost conidial surface of a previously examined strain of *A. niger* (Cole *et al.*, 1979), it was absent from the strain examined in this investigation. Based on comparisons of the two strains, lack of this amorphous wall layer had no apparent effect on viability or hydrophobicity of the conidia.

After treatment of conidia with NP-40, the rodlet fascicles appeared distended (*c.f.* Figs. 2 and 4). It is important to reemphasize, however, that rodlets were not removed from the cell surface by the detergent. One must assume, therefore, that most of the carbohydrate, protein and lipid present in the detergent soluble fraction (Table 1) escaped through the rodlet layer. Evidence was presented earlier that rodlet fascicles are capable of movement and changes in orientation resulting from alterations in turgor pressure of the cell (Cole, 1973; Cole and Samson, 1979). The rodlet layer may represent a thin, flexible and, under certain conditions, semipermeable wall component. Perhaps once the hydrophobic barrier of the conidial surface of *A. niger* is broken down, movement of certain compounds (*e.g.*, low molecular weight proteins) readily occur through the rodlet layer.

Based on electron-microscopic examinations of the sonicated outer conidial wall layer (Fig. 5), it appears that an amorphous, pebbly material is firmly attached to the undersurface of the rodlet fascicles. Comparing the chemical data for this layer (Table 1) with that of the more purified rodlet preparation derived from 0.1 N NaOH-treated conidia (Table 2), it is evident that a large percentage of the amorphous material could be protein and lipid. This suggestion is supported by an earlier analysis of the sheared outer conidial wall layer of *A. niger* (Cole *et al.*, 1979). Attempts to distinguish between protein(s) present in the rodlets and adjacent layer are in progress utilizing gel electrophoresis techniques. Glucans are also represented as structural components of the sonicated outer wall layer. The pebbly material referred to above is reminiscent of the hyphal wall component of *A. niger* and *A. awamori* described by Bobbitt *et al.* (1977) which had been identified earlier as nigeran, a hot water-soluble, linear, alternating (1→3), (1→4)-α-D-glucan (Barker *et al.*, 1953, 1957). The fibrous material (G) shown in Fig. 5 is also comparable to the fibrous glucan and chitin (β(1→3) glucan) complex described by Kreger and Kopecká (1976). Although alkali treatment removed the fibrous material, it failed to totally digest the amorphous matrix below the rodlet layer (Fig. 6). Alkali treatment has also been reported to extract galactosaminogalactan from the hyphal wall of *A. niger* (Bardalaye and Nordin, 1976). The presence of this heteropolysaccharide in the conidial wall of this fungus is suggested by the high percentage of galactose in both the detergent soluble and sonicated outer wall fractions (Table 1). Alkali followed by acid treatment of the sonicated wall preparation apparently destroyed the structural integrity of rodlets (Fig. 7), perhaps by breaking down their proteinaceous core into subunits. This proposal is also under investigation utilizing gel electrophoresis techniques.

The rodlet preparation obtained from 0.1 N NaOH-treated conidia in this study (Fig. 10) appears quite different from that of *Neurospora crassa* released from the conidial surface by vigorously shaking the cells in water (Beever *et al.*, 1979). Individual rodlets of the latter were swollen and indistinct and the fascicles were relatively thick suggesting the presence of an underlying layer. This may partly account for the differences in chemical composition between the two isolates. Lysine consisted of 4.98 mol % in the *N. crassa* rodlet preparation but was less abundant in *A. niger* (Table 2). On the other hand, while histidine represented a major component of *A. niger*, it was apparently a minor amino acid of *N. crassa* rodlets. The most dramatic difference between these two rodlet preparations was the total protein content, both determined by summation of amino acids: 79% for *N. crassa*, 8.1% protein in *A. niger*. Attempts are underway to further purify the rodlet layer of *A. niger*. On the basis of current data, it appears that the rodlets of this fungus contain less protein than reported for other species (Hashimoto *et al.*, 1976; Wu-Yuan and Hashimoto, 1977; Beever *et al.*, 1979). A large part of the remaining rodlet complex of *A. niger* may be composed of lipid (Table 1), perhaps in the form of lipoprotein(s) which is(are) highly resistant to digestive enzyme and chemical treatment.

ACKNOWLEDGMENTS

This work was supported by research grants from the National Institutes of Health (AI-15583) and the National Science Foundation (DEB76-01836). The authors are most grateful to Dr. Yoshinori Nozawa (Gifu Medical School, Japan) and Dr. James R. Brown and Philip Geis (University of Texas) for providing some of the chemical data used in this report, and to Ms. Margaret Boulanger for assistance in preparation of the manuscript.

REFERENCES

Aronson, A. I. and Fitz-James, P. (1976), Structure and morphogenesis of the bacterial spore coat, *Bacteriol. Rev*, 40, 360-402.

Bardalaye, P. C. and Nordin, J. H. (1976), Galactosaminogalactan from cell walls of *Aspergillus niger*, *J.*

Bacteriol. 125, 655-669.

Barker, S. A., Bourne, E. J., O'Mant, D. M. and Stacey, M. (1957), Studies of *A. niger*. VI. The separation and structures of oligosaccharides from nigeran. *J. Chem. Soc.* 2448-2454.

Barker, S. A., Bourne, E. J. and Stacey, M. (1953), Studies of *Aspergillus niger*. I. The structure of the polyglucosan synthesized by *Aspergillus niger* 152, *J. Chem. Soc.* 3084-3090.

Baron, C. and Thompson, T. E. (1975), Solubilization of bacterial membrane proteins using alkyl glucosides and dioctanoyl phosphatidylcholine, *Biochim. Biophys. Acta* 382, 276-285.

Bartnicki-Garcia, S. and Nickerson, W. J. (1962), Isolation, composition and structure of cell walls of filamentous and yeast-like forms of *Mucor rouxii*, *Biochim. Biophys. Acta* 58, 102-119.

Beever, R. E. and Dempsey, G. P. (1978), Function of rodlets on the surface of fungal spores, *Nature (London)* 272, 608-610.

Beever, R. E., Redgwell, R. J. and Dempsey, G. P. (1979), Purification and chemical characterization of the rodlet layer of *Neurospora crassa* conidia, *J. Bacteriol.* 140, 1063-1070.

Bligh, E. G. and Dyer, W. J. (1959), A rapid method of total lipid extraction and purification, *Canad. J. Biochem. Biophys.* 37, 911-916.

Bobbitt, T. F., Nordin, J. H., Roux, M., Revol, J. F. and Marchessault, R. H. (1977), Distribution and conformation of crystalline nigeran in hyphal walls of *Aspergillus niger* and *Aspergillus awamori*, *J. Bacteriol* 132, 691-703.

Bradley, S. G. and Ritzi, D. (1968), Composition and ultrastructure of *Streptomyces venezuelae*, *J. Bacteriol.* 95, 2358-2364.

Bronchart, R. and Demoulin, V. (1971), Ultrastructure de la paroi des basidiosporés de *Lycoperdon* et de *Scleroderma* (Gastéromycètes) comparée à celle de quelques autres spores de champignons, *Protoplasma* 72, 179-189.

Campbell, D. H., Garvey, J. S., Cremer, N. E. and Sussdorf, D. H. (1970), "Methods in Immunology", W. A. Benjamin, Inc., New York.

Cole, G. T. (1973), A correlation between rodlet orientation and conidiogenesis in Hyphomycetes, *Canad. J. Bot.* 51, 2413 -2422.

Cole, G. T. and Kendrick, B.(eds.) (1981), "Biology of Conidial Fungi" Academic Press, New York.

Cole, G. T., Sekiya, T., Kasai, R., Yokoyama, T. and Nozawa, Y. (1979), Surface ultrastructure and chemical composition of the cell walls of conidial fungi, *Exp. Mycol.* 3, 132-156.

Cole, G. T. and Samson, R. A. (1979), "Patterns of Development in Conidial Fungi" Pitman, London.

Dulley, J. R. and Grieve, P. A. (1975), A simple technique for eliminating interference by detergents in the Lowry method of protein determination, *Anal. Biochem.* 64, 136-141.

Frazier, W. and Glaser, L. (1979), Surface components and cell recognition, *Ann. Rev. Biochem.* 48, 491-523.

Gonenne, A. and Ernst, R. (1978), Solubilization of membrane proteins by sulfobetaines, novel zwitterionic surfactants, *Anal. Biochem.* 87, 28-38.

Gould, G. W., Stubbs, J. M. and King, W. L. (1970), Structure and composition of resistant layers in bacterial spore coats, *J. Gen. Microbiol.* 60, 347-355.

Hashimoto, T., Wu-Yuan, C. D. and Blumenthal, H. J. (1976), Isolation and characterization of the rodlet layer of *Trichophyton mentagrophytes* microconidial wall, *J. Bacteriol.* 127, 1543-1549.

Helenius, A., McCaslin, D. R., Fries, E. and Tanford, C. (1979), Properties of detergents. *In* "Methods in Enzymology" (Ed. S. Fleischer and L. Packer). Vol. 56, pp. 734-749. Academic Press, New York.

Hess, W. M., Bushnell, J. L., and Weber, D. J. (1972), Surface structures and unidentifed organelles of *Lycoperdon perlatum* Pers. basidiospores, *Canad. J. Microbiol.* 18, 270-271.

Hess, W. M., Sassen, M. M. A., Remsen. C. C. (1968), Surface characteristics of *Penicillium* conidia, *Mycologia* 60, 290-303.

Holt, S. C. and Leadbetter, E. R. (1969), Comparative ultrastructure of selected aerobic spore-forming bacteria: a freeze-etch study, *Bacteriol. Rev.* 33, 346-378.

Kabat, E. A. and Mayer, M. M. (1961), "Experimental Immunochemistry" Charles C. Thomas, Springfield, Ill.

Kreger, D. R. and Kopecká, M. (1976), On the nature and formation of the fibrillar nets produced by protoplasts of *Saccharomyces cerevisiae* in liquid media: an electron microscopic, X-ray diffraction and chemical study, *J. Gen. Microbiol.* 92, 207-220.

Lowry, O. H., Rosebrough, N. J., Farr, A. L. and Randall, R. J. (1951), Protein measurement with the Folin phenol reagent, *J. Biol. Chem.* 193, 265-275.

Smith, H. (1977), Microbial surfaces in relation to pathogenicity, *Bacteriol. Rev.* 41, 475-500.
Smucker, R. A. and Pfister, R. M. (1978), Characteristics of *Streptomyces coelicolor* A3(2) aerial spore rodlet mosaic, *Canad. J. Microbiol.* 24, 397-408.
Takeo, K. (1976), Existence of a surface configuration on the aerial spore and aerial mycelium of *Micropolyspora*, *J. Gen. Microbiol.* 95, 17-26.
Tanford, C., Nozaki, Y., Reynolds, J. A. and Makino, S. (1974), Molecular characterization of proteins in detergent solutions, *Biochemistry* 13, 2369-2376.
Tanner, M. J. A. and Boxer, D. H. (1972), Separation and some properties of the major proteins of the human erythrocyte membrane, *Biochem. J.* 129, 333-347.
Wehrli, E., Scherrer, P. and Kübler, O. (1980), The crystalline layers in spores of *Bacillus cereus* and *Bacillus thuringiensis* studied by high resolution electron microscopy, *Eur. J. Cell Biol.* 20, 283-289.
Wessels, J. G. H., Kreger, D. R., Marchant, R., Regensburg, B. A. and DeVries, O. M. H. (1972), Chemical and morphological characterization of the hyphal wall surface of the basidiomycete *Schizophyllum commune*, *Biochim. Biophys. Acta* 273, 346-358.
Williams, S. T., Bradshaw, R. M., Costerton, J. W. and Forge, A. (1972), Fine structure of the spore sheath of some *Streptomyces* species, *J. Gen. Microbiol.* 72, 249-258.
Williams, S. T., Sharples, G. P. and Bradshaw, R. M. (1974), Spore formation in *Actinomadura dassonvillei* (Brocq-Rousseau) Lechevalier and Lechevalier, *J. Gen. Microbiol.* 84, 415-419.
Wu-Yuan, C. and Hashimoto, T. (1977), Architecture and chemistry of microconidial walls of *Trichophyton mentagrophytes*, *J. Bacteriol.* 129, 1584-1592.
Yemm, E. W. and Cocking, E. C. (1955), The determination of amino acids with ninhydrin, *Analyst* 80, 209-214.

ADDENDUM

After preparation of the manuscript, analyses were completed on protein content of the 0.1N NaOH rodlet-free supernatant and the data are presented below in Table 3. This supernatant was obtained after the second 125,000*g* centrifugation of the surface wall components (Fig. 10) extracted by 0.1N NaOH treatment (25°C, 12 h, continuous agitation). The percentage of protein in the supernatant was also low (7.5%) but showed both qualitative and quantitative differences in amino acid composition (e.g., cystine, methionine, histidine) in comparison to the rodlet fraction (c.f. Table 2). The significance of these differences in protein composition is currently being investigated by gel electrophoresis.

TABLE 3
Amino Acid and Total Protein Composition of 0.1 N NaOH Rodlet-free Supernatant

Amino Acid	Rodlet-free Supernatant	
Aspartic acid	93.77[a]	0.013[b]
Threonine	39.04	0.002
Serine	68.77	0.007
Proline	40.91	0.008
Glutamic acid	54.57	0.006
Glycine	76.45	0.006
Alanine	42.35	0.004
Cystine	6.70	0.002
Valine	30.90	0.004
Methionine	2.83	0.0004
Isoleucine	23.21	0.003
Leucine	39.06	0.005
Tyrosine	18.88	0.003
Phenylalanine	13.35	0.002
Histidine	38.37	0.006
Lysine	12.46	0.002
Arginine	8.41	0.002
Percent Protein/mg Dry Weight[c]		7.54

[a]Values expressed as nmol amino acid per mg dry weight

[b]Values expressed as percent amino acid per mg dry weight

[c]Determined by summation of amino acids

III. NUCLEAR CONTROLS

NUCLEAR CONTROLS: DISCUSSANT'S INTRODUCTION

S. K. Dutta

Department of Botany, Howard University, Washington, D.C. 20059, U.S.A.

Fungal genomes are approximately ten-fold larger than the bacterium *Escherichia coli* (Dutta, 1974). This accounts for approximately 20 to 25,000 total genes in the haploid genomes of fungal spores. Out of these, 18 to 20,000 genes are nuclear and are located in the haploid chromosomes (Marzluf, 1977; Timberlake, 1980) of fungal spores. Dutta and Chaudhuri (1976) have shown that about 15% of total *Neurospora crassa* DNA sequences anneal with RNAs isolated from conidia. This indicates that a good portion of the nuclear genes are perhaps doing the "housekeeping" functions in the dormant conidial cells of *N. crassa*. Among the "housekeeping" genes there may be included those genes which are transcribed into rRNAs and tRNAs. Very little information is available on how many genes actually participate in morphogenetic changes in conidial and ascospore cells. Recently, Timberlake (1980) has shown in *Aspergillus nidulans* about 26% of the genome (*i.e.*, about 5600-6000 unique poly A containing RNA sequences) is expressed in vegetative cells, and 4% (*i.e.*, about 1000 new poly A containing RNA sequences) accumulate during sporulation which are absent from spores.

These and other studies (Smith, 1978) demonstrate that at sporulation there is a differential control of gene expression. Such differential gene activation depends on the rate of synthesis of mRNA which limits the rate of enzyme synthesis. A generalized scheme of gene expression can be presented as follows:

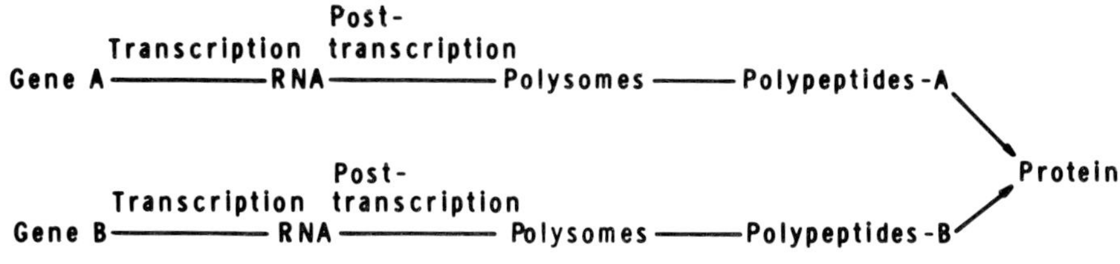

The initial stage in the gene expression is the transcription of certain nucleotide sequences (genes) by an enzyme RNA polymerase to give rise to RNA molecules whose base sequences are complementary to the transcribed strains of the DNA sequence. This RNA molecule (mRNA) may eventually pass from the nucleus to the cytoplasm and act as a template for the specification of amino acid sequences in the protein. The RNA synthesized on the DNA template may be modified before passing into the cytoplasm. If fungal RNA contain additional insertion sequences (exons), besides information sequences (introns), the former segments are likely to be spliced out during this cytoplasmic transport. Thus, it is believed that the large molecules of the nuclear RNA may be broken down into fragments until the functional molecules are formed and there may be a control mechanism that operates in such processing of nuclear RNA. The modified nuclear RNA (mRNA) moves into the cytoplasm after a series of reactions, ultimately giving rise to polypeptide chains. The bases interact by complementary codon: anticodon base pairing with the tRNA molecules each of which is specifically attached with a single amino acid residue. A number of proteins associated with the ribosomes catalyse the polymerization of the amino acids to form polypeptide chains. During this process of translation, several ribosomes are attached to the mRNA molecule. The number depends upon the length of the mRNA molecule. Aggregates of the ribosomes, mRNA and nascent polypeptide chains, are known as polysomes.

The attachment of the ribosomes to a mRNA molecule at the initial stages of protein synthesis involves certain proteins known as initiation factors, while at the end of the synthetic cycle, some factors are also needed for the release of the completed polypeptides.

All the fungal spores studied thus far (Uecker, 1979) have been shown to possess authentic chromosomes containing typical eukaryotic chromatin materials having DNA as templates. As has been stated by Maclean (1976), DNA sequences (genes) themselves exert controls at the transcriptional level which can be interpreted and termed as "gene level control". This is the subject of my topic and will be discussed later in this session.

A great deal of knowledge is already available (Schmit and Brody, 1976; Brambl *et al.*, (1978) on various morphological and biochemical events involved in germinating conidia. Axelrod *et al.* (1973) have shown genetic control of developmental competence in *A. nidulans*. According to them, fungal spores undergo an important developmental change before induction and they call this change "developmental competence" for that cell. They have identified at least "one nuclear gene mutation" which can affect the time at which a cell becomes competent. Due to the practical difficulty in analyzing developmental competence experimentally not much work was reported earlier. Dr. Champe will provide us with details on recent studies of this nature.

During the last four years (1976-1980) several powerful new methods have been developed to study DNA and RNA sequencing, molecular cloning of DNA sequences, and gene regulation. These new techniques are now being utilized to study fungal spore morphogenesis. Thus far there are few reports on this subject. Nevertheless, during this same period several reports have been presented on the role of nucleic acids on sporulation and spore germination. Details on this subject will be discussed by Dr. Van Etten in the accompanying chapter on nucleic acid.

As an introduction to these chapters on nuclear controls, I intend to present some general aspects of fungal spore morphogenesis such as (1) evidence of nuclear controls using fungal spore developmental mutants, and (2) notable examples of positive and negative controls of fungal genes.

EVIDENCE OF NUCLEAR CONTROL IN FUNGAL SPORE MORPHOGENESIS

A typical example of nuclear control in spore morphogenesis is the well-known developmental phase-specific mutants. There are at least seven different mutants that interrupt discrete steps in conidiogenesis of *N. crassa* at the following stages: (*i*) conidiophore formation, (*ii*) nuclear migration into the conidiophore tips, (*iii*) swelling of the conidiophore, (*iv*) conidiophore budding, which results in a chain of presumptive conidia, and (*v*) septation into individual spores (Selitrennikoff *et al.*, 1974). All of these seven genes are nuclear and are located on six chromosomes of *N. crassa*.

Most of the evidence of nuclear controls in fungal spore morphogenesis comes from the studies using developmental mutants of different kinds of fungal spores. An extensive review has been given by Turian (1976) on genetic controls of conidia and ascospores. Scott (1976) and Mishra (1977) have reviewed similar studies made with *Neurospora* mutants. I shall review here certain evidence that relates to nuclear control of fungal spore morphogenesis that has been reported since 1976.

Using a temperature sensitive mutant of *N. crassa*, Serna and Stadler (1978) have shown that nuclei in a population of dormant conidia are arrested at various points in the nuclear division cycle. Even when DNA synthesis is inhibited by hydroxyurea, about 30% of the fungal nuclei divide (Loo, 1976). Extensive reviews of similar studies and correlation with cell cycles and nuclear division have been presented by Ojha and Dutta (1978). The use of different inhibitors of DNA synthesis such as actinomycin-D, hydroxyurea, and cordycepin, are not always accurate since they have additional effects (Brambl *et al.*, 1978) on cellular biosynthesis or sometime do not penetrate the spore.

Horowitz *et al.* (1976) have isolated and identified a conidial germination factor in *N. crassa*. Zakharova *et al.* (1976) have studied changes in the nuclear material of the germinating conidia of *Penicillium chrysogenum*.

Most of the recent studies on this subject have been done with the yeast *Saccharomyces cerevisiae*. Rubin and Blamire (1978) have shown that a single recessive nuclear gene mutation in *S. cerevisiae* appears to be conditionally deficient in nuclear DNA metabolism. Lucchini *et al.* (1978) have reported that the inhibition of sporulation seems to occur at a very early step in meiosis. A regulator gene, different from *gal*-3, controlling growth and development of spores of *S. cerevisiae*, was reported by Bie'n *et al.* (1978). Sibirnyyi *et al.* (1977) have reported the selection of a strain of the yeast *Pichia guilliermondii* that is capable of forming a significant quantity of spores. Nasmyth (1979) has described the DNA ligase activities of a wild type and temperature sensitive lethal cde 17 mutants of *Schizosaccharomyces pombe*. He has shown that these mutations are located in the structural gene for DNA ligase. Betz (1979) has described a mutant in the yeast *S. cerevisiae* with loss of sporulation ability and low proteinase-A levels. It was suggested that the proteinase-A deficiency was due to regulatory mutation. Luong *et al.* (1978) have described mutational variability of colony morphology in the yeast *S. cerevisiae*. Esposito and Esposito (1978) proposed a working model of genetic regulation and integration processes using thermosensitive *spo* (sporulation-deficient) mutants of *S. cerevisiae*. They have studied cytological, genetical, and physiological landmark events of meiosis and ascospore formation. The presence of a complete mitochondrial genome has been shown to be required for sporulation in *S. cerevisiae* (Newlon and Hall, 1978). Johnston and Singer (1978) have given evidence that RNA metabolism plays a key role in the regulation of the cell cycle of the yeast *S. cerevisiae*.

Garber *et al.* (1978) have studied the polymorphism for sporidial colony color and sporidial nutritional requirement in the smut *Ustilago violacea*. Tani *et al.* (1977) have shown that at 28 C basidia of a mutant of *Coprinus macrorhizus* became mononucleate and meiosis stopped at meta-anaphase 1. At 20 C this mutant produced approximately 1/3 the normal size of basidiospores. Day (1979) has hypothesized that the mating type locus acts as a master developmental switch controlling entry into at least 3 distinct morphogenetic pathways, *viz.*, vegetative budding, conjugation, and sexual differentiation.

Studies on nuclear control of morphogenesis of such fungal spores as zygospores, zoospores, and myxomycete spores, conducted after 1976 are few in number. Orlowski and Sypherd (1978) have studied the regulation of macromolecular synthesis during germ tube emergence from *Mucor racemosus* sporangiospores. Genther and Borgia (1978) have studied heterokaryon formation in the same species. Studies on the genus *Allomyces* have revealed a relationship between rDNA and male differentiation (Ojha and Turian, 1978). There are also reports on nuclear control of morphogenesis in the myxomycetes and cellular slime molds (Bonner, 1969; Olive, 1975; Firtel, 1972)

NOTABLE EXAMPLES OF POSITIVE AND NEGATIVE CONTROLS OF FUNGAL GENES

Metzenberg (1979) has calculated that in *Neurospora* the concentration of each structural gene is perhaps 2×10^{-12}M when compared to 2×10^{-9}M in the bacterium *E. coli*. Since *Neurospora* has an 8-10-fold greater number of structural genes than *E. coli*, there is less regulatory demand for inactivation of genes, even though regulatory mechanisms for inactivation of genes are more complicated. This is typical for positive controls. By comparison, in negative controls regulatory macromolecules actively hold most genes in check.

Marzluf (1977) and recently Metzenberg (1979) have given excellent comprehensive reviews of several well-known control systems in fungi such as those of aromatic amino acid metabolism. As described in *Neurospora* and in yeast, the arom cluster is a single large structural gene (Marzluf, 1977) that encodes a single polypeptide chain. This large polypeptide (mol. wt. 115,000) is pentafunctional, apparently having a common control.

Genes encoding the enzymes for quinate metabolism in *Neurospora* (Vapneck *et al.*, 1977) occur as a regulated unit like those in the "lac-operon" system of *E. coli*. In this quinate system the first three enzymes for catabolism of quinic acid and shikimic acid are encoded by three tightly linked structural genes, adjacent to a fourth regulatory gene which encodes a protein that acts in a positive manner to activate transcription of the structural genes of the cluster.

In *S. cerevisiae* the three enzymes of galactose metabolism are also specified by a cluster of three tightly linked structural genes, *gal*-1, *gal*-7 and *gal*-10. Expression of those *gal* structural genes is controlled by at least two regulatory loci, *gal*-1 and *gal*-4.

In the control of the phosphorus acquisition system as described by Metzenberg (1979) and colleagues, it has been shown that at least three distinct regulatory genes control the synthesis of several genes encoding phosphorus-related enzymes. It is interesting to note that in this system there is indication of negative action of a control gene (nuc-2) because of its position in the sequential series of regulatory steps.

There are other regulatory systems of fungi which have been reported to include sequential regulatory steps with a combination of positive and negative signals. These include those for the nitrogen acquisition system (Reinert and Marzluf, 1975), acid phosphatase (Toh-e and Ishikawa, 1971), sulfur-controlled enzymes, and nucleus restricted control (Marzluf, 1977). The isoleucine and valine pathways require four enzymes which are coded by nuclear genes but are located as an aggregate in the mitochondrial matrix (Cassidy *et al.*, 1972). For these and other examples of regulatory genes in fungi, consult the excellent review of Marzluf (1977).

Thus, it is apparent that for fungi in general, including examples given which concern specifically fungal spores, the nuclear control of differentiation is beyond any question. Being primitive eukaryotic organisms, fungi do have examples of regulatory systems of both bacteria-type operons, with mostly negative controls, as well as mostly positively controlled regulatory systems which are typical of higher eukaryotes. In the following section we shall discuss a few recent reports which indicate that genomic organizations in fungi suggest that gene level controls for fungal morphogenesis are apparently different from those in higher eukaryotic control systems. Several aspects of these and other studies on nuclear controls will be covered in the next three chapters.

REFERENCES

Axelrod, D. E., Gealt, M. and Pastushok, M. (1973). Gene control of developmental competence in *Aspergillus nidulans*, *Dev. Biol.* 34:9-15.

Betz, H. (1979). Loss of sporulation ability in a yeast mutant with low proteinase A levels, *FEBS Letters* 100: 171-174.

Bie'n, M., Kotody'nski, J., Lachowicz, T. M. (1978). Respiratory mutation and galactose metabolism in yeast *Saccharomyces cerevisiae*, *Acta Microbial Pol.* 27:193-202.

Bonner, J. T. (1969). "The Cellular Slime Molds". Princeton University Press, Princeton.

Brambl, R., Dunkle, L. D. and Van Etten, J. L. (1978). Nucleic acid and protein synthesis during fungal spore germination. *In* "The Filamentous Fungi". (Eds. J. E. Smith and D. E. Berry) Vol. 3, pp. 94-118. Edward Arnold Publishers Ltd., London.

Cassidy, W. E., Leiter, E. H., Berquist, A., Wagner, R. P. (1972). Separation of mitochondrial membranes of *N. crassa* II. Submitochondrial localization of the isoleucene valine biosynthetic pathway, *J. Cell. Biol.* 53:66-72.

Day, A. W. (1979). Mating type and morphogenesis in *Ustilago violacea*, *Bot. Gaz.* 140:94-101.

Dutta, S. K. (1974). Repeated DNA sequences in fungi, *Nucleic Acid Res.* 1:1411-1419.

Dutta, S. K. and Chaudhuri, R. K. (1975). Differential transcription of nonrepeated DNA during development of *N. crassa*, *Dev. Biol.* 43:35-41.

Esposito, M. S. and Esposito, R. E. (1978). Aspects of genetic control of meiosis and ascospore development inferred from the study of Spo (Sporulation-deficient) mutant of *S. cerevisiae*, *Cell* 33:93-102.

Firtel, R. A. (1972). Changes in the expression of single-copy DNA during development of the cellular slime mold *Dictyostelium discoideum*, *J. Mol. Biol.* 66:363-377.

Garber, E. D., Baird, M. L. and Weiss, I. M. (1978). Genetics of *Ustilago violacea* II. Polymorphism of color and nutritional requirements of sporidia from natural population, *Bot. Gaz.* 139:261-265.

Genther, F. J. and Borgia, P. T. (1978). Spheroplast fusion and heterokaryon formation in *Mucor racemosus*, *J. Bact.* 134:349-352.

Horowitz, N. H., Charlang, G., Horn, G. and Williams, N. P. (1976). Isolation and identification of the conidial germination factor of *Neurospora crassa*, *J. Bact.* 127: 135-140.

Johnston, G. C. and Singer, R. A. (1978). RNA synthesis and control of cell division in the yeast *S. cerevisiae*, *Cell* 14:951-958.

Loo, M. (1976). Some required events in conidial germination of *N. crassa*, *Dev. Biol.* 54:201-213.

Lucchini, G., Biraghi, A., Carbone, M. L., de Scrilli, A., and Magni, G. E. (1978). Effect of mutation in the aromatic amino acid pathway on sporulation of *Saccharomyces cerevisiae*, *J. Bact.* 136:55-62.

Luong, L. D., Ivano, E. L. and Braced, A. A. (1978). Mutational variability of colony and cell morphology in the yeast *Saccharomyces cerevisiae*, *Genetika* 14:1543-1551.

Maclean, N. (1976). "Control of Gene Expression". Academic Press, London.

Marzluf, G. A. (1977) Regulation of gene expression in fungi. *In* "Regulatory Biology" (Eds. J. C. Copeland and G. A. Marzluf), pp. 196-242. Ohio State University Press, Columbus.

Metzenberg, R. L. (1979). Implications of some genetic mechanisms in *Neurospora*, *Microbiol. Rev.* 43:361-383.

Mishra, N. C. (1977). Genetics and biochemistry of morphogenesis in *Neurospora*, *Adv. Genetics* 19:341-405.

Nasmyth, K. A. (1979). Genetic and enzymatic characterization of conditional lethal mutants of the yeast *Schizosaccharomyces pombe* with a temperature sensitive DNA ligase, *J. Mol. Biol.* 139:272-284.

Newlon, M. C. and Hall, B. D. (1978). Inhibition of yeast sporulation by ethidium bromide, *Molec. Gen. Genet.* 165: 113-114.

Ojha, M. and Dutta, S. K. (1978). Nuclear control of differentiation. *In* "The Filamentous Fungi." (Eds. J. E. Smith and D. E. Berry) Vol. 3, pp. 8-27. Edward Arnold Publishers, Ltd., London.

Ojha, M. and Turian, G. (1978). Loss of rRNA genes during a phenotypic, masculinizing mutation in *Allomyces arbuscula*, *Molec. gen. Genet.* 166:225-227.

Olive, L.S. (1975). "The Mycetozoans". Academic Press, New York.

Orlowski, M. and Sypherd, P.S. (1978). Regulation of macromolecular synthesis during hyphal germ tube emergence from *Mucor racemosus* spore, *J. Bact.* 137:76-83.

Reinert, W.R. and Marzluf, G.A. (1975). Regulation of the purine catabolic enzymes in *N. crassa*, *Arch. Biochem. Biophys.* 166:565-574.

Rubin, B.Y. and Blamire, J. (1978). Detection of aberrant nuclear DNA metabolism in conditional mutant of *Saccharomyces cerevisiae*, *Molec. gen. Genet.* 167:177-184.

Schmit, J.C. and Brody, S. (1976). Biochemical genetics of *N. crassa* conidial germination, *Bacteriol. Rev.* 40:1-41.

Scott, W.A. (1976). Biochemical genetics of morphogenesis in *Neurospora, Annu. Rev. Microbiol.* 30:85-104.

Selitrennikoff, C.P., Nelson, R.E. and Siegel, R.W. (1974). Phase-specific genes for macroconidiation in *N. crassa, Genetics* 78:679-690.

Serna, L. and Stadler, D. (1978). Nuclear division cycle in germinating conidia of *N. crassa, J. Bact.* 136:341-351.

Sibirnyi, A.A., Zharova, V.P., Kshanovskaia, B.V. and Shavlovskii, G.M. (1977). Selection of a genetic strain of *Pichia guilliermondii* yeast capable of forming a significant quantity of spores, *Tsitol. Genets.* 11:330-333.

Smith, J.E. (1978). Asexual sporulation in the filamentous fungi. *In* "The Filamentous Fungi". (Eds. J.E. Smith and D.R. Berry) Vol. 3, pp. 214-239. Edward Arnold Publishers Ltd., London.

Tani, K., Knroiwa, T., Takemaru, T. (1977). Cytological studies on sporeless mutant in the basidiomycete *Coprinus macrorhizus, Bot. Mag.* 90:235-245.

Timberlake, W.E. (1980). Developmental gene regulation in *Aspergillus nidulans, Dev. Biol.* (In press).

Toh-e, A. and Ishikawa, T. (1971). Genetic control of the synthesis of repressible phosphatases in *N. crassa, Genetics* 69:339-351.

Turian, G. (1976). Spores in Ascomycetes : their controlled differentiation. *In* "The Fungal Spore" : Form and Function". (Eds. D.J. Weber and W.M. Hess), pp. 715-786, Wiley Interscience Publisher Inc., London.

Uecker, F.A. (1979). Development and cytology of *Anxiella endodonta, Bot. Gaz.* 140:310-317.

Vapneck, D.J., Hautala, J.A., Jacobson, J.W., Giles, N.H. and Kushner, S.R. (1977). Expression in *E. coli* K-12 of the structural gene for catabolic dehydroquinase of *N. crassa, Proc. Natl. Acad. Sci. U.S.A.* 71:4804-4812.

Zakharova, G.M., Bartoshevich, E. and Gulak, P.V. (1976). Changes in the nuclear material of the germinating conidia of *Penicillium chrysogenum, Antibiotiki* 21:881-885.

GENE LEVEL CONTROL

S. K. Dutta*, Lafayette Frederick* and W. Lena Austin#

**Department of Botany and #Department of Microbiology
Howard University, Washington, D.C. 20059 U.S.A.*

Gene regulation of fungal sporulation may be manifested in several identifiable ways. These ways include an ordering of biochemical events, the development and behavior of organelles, and the phenotypic expression of morphogenetic characters. Considerable attention has been given to the biochemical nature of gene expression during spore formation and germination. Less attention has been given to organelle genesis and behavior and the expression of morphogenetic characters. In this presentation we shall look at recent information on biochemical and certain morphological aspects of gene regulation and the nature and behavior of fungal genes.

GENE CONTROLLED MORPHOLOGICAL AND BIOCHEMICAL CHANGES

An understanding of various morphological and biochemical changes involved in fungal spore morphogenesis is essential if one is to identify gene control mechanisms involved in fungal development. Schmit and Brody (1976) and Brambl *et al.* (1978) have prepared extensive reviews of studies that have been conducted on macromolecule synthesis during spore germination. They have pointed out that in spore germination the major events are hydration and germ tube formation. The first event, hydration, occurs within a period of two minutes. Shortly thereafter, RNA synthesis begins. Both the synthesis and the processing of rRNA occurs quite rapidly. DNA synthesis is a later event. It begins 2-3 hr after the initiation of spore germination and just before germ tube formation. Neither DNA synthesis nor nuclear division appear to be essential for the initiation of spore germination.

Grange and Turian (1978) have reported that at 25 C *Neurospora crassa* germinating conidia show a maximum incorporation of ^{3}H-adenosine into DNA 6-7 hr after inoculation, when outgrowth of the germ tube is visible. When conidia were heat-treated at 46 C for 15 hr these workers found a maximum incorporation of DNA within 1 hr after inoculation followed by a decline of DNA synthesis and renewed synthesis when premature conidiation occurred. Such microcycle conidiation by heat treatment, which has also been reported in *Aspergillus niger* (Anderson and Smith, 1971), allows for an enlargement of spores while preventing germ tube formation. Such synthesis in microcyclically differentiating conidia can be sharply reduced by two specific inhibitors of DNA synthesis, *viz.*, hydroxyurea and 5-fluorodeoxyuridine (Grange and Turian, 1978).

It is interesting to note that polysome assembly and protein synthesis (Schmit and Brody, 1976) starts early (within 30 min) in the process of spore germination, *i.e.*, earlier than the initiation of DNA synthesis. Sturani *et al.* (1979) have studied the level of turnover of polyadenylate containing RNA in *N. crassa* in different steady states of growth. They found that the half lives of polyadenylated RNA vary from 30-60 min and the rate of its synthesis decreased. The rRNA increased as a function of the square of the number of duplications per hour. They concluded that in *Neurospora* the growth rate does not depend on the rate of synthesis of mRNA but rather on the rate of synthesis of rRNA. This suggests that protein synthesis is probably controlled by regulating the assembly of ribosomes into polysomes in fungi.

The regulatory mechanisms noted thus far are primarily those associated with spore germination and this important subject continues to receive considerable attention. Much less information is available on gene control mechanisms that regulate spore ontogeny, especially as it relates to the development of genetically stable morphogenetic characters. Information on some aspects of this problem is presented next.

Fungal sporulation is the end product of an intricate series of genetically regulated events. It involves organelle genesis and degradation, membrane reorganization and distribution, macromolecular assembly and accumulation, and for many fungi, cell wall formation. A distinctive morphologic unit that possesses well-defined and genetically stable phenotypic characters is commonly the result of these events. For many spores, particularly those with walls, their form or the surface features of the wall distinguishes the taxon that produces them. In such cases spore ontogeny reveals itself as a sharply circumscribed process that is obviously regulated by genes.

Fundamental to the sporulation process is membrane

activity. This activity may involve the synthesis or assembly of special macromolecules, the structural compartmentalization of cellular material within developing spores, and the transport and deposition of wall-forming materials. Although much attention has been given to biochemical mechanisms in sporulation, critical attention has not been given to membrane regulation during the spore formation event or to the ways in which membranes themselves may regulate the expression of morphogenetic characters of spores. We have been engaged in studies on this latter aspect of spore ontogeny and membrane control (Frederick and Austin, 1977) and a brief report on our observations is reported here.

The walls of many kinds of fungal spores are often furnished with distinctive patterns of surface ornamentation. Such topographic features of the spore wall frequently distinguishes the taxon. An excellent example of this situation is provided by the genus *Neurospora*. Ascospores in this genus have a longitudinally ribbed wall surface pattern (Fig. 1) and this feature distinguishes *Neurospora* from such closely related genera as *Sordaria*, which has ascospores with smooth walls, and *Gelasinospora*, which has ascospores with pitted walls. Since the report of Lindegren and Scott

Fig. 1. Scanning electron micrograph of ascospores of Neurospora terricola, a homothallic species. X4500

(1937) on the occurrence of rib-forming bodies in ascospores of *N. tetrasperma*, it has been presumed that some type of intrasporal organelle was responsible for the development of the characteristic wall surface pattern found in *Neurospora* ascospores. Although we have found that intrasporal membrane organelles are associated with the rib-forming process in ascospores of *Neurospora*, our findings do not support either Lindegren and Scott's interpretation, nor Dodge's (1957) later corroboration of the presence of rib-forming bodies. Instead, we have found that the rib-vein pattern of *Neurospora* ascospores is determined by a system of longitudinally oriented endoplasmic reticulum (ER) profiles that become appressed to the inner surface of the spore plasmalemma during the early stages of ascospore ontogeny (Figs. 2-3).

When developing asci of a homothallic species of *Neurospora*, such as *N. dodgei*, are examined under transmission electron microscopy, we have noted that soon after ascospore delimitation the spore contains an extensive system of ER profiles, abundant mitochondria with prominent cristae, numerous small vesicles, and several lipid bodies of uniform size (Fig. 2). At this stage the spore plasmalemma is covered externally by a thin electron translucent wall layer. Also, during this stage, profiles of ER cisternae migrate to the region of the spore plasmalemma and become positioned in a highly precise and regular manner at intervals along the inner surface of the membrane (Figs. 2-3). These cisternae, that we have termed intrasporal membrane plaques (IMP) (Frederick and Austin, 1977), subsequently become appressed to the inside of the plasmalemma. With IMP in place deposition of the rib layer of the spore wall commences on the outer surface of the plasmalemma in those areas on the membrane surface where there is no underlying IMP (Fig. 3). The IMP system, therefore, delineates the rib-layer configuration of *Neurospora* ascospores, serving apparently as a barrier that prevents the deposition of rib-wall material in regular and precisely spaced locations on the outer surface of the spore membrane. Those regions on the spore plasmalemma opposite IMP become the intercostal veins (Austin *et al.*, 1974) of the ascospore wall when the rib layer is fully developed (Fig. 4).

Intrasporal vesicle activity appears to be responsible for the initial deposition and build-up of ribs in the rib layer (Fig. 3). The submicroscopic dimensions of these vesicles make it unlikely that they would represent the rib-forming bodies of Lindegren and Scott (1937) or of Dodge (1957). We suspect that the numerous refractile lipid

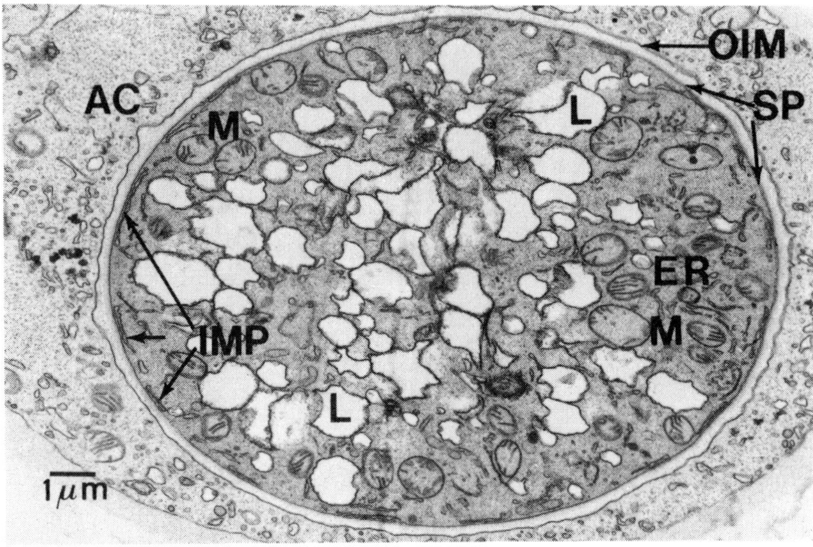

Fig. 2. Median transection of newly delimited ascospore of N. dodgei. Paraformaldehyde-potassium permanganate fixation. Ascal cytoplasm-AC; endoplasmic reticulum-ER; intrasporal membrane plaques-IMP; lipid bodies-L; mitochondria-M; outer investing membrane-OIM; spore plasmalemma-SP.

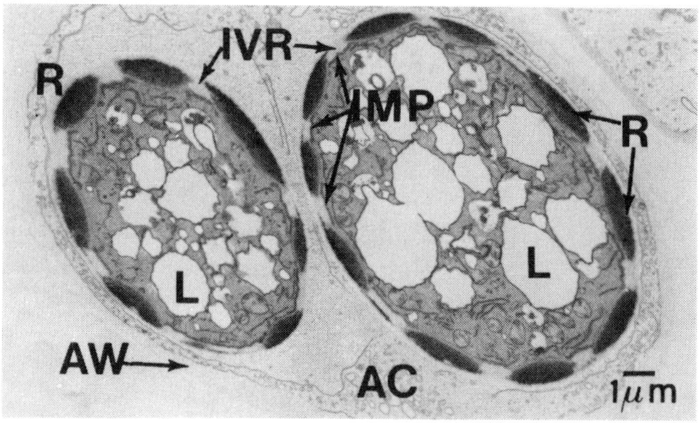

Fig. 3. Near-median transection of young ascospores of N. dodgei at early stage of rib development. Paraformaldehyde-potassium permanganate fixation. Ascal cytoplasm-AC; ascus wall-AW; intercostal vein region-IVR; intrasporal membrane plaques-IMP; lipid bodies-L; ribs-R.

bodies that are present in ascospores during the early stages of rib-layer development might have been mistaken by these investigators as intrasporal bodies from which the ribs were derived.

Our observations on rib layer development in *N. dodgei* clearly show that membrane reorganization and alignment within the spore system governs the expression of the primary taxonomic characteristic of species in the genus. In *N. dodgei*, and other homothallic species that we have studied, membrane profiles (IMP) appear with regular constancy in newly delimited ascospores, become precisely positioned at sites inside the spore plasmalemma, and persist until rib deposition has ended. Preliminary observations that we have made of ascospores at later developmental stages, when the other layers of the ascospore wall form, *i.e.*, the dense melanized layer, and the electron transparent innermost layer (Austin *et al.*, 1974; Lowry and Sussman, 1968) have revealed that IMP disappear after ribs are fully developed. This appearance of IMP, their migration within the sporoplasm, and subsequent precise positioning suggests the presence of a highly ordered genetic regulatory mechanism. In the vast body of literature on *Neurospora*, reports of variations in ascospore color, size, shape, number of germ pores, germination capability, and biochemical

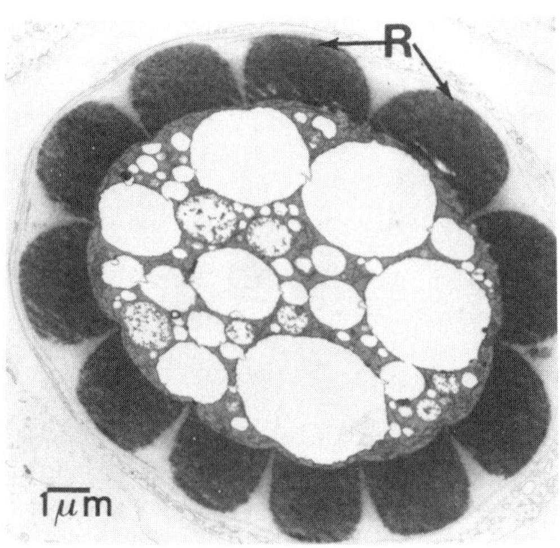

Fig. 4. Transection of ascospore of N. dodgei with fully developed ribs. Two other main wall layers form centripetal to rib layer during later stages of ascospore maturation. Paraformaldehyde-potassium permanganate fixation. Ribs-R.

processes have been shown to be regulated by genes but we are not aware of any report on any substantive variation in the wall surface pattern in the genus. The ascospore rib-vein pattern must obviously be a highly stable, genetically regulated feature whose expression is ordered by the behavior of a membrane system, the IMP.

Distinct membrane configurations have been implicated in the determination of wall surface patterns in spores of other fungi but not in the manner described here. For example, Hess and Weber (1976), have shown that the reticulate surface pattern of *Tilletia caries* and *T. controversa* teliospores corresponds to a prior invagination pattern in the plasmalemma of the young spore; Moore (1963) suggested that an internal ER system might be responsible for the reticulate markings on the wall of *Ascodesmis* ascospores; Aldrich (1974) found spine distribution on spores of the myxomycete *Didymium iridis* to be related to protrusions of the spore membrane and suggested that spine placement in these spores was no doubt genetically determined; Mims (1972) associated the development of wall ornamentations on spores of the myxomycete *Arcyria cinerea* with underlying ER cisternae; and Littlefield and Bracker (1971) implicated ER cisternae in the initiation of spines on urediospore walls of *Melampsora lini*.

ANALYSIS OF FUNGAL SPORE DNA CONSTANCY

As mentioned before, the knowledge of fungal spore DNA sequences is essential for the understanding of gene level controls. Details on fungal spore genomes are described by Van Etten *et al.* in another chapter of this volume. Here we give some of our observations on the analysis of DNA constancy studies using fungal cell types.

Recently developed procedures for the isolation of purified nuclei (Hautala *et al.*, 1977) enable us to gain more insights into genomes of fungal spores. The few data available on fungal nuclear DNAs suggest that most of the repetitive DNA sequences reported so far in fungi are perhaps nonnuclear. About 3-5% of the repetitive DNA sequences are in the chromosomes, and have high (55%) G+C mole percents. The complexity of these nuclear repetitive DNA sequences (6-9 X 10^6 daltons) are like those of the ribosomal RNA cistrons reported by Chattopadhyay *et al.* (1972). Table 1 summarizes characteristics of repetitive DNA sequences isolated from different cell fractions of *Neurospora* and *Aspergillus*.

Chemical analysis (Minagawa *et al.*, 1959) and DNA:DNA hybridization studies (Dutta and Chaudhuri, 1975) of dormant

TABLE 1

Characteristics of Repeated DNA Sequences in Cell Fractions of Fungal Genomes

Organism	Percent Repeats based on ^{3}H-cpm counts	G+C mole %	Calculated mol wt in daltons
Neurospora crassa[1] Conidia (Nuclei)	3-4	57	8-9 X 10^6
N. crassa Conidia (whole cell)	20-25	55	1-4 X 10^7
N. crassa Mycelia (Nuclei)	3-4	58	——
N. crassa Mycelia (Mitochondria)	100	32	4 X 10^7
N. crassa Mycelia (Post-Mito-chondrial)	100	60	8-9 X 10^6
Aspergillus nidulans[2] Mycelia (Nuclei)	2-3	—	6.6 X 10^6

[1]Dutta *et al.* (1980)
[2]Timberlake (1978)
Genome sizes and estimates of percents repetitive DNA sequences are calculated on the principle of DNA:DNA reassociation kinetics (Dutta, 1973). Repetitive DNA sequences were isolated using the procedure described by Dutta (1974).

conidia, germinated conidia, and mycelia of *Neurospora crassa* did not show any significant differences in nucleotide sequences. Chemical analysis of DNA in developing conidia of *Aspergillus oryzae* (Tanaka *et al.*, 1965) and *Botryodiplodia theobromae* (Dunkle and Van Etten, 1972) also confirm that DNA sequences remain the same during fungal spore morphogenesis.

To the contrary, Chaudhuri *et al.* (1973), using DNA:DNA annealing procedures, reported that DNA sequences of the *N. crassa* slime mutant was slightly different from the wild type DNA. They did not, however, identify any specific DNA sequences which might have changed in the slime mutant. Myers and Cantino (1971) showed some changes in DNA sequences during germination of zoospores, where γ-particles disappear during zoospore germination. Recently Ojha and Turian (1978) have presented evidence of rRNA gene loss in

Allomyces arbuscula. They have shown by saturation hybridization of ^{3}H-rRNA to homologous DNAs of *A. arbuscula*, and to DNA of other *Allomyces*, that approximately a 50-60% reduction in the cistrons coding for rRNA occurs in the masculinized strains.

The phenomenon of uninemy (*i.e.*, one DNA duplex molecule per chromosome) seems to be the property of each chromosome of higher eukaryotes. Extensive data in support of this concept come from the work of Petes *et al.* (1973) and Cryer *et al.* (1973) using the yeast *Saccharomyces cerevisiae*. It is estimated that the average size of the DNA molecule in each of the 17 genetic linkage groups (or chromosomes) is 6 X 10^8 daltons. Most of the rRNA genes (140 of them) in yeast, each being 2.5 X 10^6 dalton size, are in chromosome #12. This is based on the fact that most of the mapped genetic markers are located in other chromosomes (Cryer *et al.*, 1973).

ORGANIZATION OF SPECIFIC GENES AS REVEALED BY CLONING AND RESTRICTION ANALYSIS

The amount of fungal nuclear DNA that codes for rRNA, tRNA, and mRNA are respectively 1 to 3%, 0.1 to 0.3%, and 97 to 98% (Chattopadhyay *et al.*, 1972; Ray and Dutta, 1972; Dutta, 1973; Dutta and Chaudhuri, 1975; Ojha and Turian, 1977; Cryer *et al.*, 1973). Most of these estimates were based on DNA:RNA reactions using hydroxyapatite chromatography. This procedure not only gave quantitative estimates of DNA sequences transcribed (Dutta, 1973; Dutta and Chaudhuri, 1975) but also enabled the isolation of DNA transcripts. These isolated DNA transcripts have been able to identify specific DNA sequences by hybridizing these DNA transcripts with respective RNA molecules (Dutta and Chaudhuri, 1978). These isolated transcripts or the specific RNA cistrons have enabled us to understand certain genes whose characteristics were reviewed earlier (Ojha and Dutta, 1978). The extensive information available now suggest that most of the rDNAs have high G:C sequences except the zoosporic fungus *Allomyces* with high A:T sequences (Ojha and Turian, 1977).

Unfortunately DNA:DNA reassociation kinetics data alone cannot answer many vital questions for the complete understanding of the nature of genes. Some of these questions are: (1) How many types of specific RNA genes are present in fungal spore genomes? (2) What is the organization and/or location of specific RNA genes? (3) What is the nature of the structural configuration of regulatory sequences, *i.e.*, the promoter and/or initiator DNA sequences? It is now possible to answer these questions by cloning the specific DNA sequences using the highly successful recombinant-DNA technology. In the subsequent sections a few fungal genes which have been cloned are briefly described. Details of these procedures are in two books on recombinant-DNA molecules (Wu, 1979; Beers and Bassett, 1976).

Organization of Yeast rRNA Genes

Figure 5A, adapted from the work of Nath and Bollon (1978) shows the organization of a yeast rRNA gene. In the figure it may be noted first that genes coding for the yeast 5.8S, 18S and 25S rRNAs are present in a unit length of DNA of about 5.0 X 10^6 daltons (excluding the 5S gene) and approximately 9000 bases; second that the 9Kb repeat unit consists of seven sites for Eco R_1 (5^1GAATTC); and 5^1-AAGCTT base sequence at two sites for Hind III; third that one of the base sequences 5^1GAATTC is situated at or very close to the 18S coding terminus of the 35S precursor rRNA transcription unit; and fourth that two 35S repeat units are separated from each other by a region of DNA sequence unit of 1 X 10^6 dalton size which contains the separate transcription unit for the 5S RNA. They concluded furthermore that by extensive restriction analysis (Nath and Bollon, 1978) that out of 17 DNA restriction sites present in the 5 X 10^6 dalton "repeat unit", 9 are located at or near the 3' transcribed spacer regions. Several other workers (Valenzuela *et al.*, 1977; Maxam *et al.*, 1977) studied the physical structure and sequencing of 5S rRNA gene and the promoter regions. An analysis of the promoter region for the yeast 5S rRNA is helpful to understand the control of transcription by eukaryotic promoters. The yeast 5S RNA has a 5' triphosphate, unmodified by maturation at the 5' end. This suggests that the DNA sequence immediately before this region must contain the recognition sequences where the RNA polymerase binds to begin synthesis.

Petes *et al.* (1978) have shown recently by restriction analysis of recombinant DNA molecules the existence of two types of yeast rDNA genes in approximately equal amounts in the yeast *Saccharomyces carlsbergensis*. The first (type 1) group, having seven Eco R_1 sites (GBEFADC), were sequenced as reported earlier by Cramer *et al.* (1972). The type II (second) was found to contain only six sites (GX'FADC) recognized by Eco R_1. These results are different from those reported by Nath and Bollon (1978) who used different strains of the yeast *Saccharomyces cerevisiae*. It is possible that the observed differences are strain related.

Organization of Neurospora rDNAs

The *N. crassa* rDNA repeats (Free *et al.*, 1979) are also about 6 X 10^6 daltons in size, the same size as that of the yeast rDNA repeats and are tandemly arranged. *Neurospora crassa* rDNA repeats, however, differ from yeast rDNA repeats in that they do not contain the 5S (Fig. 5B) rRNA coding sequence. Free *et al.* (1979) suggest that 5S rDNA genes of *N. crassa* are not physically tightly clustered. It will be interesting to ascertain whether *N. crassa* 5S rRNA genes are homogenous, since they are found (unlike yeast 5S rDNAs) to be surrounded by heterogenous flanking sequences (Free *et al.*, 1979). Regulation of transcription and description of promoter regions for *N. crassa* 5S genes are yet to be studied.

Studies on arrangements of rDNAs in the genomes of five other lower eukaryotes reveal other distinctive features. The rDNAs of the true slime mold *Physarum polycephalum*

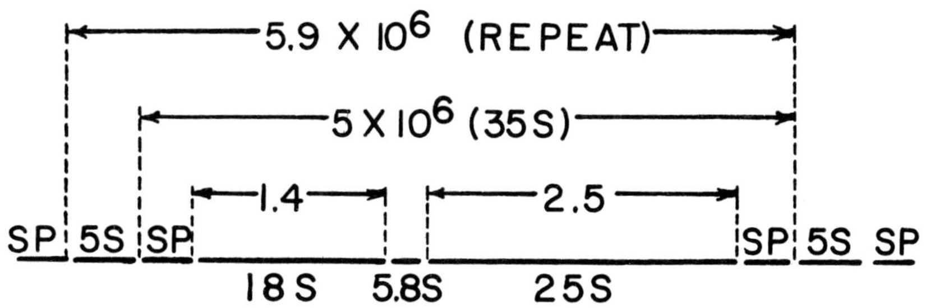

Fig. 5A. Yeast rRNA gene organization. (after Nath and Bollon, 1978)

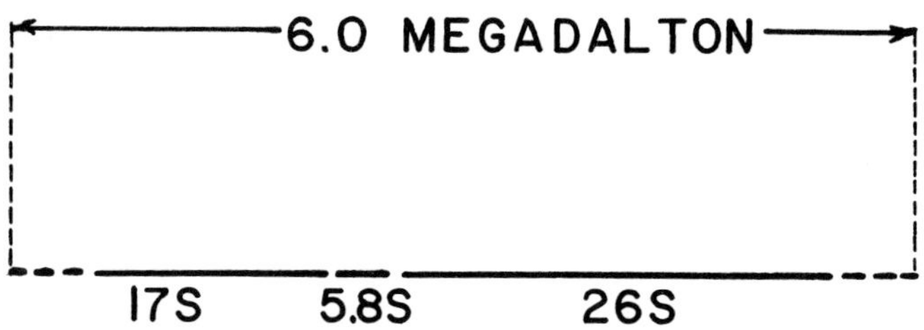

Fig. 5B. Neurospora rRNA gene organization. (after Free et al., 1979)

(Seebeck *et al.*, 1979) and of the cellular slime mold *Dictyostelium discoideum* (Taylor *et al.*, 1977) have all four (5S, 5.8S, 17S, and 25S rRNAs) of the rRNA coding region contained within a large extrachromosomal palindrome. The term "palindrome" refers to two "rDNA repeats: paired head to head instead of a tandem arrangement of repeats" (Krumlauf and Marzluf, 1980).

Although sequence arrangements in DNA sequences coding for mRNAs in fungi have not been studied, several excellent studies on these are in progress at the laboratories of Phil Leder of the U. S. National Institutes of Health, Bethesda, Maryland, and in the laboratories of O'Mally, Baylor University College of Medicine, Houston, Texas. Since fungi have very little, if any, repeated (except rDNA and tDNA) sequences in the nuclear DNAs, these models in higher eukaryotes may not be useful for understanding fungal gene regulation.

INTERVENING SEQUENCES IN FUNGAL GENES

The concept of one continuous polynucleotide (*i.e.*, one gene), one polypeptide (*i.e.*, one enzyme) in eukaryotes has undergone a drastic change during the last 2-3 years. Several eukaryotic genes are now known to be interrupted in their coding regions by DNA sequences. These are called intervening DNA sequences and are not represented in the mature RNA transcripts. Such interrupting DNA sequences have been reported in both structural and messenger RNA; for example, genes for tRNA of yeast (Valenzuela *et al.*,1977), 25S ribosomal RNA of *Tetrahymena pigmentosa* (Wild and Gall, 1979), 24S mitochondrial ribosomal RNA of *N. crassa* (Hahn *et al.*, 1979), 28S rRNA genes of higher eukaryotes (Glover and Hogness, 1977).

These observations naturally raise many vital questions such as: (1) Do all of these intervening sequences transcribe? (2) Since these intervening sequences disappear in mature RNA transcripts, then how are these sequences processed? (3) What kind of controls may these intervening sequences have on cell development, particularly in fungal morphogenetic events? Several workers have already addressed some aspects of these questions in higher eukaryotes. Wild and Gall (1979) have observed that intervening sequences of 25S rRNA genes were not present in all strains and species of *Tetrahymena* from different geographical locations. This is considered interesting from an evolutionary point of view, particularly because some *Tetrahymena* strains mate perfectly well and grow normally without the intervening sequences. It is therefore difficult at present to assess regulatory and controlling functions for these intervening sequences. Are these intervening sequences present on other parts of the genome? These and other questions need serious future consideration to understand gene level controls of fungal spore morphogenesis.

GENE AMPLIFICATION

Differential replication of only a part of a specific genome is known as "amplification." Several lines of evidence (Swift, 1973) have provided documentation of this phenomenon, such as DNA puffs of *Rhynochosiiara* and a few other insect species, in *Tetrahymena*, and clearly in the amphibian, *Xenopus* (oocytes). A similar term "magnification" has been used by Ritossa (see Swift, 1973) to designate the increase in ribosomal cistrons found in different eukaryotes, such as is found in *Drosophila melanogaster*, in most of the oocytes of *Xenopus laevis* (Bird, 1977), and in nuclear and extrachromosomal rRNA cistrons (rDNAs) in *Tetrahymena* (Yao *et al.*, 1978). A good summary of rDNA magnification in different species of mammals and plants, including yeast, has been given by Grierson *et al.* (1970). There are examples of "demagnification" or loss of rDNAs in female *Drosophila* flies (Swift, 1973). The wild type (bb+) *Drosophila* has 150 rDNA cistrons but bobbed (bb-) flies contain only 30 rDNA cistrons (Strausbaugh and Kiefer, 1979). The most striking example of a similar nature has been given by Bird (1977) who has shown that in the male *Xenopus laevis* all detectable amplified rDNA is lost but in the female a dramatic increase (2500 times) in chromosomal rDNA occurs after a transient reduction in rDNA.

Gene Amplification vs. Differential Gene Expression in Fungi

All of the fungal species studied so far, *viz.*, *Allomyces*, *Saccharomyces*, *Neurospora*, and *Dictyostelium*, have been reported to have 80 to 240 rRNA genes per haploid cell, and 435 copies of rDNA as reported for a diploid strain of *Achlya bisexualis* (Jaworski and Thomson, 1980). Lovett (1975) has provided evidence for the differential synthesis of rDNA in *Blastocladiella emersonii*. Differential rDNA replication has been reported recently by Mukhopadhyay and Dutta (1979) in germinating conidia of *N. crassa*. They have reported 200 copies of rDNA cistrons in germinating conidia and only 100 such copies in ungerminated conidia. The correlation of differential rDNA copies in the sexual differentiation of *Allomyces* (Ojha and Turian, 1978) and in *Xenopus* (Bird, 1977) is well established.

In *N. crassa* (Mukhopadhyay, Gbadamosi and Dutta, unpublished observations) we have detected highly repetitive rDNA-like molecules in the post-mitochondrial fraction of *N. crassa* germinating conidia. This observation raises the question of the origin of extra-chromosomal rDNAs. Hourcade *et al.* (1973) have given very convincing cytological evidence, using the electron microscope, to demonstrate the involvement of "rolling circle" intermediates in the thousand-fold amplification of the rDNA genes during development of *Xenopus* oocytes. This theory has, however given no definite answer regarding the origin of the first extra-chromosomal rDNA copies, based on the restriction enzyme analysis and by the electron microscope studies of rDNAs. Based on these studies, Bird (1977) suggests that there is more than one "excision" event, or that the "excised" segment contains a number of repeats. Hence, certain repeats that contain unique spacer regions in a cluster, are preferentially amplified.

Ribosomal RNA synthesis and rRNA gene amplification perhaps provide the best system (Grierson *et al.*, 1970) to study the activity of a gene and its products directly. This is the only case in which the behavior of the corresponding chromosomal region, the nucleolar organizer (McClintock, 1934), the processing of precursor rRNA to mature RNA, and their transport into the cytoplasm can be followed in great detail. Uecker (1979) and his associates have shown that in *Sordaria humana* nuclear structures enlarge soon after nuclear fusion.

Heterogeneity of rRNA Cistrons

Schibler *et al.* (1975) have presented excellent cytological evidence to prove that 28S and 18S components of the rRNA cistron are highly conserved in diverse eukaryotic vertebrate organisms. There is, however, a remarkable difference in the spacer regions which contribute to the heterogeneity of rRNA genes. The amount of spacer regions is largest in the warm-blooded animals. In lower animals and in plants (Grierson *et al.*, 1970) these spacer regions are less than half that in mammals and can very even between closely related species. Very few studies have been conducted to determine whether this is true for rRNA cistrons of very closely related species of fungi. In order to test this possibility we chose four closely related species of homothallic species of *Neurospora* which differ only slightly morphologically. We have (Mukhopadhyay, Mimiko and Dutta, unpublished, 1980) observed 2-3% differences among rDNA sequences in three homothallic species of *Neurospora, viz.*, *N. dodgei*, *N. africana*, and *N. lineolata*, which could be due to differences in spacer regions of their respective DNAs.

Bird (1977) has shown that out of 21 oocytes of *Xenopus* examined, 18 different spacer combinations were present. Using the restriction enzyme Eco R_1 digestion patterns of rDNAs from different oocytes he finds that certain repeats in a cluster are preferentially amplified, perhaps by virtue of their position within the tandem gene array.

CONTROL OF FUNGAL GENE EXPRESSION

The fungal spore nuclear genomes contain very little (2-5%) repetitive DNAs. The role of these repetitive DNAs in gene regulation has been of great interest, particularly in the nature of their interspersions with the non-repetitive DNA sequences.

Several workers have attempted to study the relationship different patterns of DNA sequences (*i.e.*, patterns of interspersions of repetitive sequences in the genome) have on the control of gene regulation in fungal species (Krumlauf and Marzluf, 1980). Repetitive and single copy sequence could be arranged in a long-period interspersion pattern. In this pattern, 4000 nucleotide pairs (Nt pr) are linked to single copy DNA sequences of 10,000 Nt pr. It is difficult at present to ascertain regulatory functions of these long repetitive DNA sequences. On the contrary, short-period interspersion patterns comprise 200-400 nucleotide base pairs, linked to single copy DNA sequences at intervals of 1000-2000 Nt pr (Britten and Davidson, 1969). This short period interspersion pattern seems to be typical for higher eukaryotic organisms. Davidson and Britten (1979) have suggested that these short-period repeats could play a role in eukaryotic gene regulation, however.

Recently, attempts have been made by several workers to study the pattern of DNA sequence interspersions in the water mold *Achlya* (Hudspeth *et al.*, 1977), *N. crassa* (Krumlauf and Marzluf, 1980), and *Aspergillus niger* (Timberlake, 1978). All of these workers have suggested that in fungi, perhaps, repetitive DNA sequences identified by renaturation experiments, do not serve a regulatory function due to their long interspersion patterns. However, it has been suggested (Krumlauf and Marzluf, 1980) that very short repeated sequences, undetected by DNA renaturation experiments, may serve as recognition sites. Perhaps cloned repetitive DNA sequences of fungal species might give insights to these important questions on fungal gene regulation. It is rather early to conclude on the nature of fungal repeats on gene regulation.

Is Control Transcriptional?

There are several questions that remain unanswered in this context. It is reasonable to conclude, however, that in most regulatory genes, as described in the introductory section, the level of controls of known fungal genes are probably not posttranslational (Metzenberg, 1979). An excellent example of gene level controls in fungi is that of the quinic acid (qa) catabolic system of *N. crassa* described by Norman H. Giles and his collaborators of the University of Georgia, Athens, GA. The *qa* gene cluster of *N. crassa* (Vapneck *et al.*, 1977) consists of three loci which are structural genes for the individual *qa* enzyme: *qa*-2 catabolic dehydroquinase; *qa*-3 quinate dehydrogenase; and *qa*-4 dehydroshikimate dehydrase. Another tightly linked regulatory locus, *qa*-1, encodes a regulatory protein which in conjunction with the inducer, quinic acid, controls the expression of the three structural genes. Thus the *qa* gene cluster proved to be an ideal DNA sequence for cloning on a recombinant plasmid. Cloning of the entire *N. crassa qa* gene cluster in *E. coli* plasmids may provide many opportunities to study controls of expressions of the component genes in the future.

In this system it is not yet known whether the three structural genes *qa*-3, *qa*-2, and *qa*-4 are all transcribed (Metzenberg, 1979) together (*i.e.*, one precursor mRNA) and processed, neither is it clear yet whether these individual genes are transcribed separately from their one promoter. One group (Vapneck *et al.*, 1977) has cloned *qa*-2^+ genes in the plasmid pBR 322 and have studied the expression of the catabolic dehydroquinase of *Neurospora* in *E. coli* aromatic amino acid auxotrophs. It is possible that the RNA synthesized on the *qa*-2 gene template was transcribed from a *Neurospora* DNA promoter. Confirmation of this information is now expected using the powerful recombinant-DNA technology that Giles and his group are currently using.

A few other regulatory circuits have been described in fungi. For example, in *Aspergillus* (Maclean, 1976) there is an example of two closely linked *ma* D and *nu* A structural genes which are controlled by a powerful regulator gene distinct from them but on the same linkage group VII. These genes are involved in the complete reduction of nitrate to ammonium. Reinert and Marzluf (1975) have described another positive control gene, *nit*-2-gene, which is considered to regulate unlinked structural genes controlling the synthesis of many enzymes involved in nitrogen metabolism. They have postulated that *nit*-2 specifies a regulatory element which binds to control regions adjacent to structural genes to "turn on" their expression. Not much is known about the nature of transcription of these gene clusters, and the question cannot be resolved.

PROPOSED EXPERIMENTAL APPROACHES

In the preceding examples we have seen that fungal systems provide many examples of regulatory circuits or gene clusters. Most of these genes are positively controlled and a few are negatively controlled. Unfortunately not much information on gene level controls is currently available for the fungi although morphogenetic changes in fungal spore germination or sporulation no doubt provide many opportunities for future experimentation. In order to understand the control mechanisms of fungal spore morphogenesis at the gene level, the following areas of research may be of interest.

Extensive Molecular Analysis of Fungal Spore Genomes

Most of the examples discussed are taken from a small number of kinds of filamentous fungi. One can easily see the need for this kind of work using spores of essentially unexplored species of basidiomycetes, ascomycetes, oomycetes, and other groups of fungi. Extensive knowledge on the sequence, complexity and genomic organization of various fungal spores and related developmental cell types will be essential to understand gene level controls. These studies may include: (a) use of restriction enzymes to detect any possible rearrangements in DNA sequences of specific genes which might have occurred in morphogenesis, (b) the nature of foldback DNA sequences (inverted repeats, sizes, whether free in cytoplasm or only in the chromosomes), and (c) a search for short repeated DNA sequences, as found in higher eukaryotes (Davidson and Britten, 1979). This can best be studied by molecular cloning of repetitive and non-repetitive DNA sequences. The basis for these suggestions is that each control circuit should contain a family of repeated DNA sequences depending on the number of structural genes within the circuit (Davidson and Britten, 1979).

Fungi seem to be strikingly different from higher eukaryotes in that short (200-400 Nt bp) repetitive DNA sequences do not exist (Krumlauf and Martzluf, 1980; Timberlake, 1978). This does not rule out the presence of very short sequences (10-25 Nt bp) that might exist contiguous with structural genes (Krumlauf and Marzluf, 1980). Answers to these questions can now be obtained by cloning and partial sequencing of unique sequences of known structural genes. This step has already been initiated by Giles and his associates at the University of Georgia using the fungus *N. crassa*.

Analysis of DNA Constancy by Restriction Analysis and DNA Sequencing

In order to test the occurrence of rearrangements in DNA during development and to assess the rate of DNA divergence, if any, during fungal spore morphogenesis, the analysis of restriction fragments should be ideal. Restriction endonucleases allow manipulation of specific fragments of DNA in a manner that has not been possible. These enzymes can recognize specific sites on the DNA molecule, and cut out fragments containing one or many genes. There are more than 40 different specific endonucleases that have been discovered and recognition sequences for more than 23 of these have been deduced (Roberts, 1976). Each of these enzymes cleave DNAs of different sizes and at different recognition sequences. The best example of detecting minute differences in any possible rearrangements of genomes in developmental cell types comes, however, from studies with the nematode *Caenorhabditis elegans* (Emmons *et al.*, 1979). These authors found a surprisingly great incidence of DNA rearrangements in sperms and somatic cells of two strains. Similar studies are in progress in our laboratory (Dutta) using different cell types of *Neurospora*. In addition, it is likely that the existing concept of one continuous polynucleotide sequence coding for one polypeptide may not be true anymore. Several examples of intervening sequences in fungal rDNAs have been discussed before. One recent example of intervening sequences in the actin gene of *S. cerevisiae* has generated considerable interest. It appears that a broader search for intervening sequences is definitely needed. These studies will give much insight towards understanding gene level controls in fungal spore morphogenesis.

Use of DNA Insertion Mutants Within the Regulatory Circuits

The galactose regulatory circuit of *E. coli* is one of the best examples for understanding gene level controls using DNA insertions (Adhya and Miller, 1979). Das *et al.* (1976) have described three types of DNA insertions containing 700, 1400, and 1200 nucleotide pairs. The galactose (*gal*) operon of *E. coli*, as described below, contains three structural genes (K, T, and E) and the promoter-operator (P-O) located at the E end of the transcriptional unit (Adhya and Miller, 1979).

K   T   E   DNA   P-O

Insertions

These insertion mutants have been shown to have polar effects, *i.e.*, decrease in the levels of the products of genes located distal to the polar mutations (Das *et al.*, 1976). These insertion mutations, which can now be made *in vitro* using recombinant-DNA technology, should also help to increase the depth of our understanding of promoters or operators of regulatory circuits. Until more fungal genes are cloned and are analysed this suggestion should be taken as speculatory.

SUMMARY

In this chapter our discussion has been on gene level controls of fungal morphogenesis, in general, with special reference to fungal spores, where possible. Our discussion has been primarily on the involvement of DNA sequences (genes) themselves in fungal spore morphogenetic controls. Numerous evidence now exists to demonstrate that fungal spore morphogenetic events can be identified conclusively as under control of specific gene or regulatory circuits. In addition, we have provided electron microscopic evidence that the regulation of the rib-vein patterns of ascospores of *Neurospora* is effected by a lineally aligned complex of cisternae of the endoplasmic reticulum. This finding delineates the basis for the taxonomically stable wall surface characteristic of fungal spores that is typical for such a well-defined taxon.

Several regulatory circuits reported by others in fungi have been discussed (see Discussant's Introduction to this session on nuclear controls). Most of these gene clusters are positively controlled such as: "arom" cluster regulating aromatic amino acid metabolism; genes encoding the enzymes for quinate metabolism; three tightly linked genes controlling galactose metabolism; and nitrogen acquisition system. The control of the phosphorus acquisition system is perhaps under negative control action. In a negatively controlled regulatory circuit, macromolecular activity holds most genes in check.

Amplification of rRNA genes in fungal spores have been

discussed. The "rolling circle mechanism" of gene amplification is perhaps the basis of rDNA amplification in fungal spores. The rDNA amplification and differential gene expression in fungal spore development has been discussed.

A summary of the studies made with the molecular analysis of the fungal spore genomes, organization of specific rRNA genes of the yeast *S. cerevisiae* and *N. crassa* is given. It appears that nuclear genomes of fungal spores contain only 3 to 6% repetitive DNA sequences of the rDNA size (3-6 X 10^6 daltons) when compared to 10-20% repetitive DNA sequences of whole vegetative cell DNAs, indicating that most of the repetitive DNAs in fungal spores are mitochondrial. Evidence that post-mitochondrial DNAs have the same size as those of rDNAs was obtained. All of these post-mitochondrial DNAs were repetitive. Both yeast and *Neurospora* cloned rDNAs had 17S, 5.8S, and 26S rRNA genes. The 5S rRNA gene was, however, absent in the total cloned rRNA gene in *N. crassa*. Our studies have established the heterogeneity of rDNAs of closely related fungal species which could be attributed to differences in spacer regions of fungal rDNAs.

It is likely that the old concept of one continuous polynucleotide sequence code for one polypeptide may not be true any more. Studies are underway to test the significance of these findings in fungal morphogenesis.

Finally, these suggestions on proposed experimental approaches for the further elucidation of gene level control in fungal morphogenesis, represent, of course, our own personal biases. The need for extensive studies for very short repetitive DNA interspersions with the unique sequences in fungal spores should be explored. The use of DNA insertion mutants with the regulatory circuits (operons) of fungal spores using recombinant-DNA technology will help to elucidate morphogenetic controls.

ACKNOWLEDGMENTS

Support for different aspects of the studies reported here have come from the Brown-Hazen Fund of the Research Corporation, the Department of Energy and the National Institutes of Health, Division of Research Resources. The critical reviews of Dr. J. C. Schmit, of the Southern Illinois University, Dr. S. Adhya of the National Institutes of Health, and Professor G. Turian of the University of Geneva are gratefully acknowledged. We also wish to express grateful appreciation to Mrs. A. Frederick for typing the manuscript.

REFERENCES

Adhya, S. and Miller, W. (1979). Modulation of the two promoters of the galactose operon of *Escherichia coli*, *Nature* 279:492-494.

Aldrich, H. C. (1974). Spore cleavage and the development of wall ornamentation in two myxomycetes, *Proc. Natl. Acad. Sci.* 81:28-36.

Anderson, J. G. and Smith, J. E. (1971). The production of conidiophores and conidia by newly germinated conidia of *Aspergillus niger*, *J. Gen. Microbiol.* 69:185-197.

Austin, W. Lena, Frederick, L. and Roth, I. L. (1974). Scanning electron microscope studies on ascospores of homothallic species of *Neurospora*, *Mycologia* 66:130-138.

Beers, R. F. and Bassett, E. G. (Eds.) (1976). "Recombinant Molecules: Impact on Science and Society." Raven Press, New York.

Berget, S. M., Moore, C. and Sharp, P. A. (1977). Spliced segments at the 5' terminus of adenovirus 2 late mRNA, *Proc. Natl. Acad. Sci.* 74:3171-3175.

Bird, A. P. (1977). A study of early events of ribosomal gene amplification, *Symposia in Quantitative Biology* 42: 1179-1183. Cold Spring Harbor Laboratory, N. Y.

Brambl, R., Dunkle, L. D. and Van Etten, J. L. (1978). Nucleic acid and protein synthesis during fungal spore germination. *In* "Filamentous Fungi" (Eds. J. E. Smith and D. R. Berry). Vol. 3, pp. 94-118. Edward Arnold Publishers Ltd., London.

Britten, R. J. and Davidson, E. H. (1969). Gene regulation for higher cells: a theory, *Science* 165:349-357.

Chattopadhyay, S. K., Kohne, D. E. and Dutta, S. K. (1972). Ribosomal RNA genes of *Neurospora*: Isolation and characterization, *Proc. Natl. Acad. Sci.* 69:3256-3259.

Chaudhuri, R. K., Ojha, M. and Dutta, S. K. (1973). Morphological mutants of *N. crassa*: Possible evidence of abnormal morphology due to change in DNA composition, *Oncology* 28:553-566.

Cramer, J. H., Bhargava, M. M. and Halvorson, H. O. (1972). Isolation and characterization of rDNA in *Saccharomyces cerevisiae*, *Jour. Mol. Biol.* 71:11-20.

Crippa, M., Meza, I. and Dina, D. (1973). Sequence arrangement in mRNA: Presence of Poly (A) and identification of a repetitive fragment at 5' end, *Symposia in Quantitative Biology* 38:933-942. Cold Spring Harbor Laboratory, New York.

Cryer, D. R., Goldthwaite, C. D., Zinker, S., Lam, K. B., Storm, E., Hischberg, R., Blaimire, J., Finkelstein, D.

B. and Marmur, J. (1973). Studies on nuclear and mitochondrial DNA of *S. cerevisiae*, *Symposia in Quantitative Biology* 38:17-29. Cold Spring Harbor Laboratory, New York.

Das, Asis, Court, Don and Adhya, S. (1976). Isolation and characterization of conditional lethal mutants of *Escherichia coli* defective in transcription factor rho, *Proc. Natl. Acad. Sci.* 73:1959-1963.

Davidson, E. H. and Britten, R. J. (1979). Regulation of gene expression in animal cells. A new look at the possible role of interspersed repetitive sequences, *Science* 204:1052-1059.

Dodge, B. O. (1957). Rib formation in ascospores of *Neurospora* and questions of terminology, *Bull. Torr. Bot. Club* 84:182-188.

Dunkle, L. D. and Van Etten, J. L. (1972). Characteristics and synthesis of DNA during fungal spore germination. *In* "Spores" (Eds. H. O. Halvorson, R. Hanson, and L. L. Campbell) pp. 283-298. Amer. Soc. Microbiol., Washington, D. C.

Dutta, S. K. (1973). Transcription of non-repeated DNA in *Neurospora crassa*, *Biochem. Biophys. Acta* 324:482-487.

Dutta, S. K. (1974). Repeated DNA sequences in fungi, *Nucleic Acid Research* 1:1411-1419.

Dutta, S. K. and Chaudhuri, R. K. (1975). Differential transcription on nonrepeated DNAs in *Neurospora crassa* mutants, *Mol. Gen. Genet.* 164:45-50.

Dutta, S. K., Gbadamosi, I., Chambers, C. and Mukhopadhyay, D. K. (1980). Comparison of nuclear DNA with whole cell DNA isolated from *Neurospora crassa*, *Neurospora Newsletter* 27:18-19.

Emmons, S. W., Klass, M. R. and Hirsh, D. (1979). Analysis of the constancy of DNA sequences during development and evolution of the nematode *Caenorhabditis elegans*, *Proc. Natl. Acad. Sci.* 76:1333-1337.

Frederick, L. and Austin, W. Lena (1977). The structure and development of the rib layer in *Neurospora* ascospores, *Abstracts of Second International Mycological Congress*, Vol. A-L, p. 204.

Free, S. J., Rice, P. W. and Metzenberg, R. L. (1979). Arrangement of the genes coding for rRNA in *Neurospora crassa*, *Jour. Bact.* 137:1219-1226.

Firtel, R. A. and Bonner, J. (1972). Characterization of the genome of the cellular slime mold *Dictyostelium discoideum*, *Jour. Mol. Biol.* 66:339-361.

Gallwitz, D. and Sures, I. (1980). Structure of a split gene: complete nucleotide sequence of the actin gene in

S. cerevisiae, Proc. Natl. Acad. Sci. 77:2546-2550.

Glover, D. M. and Hogness, D. S. (1977). A novel arrangement of the 18S and 28S sequences in a replicating unit of *Drosophila melanogaster* DNA, *Cell* 10:167-176.

Grange, F. and Turian, G. (1978). Differential DNA synthesis during microcycle conidiation in *Neurospora crassa, Arch. Microbiol.* 119:257-261.

Grierson, D., Rogers, M. E., Sartirana, M. L. and Loening, U. E. (1970). The synthesis of rRNA in different organisms: Structure and evolution of the rRNA precursor, *Symposia in Quantitative Biology*, Vol. 35, pp. 589-597. Cold Spring Harbor Laboratory, New York.

Hahn, U., Lazarus, C. M., Lunsodorf, H. and Kuntzel, Hans (1979). Split gene for mitochondrial 24S rRNA of *Neurospora crassa, Cell* 17:191-200.

Hautala, J. A., Conner, B. H., Jacobson, J. W., Patel, G. L. and Giles, N. H. (1977). Isolation and characterization of nuclei from *Neurospora crassa, Jour. Bact.* 130:704-713.

Hess, W. M. and Weber, D. J. (1976). Form and function in basidiomycete spores. *In* "The Fungal Spore." (Eds. D. J. Weber and W. M. Hess), pp. 643-714. Wiley Interscience, London.

Hourcade, D., Dresster, D. and Wolfson, J. (1973). The nucleolus and the rolling cycle, *Symposia in Quantitative Biology* 38:537-571. Cold Spring Harbor Laboratory, New York.

Hudspeth, N. E. S., Timberlake, W. E. and Goldsberg, R. B. (1977). DNA sequence organization in the water mold *Achlya, Proc. Natl. Acad. Sci.* 74:4332-4336.

Jaworski, A. J. and Thomson, K. (1980). A temporal analysis of the synthesis of the mRNA sequestered in zoospores of *Blastocladiella emersonii, Develop. Biol.* 75:343-347.

Krumlauf, R. and Marzluf, G. A. (1979). Characterization of the sequence complexity and organization of the *Neurospora crassa* genome, *Biochemistry* 18:3705-3713.

Krumlauf, R. and Marzluf, G. A. (1980). Genome organization and characterization of the repetitive and inverted repeat DNA sequences in *Neurospora crassa, Jour. Biol. Chem.* 225:1135-1145.

Lee, A. S., Britten, R. J. and Davidson, E. W. (1977). Short period repetitive-sequence interspersion in cloned fragments of sea urchin DNA, *Symposia in Quantitative Biology*, Vol. 42:1065-1076. Cold Spring Harbor Laboratory, New York.

Lindegren, C. C. and Scott, M. A. (1937). Formation of the ascospore wall in *Neurospora, Cellule* 15:360-373.

Littlefield, L. J. and Bracker, C. E. (1971). Ultrastructure and development of urediospore ornamentation in *Melanospora lini*, *Can. J. Bot.* 49:2067-2073.

Lovett, J. S. (1975). Growth and differentiation of the water mold *Blastocladiella emersonii:* Cytodifferentiation and the role of RNA and protein synthesis, *Bact. Rev.* 39: 345-404.

Lowry, R. J. and Sussman, A. S. (1968). Ultrastructural changes during germination of ascospores of *Neurospora tetrasperma*, *J. Gen. Microbiol.* 51:403-409.

Maclean, N. (1976). "Control of Gene Expression". Academic Press, London.

Marzluf, G. A. (1977). Regulation of gene expression in fungi. *In* "Regulatory Biology." (Eds. J. C. Copeland and G. A. Marzluf), pp. 196-242. Ohio State University Press, Columbus.

Maxam, A. M., Tizard, R., Skryabrin, K. G. and Gilbert, W. (1977). Promoter region for yeast 5S ribosomal RNA, *Nature* 267:643-644.

McClintock, B. (1934). The relationship of a particular chromosomal element to the development of the nucleoli in *Zea mays*, *Zeitschrift Zellforsch. Mikrosk. Anat.* 21: 294-298.

Metzenberg, R. L. (1979). Implications of some genetic mechanisms in *Neurospora*, *Microbiol. Reviews* 43:361-383.

Mims, C. W. (1972). Spore-wall formation in the myxomycete *Arcyria cinerea*, *Trans. Br. Mycol. Soc.* 59:477-481.

Minagawa, T., Wagner, B. and Strauss, B. (1959). The nuclei acid content of *N. crassa*, *Arch. Biochem. Biophysic.* 80:442-445.

Moore, R. T. (1963). Fine structure of Mycota. I. Electron microscopy of the discomycete *Ascodesmis*, *Nova Hedwg.* 5:263-298.

Mukhopadhyay, D. K. and Dutta, S. K. (1979). Evidence of increased rRNA genes in germinating conidia of *N. crassa*, *Ind. J. Exptl. Biol.* 17:620-621.

Myers, R. B. and Cantino, E. C. (1971). DNA profile of the spore of *Blastocladiella emersonii*: Evidence of γ-particle DNA, *Archiv Mikrobiol.* 78:252-267.

Nath, K. and Bollon, A. P. (1978). Organization of the yeast rRNA gene cluster via cloning and restriction analysis, *Molec. Gen. Genet.* 160:235-245.

Ojha, M. and Dutta, S. K. (1978). Nuclear control of differentiation. *In* "The Filamentous Fungi" (Eds. J. E. Smith and D. E. Berry). Vol. 3, pp. 8-27. Edward Arnold Publishers Ltd., London.

Ojha, M. and Turian, G. (1977). Ribosomal RNA cistrons in

Allomyces arbuscula, Biochem. Biophysica Acta 478:392-406.
Ojha, M. and Turian, G. (1978). Loss of rRNA genes during a phenotypic, masculinizing mutation in *Allomyces arbuscula, Molec. Gen. Genet.* 166:225-227.
Petes, T. D. (1979). Yeast ribosomal DNA genes are located on chromosome XII, *Proc. Natl. Acad. Sci.* 76:410-414.
Petes, T. D., Hereford, L. M. and Skyrabin, K. G. (1978). Characterization of two types of yeast rDNA genes, *Jour. Bact.* 134:295-305.
Petes, T. D., Newlon, C. S., Byers, B. and Fangman, W. L. (1973). Yeast chromosomal DNA: size, structure and replication, *Symposia in Quantitative Biology* 38:1-29. Cold Spring Harbor Laboratory, New York.
Ray, R. and Dutta, S. K. (1972). Isolation of transfer RNA genes from *Neurospora crassa, Biophys. Biochem. Res. Comm.* 47:1458-1463.
Reinert, W. R. and Marzluf, G. A. (1975). Regulation of the purine catabolic enzymes in *Neurospora crassa, Arch. Biochem. Biophysic.* 166:565-574.
Roberts, R. J. (1976). The role of restriction endonucleases in genetic engineering in recombinant moles. *In* "Recombinant Molecules: Impact on Science and Society" (Eds. R. F. Beers and E. G. Bassett), pp. 21-32. Raven Press, New York.
Schibler, V., Wyler, T. and Hagenbuchle, O. (1975). Changes in size and secondary structure of the ribosomal transcription unit during vertebrate evolution, *Jour. Mol. Biol.* 94:503-517.
Schmit, J. C. and Brody, S. (1976). Biochemical genetics of *Neurospora crassa* conidial germination, *Bacteriol. Reviews* 40:1-41.
Schweizer, E., Mackechnie, C. and Halvorson, H. O. (1969). The redundancy of ribosomal and transfer RNA genes in *S. cerevisiae, J. Mol. Biol.* 40:261-277.
Seebeck, T., Stadler, J. and Braun, R. (1979). Isolation of minichromosome containing the ribosomal genes from *Physarum polycephalum, Biochemistry* 18:484-490.
Serna, L. and Stadler, D. (1978). Nuclear division cycle in germinating conidia of *Neurospora crassa, Jour. Bact.* 136:341-351.
Strausbaugh, L. D. and Kiefer, B. I. (1979). Genetic modulation of RNA metabolism in Drosophila III. Requirement for an rDNA-deficient X-chromosome in Y^{bb}SuVar-5 mediated increases in RNA synthesis, *Genetics* 93:411-422.
Sturani, E., Costantini, M. G., Bartigani, E. and Alberghina, L. (1979). Level of turnover of polyadenylate-contain-

ing RNA in *N. crassa* in different steady states of growth. *Eur. J. Biochem.* 99:1-7.

Srere, P. A. and Mosbach, K. (1974). Metabolic compartmentation: symbiotic, organellar, multienzymic, and microenvironmental, *Annu. Rev. Microbiol.* 28:61-83.

Swift, H. (1973). The organization of genetic material in eukaryotes: progress and prospects, *Symposia in Quantitative Biology* 38:963-979. Cold Spring Harbor Laboratory, New York.

Tanaka, K., Kogane, F. and Yanagita, T. (1965). Is DNA of *Aspergillus oryzae* conidia modified chemically in the early period of germination? *J. Gen. Appl. Microbiol.* 11:85-90.

Taylor, W. C., Cockburn, A. F., Frankel, G. A., Newkirk, M. J. and Firtel, R. A. (1977). Organization of ribosomal and 5S coding regions in *Dictyostelium discoideum*. *In* "Symposium on Molecular and Cellular Biology" (Eds. J. Abelson and J. C. Copeland), pp. 309-313. Academic Press, New York.

Timberlake, W. E. (1978). Low repetitive DNA content in *Aspergillus nidulans*, *Science* 202:973-982.

Turian, G. (1976). Spores in Ascomycetes, their controlled differentiation. *In* "The Fungal Spore: Form and Function" (Eds. D. J. Weber and W. M. Hess), pp. 715-786. Wiley Interscience Publishers, Inc., London.

Uecker, F. A. (1979). Development and cytology of *Anixiella endodonta*, *Bot. Gaz.* 140:310-317.

Valenzuela, P., Bell, G. I., Venegas, A., Sewell, E. T., Masiarz, F. R., DeGennaro, L. J., Weinberg, F. and Rutter, W. J. (1977). rRNA genes of *Saccharomyces cerevisiae* II. Physical map and nucleotide sequence of the 5S rRNA gene and adjacent intergenic regions, *Jour. Biol. Chem.* 252:8126-8135.

Van Etten, J. L. and Rawn, C. D. (1979). Polyadenylate-containing RNA in dormant and germinating *Botryodiplodia theobromae* pycnidiospores, *Can. J. Microbiol.* 25:375-379.

Vapnek, D., Hautala, J. A., Jacobson, J. W., Giles, N. H. and Kushuer, S. R. (1977). Expression of *Escherichia coli* K-12 of the structural gene for catabolic dehydroquinase of *Neurospora crassa*, *Proc. Natl. Acad. Sci.* 74: 3508-3512.

Wild, M. A. and Gall, J. G. (1979). An intervening sequence in the gene coding for 25S rRNA of *Tetrahymena pigmentosa*, *Cell* 16:565-573.

Wu, Ray (1979). Recombinant DNA. *In* "Methods in Enzymology" (Eds. S. P. Colowick and N. O. Kaplan). Vol. 68. Academic Press, New York.

SPORE FORMATION IN *ASPERGILLUS NIDULANS*: COMPETENCE AND OTHER DEVELOPMENTAL PROCESSES

Sewell P. Champe*, Myra B. Kurtz*, Lawrence N. Yager*, Nancy J. Butnick*, and David E. Axelrod#

**Waksman Institute of Microbiology, Rutgers University, New Brunswick, New Jersey 08903, U.S.A.*

#*Department of Biological Sciences, Douglass College, Rutgers University, New Brunswick, New Jersey 08903, U.S.A.*

INTRODUCTION

Fungi have long been recognized as useful experimental organisms for the study of eukaryotic development. Their phylogenetic position predicts that they will have retained some of the primitive control mechanisms of their prokaryotic ancestors, but may have elaborated some of the more sophisticated control mechanisms of higher eukaryotes. The homothallic ascomycete *Aspergillus nidulans*, in particular, has a number of experimentally advantageous features. It grows rapidly on simple, defined media, it is amenable to genetic and biochemical analysis, and it has a relatively simple life cycle during which well-defined differentiated structures are formed.

This review will discuss primarily the formation of asexual spores in *A. nidulans* with emphasis on the genetic elements that control the inducibility of this developmental process. In higher eukaryotes the acquired responsiveness to a stimulus that induces development has been termed "competence" (Nöthiger, 1972). Some features of spore formation in *A. nidulans*, to be described, lead to the view that a growing mycelium must, likewise, become competent before sporulation is inducible. A study of the genetic control of sporulation in this organism thus has the prospect of identifying genes which may function as master switches for developmental processes.

THE SPORULATION SCHEDULE

Asexual sporulation

If a single conidium (asexual spore) is placed on a nutrient agar surface it will germinate in several hours as evidenced by the emergence of a hyphal germ tube. This tube elongates and repeatedly branches to form the mycelial mesh which at early times grows parallel to the agar surface. After a fixed period of vegetative growth the walls of some of the hyphae become thickened. From these specialized hyphae (foot cells) similarly thick-walled hyphae (conidiophores) branch and elongate aerially. Suddenly, within a period of 30 to 60 minutes for a given colony, the tips of the conidiophores swell to form a spherical bulb (the vesicle) containing many nuclei. Shortly thereafter, the surface of the vesicle becomes covered with small mononucleate bud cells (primary sterigmata or metulae) each of which divides once to form the apical conidiogenic cells (secondary sterigmata or phialides). Repeated mitotic divisions of the secondary sterigmata result in long unbranched chains of progeny cells which mature into haploid conidia genetically identical to the initiating conidium. The conidial apparatus is illustrated in Figure 1.

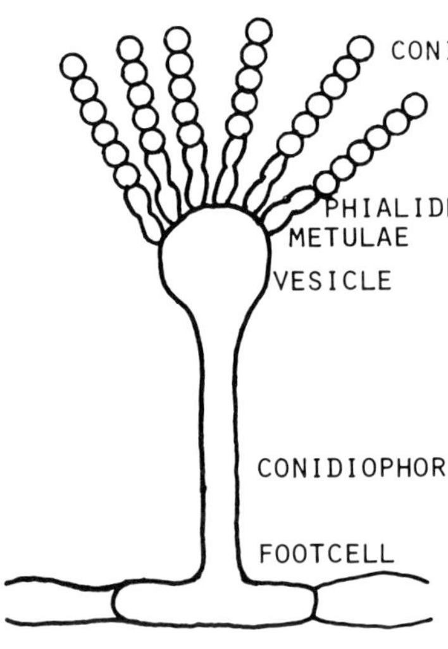

Fig. 1. The conidial apparatus of Aspergillus nidulans

The formation of the conidiophore vesicle, a rapid and easily observable process, provides a convenient marker event for the onset of conidiation which, at 37°C, occurs 24 hours after conidial inoculation (Axelrod, 1972). The synchrony is remarkably precise from colony to colony and the medium has little effect on the schedule (Pastushok and Axelrod, 1976). The time of conidiation is, however, temperature-dependent and varies inversely with colony density, being about 6 hours earlier for a confluent lawn than for a single colony at 37°C. This developmental regularity makes the conidiation process of *A. nidulans* particularly amenable to experimental study.

Sexual sporulation

Compared to conidiation, sexual sporulation is a late event occurring, depending on the genetic background and cultural conditions, several days to more than a week after conidiation is complete. Sexual spores develop within asci located throughout the interior of closed spherical shells (cleistothecia) composed of fused, interwoven, sterile hyphae. When mature, cleistothecia, which attain a diameter of up to 0.5 mm, are firm, resilient structures, the walls of which are infused with a dark-red, insoluble pigment. The number of cleistothecia per unit area, which can be several thousand per cm^2, can be used as an index of sexual sporulation.

A. nidulans and several other *Aspergillus* species elaborate clusters of large, thick-walled cells, called Hülle cells (Thom and Raper, 1945), which arise by the enlargement of terminal and intercalary hyphal segments (Benjamin, 1955; Ellis *et al.*, 1973; Zonneveld, 1977). In *A. nidulans* these Hülle cells surround the developing cleistothecia and thus identify the loci at which sexual development is taking place. Quantitation of these loci at early stages is difficult, however, because most are obscured by overlying conidial heads and the surrounding mycelial matrix. As described in a later section, we have discovered that the cleistothecial primordia and the surrounding Hülle cells produce a phenol oxidase which allows specific staining of the primordia with a chromogenic substrate. Although sexual development is notoriously hypersensitive to cultural conditions, quantitation of the primordia by this method shows that initiation of the process can be as synchronous as conidiation.

THE INDUCTION OF CONIDIATION

Surface versus submerged growth

The schedule for conidiation described above applies to a colony arising from a single spore plated on a nutrient agar surface. If liquid medium is inoculated with spores, submerged mycelial colonies will develop but may remain undifferentiated indefinitely. The suppression of conidiation in submerged culture is not absolute; it is medium-dependent and strain-dependent. Conditions which promote submerged conidiation have been extensively studied in several species (Smith and Anderson, 1973; Smith, 1978), including *A. nidulans* (Saxena and Sinha, 1973; Martinelli, 1976). For the medium[1] and the strain (FGSC-4) employed in the following studies, however, conidiation is completely suppressed during the time period under consideration. In a later section we will discuss the identification of a genetic locus involved in the control of submerged conidiation.

The acquisition of competence

If a spore-originated colony growing in liquid medium is transferred to a solid nutrient surface, the colony will begin to conidiate after a certain time. The transfer from liquid to solid medium can thus be considered an inducing stimulus for conidiation. As shown in Figure 2, if transfer is made at any time before 20 hours, conidiation (as judged by the appearance of conidiophore vesicles) always begins at 24 hours. If, however, transfer is made any time after 20 hours, conidiation commences 4 hours later. This experiment shows that responsiveness to induction, or competence, is acquired at the critical time of 20 hours. Thereafter conidiation follows induction after a fixed maturation period of 4 hours (Axelrod *et al.*, 1973). The decision to conidiate at 24 hours (and to become competent at 20 hours) appears to be endogenously controlled since this time is unaffected by continuous replacement of the medium (Pastushok and Axelrod, 1976).

It might be argued that the requirement for a competent state prior to conidiation is inconsistent with the phenomenon of microcycle conidiation whereby, under special conditions, conidiophores develop directly from germinating spores without an intervening period of vegetative growth. However, the various regimens used to induce microcycle development require times at least as long as those for normal conidiation (Anderson, 1978; Smith, 1978). Under such

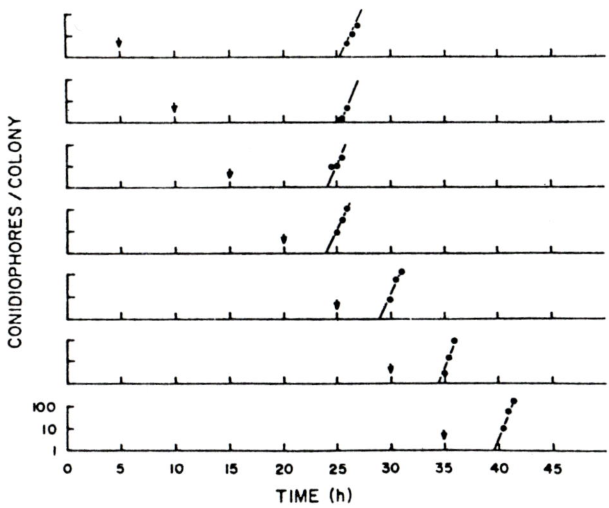

Fig. 2. Definition of the time of competence. Submerged colonies growing in liquid medium were induced by transfer to solid medium at the times indicated (↓). From Axelrod et al. (1973), by permission.

conditions, competence may be acquired by the germinating spore. Alternatively, it is conceivable that the competent state is bypassed in the microcycle.

It should be noted that if spore germination is very asynchronous or if competence is acquired very early after spore germination, the time of competence would be difficult to define. Thus the failure to observe competence in other fungi may be for purely technical reasons and does not necessarily imply that the phenomenon does not exist.

Precocious mutants

Mutants which conidiate earlier than wild-type have been isolated after UV-irradiation of conidia (Axelrod *et al.*, 1973). Such mutants could have an altered time of competence, an altered maturation period, or both. One such precocious mutant, BB142, was studied in some detail. As compared to wild-type it has the same maturation period and produces conidia at the same rate, but acquires competence 2 hours earlier. The mutation controlling the precocious phenotype segregates from the wild-type allele in the ratio of 1:1 and assorts independently of several chromosomal markers in sexual crosses. Although the precocious mutation has not been mapped, its behavior is consistent with it being a chromosomal locus.

Conidia of the precocious mutant BB142 were subjected to further UV mutagenesis and plated for the selection of a derivative showing more extreme precocity. The derivative so obtained (designated BP3) becomes competent 3.5 hours earlier than wild-type but again has the same maturation period (Kurtz and Champe, 1979). This multi-step approach might yield even more extreme derivatives, but the rarity of the second-step derivative suggests that there is a limit on the degree to which the precompetent state can be shortened by mutation.

PHYSIOLOGICAL CHANGES DURING PRECOMPETENT GROWTH

If the time of competence is endogenously determined, some property of the growing mycelium must change to record the passage of time. The time-keeping mechanism could be a set of sequentially dependent events, or, alternatively, a single parameter which changes with age but which has no physiological effect until a threshold value is reached. Several studies with *A. nidulans* have revealed changes in vegetative mycelia which are coincident with or preceed the acquisition of competence.

Enzyme inducibility

If the transition to the competent state potentiates the expression of genes involved in conidiation, it is conceivable that the expressibility of other genes might also be affected. This possibility was explored by measuring the biochemical induction of certain enzymes as a function of culture age (Gealt and Axelrod, 1974). Nitrate reductase (induced by addition of nitrate) and extracellular neutral protease (induced by sulfate deprivation) were chosen because they are metabolically unrelated and are not required for growth or conidiation. Both enzymes, which are strongly inducible in precompetent mycelia, were found to become only weakly inducible as competence is acquired. For nitrate reductase, at least, the loss of inducibility is gradual, beginning some 6 hours before the transition to the competent state, as shown in Figure 3. By contrast, the constitutive enzyme glucose-6-phosphate dehydrogenase was found to be essentially the same in pre- and postcompetent cultures. Thus the decrease of inducibility is not due to some general deterioration of the protein synthetic machinery. These results

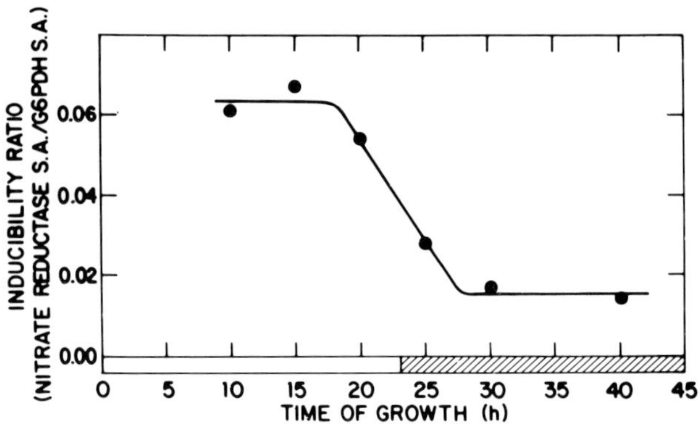

Fig. 3. Enzyme inducibility as a function of culture age. Under the conditions employed the culture became competent for induction of conidiation at 23 hours. From Gealt and Axelrod (1974), by permission.

suggest that rather general metabolic changes accompany the transition to the competent state as is further indicated by a concomitant decrease in the level of cyclic AMP (Clutterbuck, 1975).

DNA extractability

A commonly used procedure for the estimation of DNA in biological samples involves the extraction of the tissue with perchloric acid. When this method was used to measure the DNA content of *A. nidulans* during submerged growth, a dramatic change was observed in the efficiency of extraction: the extractability of DNA from a young culture was found to be only 15% that of an old culture and the transition occurred in close coincidence with the time of competence (Hall and Axelrod, 1978). The poor extractability of DNA from young mycelia was traced to an interference by a low molecular weight weight component which also inhibits the hydrolysis of added calf thymus DNA. The interferring component was identified as iron (Hall and Axelrod, 1977).

Measurements of mycelial iron content showed that young mycelia have sufficient iron to explain the low efficiency of DNA extraction. Although the specific iron content decreases with mycelial age (by cessation of uptake and dilution due to growth) it does not decrease sufficiently to explain the complete extractability of DNA from old mycelia. It was proposed that postcompetent mycelia contain an iron-binding component, absent from precompetent mycelia, which renders the iron unavailable for interaction with the DNA in the extraction process. Although such a component has not yet been directly demonstrated in *A. nidulans*, a number of physiological iron chelators are well known in other fungi, and some iron binding components have been shown to change during development of various organisms (Brown and Caston, 1962; David, 1974). A recent report that RNA polymerases can be both activated and inhibited (depending on the particular polymerase) by ferrous iron (Vaisius and Horgen, 1979) lends some credence to the idea that the availability of cellular iron could have direct and profound effects on gene expression.

Carbohydrate transport

A third developmental change manifested by vegetative mycelia of *A. nidulans* is the rate of carbohydrate uptake. The rate of glucose uptake per unit dry weight increases during spore germination until a maximum rate is attained at about 11 hours. Thereafter the uptake rate declines exponentially with age (Kurtz and Champe, 1979). During the period of decline, mycelial mass continues to increase at a constant rate until at least 30 hours after inoculation. The fact that the uptake of the non-metabolizable glucose analog, 2-deoxy-glucose, also decreases and that hexokinase activity changes very little as the culture ages argues strongly that the decrease in glucose uptake reflects an alteration in transport rather than utilization.

Under the growth conditions usually employed for these experiments the mycelia aggregate into tight pellets which could limit the access to nutrients; however, the suppression of pellet formation (with Carbopol) does not change the rate of transport loss. In addition, if mycelia of a certain age and uptake rate are either starved or transferred to fresh growth medium, the uptake rate remains essentially unchanged. This parameter thus appears to be strictly a function of culture age rather than nutritional state and could function as a clock to measure time from spore germination.

Support for this hypothesis is provided by uptake measurements on mutants which become competent earlier than wild-type. As described earlier, two precocious mutants, BB142 and BP3, become competent respectively 2 hours and 3.5 hours sooner than wild-type. As shown in Figure 4, both of these mutants lose transport capacity at a greater rate than wild-type, and although it is dangerous to generalize from two mutants, the rate of decline appears to vary with the degree of precocity.

The loss of transport capacity with age is not specific

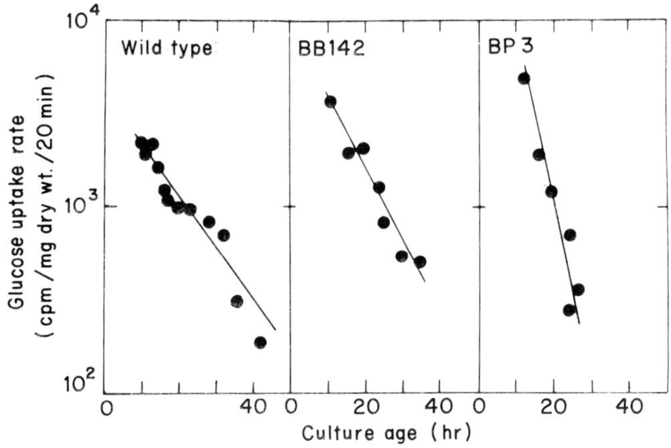

Fig. 4. Glucose uptake rate as a function of culture age for wild-type A. nidulans and two precocious mutants. From Kurtz and Champe (1979), by permission.

for glucose but has been observed also for fructose, sucrose, and for at least some amino acids, suggesting a general shut-off of transport with mycelial age. In an extension of these studies (Kurtz, 1980), the fructose uptake system was chosen because uptake of this sugar can be measured without the complication of subsequent metabolism by using a fructokinase deficient mutant. Transport loss with age could not be attributed to energy limitation nor to a physical modification or replacement of the fructose carrier. Rather, the transport activity appears to be regulated both by the rate and extent of synthesis of a limiting component of the transport system. The rate of synthesis of the activity was found to be almost 5 times greater for a 12-hour culture than for a 20-hour culture, and the maximum activity attained by the younger culture was some 3 times that of the older culture. As a tentative model it is proposed that the number of sites available for insertion of the carrier in the membrane decreases with mycelial age which could be caused by a changing composition of either lipid or protein membrane components.

ACONIDIAL MUTANTS

Abortive mutants

Mutants of *A. nidulans* defective in conidiation have been studied in several laboratories (Clutterbuck, 1969; Dorn, 1970; Martinelli and Clutterbuck, 1971; Barbata *et al.*, 1973; Martinelli, 1979). In a rather exhaustive search for mutants blocked after conidiophore formation, Clutterbuck (1969) found only two morphological variants called *bristle* and *abacus*. *Bristle* mutants are blocked in the formation of conidiophore vesicles and all subsequent structures, but the conidiophore itself continues to elongate to many times its normal length. *Abacus* mutants have normal vesicles, but the sterigmata develop abnormally into rods with periodic swellings instead of the usual chains of conidia. Allelism tests between many independent mutant isolates indicate a single *bristle* locus and a single *abacus* locus.

Clutterbuck (1969) isolated three additional variants which, although developmentally aberrant, manage to produce conidia: *medusa* mutants produce chains of metulae (primary sterigma) four or more units long from which conidia eventually bud; *stunted* mutants produce conidia directly on the surface of the conidiophore vesicle, dispensing with the sterigma state; *wet* mutants carry out an apparently normal conidiation process, but the conidia autolyze within a few days. Interestingly, *medusa* and *stunted* mutants are also blocked in cleistothecial development indicating a coupling between asexual and sexual sporulation. Certain thermosensitive aconidial mutants, discussed below, are also acleistothecial.

In a recent study, Martinelli (1979) sought to deduce the order of expression of genes in the conidiation pathway from the phenotypes of double mutants. The rationale for this approach is that mutations in early-acting genes should be epistatic over mutations in late-acting genes. The deductions essentially confirm the expectations based on the morphological defects of single mutants, but also indicate that the time periods for gene expression in the conidiation pathway can be overlapping.

Martinelli and Clutterbuck (1971) estimated the number of genes involved in conidiation to be 45-150 by comparing the frequency of aconidial mutants with the frequency of a class of auxotrophic mutants. Since only two late-functioning genes had been identified, it was concluded that the vast majority of genes involved in conidiogenesis function before conidiophore formation. The genetics of the conidiation process has been recently reviewed by Clutterbuck (1977, 1978).

Thermosensitive mutants[2]

Another approach to determine the order of gene expression is by use of thermosensitive aconidial mutants. If colonies growing in liquid medium at the permissive temperature (28°) are induced after competence is acquired and then, at various times, are shifted to the restrictive temperature (42°), conidiation will result if the thermosensitive function has been completed before the time of the temperature upshift.

We have applied this approach to a number of thermosensitive aconidial mutants (*aco*^{ts}) the results of which are shown in Figure 5. Most of the mutants do, in fact, lose their thermosensitivity at various times during the maturation period, which is 7 hours at 28°C. Two mutants, *aco*A14 and *aco*A102, retain their thermosensitivity until the end of the maturation period (i.e. the earliest time that a temperature upshift results in conidiation is indistinguishable from the time at which an unshifted control plate conidiates). For these mutants, the thermosensitive function is either performed very late or is required throughout the maturation period. Since this experiment determines only the end of the thermosensitive period, we cannot distinguish between these alternatives. In some developmental systems temperature downshift experiments can be used to define the beginning of the thermosensitive period. Downshift experiments are not feasible with these mutants, however, because they recover their ability to conidiate when returned to the permissive temperature.

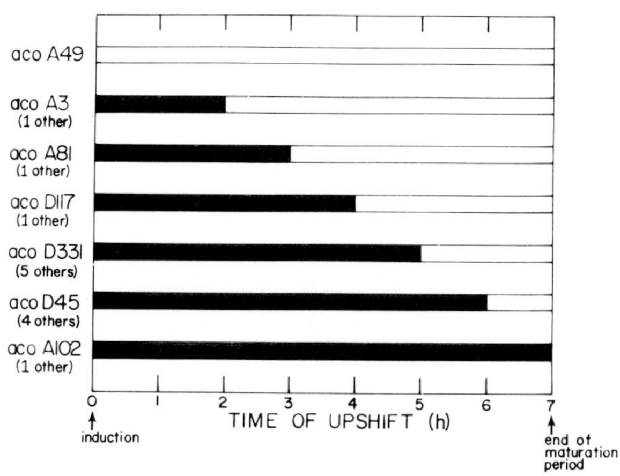

Fig. 5. Time of completion of the thermosensitive functions of $\underline{aco}^{ts}$ *mutants. See text for details.*

Identification of a competence mutant

Nineteen of the twenty mutants listed in Fig. 5 clearly require the thermosensitive function to be expressed after induction and are thus termed maturation mutants. One exceptional mutant, *aco*A49, conidiates even when the temperature upshift is performed at the time of induction. However, since temperature equilibration is not instantaneous, the apparent lack of thermosensitivity could result merely from the rapid completion of the thermosensitive function immediately after induction. Alternatively, the thermosensitive function of this mutant might have been completed before induction.

These alternatives can be distinguished by an experiment in which induction and temperature upshift are carried out simultaneously but at various times, as in the experiment of Figure 6. It is seen that colonies of *aco*A49 are able to conidiate only if the induction-upshift is after a certain time, in contrast with the control maturation mutant *aco*D331, which fails to conidiate regardless of the time. Since its thermosensitive function is completed before induction, *aco*A49 has the properties expected of a mutant blocked in the acquisition of competence.

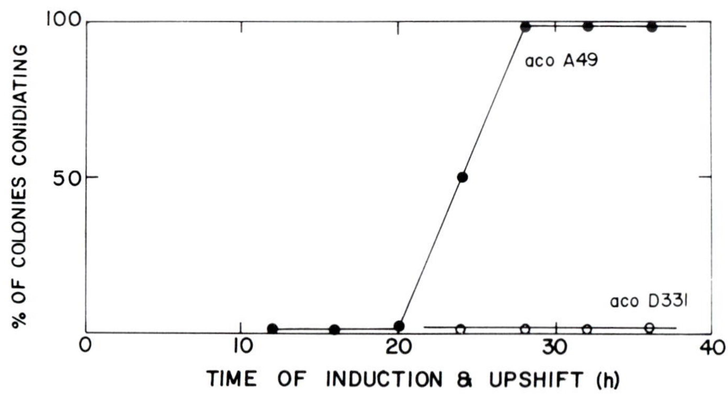

Fig. 6. The distinction between a maturation mutant, acoD331, and a competence mutant, acoA49. Colonies growing in liquid medium at 28°C were induced and simultaneously shifted to 42°C. Conidiation was scored 24 hours later.

It is to be noted that the time of induction-upshift for which half of the *aco*A49 colonies conidiate is about 23 hours. This is some 11 hours before *aco*A49 and wild-type become competent (32 ± 1 hr) at the permissive temperature employed (28°C), and is not an unexpected result since it is reasonable to assume that the acquisition of competence could require a sequence of steps, extending back to the time of spore germination, any one of which could be blocked by mutation.

A protein associated with conidiation

By examining the spectrum of proteins synthesized during

development, and changes in this spectrum resulting from mutation, it is sometimes possible to deduce the sequential dependency of the protein synthetic events as described by Francis (1977) using mutants of *Polysphondylium pallidum*. This sequential dependency is relevant to the manner by which the synthetic and developmental events are regulated.

In an initial attempt to explore this approach in *A. nidulans* the proteins of confluent surface-grown wild-type cultures were examined by SDS-acrylamide gel electrophoresis as a function of culture age. As seen in Figure 7, the most striking change that occurs with age is the appearance of a 36,000 dalton protein (designated p36) at about the time that conidiophores first develop (16-18 hr in these confluent cultures). By 24 hours this protein is present at its maximum amount, accounting for some 1% of the total soluble protein, and remains relatively constant until at least 40 hours.

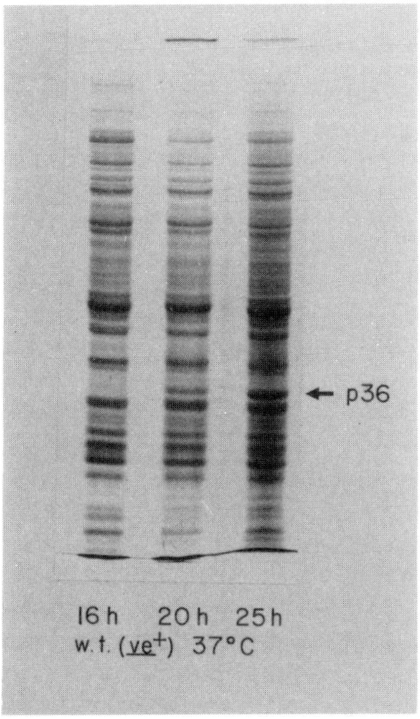

Fig. 7. The appearance of a condidiation-associated protein during development of A. nidulans.

The protein spectrum of 20 thermosensitive mutants was examined for surface cultures grown at the permissive temperature of 28°C and at 42°C where the aconidial phenotype is expressed. At 28°C, as expected, all of the mutants exhibit a pattern identical to the wild-type. At 42°C the pattern for wild-type and the majority of the thermosensitive mutants is little different from that at 28°C with the notable exception that the protein p36 is absent. Certain of the mutants do, however, clearly synthesize p36 at the restrictive temperature. The absence of p36 in the wild-type and some mutants, but not others, could be due merely to a slight mutant-dependent shift in the temperature threshold for whatever process that inhibits the formation of p36 in the wild-type. On the other hand, the pattern may reflect a negative control of p36 synthesis by conidiation functions that are expressed by some aconidial mutants but not by others.

One of the aconidial mutants, *aco*A49, is exceptional in that it produces several times more p36 than other aconidial mutants at the restrictive temperature. The overproduction of p36 by *aco*A49 could be due to enhanced synthesis, to inhibition of turnover, or to inhibition of some more specific post-translational modification that changes its physical form. This question is being studied by pulse-labeling experiments. At whatever level this protein is regulated, it is clear that the mechanism has gone amiss in *aco*A49.

We have described p36 as a protein associated with conidiation from the fact that its appearance coincides with the appearance of conidiophores. On the other hand, conidiation can occur in the absence of detectable p36 (wild-type at 42°C), and p36 can be present, even in excess, without the occurrence of conidiation (*aco*A49 at 42°C). It is thus unlikely that p36 has any causitive role in the conidiation process. It is nevertheless curious that, among the twenty aconidial mutants studied, the one blocked in the acquisition of competence also has an altered regulation of the amount of the one protein associated with conidiation.

CONTROL OF SPORULATION BY THE *VELVET* LOCUS

The velvet phenotype

The original Glasgow isolate of *A. nidulans* (FGSC-4), which we have used as our wild-type, produces a profuse growth of aerial hyphae which, in older colonies, tower over and partially obscure the underlying conidial heads, giving the colony a velvety appearance. Käfer (1965) described a derivative which lacks the aerial hyphae and showed that the velvet phenotype is controlled by a locus on chromosome VIII. The parent and mutant alleles are designated ve^+ and ve^- respectively. Käfer also noted that ve^- strains conidiate better than ve^+ strains, whereas ve^+ strains appeared to be more sexually precocious than ve^- strains. All of our thermosensitive mutants, being derived from FGSC-4, are *ve*$^{+}$, whereas almost all other strains in use are *ve*$^{-}$. Since its description by Käfer, the *velvet* locus has been completely ignored except as a mapping marker.

We have found that the allelic state of the *velvet* locus also affects the conidiation in submerged liquid culture under certain conditions. In Cove's minimal medium at pH 5.3 with 10 mM glutamate as a nitrogen source, *ve*$^{-}$ strains begin to form distinct conidiophores between 28 and 30 hours, whereas under identical conditions, *ve*$^{+}$ strains do not show similar differentiation until after 48 hours. This difference, however, is observed only at high inocula densities (above 10^5/ml) and only in certain media.

The above observation and those described below were made with various *ve*$^{-}$ strains whose isogenicity with FGSC-4 might be questionable. For this reason an isogenic pair was obtained by reisolating a *ve*$^{-}$ derivative from a *ve*$^{+}$ parent. The differences observed between *ve*$^{+}$ and *ve*$^{-}$ strains were confirmed using this pair.

Cleistothecial development in ve^{+} and ve^{-} strains

Studies on the kinetics of cleistothecial development have been limited to qualitative observations because, until these bodies mature, most are invisibly buried in the mycelial mat and under conidial heads making quantitation difficult. This difficulty can be overcome by taking advantage of the fact, recently discovered in our laboratory, that cleistothecial primordia and the surrounding Hülle cells produce a *p*-diphenol oxidase (laccase). The primordia can thus be observed and quantified from early stages of development by applying a chromogenic laccase substrate (such as dimethyl phenylenediamine) to the mycelial mat. The resulting darkly pigmented foci are best observed in yellow conidial strains in which the conidial laccase is absent (Clutterbuck, 1972). Staining and observation is facilitated by first treating the mat with chloroform which flattens the mat and renders it wettable.

Using this technique, we have confirmed quantitatively Käfer's observation that sexual development occurs more readily in *ve*$^{+}$ strains relative to *ve*$^{-}$ strains. As shown in Figure 8, cleistothecial development of *ve*$^{-}$ at 37°C is not only delayed but attains a lower final density of cleistothecia relative to *ve*$^{+}$. At 42°C cleistothecial initiation is completely blocked in the *ve*$^{-}$ strains but still occurs in the *ve*$^{+}$ strain, albeit very late. Strains carrying the *ve*$^{-}$ allele thus behave as thermosensitive acleistothecial mutants.

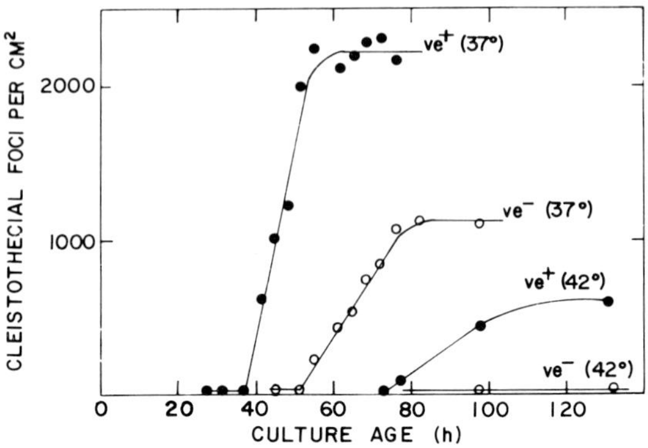

Fig. 8. Cleistothecial development in ve^+ *and* ve^- *strains of A. nidulans. A series of plates were inoculated with* 10^5 *conidia to obtain a confluent lawn. At the times indicated, the plates were stained for laccase, as described in the text, and the laccase-positive foci were counted by transillumination.*

By transferring confluent surface cultures of ve^- from 37°C to 42°C at various times it is possible to determine whether there is a particular time at which the thermosensitivity is relieved. As shown in Figure 9, cleistothecium formation becomes thermoinsensitive between 40 and 50 hours which is some 10-15 hours earlier than the first primordia can be observed by the laccase staining method. In addition to identifying a critical period for sexual development, this experiment shows that the initiation of sexual development is precisely timed.

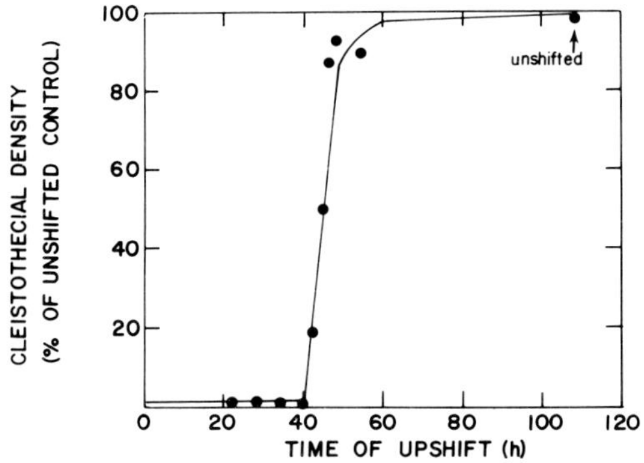

Fig. 9. Cleistothecial development of ve^- *following temperature shift from 37°C to 42°C. The experiment was performed identically to that of Fig. 8 except that plates were upshifted at the indicated times and cleistothecial densities were measured at 108 hours.*

The protein spectra of *ve*$^+$ *and* *ve*$^-$

The soluble protein from confluent plate cultures of *ve*$^+$ and *ve*$^-$ grown at 37°C were compared by SDS acrylamide gel electrophoresis as shown in Figure 10. The patterns are identical with the exception of one band present in *ve*$^+$ but absent in *ve*$^-$. The band missing in *ve*$^-$ corresponds to the conidiation-associated protein p36 of Figure 7 as shown by a separate experiment. (The banding patterns of Figures 7 and 10 differ because of slightly different gel formulations.) To confirm that the absence of p36 is, in fact, due to the *ve*$^-$ mutation, 10 *ve*$^+$ and 10 *ve*$^-$ recombinants from a *ve*$^+$ × *ve*$^-$ cross were examined for the presence of p36. The protein was present only in those recombinants which exhibited the *ve*$^+$ phenotype.

The experiments of Figures 9 and 10 delimit the time period during which the *velvet* gene functions. Since p36 begins to appear at about 18 hours in *ve*$^+$ but not in *ve*$^-$, the *velvet* gene must be expressed prior to this time. The temperature shift experiment of Figure 9 shows, on the other hand, that the function of the *velvet* gene is required until 40 hours.

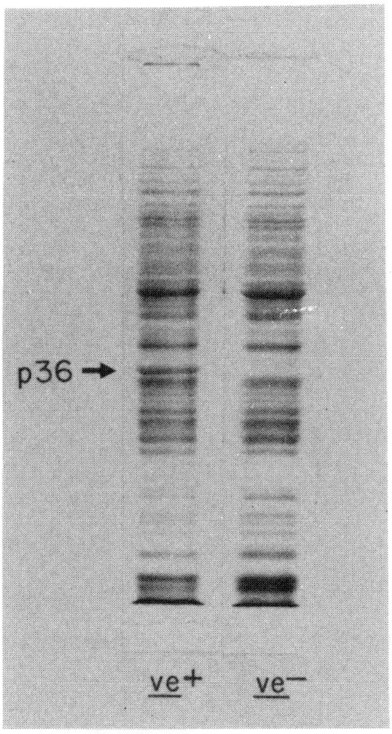

Fig. 10. The protein spectra of *ve*$^+$ *and* *ve*$^-$.

SUMMARY AND PROSPECTS

The studies reviewed here show that a spore-originated colony of *Aspergillus nidulans* must grow vegetatively for a certain length of time before it is inducible to conidiate. By analogy with similar phenomena in higher organisms the inducible state has been termed competence. It is not assumed, however, that competence in *Aspergillus* and competence in *Drosophila* have the same, or even a similar, mechanism.

Vegetative mycelia exhibit several changes which may be related to the acquisition of competence: inducible enzymes become less inducible, DNA extractability (inversely related to mycelial iron content) increases, and the transport of carbohydrates and other metabolites progressively decreases. For the latter phenomenon, mutants which become competent earlier than wild-type lose transport capacity more rapidly than wild-type, indicating a relationship between transport and competence, though not necessarily a causal one. The loss of transport capacity with mycelial age and its irreversibility by medium replacement suggests that this parameter may be endogenously programmed and could thus serve as a timing mechanism.

A set of thermosensitive aconidial mutants have been used in a mutational analysis of the conidiation process. Temperature shift experiments show that most mutants are blocked at various times during the maturation period, i.e. after induction. One exceptional mutant, *aco*A49 was found to be blocked before induction and is, by definition, defective in the acquisition of competence. Genetic analysis of *aco*A49 is in progress. Preliminary indications are that this mutant differs from the wild-type by more than one mutation.

Electrophoretic examination of the protein spectrum of wild-type as a function of culture age at 37°C shows that one prominent protein, p36, appears at a time closely coincident with the initiation of conidiation. This protein, however, is not necessary for conidiation and is not present in mature conidia. At 42°C, p36 is not produced by wild-type, but is produced by certain aconidial mutants and is overproduced by *aco*A49. In addition, *ve*$^-$ strains, which are acleistothecial at elevated temperature, do not produce p36 at any temperature. Although the significance of this pattern is unclear, it is noteworthy that the presence or absence of this protein depends on the function of genes of both the asexual and sexual pathways. Even without knowledge of its function, p36 may prove useful as an event marker to define the points of coupling between the two sporulation pathways. To this end, similar studies on acleistothecial mutants are being undertaken.

The *velvet* gene has a strong effect on the choice of developmental pathways. Strains carrying the *ve*$^-$ allele are enhanced over *ve*$^+$ in conidiation both on solid medium and, under certain conditions, in submerged culture. By contrast, the *ve*$^-$ allele is antagonistic to sexual sporulation: cleistothecial development at 37°C is delayed relative to *ve*$^+$ and is completely blocked at 42°C. The tacit assumption that there is only one *velvet* gene is being tested by genetic analysis of additional *ve*$^-$ isolates. The one new isolate we have studied behaves qualitatively like the original *ve*$^-$A of Käfer, but differs from it quantitatively in its effect on sexual development.

The discovery that cleistothecial primordia contain laccase opens the way to a quantitative analysis of sexual development which has heretofore not been feasible. In addition, the fact that Hülle cells also manifest this enzyme activity suggests a possible function for these cells: Hülle cells may supply the developing cleistothecium with laccase which, in turn, may be responsible for melanization of the cleistothecial wall.

Our initial studies of cleistothecium formation show that a function controlled by the *velvet* gene is required considerably earlier than the onset of morphological differentiation. Zonneveld (1977) has recently reviewed the metabolic requirements for sexual development in which he points out the importance of α-1,3 glucan, the synthesis or degradation of which could be the process which is thermosensitive in *ve*$^-$.

The initiating event for sexual development is, of course, the mysterious fusion of specialized hyphae resulting in the dikaryotic cell from which all of the ascospores of a cleistothecium arise. The time of this fusion event in the developmental schedule is, however, not known precisely.

Analysis of these metabolic and genetic events during development of *ve*$^-$ and other acleistothecial mutants should help identify the functions necessary for sexual development. Of particular interest will be whether such a mutational analysis can define a state of competence for sexual development as *aco*A49 does for asexual development. In this context, it is noteworthy that, although many of the thermosensitive aconidial mutants are normal in cleistothecium formation at 42°C, some, including *aco*A49, are completely acleistothecial. Further studies on mutants of the latter type may help to identify control functions which operate on or between the asexual and sexual pathways.

FOOTNOTES

[1]The growth medium employed for the experiment of Figure 2 was composed of 0.5% yeast extract (Difco) and 2% glucose. For plates, the medium was solidified with 1.5% agar (Difco). For the experiments of Figures 5-10 this medium was supplemented with trace elements at the concentrations used by Cove (1966). The media for other experiments reviewed are described in the referenced papers.

[2]This and subsequent sections are summary descriptions of as yet unpublished studies from our laboratory.

ACKNOWLEDGEMENTS

We would like to express our appreciation to Amy Chang for able technical assistance, and to Dr. Pamela Champe for useful suggestions and editorial assistance in preparing this manuscript. This work was supported by grant GM 17020 from the National Institute of General Medical Sciences and the Charles and Johanna Busch Memorial Fund.

REFERENCES

Anderson, J.G. (1978), Temperature-induced fungal development. *In* "The Filamentous Fungi" (Ed. J.E. Smith and D.R. Berry). Vol. 3, pp. 358-375. Edward Arnold, London.

Axelrod, D.E. (1972). Kinetics of differentiation of conidiophores and conidia by colonies of *Aspergillus nidulans*. *J. Gen. Microbiol.* 73, 181-184.

Axelrod, D.E., Gealt, M. and Pastushok, M. (1973). Gene control of developmental competence in *Aspergillus nidulans*. *Develop. Biol.* 34, 9-15.

Barbata, G., Valdes, L. and Sermonti, G. (1973). Complementation among developmental mutants in *Aspergillus nidulans*. *Mol. Gen. Genet.* 126, 227-232.

Benjamin, C.R. (1955). Ascocarps of *Aspergillus* and *Penicillium*. *Mycologia* 47, 669-687.

Brown, G.R. and Caston, J.D. (1962). Biochemistry of amphibian development III. Identification of ferritin in the egg and early embryos of *Rana pipiens*. *Develop. Biol.* 5, 445-451.

Clutterbuck, A.J. (1969). A mutational analysis of conidial development in *Aspergillus nidulans*. *Genetics* 63, 317-327.

Clutterbuck, A.J. (1972). Absence of laccase from yellow-spored mutants of *Aspergillus nidulans*. *J. Gen. Microbiol.* 70, 423-435.

Clutterbuck, A.J. (1975). Cyclic AMP levels during growth and conidiation. *Aspergillus News Letter* 12, 13-15.

Clutterbuck, A.J. (1977). Genetics of Conidiation. *In* "Genetics and Physiology of *Aspergillus*" (Ed. J.E. Smith and J.A. Pateman). pp. 305-317. Academic Press, London.

Clutterbuck, A.J. (1978). Genetics of vegetative growth and asexual reproduction. *In* "The Filamentous Fungi" (Ed. J.E. Smith and D.R. Berry). Vol. 3, pp. 240-274. Edward Arnold, London.

Cove, D.J. (1966). The induction and repression of nitrate reductase in the fungus *Aspergillus nidulans*. *Biochem. Biophys. Acta* 113, 51-56.

David, C.N. (1974). Ferritin and iron metabolism in *Phycomyces*. *In* "Microbial Iron Metabolism: A Comprehensive Treatise" (Ed. J.B. Neilands). pp. 149-158. Academic Press, New York.

Dorn, G.L. (1970). Genetic and morphological properties of undifferentiated and invasive variants of *Aspergillus nidulans*. *Genetics* 66, 267-279.

Ellis, T.T., Reynolds, D.R. and Alexopoulos, C.J. (1973). Hülle cell development in *Emericella nidulans*. *Mycologia* 65, 1028-1035.

Francis, D. (1977). Synthesis of developmental proteins in morphogenetic mutants of *Polysphondylium pallidum*. *Develop. Biol.* 55, 339-346.

Gealt, M.A. and Axelrod, D.E. (1974). Coordinate regulation of enzyme inducibility and developmental competence in *Aspergillus nidulans*. *Develop. Biol.* 41, 224-232.

Hall, N.E.L. and Axelrod, D.E. (1977). Interference of cellular ferric ions with DNA extraction and the application to methods of DNA determination. *Analyt. Biochem.* 79, 425-430.

Hall, N.E.L. and Axelrod, D.E. (1978). Sporulation competence in *Aspergillus nidulans*: a role for iron in development. *Cell Differen.* 7, 73-82.

Käfer, E. (1965). The origin of translocations in *Aspergillus nidulans*. *Genetics* 52, 217-232.

Kurtz, M.B. (1980). Regulation of fructose transport during growth of *Aspergillus nidulans*. *J. Gen. Microbiol.* 118, 389-396.

Kurtz, M.B. and Champe, S.P. (1979). Genetic control of transport loss during development of *Aspergillus nidulans*. *Develop. Biol.* 70, 82-88.

Martinelli, S.D. (1976). Conidiation of *Aspergillus nidulans* in submerged culture. *Trans. Brit. Mycolog. Soc.* 67, 121-128.

Martinelli, S.D. (1979). Phenotypes of double conidiation mutants of *Aspergillus nidulans*. *J. Gen. Microbiol.* 114, 277-287.

Martinelli, S.D. and Clutterbuck, A.J. (1971). A quantitative survey of conidiation mutants in *Aspergillus nidulans*. *J. Gen. Microbiol.* 69, 261-268.

Nöthiger, R. (1972). The larval development of imaginal disks. *In* "The Biology of Imaginal Disks" (Ed. H. Ursprung and R. Nöthiger). pp. 1-34. Springer-Verlag, New York.

Pastushok, M. and Axelrod, D.E. (1976). Effect of glucose, ammonium and media maintenance on the time of conidiophore initiation by surface colonies of *Aspergillus nidulans*. *J. Gen. Microbiol.* 94, 221-224.

Saxena, R.K. and Sinha, U. (1973). Conidiation of *Aspergillus nidulans* in submerged liquid culture. *J. Gen. Appl. Microbiol.* 19, 141-146.

Smith, J.E. (1978). Asexual sporulation in filamentous fungi. *In* "The Filamentous Fungi" (Ed. J.E. Smith and D.R. Berry). Vol. 3, pp. 214-239. Edward Arnold, London.

Smith, J.E. and Anderson, J.G. (1973). Differentiation in the *Aspergilli*. *Symp. Soc. Gen. Microbiol.* 23, 295-337.

Thom, C. and Raper, K.B. (1945). "A Manual of the *Aspergilli*" Williams and Wilkins Co., Baltimore.

Vaisius, A.C. and Horgen, P.A. (1979). The effects of several divalent cations on the activation or inhibition of highly purified RNA polymerases II from *Agaricus bisporius* and from calf thymus. *J. Cell Biol.* 83 No. 2, Part 2, 405a.

Zonneveld, B.J.M. (1977). Biochemistry and ultrastructure of sexual development in *Aspergillus*. *In* "Genetics and Physiology of *Aspergillus*". (Ed. J.E. Smith and J.A. Pateman). pp. 59-80. Academic Press, London.

NUCLEIC ACIDS

J.L. Van Etten, K.R. Dahlberg, and G.M. Russo

Department of Plant Pathology, University of Nebraska, Lincoln, Nebraska 68583, U.S.A.

Since the last symposium on fungal spores many new techniques have been developed for studying DNA and RNA. The application of these techniques suggests that there are certain differences in how DNA is organized and transcribed in fungi as compared to higher eukaryotic organisms. This review will briefly consider some of the new developments in nucleic acid research, describe the current information on the organization and transcription of the fungal genome, and mention apparent differences between fungi and higher eukaryotes. Finally, new information on the role of nucleic acids in sporulation and spore germination will be discussed. Earlier work describing the general properties of DNA and RNA in fungi and the molecular events which accompany spore germination has been detailed in other review articles (Brambl *et al.*, 1978; Lovett, 1975; 1976; Lovett *et al.*, 1977; Schmit and Brody, 1976; Van Etten *et al.*, 1976; 1977) and will not be covered in this review. Because of page limitations much of the information cited in this article will be from review articles or a few examples of primary articles and should not be considered an exhaustive review of the literature. We beg your indulgence if we prove to be guilty of gross generalizations and oversimplifications in this rapidly expanding field of research.

NEW TECHNIQUES AND DEVELOPMENTS IN NUCLEIC ACID RESEARCH.

Probably the single most important discovery in DNA analysis in the past few years was the identification and subsequent isolation and characterization of enzymes (restriction endonucleases) from bacteria which cleave DNA into fragments of defined size and genetic information. Furthermore, these fragments can be joined (with ligase enzymes) to DNA from other sources, such as bacterial plasmid DNA or λ phage DNA, and inserted into bacteria (usually *Escherichia coli*). With appropriate selection techniques colonies of bacteria which contain the DNA of interest can be isolated and propagated (cloned). The net result is that large quantities of specific DNA fragments can be obtained for chemical and biological analyses.

The identification of specific DNA fragments relies on other recently developed techniques such as the synthesis of complementary (c) DNA from RNA. Complementary DNA can be used to detect genomic DNA fragments after electrophoresis on gels (Southern blots) and for the *in situ* selection of bacterial colonies containing cloned DNA. Concurrently the development of efficient *in vitro* protein synthesizing systems (e.g. from wheat germ and reticulocytes) allows the identification of mRNAs isolated from cells.

While procedures were becoming available for the isolation and identification of specific fragments of DNA, rapid progress was made in the sequencing of DNA molecules. It is now possible to sequence DNAs of ca. 250 bases in a few hours. Techniques for sequencing RNA have not developed as fast as those for DNA, but progress is occurring and it is only a matter of time before RNA sequencing becomes as commonplace as DNA sequencing. Consequently, one can anticipate the sequencing of selected fungal genes in the near future.

There are two other recent developments which will have a long range effect on the study of genes and gene expression in eukaryotes and especially fungi. First, transformation systems have been developed for *Saccharomyces cerevisiae* (Hinnen *et al.*, 1978; Struhl *et al.*, 1979) and *Neurospora crassa* (Case *et al.*, 1979), and these open up new possibilities for studying the genes involved in development. Second, there is evidence that yeast and probably other fungi, like bacteria, corn, and *Drosophila*, contain transposing elements (Greer and Fink, 1979; Cameron *et al.*, 1979). The role, if any, that these might play in sporulation and spore germination is not known, although they are involved in determining mating types of yeast (e.g. Strathern *et al.*, 1980).

NATURE OF THE FUNGAL GENOME.

Knowledge of the composition and compartmentalization of cellular DNA is essential to understanding the mechanisms by which nucleic acids may regulate the sporulation and germination processes. Most properties of the fungal genome have been determined from vegetative cells; however, the general properties of DNA from vegetative cells are probably identical to those of DNA from spores.

Nuclear chromosomal DNA comprises ca. 80-99% of the total cellular DNA in fungi. Studies on yeast chromosomes (Petes and Fangman, 1972; Petes *et al.*, 1973) indicate that each chromosome contains a single DNA molecule ranging in size from 0.5 to 12 x 10^8 daltons, with an average mass of 5 to 7 x 10^8 daltons (four times smaller than the *Escherichia coli* DNA). The majority of the remaining DNA is located in mitochondria (Cummings *et al.*, 1979). In addition, *Saccharomyces cerevisiae*, *Podospora anserina*, and probably other fungi contain a small covalently linked closed circular double stranded (ds) DNA (called 2 μm DNA or plasmid DNA). The properties and characteristics of these DNAs as well as virus like particle nucleic acids found in fungi have been the subject of a recent book (Lemke, 1979) and will not be discussed further.

Fungal Chromatin.

Basic proteins, called histones, are associated with the chromosomal material of higher plants and animals. Five proteins, termed H1, H2A, H2B, H3 and H4, comprise the histone group, although other histones are present in some specialized cells. A controversy existed for years as to whether histones were present in fungal chromatin (reviewed by Horgen and Silver, 1978; Morris, 1980). With the development of better isolation techniques, it is clear that fungal chromatin, e.g. *N. crassa* (Goff, 1976) and *A. nidulans* (Felden *et al.*, 1976), contains a complete complement of histone-like proteins that are similar but probably not identical to those of higher organisms.

In higher eukaryotes, two copies each of histones H2A, H2B, H3, and H4 are present in clusters (called nucleosomes) at regularly spaced intervals along the DNA [ca. 200 base pairs (bp) apart in most organisms], such that chromatin has the appearance of "beads on a string" when viewed under the electron microscope. This basic structure exists in fungi, although the repeat unit is only 150-160 bp. This shorter repeat unit is due to a reduced distance between nucleosomes (10-20 bp in fungi versus 50-60 bp in higher eukaryotes) rather than due to a reduced nucleosome size. The role of histone H1 is less well understood, but it is believed to be associated with linker DNA between the nucleosomes and may be involved in condensing the chromatin material. Nonhistone proteins (acidic proteins) are also associated with chromatin in eukaryotic organisms, including fungi. However, their function is not known.

Size and organization of fungal DNA.

The size of the haploid fungal genome ranges from about 0.5×10^{10} in *Mucor rouxii* to 3×10^{10} in *Coprinus lagopus* (Storck, 1974). Some recent papers report values of 2.7×10^{10} (4.3×10^7 bp) for *Achlya bisexualis* (Hudspeth *et al.*, 1977), 2.3×10^{10} (3.7×10^7 bp) for *Schizophyllum commune* (Dons *et al.*, 1979), 1.7×10^{10} (2.7×10^7 bp) for *Aspergillus nidulans* (Timberlake, 1978) and *Neurospora crassa* (Krumlauf and Marzluf, 1979), and 0.9×10^{10} (1.4×10^7 bp) for *Saccharomyces cerevisiae* (Lauer *et al.*, 1977). Thus, the fungal genome is roughly 2 to 10 times more complex than the bacterial genome and ca. 1% the size of the genome in man.

Fifty percent or more of the DNA in higher eukaryotic organisms contains repetitive sequences, with the remainder consisting of single copy or unique sequences. In contrast, the amount of repetitive DNA in fungi is much less. For example, 16% of the DNA in *Achlya* (Hudspeth *et al.*, 1977), 8% in *N. crassa* (Krumlauf and Marzluf, 1979), and 2-3% in *A. nidulans* (Timberlake, 1978) is reiterated. Earlier studies on fungi (Ojha and Dutta, 1978) also reported values of 20% or less. In the majority of eukaryotes the arrangement of repetitive and single copy DNA is characterized by the interspersion of repetitive sequences, 200-400 bp in length, throughout a major fraction of the genome. These short repetitive sequences are linked to single copy DNA at intervals averaging 1000-2000 bp (Davidson *et al.*, 1975); this arrangement is called a "short period interspersion pattern". Since the repetitive DNA is organized in such highly ordered patterns in nearly all eukaryotes it has been postulated that interspersed repetitive sequences play a central role in the coordinated regulation of sets of unlinked structural genes (Davidson and Britten, 1979). A few organisms, mainly insects, exhibit a "long-period interspersion pattern" (Crain *et al.*, 1976). That is, the repetitive DNA is organized in long stretches (4000-10,000 bp) that are interspersed with single copy DNA at intervals of at least 10,000 bp.

Recent studies on the organization of fungal DNAs indicate that fungi do not fit either pattern. For example, repetitive and single copy DNA sequences in *Achlya* consist of 2.7×10^4 and 1.35×10^5 bp, respectively (Hudspeth *et al.*, 1977). Likewise, the repetitive sequences in *N. crassa* are estimated to be 7.3×10^4 bp long (Krumlauf and Marzluf, 1979). Since 8% of the 2.7×10^7 bp of DNA in *N. crassa* is repetitive, only about 30 clusters of repetitive DNA, as a maximum, can exist in the entire genome. Other experiments suggest that ca. 90% of the reiterated DNA in *N. crassa* (Krumlauf and Marzluf, 1980) and possibly all of the repetitive DNA in *A. nidulans* (Timberlake, 1978), codes for ribosomal (r) RNA.

The net result is that fungi, which exhibit coordinated gene control, do so in the absence of extensive amounts of reiterated DNA. It should be noted, however, that the limits of detection in these experiments is ca. 15-20 bp. That is, if fungi have reiterated sequences less than 20 bp long interspersed between single copy DNA, they would not be detected.

The number of potential genes in fungi can be estimated from the amount of single copy DNA, assuming the average gene is 1200 bp long. Thus *N. crassa* contains ca. 20,700 genes (Krumlauf and Marzluf, 1979), *Achlya* ca. 30,000 (Hudspeth *et al.*, 1977), and *A. nidulans* ca. 22,000 (Timberlake, 1978). Ten to 25% (3000 to 6000) of these genes are expressed during vegetative growth. However, not all genes are expressed to the same extent. That is, a few genes (20 to 30) are expressed 200 to 1000 times per cell (high frequency), 200 to 400 genes are expressed 20 to 150 times per cell (medium frequency), and the remaining genes 1 to 11 times per cell (low frequency) (Hereford and Rosbash, 1977a; Rozek *et al.*, 1978). Hereford and Rosbash (1977b) have suggested that ribosomal proteins in yeast may be translated from the medium and high frequency expressed gene products.

Detailed studies of fungal ribosomal RNA genes.

Fungi, like other eukaryotic organisms, contain multiple genes for rRNAs (100-300 copies per nucleus). Probably all fungi, like other eukaryotes, synthesize the three rRNAs (5.8S, 18S, and 25S) from a single large precursor RNA. Furthermore, these rRNA genes are arranged in the order of 5'-18S, 5.8S, 25S-3' in *S. cerevisiae* (Bell *et al.*, 1977), *N. crassa* (Free *et al.*, 1979), and *Mucor racemosus* (Cihlar and Sypherd, 1980), with noncoding sequences between the three genes. This arrangement may be common to all fungi. Beyond this basic similarity, however, differences exist among fungi with regard to the molecular details of the structural organization. For example, the size of the basic unit ranges from 5.8×10^6 in *S. cerevisiae* (Nath and Bollon, 1977) to approximately 30×10^6 in *Dictyostelium* (Taylor *et al.*, 1977). The gene cluster may exist as a tandem repeat as in *N. crassa* (Free *et al.*, 1979), or in an extrachromosomal palindrome arrangement as in *Physarum polycephalum* (Molgaard *et al.*, 1976; Seebeck *et al.*, 1979). Intervening sequences have been reported in the 25S gene in *Physarum* (Campbell *et al.*, 1979) but are absent in *S. cerevisiae* (Philippsen *et al.*, 1978). Finally the 5S RNA gene is closely associated with the other rRNA genes in *S. cerevisiae* (Nath and Bollon, 1977), *Achlya ambisexualis* (Rozek and Timberlake, 1979), and *M. racemosus*

(Cihlar and Sypherd, 1980), but this same gene is scattered throughout the genome in *N. crassa* (Free *et al.*, 1979). In *S. cerevisiae* the 5S RNA gene is transcribed from the opposite (complementary)strand from that transcribing the other rRNAs (Valenzuela *et al.*, 1977). The major conclusion from these experiments is that differences in the arrangement of the rRNA genes exist among fungi.

While the most detailed studies on fungal genes have been conducted on rRNA genes, other genes are beginning to be examined. For example, the yeast structural gene for imidazole-glycerolphosphate dehydratase (Struhl and Davis, 1977) and the *N. crassa* gene for catabolic dehydroquinase (Hautala *et al.*, 1979) have been isolated and cloned in *E. coli*. In both cases the gene is expressed in the bacterium. Consequently, one can expect rapid progress in our understanding of selected fungal genes in the near future.

Transcription

Most types of RNA, with the possible exception of 5S RNA, undergo some type of post-transcriptional modifications in eukaryotic organisms. As noted in the previous section, rRNAs are initially transcribed as a large precursor molecule which is then cleaved to yield 5.8S, 18S, and 25S RNAs. Likewise many, but not all, yeast transfer (t) RNAs are initially transcribed as molecules of ca. 100 nucleotides (e.g. Johnson *et al.*, 1980); these molecules are cleaved and some individual bases modified to yield a final product of about 75 bases.

Extensive post transcriptional modification occurs during the synthesis of messenger (m) RNA in higher eukaryotes. The initial transcriptional product (termed heterogeneous nuclear RNA or HnRNA) may be several times larger than the final product. The processing of HnRNA to mRNA involves the removal of internal regions (called intervening sequences) and the subsequent rejoining of the RNA fragments to yield the final product (Crick, 1979).

At present there is no evidence that fungal mRNAs are derived from large HnRNA like precursors (Timberlake *et al.*, 1977; Brambl *et al.*, 1978). If they are initially transcribed as a larger molecule, the initial transcript, on the average, is only 10 to 20% larger than the final product (Firtel and Lodish, 1973; Freer *et al.*, 1977). However, the coding region of at least one fungal gene (actin gene in *S. cerevisiae*) is interrupted by a 304 bp intervening sequence (Gallwitz and Sures, 1980). Based on this one example it is too early to tell if intervening sequences will be common in fungal genes or if they only occur in a few genes.

The average size of fungal mRNAs has been estimated to be 750 nucleotides long in *S. cerevisiae* (McLaughlin *et al.*, 1973), 870 nucleotides in *Rhizopus stolonifer* (Freer *et al.*, 1977), 1140 nucleotides in *Achlya* (Timberlake *et al.*, 1977) and 1275 nucleotides in *Botryodiplodia theobromae* (Van Etten and Rawn, 1979). It is unlikely that these values reflect real differences among fungi but rather vary because of the way the values were obtained. For discussion purposes we will assume that fungal mRNAs average about 1200 nucleotides.

The synthesis of functional mRNAs in fungi involves two other post transcriptional modifications common to most eukaryotic organisms. Adenylate residues are added to the 3' hydroxyl end of the RNA, giving rise to poly adenylate (A) segments as long as 100 nucleotides (Freer *et al.*, 1977) which are rapidly processed to a length of 30-50 nucleotides (Brambl *et al.*, 1978). This contrasts with poly A segments of 100 to 200 nucleotides attached to mRNA from higher eukaryotes (Brawerman, 1976). The second post transcriptional event is the addition of a 7-methylguanosine triphosphate to the 5' hydroxyl end of the mRNA (cap structure). Cap structures have been identified in mRNAs from *S. cerevisiae* and *N. crassa* (Banerjee, 1980) and presumably are present in all fungi.

To summarize, studies on the organization of the fungal genome suggest that fungal DNAs and RNAs are in many ways identical to those found in higher eukaryotes; however, some notable differences do exist. The lack of large amounts of repetitive DNA in fungi may be especially significant in light of recent models on coordinated regulation of gene expression (Davidson & Britten, 1979). Detailed studies on the organization and function of the fungal genome may be tremendously important in contributing to our understanding of the regulation of eukaryotic gene expression in general, since such differences in genome organization between fungi and higher eukaryotes must be accounted for.

DEVELOPMENTAL GENE EXPRESSION IN FUNGI.

A widely held view among developmental biologists is that development, or the formation of a specialized cell from an unspecialized cell, is the result of differential expression of genes. This maxim is largely supported among workers studying fungal sporulation and spore germination. Although it has been suggested that "this dogma is not realistic and, therefore, is detrimental to our thinking, teaching and research" (Wright, 1979), fungal sporulation and spore germination have been studied in this light.

This section will briefly consider evidence suggesting that development in fungi is characterized by differential changes in gene expression, but will not attempt to ascribe causality to it. Rather, it will focus on attempts to detect and identify the differential expression of functional genes during fungal sporulation and spore germination. The search for changes in expression of genes during both spore germination and sporulation has largely been done by searching for gross changes in protein content or enzyme activity, by screening for changes in the cellular content of messenger RNAs, and, to a much lesser extent, by using the newer DNA techniques. The following summary of the work done in this area suggests that the newer recombinant DNA techniques hold much promise for understanding the differential activation of genes during fungal development.

Sporulation

The subject of nucleic acid metabolism during fungal sporulation is difficult to study because many fungi possess eucarpic fruiting bodies in which nucleic acid synthesis presumably occurs in both the nascent spores and accessory fruiting body structures. Removal of nascent spores from the developing thalli may cause abortion of development. For instance, in *Botryodiplodia theobromae*, preliminary results suggest that removal of spores from pycnidia early in the sporulation process terminates the synthesis of a spore-specific protein (Dahlberg and Van Etten, unpublished data). Thus, most investigations of fungal sporulation have been on holocarpic organisms, including the yeast, *S. cerevisiae*, and the water mold, *Blastocladiella emersonii*.

Another problem frequently encountered in studying fungal sporulation was noted by Lovett (1975) in his review of *Blastocladiella* development. That is, fungal sporulation is often triggered by dramatic shifts in the nutritional status of the organism, and these nutritional changes themselves cause alterations in nucleic acid metabolism. This problem can in part be obviated by a judicious choice of the organism. *S. cerevisiae* sporulates only in the heterozygous (a/α) state, while homozygotes (a/a and α/α) do not sporulate. Subjecting homozygotes to similar treatments as the sporulating heterozygotes provides a control to distinguish sporulation specific rather than nutritionally specific changes in nucleic acids.

Probably the best characterized system is sporulation in *S. cerevisiae* (most recently reviewed by Haber *et al.*, 1977). Immediately following transfer of cells to sporulating conditions, cells complete mitotic DNA synthesis already in progress, later followed by a round of meiotic DNA synthesis which is required for ascus formation. A functional mitochondrial genome is not required for sporulation (Küenzi *et al.*, 1974).

After an initial increase in total RNA levels, cellular RNA concentrations decrease throughout the course of sporulation. RNA turnover is high, with active synthesis and degradation occurring throughout sporulation. Most of the RNA synthesized during yeast sporulation appears to be rRNA, and up to 50% of the rRNA present at the end of sporulation is newly synthesized. It is not known if this rRNA synthesis is required for sporulation, but no obvious differences in the 18S and 26S rRNAs have been detected between vegetative and sporulating cells.

A methylated 20S RNA species is normally synthesized during yeast sporulation. This RNA was originally thought to be a precursor of the 18S rRNA, but was subsequently shown to be encoded by a cytoplasmically inherited element and is not required for yeast sporulation (Garvik and Haber, 1978; Wejksnora and Haber, 1978).

Circumstantial evidence for the existence of sporulation specific proteins in yeast ascospores is abundant. Protein synthesis is required for the completion of sporulation and inhibition of protein synthesis by cycloheximide prevents ascus formation. Also, temperature sensitive mutants for protein synthesis initiation fail to complete ascus formation at the restrictive temperature. The meiotic apparatus of the cells appears during sporulation, but is virtually absent from vegetative cells. An apparently specific protease and α-1, 4-glucosidase (Colonna and Magee, 1978) activities appear during sporulation. Attempts to identify other sporulation specific proteins have met with limited success. Hopper *et al.* (1974) failed to detect major differences in protein patterns specific to sporulation by autoradiography of one-dimensional gels. Labelling of cells with ^{35}S-methionine followed by two-dimensional gel electrophoresis also failed to resolve differences in protein patterns specific to yeast sporulation (Litske Petersen *et al.*, 1979; Trew *et al.*, 1979). Although numerous changes occurred as a result of the nutritional shiftdown to sporulating conditions, all of these changes could be attributed to the shiftdown itself, or also occurred in nonsporulating strains.

One possible explanation for the failure to detect sporulation specific proteins is that gene products crucial for yeast sporulation are below the resolving power of the two-dimensional gels. The possibility also exists, however, that some of the protein changes common to sporulating and non-sporulating cells are actually sporulation specific, but are not detected as such. The physiological or biochemical nature of the block to sporulation in homozygous yeast cells is not understood. It is conceivable that the block to ascus formation in these cells is rather late in a series of steps leading to sporulation. Under this condition, a sporulation specific product might be expressed despite the failure of the cells to complete ascus formation (Trew *et al.*, 1979).

One report suggests that 45 significant changes occur among the 400 most prominent polypeptides detected in yeast after 24 h under sporulating conditions (Wright and Dawes, 1979). Thirteen new polypeptides appeared, and seven polypeptides increased in intensity among the sporulation competent cells, suggesting the existence of sporulation specific proteins. One might note, however, that these workers pulse-labelled yeast cells with $^{35}SO_4{=}$ rather than ^{35}S-methionine, and apparently took no precautions to ensure that only proteins were added to their two-dimensional gels.

Another system in which an appreciable amount of work has been done on nucleic acid metabolism during sporulation is zoosporogenesis in *B. emersonii* (Lovett, 1975). The primary features of this system include zoosporangium and zoospore differentiation. While DNA synthesis occurs exponentially during vegetative growth, levels remain constant during zoosporogenesis. Total RNA levels initially increase during zoosporangium differentiation, but then decrease during subsequent development. While rRNA synthesis virtually ceases during zoosporogenesis some mRNA is still synthesized during the process.

RNA synthesis is required for zoospore differentiation. The process is inhibited by actinomycin D, polysomes persist, and zoospores contain poly A(+) mRNA (Johnson *et al.*, 1977). Poly A(+) RNA synthesized early during sporulation is preferentially degraded, while that synthesized late in sporulation is packaged into zoospores. As much as one-third of the poly A(+) RNA packaged into the zoospore is synthesized during the final 15-20 min of sporulation (Jaworski and Thomson, 1980). Thus, it appears that quantitative differences in poly A(+) RNA degradation occur at different phases of *Blastocladiella* sporulation. Whether the RNAs sequestered in the zoospores are qualitatively different from those synthesized during early sporulation is unknown.

In *D. discoideum*, there are quantitative and qualitative differences in the numbers of genes expressed at various stages of the life cycle (for review, see Loomis, 1975). Based on hybridization studies, vegetative cells express 30.2% of their genome as RNA transcripts, while sporulating cells express 35.6% of the genome. Some of these transcripts are unique to vegetative and sporulation stages. A variety of changes in enzyme activities are also detected during sporulation (culmination) in *D. discoideum*. A number of these changes cannot be attributed to changes in isozyme patterns or to unmasking of enzyme activities, but rather result from *de novo* synthesis of the enzymes. The *Dictyostelium* sporulation system is also noteworthy in that differential gene expression during sporulation appears to be under the direction of an underlying temporal control mechanism. Enzyme activities appear sequentially even when sporulation is perturbed by mutations causing rapid or delayed development.

Major proteins associated with perithecium development in various *Neurospora* species and related Ascomycetes have been reported (Nasrallah and Srb, 1973, 1977). These proteins fall into several serological groups and have varying electrophoretic mobilities, but all appear as major components of the fruiting bodies, and are not detected in vegetative mycelia. The *N. tetrasperma* protein is localized in the noncellular matrix surrounding the asci, and may function to maintain a proper nutritional or osmotic environment for ascus development (Nasrallah and Srb, 1978).

A sporulation specific major protein has also been detected in pycnidiospores of *B. theobromae* (Van Etten *et al.*, 1979). This putative storage protein comprises approximately 23% of the total spore protein, appears to be degraded during spore germination, and is absent or undetectable in the vegetative mycelium. The protein is specifically synthesized during sporulation (Dahlberg and Van Etten, unpublished data).

No significant qualitative differences were found between poly A(+) RNA fractions from vegetative and sporulating stages of *Schizophyllum commune* (Zantinge *et al.*, 1979). No changes in RNA transcription were detected after photoinduction of conidiation in *Trichoderma viride* (Stavy *et al.*, 1972). However, a failure to detect minor changes in nucleic acids during sporulation by techniques commonly available may not be surprising. Rozek and Timberlake (1980) use a variety of approaches with a wide range of sensitivities to detect changes in mRNA populations associated with antheridiol induced morphogenesis in *Achlya ambisexualis*. No changes were found even using DNA techniques potentially capable of detecting a single RNA sequence difference. These workers concluded that alterations in gene expression during hormone induced *Achlya* differentiation are subtle.

Some of the most interesting work on gene expression during fungal development is being done with *Aspergillus nid-*

ulans conidiation. Timberlake (1980) has shown by DNA:RNA hybridization that vegetative *Aspergillus* cells contain 5600-6000 diverse poly A(+) RNA sequences. During conidiation about 1300 new poly A(+) RNA sequences are expressed. Of these, 300 are present only in conidia. Thus, in *Aspergillus*, it is possible to detect some RNA sequences which are specific only to spores. Most of the poly A(+) RNA sequences present in vegetative cells are also present during development.

In further work Zimmerman *et al.* (1980) cleaved *Aspergillus* nuclear DNA with specific restriction endonucleases, inserted fragments into a lambda phage, and amplified them by growing the phage in *E. coli*. Complementary DNAs specific for conidiation were selected for by repeatedly hybridizing the cDNA to vegetative poly A(+) RNA; conidiation specific cDNAs fail to hybridize under these conditions. This cDNA was then used to select phage clones containing sporulation specific *Aspergillus* genes. One of the clones contained five coding regions, but did not exhibit an operon type of arrangement. Five peptides were coded for by this clone, but the biological function of these products remains to be determined (Timberlake, personal communication).

Germination

In the majority of fungi total DNA synthesis begins after the initiation of RNA and protein syntheses but prior to or at about the time of germ tube emergence. However, not all fungi exhibit concomitant initiation of nuclear and mitochrondrial DNA syntheses. Both nuclear and mitochondrial DNA syntheses are probably not essential for germination of many fungal spores.

The bean rust fungus, *Uromyces phaseoli*, exhibits somewhat unusual DNA metabolism in that mitochondrial and nuclear DNA syntheses appear as mutually exclusive processes (Staples, 1974). Only mitochondrial DNA is synthesized during germination of uredospores and undifferentiated germ tube growth. However, when germ tubes are induced to differentiate infection structures, mitochondrial DNA synthesis stops and nuclear DNA synthesis begins. Inhibitor studies indicate that nuclear DNA synthesis is required for differentiation of infection structures by this fungus (Staples *et al.*, 1975).

The relationship between RNA synthesis and spore germination is more variable among different species of fungi. Each fungus is classified into one of three categories depending on whether RNA and protein synthesis begin simultaneously during spore germination (category 1), or whether RNA synthesis (category 2) or protein synthesis (category 3) is the first to commence. Furthermore, there are two subgroups within each category. Fungi belonging to one subgroup initiate the synthesis of all species of RNA concomitantly, whereas in the other subgroup, syntheses of the various species of RNA begin sequentially. Variability also exists among fungi as to whether RNA synthesis is required for germ tube formation. Several reports indicate that poly A(+) RNA synthesized early in germination is rapidly incorporated into polysomes (Mirkes and McCalley, 1976; Johnson *et al.*, 1977; Freer *et al.*, 1977; Van Etten and Rawn, 1979). However, translation of these newly formed mRNAs does not necessarily indicate that the protein products are unique and/or essential for completion of the germination process.

Prepackaging of mRNA in polysomes within the dormant spore has been demonstrated for several fungi (see reviews: Brambl *et al.*, 1978; Van Etten *et al.*, 1976). In other fungi, such as *D. discoideum* (Giri and Ennis, 1978), *Polysphondylium pallidum* (Ennis *et al.*, 1978), and *Rhizopus stolonifer* (Van Etten and Freer, 1978) evidence for polysomes in dormant spores or microcysts is lacking. However, hybridization of total RNA with [^{3}H] poly (U) or binding of total RNA to oligo (dT)-cellulose indicates that dormant spores or microcysts of these fungi contain poly A(+) RNA. The form in which this mRNA or its precursor is stored within the dormant spore is unknown. Estimates of the number of preformed poly A(+) RNA molecules packaged within each dormant spore are 6.6×10^5 for *R. stolonifer* sporangiospores (Van Etten and Freer, 1978) and 6.25×10^5 for *Botryodiplodia theobromae* pycnidiospores (Van Etten and Rawn, 1979). The similarity of these numbers, in spite of the vast difference in size of the spores of these two unrelated fungi, suggests that the number of prepackaged poly A(+) RNA molecules may be similar for a variety of fungi. Many, if not all, of these preformed mRNAs are functional, since they stimulate the incorporation of labeled amino acids into proteins in cell-free protein synthesizing systems (Van Etten and Freer, 1978; Wenzler and Brambl, 1978).

In *B. theobromae* protein synthesis begins in the absence of RNA synthesis immediately upon initiation of germination (Brambl and Van Etten, 1970; Knight and Van Etten, 1976). Thus, the proteins synthesized early in germination are translated from mRNA contained in the dormant spore. The early translation of this preformed mRNA by cytoplasmic ribosomes is apparently essential for the subsequent synthesis of nucleic acids, germination, and the activation of a cyanide-sensitive respiratory system (Brambl and Van Etten, 1970; Brambl, 1975).

Wenzler and Brambl (1978) isolated poly A(+)RNA from dormant and germinating *B. theobromae* spores and translated these RNAs in a wheat germ cell-free system. Their results suggested that the mRNA prepackaged in dormant spores contains genetic information which is qualitatively different from that of germinated spores. These results corroborate an earlier study (Van Etten *et al.*, 1972) in which differences were found in the relative distribution of soluble polypeptides synthesized at early and late stages of germination of *B. theobromae* spores.

As will be discussed in another review from this symposium, (Brambl) experiments using immunochemical techniques indicate that mRNA from *B. theobromae* spores codes for certain cytoplasmic polypeptides of the multiple-subunit enzymes F-adenosine triphosphatase and cytochrome C oxidase (Wenzler and Brambl, 1978). Previous studies had shown that mitochondria from dormant *B. theobromae* spores lacked a functional cytochrome C oxidase (Brambl, 1977; Brambl and Josephson, 1977).

Several investigators have looked for the synthesis of stage-specific proteins at intervals during germination of *R. stolonifer* (Freer and Van Etten, 1978) and *D. discoideum* (Giri and Ennis, 1978; Dowbenko and Ennis, 1980) spores and *P. pallidum* microcysts (Ennis *et al.*, 1978), as a means of identifying developmentally regulated genes. Proteins pulse labeled with ^{35}S-methionine *in vivo* at specified intervals during germination were extracted and electrophoresed on SDS-polyacrylamide gels. Protein bands were detected by staining and/or fluorography. Using two dimensional gel electrophoresis, Dowbenko and Ennis (1980) distinguished six classes of proteins in germinating spores of *D. discoideum*, based on the time of onset and duration of their synthesis. Freer and Van Etten (1978) reported that a few proteins were synthesized only at discrete time periods during germination of *R. stolonifer* sporangiospores, but at no time during germination was a germination-specific protein synthesized which represented a major fraction of the total proteins synthesized. Likewise, Ennis *et al.*, (1978) reported changes in the pattern of proteins synthesized at various times during germination of *P. pallidum* microcysts.

A comparison of the proteins synthesized *in vivo* with those synthesized *in vitro* in a cell-free system, using either total messenger RNA or polysomal messenger RNA as templates, indicated that the majority of proteins synthesized during germination are under transcriptional control (Freer and Van Etten, 1978; Giri and Ennis, 1978; Dowbenko and Ennis, 1980). That is, qualitative changes in the population of mRNAs at specified intervals during germination correlate well with the majority of the changes in the proteins synthesized *in vivo*. Likewise Hansberg *et al*. (1979) reported that glutamine synthetase activity in *N. crassa* conidia is regulated at the transcriptional level.

Other types of control mechanisms are probably also involved in the regulation of protein synthesis during germination, but to a lesser extent. Messenger ribonucleoprotein (mRNP) complexes released from EDTA-dissociated polysomes of *Neurospora crassa* (Mirkes, 1977) are presently being studied in terms of their possible involvement in regulating protein synthesis during the initial stages of spore germination. It is possible that the addition, deletion, or modification of proteins associated with the mRNP complexes might regulate the selection of mRNA for translation, and/or its degradation. Freer and Van Etten (1978) reported that in *R. stolonifer*, mRNAs for at least five proteins appear to be masked during germination. The mRNAs for these proteins could be detected *in vitro* several hours before the synthesis of these proteins could be detected *in vivo*. Both *D. discoideum* (Alton and Lodish, 1977) and *R. stolonifer* (Freer and Van Etten, 1978) possess specific proteins which are synthesized *in vivo* only early in germination; however, the mRNA for these proteins remains in a translatable form in the cells for several additional hours. Furthermore, certain mRNAs may have different affinities for the ribosomes. For example, Freer and Van Etten (1978) reported that during *R. stolonifer* germination one protein is easily detected in the *in vivo* sample. The mRNA for this protein was detected in total RNA, but not polysomal RNA. Another polypeptide appears to be under the converse type of regulation. This protein is synthesized sparingly in the *in vivo* sample, but its mRNA is not detected until 3 h in total RNA extracts. However, this mRNA is abundant in the polysomal RNA extracts.

CONCLUSION

As the literature reviewed in this chapter indicates, notable advances have been made in the development of techniques used for analysis of nucleic acids and proteins. The application of these techniques has led to the unraveling of many important facts about the structural organization of the fungal genome, and the ways in which it resembles or differs from that of higher eukaryotes. Researchers have begun to apply some of these techniques to the study of fungal development. By the time of the next fungal spore symposium, one can anticipate a large increase in our understanding of gene expression in this fascinating group of organisms.

ACKNOWLEDGMENTS

We thank Bill Timberlake for allowing us to cite results prior to publication. Research from our laboratory was supported by grants from the National Institutes of Health and the National Science Foundation. This paper is published with the approval of the Director, Nebraska Agricultural Experiment Station. The work was conducted under Nebraska Agricultural Experiment Station Project No. 21-17.

REFERENCES

Alton, T.H. and Lodish, H.F. (1977). Translational control of protein synthesis during the early stages of differentiation of the slime mold *Dictyostelium discoideum*. *Cell* 12:301-310.

Banerjee, A.K. (1980). 5'-terminal cap structure in eukaryotic messenger ribonucleic acids. *Microbiol. Revs.* 44:175-205.

Bell, G.I., DeGennaro, L.J., Gelfond, D.H., Bishop, R.J., Valenzuela, P. and Rutter, W.J. (1977). Ribosomal RNA genes of *Saccharomyces cerevisiae*. 1. Physical map of the repeating unit and location of the regions coding for 5S, 5.8S, 18S, and 25S ribosomal RNAs. *J. Biol. Chem.* 252: 8118-8125.

Brambl, R. (1975). Characteristics of developing mitochondrial genetic and respiratory functions in germinating fungal spores. *Biochim. Biophys. Acta* 396:175-186.

Brambl, R. (1977). Mitochondrial biogenesis during fungal spore germination. Development of cytochrome c oxidase activity. *Arch. Biochem. Biophys.* 182:273-281.

Brambl, R., Dunkle, L.D. and Van Etten, J.L. (1978). Nucleic acid and protein synthesis during fungal spore germination. *In* "The Filamentous Fungi" (Ed. J.E. Smith and D.R. Berry). Vol. 3, p. 94-118. Edward Arnold (Publishers) Limited, London.

Brambl, R. and Josephson, M. (1977). Mitochondrial biogenesis during fungal spore germination: respiratory cytochromes of dormant and germinating spores of *Botryodiplodia*. *J. Bacteriol.* 129:291-297.

Brambl, R.M. and Van Etten, J.L. (1970). Protein synthesis during fungal spore germination. V. Evidence that the ungerminated conidiospores of *Botryodiplodia theobromae* contain messenger ribonucleic acid. *Arch. Biochem. Biophys.* 137:442-452.

Brawerman, G. (1976). Characteristics and significance of the polyadenylate sequence in mammalian messenger RNA. *In* "Progress in Nucleic Acid Research and Molecular Biology" (Ed. W.E. Cohn). Vol. 17, p. 117-148. Academic Press, New York.

Cameron, J.R., Loh, E.Y., and Davis, R.W. (1979). Evidence for transposition of dispersed repetitive DNA families in yeast. *Cell* 16:739-751.

Campbell, G.R., Littau, V.C., Melera, P.W., Allfrey, V.G, and Johnson, E.M. (1979). Unique sequence arrangement of ribosomal genes in the palindromic rDNA molecule of *Physarum polycephalum*. *Nuc. Acid Res.* 6:1433-1447.

Case, M.E., Schweizer, M., Kushner, S.R. and Giles, N.H. (1979). Efficient transformation of *Neurospora crassa* by utilizing hybrid plasmid DNA. *Proc. Natl. Acad. Sci. USA*. 76:5259-5263.

Cihlar, R.L. and Sypherd, P.S. (1980). The organization of the ribosomal RNA genes in the fungus *Mucor racemosus*. *Nuc. Acid Res.* 8:793-804.

Colonna, W.J. and Magee, P.T. (1978). Glycogenolytic enzymes in sporulating yeast. *J. Bacteriol.* 134:844-853.

Crain, W.R., Eden, F. C., Pearson, W.R., Davidson, E.H. and Britten, R.J. (1976). Absence of short period interspersion of repetitive and nonrepetitive sequences in the DNA of *Drosophila melanogaster*. *Chromosoma* 56:309-326.

Crick, F. (1979). Split genes and RNA splicing. *Science* 204:264-271.

Cummings, D.J., Borst, P., Dawid, I.B., Weissman, S.M., and Fox, C. F. (Eds). (1979). Extrachromosomal DNA. Academic Press, New York.

Davidson, E.H. and Britten, R.J. (1979). Regulation of gene expression: possible role of repetitive sequences. *Science* 204: 1052-1059.

Davidson, E.H., Galau, G.A., Angevev, R.C. and Britten, R.J. (1975). Comparative aspects of DNA organization in Metazoa. *Chromosoma* 51:253-259.

Dons, J.J.M., DeVries, O.M.H., and Wessels, J.G.H. (1979). Characterization of the genome of the basidiomycete *Schizophyllum commune*. *Biochim. Biophys. Acta* 563: 100-112.

Dowbenko, D.J. and Ennis, H.L. (1980). Regulation of protein synthesis during spore germination in *Dictyostelium discoideum*. *Proc. Natl. Acad. Sci. USA* 77:1791-1795.

Ennis, H.L., Pennica, D., and Hill, J.M. (1978). Synthesis of macromolecules during microcyst germination in the cellular slime mold *Polysphondylium pallidum*. *Devel. Biol.* 65:251-259.

Felden, R.A., Sanders, M.M., and Morris, N.R. (1976). Presence of histones in *Aspergillus nidulans*. *J. Cell. Biol.* 68: 430-439.

Firtel, R.A. and Lodish, H.F. (1973). A small nuclear precursor of messenger RNA in the cellular slime mold *Dictyostelium discoideum*. *J. Mol. Biol.* 79:295-314.

Free, S.J., Rice, P.W. and Metzenberg, R.L. (1979). Arrangement of the genes coding for ribosomal ribonucleic acids in *Neurospora crassa*. *J. Bacteriol.* 137:1219-1226.

Freer, S.N., Mayama, M., and Van Etten, J.L. (1977). Synthesis of polyadenylate-containing RNA during the germination of *Rhizopus stolonifer* sporangiospores. *Exp. Mycol.* 1: 116-127.

Freer, S.N. and Van Etten, J.L. (1978). Changes in messenger RNAs and protein synthesis during germination of *Rhizopus stolonifer* sporangiospores. *Exp. Mycol.* 2:313-325.

Gallwitz, D. and Sures, I. (1980). Structure of a split yeast gene: Complete nucleotide sequence of the actin gene in *Saccharomyces cerevisiae*. *Proc. Natl. Acad. Sci. USA* 77: 2546-2550.

Garvik, B., and Haber, J.E. (1978). New cytoplasmic genetic element that controls 20S RNA synthesis during sporulation in yeast. *J. Bacteriol.* 134:261-269.

Giri, J.G. and Ennis, H.L. (1978). Developmental changes in RNA and protein synthesis during germination of *Dictyostelium discoideum* spores. *Devel. Biol.* 67:189-201.

Goff, C.G. (1976). Histones of *Neurospora crassa*. *J. Biol. Chem.* 251:4131-4138.

Greer, H. and Fink, G.R., (1979). Unstable transpositions of his 4 in yeast. *Proc. Natl. Acad. Sci. USA* 76:4006-4010.

Haber, J.E., Wejksnora, P.J., Wygal, D.D. and Lai, E.Y. (1977). *In* "Eucaryotic microbes as model developmental systems" (Ed. D.H. O'Day and P.A. Horgen). p. 129-154. Marcell Dekker, Inc., New York.

Hansberg, W., Espin, G., Palacios, R. and Sanchez, F. (1979). Regulation of glutamine synthetase messenger ribonucleic acid in conidia of *Neurospora crassa*. *Devel. Biol.* 73:68-75.

Hautala, J.A., Bassett, C.L., Giles, N.H. and Kushner, S.R. (1979). Increased expression of a eukaryotic gene in *Escherichia coli* through stabilization of its messenger RNA. *Proc. Natl. Acad. Sci. USA* 76:5774-5778.

Hereford, L.M. and Rosbash, M. (1977a). Number and distribution of polyadenylated RNA sequences in yeast. *Cell* 10: 453-462.

Hereford, L.M. and Rosbash, M. (1977b). Regulation of a set of abundant mRNA sequences. *Cell* 10:463-467.

Hinnen, A., Hicks, J.B. and Fink, G.R. (1978). Transformation of yeast. *Proc. Natl. Acad. Sci. USA* 75:1929-1933.

Hopper, A.K., Magee, P.T., Welch, S.K., Friedman, M. and Hall, B.D. (1974). Macromolecular synthesis and breakdown in relation to sporulation and mitosis in yeast. *J. Bacteriol.* 119:619-628.

Horgen, P.A. and Silver, J.C. (1978). Chromatin in eukaryotic microbes. *Ann. Rev. Microbiol.* 32:249-284.

Hudspeth, M.E.S., Timberlake, W.E., and Goldberg, R.B. (1977). DNA sequence organization in the water mold *Achlya*. *Proc. Natl. Acad. Sci. USA* 74:4332-4336.

Jaworski, A.J. and Thomson, K. (1980). A temporal analysis of the synthesis of the mRNA sequested in zoospores of *Blastocladiella emersonii*. *Devel. Biol.* 75:343-357.

Johnson, J.D., Ogden, R., Johnson, P., Abelson, J., Dembeck, P. and Itakura, K. (1980). Transcription and processing of a yeast tRNA gene containing a modified intervening sequence. *Proc. Natl. Acad. Sci. USA* 77:2564-2568.

Johnson, S.A., Lovett, J.S., and Wilt, F.H. (1977). The polyadenylated RNA of zoospores and growth phase cells of the aquatic fungus, *Blastocladiella*. *Devel. Biol.* 56:329-342.

Knight, R.H. and Van Etten, J.L.(1976). Synthesis of ribonucleic acids during the germination of *Botryodiplodia theobromae* pycnidiospores. *J. Gen. Microbiol.* 95:257-267.

Krumlauf, R. and Marzluf, G.A. (1979). Characterization of the sequence complexity and organization of the *Neurospora crassa* genome. *Biochemistry* 18:3705-3713.

Krumlauf, R. and Marzluf, G.A. (1980). Genome organization and characterization of the repetitive and inverted repeat DNA sequences in *Neurospora crassa*. *J. Biol. Chem.* 255: 1138-1145.

Küenzi, M.T., Tingle, M.A. and Halvorson, H.O. (1974). Sporulation of *Saccharomyces cerevisiae* in the absence of a functional mitochondrial genome. *J. Bacteriol.* 117: 80-88.

Lauer, G.D., Roberts, T.M. and Klotz, L.C. (1977). Determination of the nuclear DNA content of *Saccharomyces cerevisiae* and implications for the organization of DNA in yeast chromosomes. *J. Mol. Biol.* 114:507-526.

Lemke, P.A. (1979). "Viruses and Plasmids in Fungi". Marcel Dekker Inc., New York.

Litske Petersen, J.G., Kielland-Brandt, M.C. and Nilsson-Tillgren, T. (1979). Protein patterns during yeast sporulation. *Carlsberg Res. Commun.* 44:149-162.

Loomis, W.F. (1975). "*Dictyostelium discoideum: A Developmental System*". Academic Press, New York.

Lovett, J.S. (1975). Growth and differentiation of the water mold *Blastocladiella emersonii:* cytodifferentiation and the role of ribonucleic acid and protein synthesis. *Bacteriol. Revs.* 39:345-404.

Lovett, J.S. (1976). Regulation of protein metabolism during spore germination. *In* "The Fungal Spore: Form and Function" (Ed. D.J. Weber and W.H. Hess). p. 189-242. John Wiley and Sons, Inc., New York.

Lovett, J.S., Gong, C.-S., and Johnson, S.A. (1977). *In* "Eukaryotic Microbes as Model Developmental Systems" (Ed. D.M. O'Day & P.A. Horgen). p. 402-424. Marcel Dekker, New York.

McLaughlin, C.S., Warner, J.R., Edmonds, M., Nakazato, H. and Vaughan, M.H. (1973). Polyadenylic acid sequences in yeast messenger ribonucleic acid. *J. Biol. Chem.* 248:1466-1471.

Mirkes, P.E. (1977). Messenger ribonucleoprotein complexes isolated by oligodeoxythymidylate-cellulose chromatography from *Neurospora crassa* polysomes. *J. Bacteriol.* 131: 240-246.

Mirkes, P.E. and McCalley, B. (1976). Synthesis of polyadenylic acid-containing ribonucleic acid during the germination of *Neurospora crassa* conidia. *J. Bacteriol.* 125:174-180.

Molgaard, H.V., Matthews, H. R. and Bradbury, E.M. (1976). Organization of genes for ribosomal RNA in *Physarum polycephalum. Eur. J. Biochem.* 68:541-549.

Morris, N.R. (1980). *In* "The Eukaryotic Microbial Cell" (Ed. G.W. Gooday, D. Lloyd, and A.P.J. Trinci). p. 41-76. Cambridge University Press, Cambridge.

Nasrallah, J.B. and Srb, A.M. (1973). Genetically related protein variants specifically associated with fruiting body maturation in *Neurospora. Proc. Natl. Acad. Sci. USA* 70:1891-1893.

Nasrallah, J.B. and Srb, A.M. (1977). Occurrence of a major protein associated with fruiting body development in *Neurospora. Proc. Natl. Acad. Sci. USA* 74:3831-3834.

Nasrallah, J.B. and Srb, A.M. (1978). Immunofluorescent localization of a phase-specific protein in *Neurospora tetrasperma* perithecia. *Exp. Mycol.* 2:211-215.

Nath, K. and Bollon, A.P. (1977). Organization of the yeast ribosomal RNA gene cluster via cloning and restriction analysis. *J. Biol. Chem.* 252:6562-6571.

Ojha, M. and Dutta, S.K. (1978). Nuclear control of differentiation. *In* "The Filamentous Fungi". (Ed. J.E. Smith and D.R. Berry). Vol. 3, p. 8-27. Edward Arnold (Publishers) Limited, London.

Petes, T.D., Byers, B. and Fangman, W.L. (1973). Size and structure of yeast chromosomal DNA. *Proc. Natl. Acad. Sci. USA* 70:3072-3076.

Petes, T.D. and Fangman, W.L. (1972). Sedimentation properties of yeast chromosomal DNA. *Proc. Natl. Acad. Sci. USA* 69:1188-1191.

Philippsen, P., Thomas, M., Kramer, R.A. and Davis, R. (1978). Unique arrangement of coding sequences for 5S, 5.8S, 18S, and 25S ribosomal RNA in *Saccharomyces cerevisiae* as determined by R-Loop and hybridization analysis. *J. Mol. Biol.* 123:387-404.

Rozek, C.E., Orr, W.C. and Timberlake, W.E. (1978). Diversity and abundance of polyadenylated RNA from *Achlya ambisexualis*. *Biochemistry* 17:716-722.

Rozek, C.E. and Timberlake, W.E. (1979). Restriction endonuclease mapping by crossed contact hybridization: the ribosomal RNA genes of *Achyla ambisexualis*. *Nuc. Acid. Res.* 7:1567-1578.

Rozek, C.E. and Timberlake, W.E. (1980). Absence of evidence for changes in messenger RNA populations during steroid hormone-induced cell differentiation in *Achlya*. *Exp. Mycol.* 4:33-47.

Schmit, J.C. and Brody, S. (1976). Biochemical genetics of *Neurospora crassa* conidial germination. *Bacteriol. Revs.* 40:1-41.

Seebeck, T., Stalder, J. and Braun, R. (1979). Isolation of a minichromosome containing the ribosomal genes from *Physarum polycephalum*. *Biochemistry* 18:484-490.

Staples, R.C. (1974). Synthesis of DNA during differentiation of bean rust uredospores. *Physiol. Plant Pathol.* 4:415-424.

Staples, R.C., App, A.A. and Ricci, P. (1975). DNA synthesis and nuclear division during formation of infection structures by bean rust uredospore germlings. *Arch. Microbiol.* 104:123-127.

Stavy, R., Galun, E., and Gressel, J. (1972). Morphogenesis in *Trichoderma:* RNA-DNA hybridization studies. *Biochim. Biophys. Acta* 259:321-329.

Storck, R. (1974). Molecular Mycology. *In* "Molecular Microbiology" (Ed. J.B.G. Kwapinski). p. 423-477. John Wiley and Sons, Inc., New York.

Strathern, J.N., Spatola, E., McGill, C. and Hicks, J.B. (1980). Structure and organization of transposable mating type cassettes in *Saccharomyces* yeasts. *Proc. Natl. Acad. Sci. USA* 77:2839-2843.

Struhl, K. and Davis, R.W. (1977). Production of a functional eukaryotic enzyme in *Escherichia coli:* cloning and expression of the yeast structural gene for imidazoleglycerol phosphate dehydratase (his 3). *Proc. Natl. Acad. Sci. USA* 74:5255-5259.

Struhl, K., Stinchcomb, D.T., Scherer, S. and Davis, R.W. (1979). High-frequency transformation of yeast autonomous replication of hybrid DNA molecules. *Proc. Natl. Acad. Sci. USA* 76:1035-1039.

Taylor, W.C., Cockburn, A.F., Frankel, G.A., Newkirk, M.J. and Firtel, R.A. (1977). Organization of ribosomal and 5S RNA coding regions in *Dictyostelium discoideum. In* "Molecular Approaches to Eukaryotic Genetic Systems" (Ed. G. Wilcox, J. Abelson and C.F. Fox). Vol. 8, p. 309-313. Academic Press, New York.

Timberlake, W.E. (1978). Low repetitive DNA content in *Aspergillus nidulans. Science* 202:973-975.

Timberlake, W.E. (1980). Developmental gene regulation in *Aspergillus nidulans. Devel. Biol.* (in press).

Timberlake, W.E., Shumard, D.S. and Goldberg, R.B. (1977). Relationship between nuclear and polysomal RNA populations of *Achlya:* a simple eukaryotic system. *Cell* 10: 623-632.

Trew, B.J., Friesen, J.D. and Moens, P.B. (1979). Two-dimensional protein patterns during growth and sporulation in *Saccharomyces cerevisiae. J. Bacteriol.* 138:60-69.

Valenzuela, P., Bell, G.I., Venegas, A., Sewell, E.T., Masiarz, F.R., DeGennaro, L.J., Weinberg, F. and Rutter, W.J. (1977). Ribosomal RNA genes of *Saccharomyces cerevisiae.* 2. Physical map and nucleotide sequence of the 5S ribosomal RNA gene and adjacent intergenic regions. *J. Biol. Chem.* 252:8126-8135.

Van Etten, J.L., Dunkle, L.D. and Knight, R.H. (1976). Nucleic acids and fungal spore germination. *In* "The Fungal Spore: Form and Function" (Ed. D.J. Weber and W. M. Hess). p. 243-300. John Wiley and Sons, Inc., New York.

Van Etten, J.L., Dunkle, L.D. and Freer, S.N. (1977). Germination of *Rhizopus stolonifer* sporangiospores. *In* "Eukaryotic Microbes as Model Developmental Systems" (Ed. D.M. O'Day and P.A. Horgen). p. 372-401. Marcell Dekker, New York.

Van Etten, J.L. and Freer, S.N. (1978). Polyadenylate-containing RNA in dormant and germinating sporangiospores of *Rhizopus stolonifer. Exp. Mycol.* 2:301-312.

Van Etten, J.L., Freer, S.N. and McCune, B.K. (1979). Presence of a major (storage?) protein in dormant spores of the fungus *Botryodiplodia theobromae. J. Bacteriol.* 138: 650-652.

Van Etten, J.L. and Rawn, C.D. (1979). Polyadenylate-containing RNA in dormant and germinating *Botryodiplodia theobromae* pycnidiospores. *Can. J. Microbiol.* 25: 375-379.

Van Etten, J.L., Roker, H.R. and Davies, E. (1972). Protein synthesis during fungal spore germination: differential protein synthesis during germination of *Botryodiplodia theobromae* spores. *J. Bacteriol.* 112:1029-1031.

Wejksnora, P.J. and Haber, J.E. (1978). Ribonucleoprotein particle appearing during sporulation in yeast. *J. Bacteriol.* 134:246-260.

Wenzler, H. and Brambl, R. (1978). *In vitro* translation of polyadenylate-containing RNAs from dormant and germinating spores of the fungus *Botryodiplodia theobromae*. *J. Bacteriol.* 135:1-9.

Wright, B. (1979). Causality in biological systems. *Trends in Biochem. Sci.* 4:N110-N111.

Wright, J.F. and Dawes, I.W. (1979). Sporulation-specific protein changes in yeast. *FEBS Lett.* 104:183-186.

Zantinge, B., Dons, H. and Wessels, J.G.H. (1979). Comparison of poly (A)-containing RNAs in different cell types of the lower eukaryote *Schizophyllum commune*. *Eur. J. Biochem.* 101:251-260.

Zimmerman, C.R., Orr, W.C., Leclerc, R.F., Barnard, E.C. and Timberlake, W.E. (1980). Molecular cloning and selection of genes regulated in *Aspergillus* development. (submitted).

IV. INFORMATIONAL EXPRESSION

INFORMATIONAL EXPRESSION : DISCUSSANT'S INTRODUCTION

D. J. Weber

Dept. of Botany and Range Science
Brigham Young University
Provo, Utah, USA 84602

Informational expression of the genetic message from fungi involves the translation of the genetic message into protein. First, the genetic code must be transcribed into various classes of RNA (mRNA, tRNA, and rRNA). Then, the information in mRNA must be translated into protein.

Protein synthesis requires a number of macromolecules namely: ribosomes, tRNA, mRNA, specific enzymes (amino acyl-tRNA synthetases) and specialized protein factors (Mazumder and Szer, 1977). Most of our understanding of the reactions in protein synthesis have been obtained with prokaryotic organisms, although more eukaryotic organisms are now being investigated. Of all the protein synthesis papers (6768) listed in biological abstracts from 1969 to the present, fungal studies represented 23% of the total. Generally, the steps in protein synthesis are considered to be similar in fungi as compared to other eukaryotic organisms. The following is a short review of the basic aspects of protein synthesis in eukaryotic organisms particularly as they relate to fungi.

Ribosomes. The molecular weights of ribosomes in eukaryotic cells are about 4.0 to 4.6 million with a sedimentation coefficient of 80S. Ribosomes are also present in the organelles as well as in the cytoplasm. Organelle ribosomes have structural features that are distinctive from cytoplasmic ribosomes and tend to resemble prokaryotic ribosomes (Nomura et al., 1974).

All ribosomes appear to be made of two non-identical subunits (the small and large ribosomal subunits). In eukaryotic cells, the 80S ribosomes have two subunits (40S and 60S). The 40S ribosome subunit is composed of 18S ribosome RNA and about 33 proteins, while the 60S subunit is composed of 5S and 28S RNA and about 49 proteins. Eukaryotic ribosomes contain about 16,000 amino acids. The task of determining the amino acid sequence is monumental, however, the amino-terminal sequences of the separated ribosomal proteins up to 20 to 25 amino acids have been determined for rat liver ribosomal proteins (Wool, 1979). Generally a large percentage of the ribosomes present in the cytoplasm of eukaryotic cells are attached to the endoplasmic reticulum.

Kellems *et al.* (1974) found cytoplasmic 80S ribosomes associated with the outer membranes of mitochondria of *Saccharomyces cerevisiae*. The cytoplasmic polysomes could be released from the mitochondria by Triton X-100. The amount of cytoplasmic 80S ribosomes associated with the mitochondria was greater on mitochondria obtained from starved cells than normal mitochondria (Kellems and Butow, 1974).

Transfer RNA. The tRNAs function as adaptors which form covalent bonds with amino acids. The tRNAs also interact with the codons on mRNA by complement base pairing (anticodons) (Crick, 1966). There is at least one unique tRNA for each given amino acid. On the other hand, with few exceptions there are several tRNAs which can accept the same amino acid. They are called isoaccepting species. The complete nucleotide sequence of several yeast tRNAs have been determined. Each tRNA has a single strand of about 80 nucleotides with a molecular weight of about 25,000. They all share a common secondary structure of a clover leaf type of arrangement (Holley *et al.*, 1965, Heckman *et al.*, 1978). X-ray crystallography has indicated that the clover leaf arrangement folds upon itself to form a L-shape tertiary structure (Kim *et al.*, 1973).

Amino Acyl-tRNA Synthetases. Amino acyl-tRNA synthetases are enzymes which serve the function of translators. They recognize both a unique tRNA and its cognate amino acid. The general reaction involves:

$$\text{amino acid} + \text{tRNA} + \text{ATP} \rightleftharpoons \text{amino acyl-tRNA} + \text{AMP} + \text{PPi}.$$

Within a given species, amino acyl-tRNA synthetases display specificity with respect both to the amino acid and the tRNA. Generally a specific synthetase will transfer only one amino acid to its cognate tRNA. Evidence suggests that the loading of an amino acid to its cognate tRNA is a multiple step process which makes possible the rejection of the wrong amino acid in each step (Soll and Schimmel,

1974). There are at least 20 different and specific synthetases for the 20 different amino acids found in protein. In the bacteria there appears to be with one exception only one amino acyl-tRNA synthetase for each amino acid. In contrast in eukaryotic cells there appears to be multiple forms of the same synthetase in the cytoplasm. The business end of the tRNA which accepts the amino acid is the adenosine moiety located at its 3' end. This end has a common base sequence namely CCA_{OH}. As a result of the synthetase catalyzed reaction, the amino acid becomes covalently linked to the 3' terminal adenosine of tRNAs through an esterlinkage of the carboxyl group of the amino acid and either the 2' or 3' hydroxyl group of the adenosine moiety. The synthetase catalyzed reaction activates the carboxyl group of amino acids. The activation step requires ATP, but subsequent reactions with amino groups of other amino acids do not require additional energy to form the peptide bond.

Protein Factors. There are specific protein factors which ensure that the message is translated with the utmost fidelity and at optimum rates. Several distinct protein factors are involved in the initiation, elongation and termination (release of polypeptide chain). For example, in *E. coli* there is evidence of functional participation of three initiation factors (IF-1, IF-2, and IF-3), three elongation factors (EF-TU, EF-TS, and EF-G) and three release factors, (RF-1, RF-2, and RF-3) (Ochoa and Mazumder, 1974). The suggested factors in eukaryotic cells are: initiation (eIF-2, eIF-3, eIF-5), elongation (eEF-1, eEF-3, eEF-5) and termination (eRF-1, eRF-2) (Mazumder and Szer, 1977). The number of eukaryotic initiation factors and their role in eukaryotic protein synthesis is not as clear as prokaryotic factors. The factors have been found in fungi, but not all of the factors have been well characterized. However, the factors appear to function in the following manner (Mazumder and Szer, 1977).

The eIF-2 promotes the GTP-dependent binding of MET-tRNA to 40S subunits. The eIF-5 (possessing ribosome-dependent GTPase activity) is involved in joining 60S subunits to 40S initiation complexes. The joining of these subunits to form the 80S ribosome is a complex process with a stringent requirement for GTP and additional specialized proteins. McNeil and McLaughlin (1974) found this factor in yeast.

The eIF-3 is involved in the translation of natural mRNAs. The elongation factors in eukaryotic organisms are not as well documented as in prokaryotic organisms.

The eEF-1 brings an amino acyl-tRNA to the A site of the ribosome as specified by the codon on the mRNA. The transfer of the amino acyl-tRNA from the complex to site A on the ribosome results in the hydrolysis of GTP and the release of eEF-1 as eEF-1-GDP.

The eEF-3 appears to function to regenerate eEF-1 by releasing it from eEF-1-GDP.

The eEF-5 facilitates the translocation of the amino acyl-tRNA from site A to the P site after the peptide bond has been formed. Then the deacylated tRNA is ejected from P site, and the ribosome moves along the mRNA to a new codon site.

The three termination codons are UAG, UAA, and UGA. Termination factors are required for the release of the completed polypeptide chain from the ribosome and the dissociation of the ribosome.

The eRF-1 binds to ribosomes in response to UAA and UAG to release the peptides.

The eRF-2 binds to ribosomes in response to UAA and UGA to release the peptides.

The RF-3 increases the rate of release of the peptide from the ribosome in prokaryotes. However, no factor similar to RF-3 has been found in mammalian eRF factors. There is also a requirement in mammals that the nonsense codon be followed by a purine nucleotide at the 3' end.

Sequential Assembly of Polypeptides Chain. The reading of the mRNA proceeds from the 5' to the 3' end, while the polypeptide chains are initiated from the NH_2 terminal end and proceed to the COOH terminal end. The initiation, elongation, and the termination factors are all involved; GTP is involved in the dependent binding of these factors and appears to be hydrolyzed in the initiation, elongation, and the termination steps of protein synthesis (Mazumder and Szer, 1977).

Regulation of Eukaryotic Protein Synthesis. Gene expression in prokaryotes is controlled mainly at the level of transcription, whereas in eukaryotes, with more stable mRNAs, there is control also at the translation level (Ochoa and De Haro, 1979). One method of translational control of polypeptide initiation occurs by the blocking of the function of the initiation factor eIF-2 by phosphorylation of one of its three subunits. This phosphorylation is catalyzed by a cAMP-independent protein kinase (eIF-2 kinase). The second mechanism of translation control is by degrading mRNA by endonucleases. These actions (blocking of the eIF-2 function and mRNA degradation) appear to act in a non-selective fashion.

Protein synthesis control in yeast. Kraft-Creech and Lochmann (1978) analyzed the rate of protein synthesis on membrane-bound and membrane-free ribosomes and suggested that protein synthesis in yeast occurred mainly on membrane-bound ribosomes. The free ribosomes were considered to be a ribosome pool that allowed the cells to increase its rate of protein synthesis rapidly. During the cell cycle of yeast, DNA synthesis occurred periodically but RNA and protein synthesis increased at an expotential rate (Elliott and McLaughlin, 1978). Fraser (1975) investigated turn over with labeling experiments in *Schizosaccharomyces pombe*, and suggested that protein synthesis control was at the level of mRNA. Shilo *et al.* (1978) concluded from starvation and inhibition studies that the rate of the cycle initiation was a more sensitive control factor than protein synthesis. He suggested that the initiation of the cell cycle depended upon the synthesis of particular proteins whose production or function requires the limiting nutrients. The high sensitivity of cell initiation to decrease the rate of protein synthesis could explain the ability of yeast cells to complete the cycle and arrest its stationary phase upon depletion of medium components.

Effect of light on protein synthesis. Rubin *et al.*, (1973) suggested that light influenced RNA and protein synthesis and that light sensitive regulatory systems were present in fungi. In the case of yeast, Ulaszewski *et al.* (1979) suggested cytochrome b and a/a_3 were the primary light sensitive receptors for the inhibitory effects of light on fungi.

Determination of informational expression by protein analysis. Protein represents the expression of the genetic message and is the normal means of measuring informational expression. A major problem in analyzing proteins is the need to separate individual proteins from protein mixtures. In the past, various protein fractions were obtained by precipitation with salts, differential centrifugation, and low pressure liquid chromatography. The proteins were separated on the basis of their charges using weak ion exchange columns (DEAE cellulose and CM cellulose) and by size using steric exclusion (Sephadex). The process of isolation with classic liquid chromatography was slow and resolution was often poor.

Analysis of Proteins Using High Pressure Liquid Chromatography. With the development of high pressure liquid chromatography (HPLC), high resolution and rapid separation of small molecular weight compounds were readily obtained in 10 to 15 minutes. The application of HPLC to protein separation was limited because the high pressure packed the DEAE cellulose and the Sephadex gels with such force that limited or no flow rate occurred. Recent developments in HPLC column packings now permit separation of peptides and proteins (Walton, 1980)

Rubinstein (1979) reviewed the application of preparative HPLC and reported that proteins with molecular weights above 10,000 have been fractionated. The procedures included silica absorbent, reverse phase chromatography, steric exclusion, and ion exchange columns. He suggested the biggest limitation was that generally some organic solvent was involved which could denature proteins. The use of reverse phase columns and more polar solvents such as water has reduced this problem. Gruber *et al.* (1979) used a HPLC column (synchropak GPC-100) to determine the molecular weight of proteins between 5,000 and 7,000. Ui (1979) also used HPLC to estimate molecular weights of proteins.

The concept of paired ion chromatography in which the stationary phase is a hydrophobic packing and the mobile phase contains long chained hydrophilic ions of opposite charge to the ions being separated has been a very successful method. It appears that the highly hydrophobic counter ions are incorporated into the surface of the packing making it essentially an ion exchanger. The interest in HPLC is illustrated by the publication of over 170 papers in 1979 on HPLC protein separation representing the resolution of some 2,000 different proteins.

Protein Separation Using two-Dimensitional Gel Electrophoresis. Gel electrophoresis has been a useful technique for obtaining resolution of protein mixtures (Allen and Maurer, 1972). The separation was due to a combination of protein movement in a charged field in relation to a sieving effect (amount of cross-linking of the acrylamide). The smallest protein with the strongest charge moves the fastest through the acrylamide gel. Another excellent protein separation technique is isoelectric focusing which involves a pH gradient in acrylamide gel. The protein moves in the electrical field in relation to the charge on the protein until it reaches its isoelectric point (the specific pH where the protein charge is neutral), then the protein movement is stopped. This method provides excellent resolution of proteins (Saravis and Zamcheck, 1979).

These two methods were combined by O'Farrell in 1975 in a technique called two-dimensional gel electrophoresis. This method provides excellent resolution of a large number of proteins. The proteins are first separated by isoelectric focusing in a tube. Then the gel is removed and placed on top of a regular acrylamide slab and the proteins separate in a second direction by regular gel electrophoresis. The proteins obtained can be detected by protein staining. The greatest number of proteins have been detected by labeling the proteins with a specific isotope and then placing x-ray film over the gel slab to detect the radioactivity of the proteins. Using this technique, over 1000 proteins from *E. coli* have been detected. This is a great improvement in the potential of identifying the products of the genone in terms of informational expression. Two-dimensional gel electrophoresis has been applied in a number of different ways. The method has been used as a finger print for various strains of organisms (Piperno *et al.*, 1977). Roberts *et al.*, (1980) used the two-dimensional gel electrophoresis to compare 57 species and strains of *Rhizobium*. We have used the method to compare species of smut (Hemming *et al.*, 1978). O'Farrell (1975) analyzed the proteins in human blood plasma. A map of these blood proteins were made by Anderson and Anderson (1977). Recently Bloch *et al.*, (1980) undertook the identification of proteins from *E. coli*. They obtained 81 purified proteins and were able to identify them in specific locations in the pattern. The indications are that this technique has considerable potential for analyzing proteins.

Elliott and McLaughlin (1978) used two-dimensional electrophoresis to determine the rate of synthesis of individual proteins in yeast. They investigated 111 individual proteins out of over 1000 proteins that they separated by electrophoresis.

Two dimensional gel electrophoresis has been used to follow changes in protein during the sporulation of *Saccharomyces cerevisiae*. Petersen *et al.* (1979) detected approximately 250 polypeptides during the sporulation period. Only a few of the proteins increased or decreased in concentration during sporulation. Five new proteins were detected in stationary and sporulating yeast cells that were absent in the exponentially growing cultures.

Two dimensional autoradiograms using ^{35}S methionine to label the yeast protein revealed some proteins were preferentially labeled early during sporulation while other proteins were labeled at later sporulation stages.

Protein Purification Using Affinity Chromatography. The size, shape, and charge of proteins have been important properties in the separation of proteins. Affinity chromatography is an effective method that used the unique biological properties of the specific protein in relation to binding to a ligand to isolate specific proteins (Gray, 1980). The ligand is bound to an insoluble support and packed in a chromatography column. Supports such as agarose, sepharose, cellulose, polyacrylamide,and silica have been used. In principle only the proteins with appreciable affinity for these ligands will be retained on the column. Other proteins will pass through unretarded. In some cases the ligand has to be placed on a spacer arm in order to permit the proper stereochemistry for the protein to attach. The basic principle in affinity chromatography can apply when any practically ligand interacts specifically with a biomolecule. Affinity chromatography has been used to purify enzymes, antibodies, nucleic acids, co-factors, hormones, vitamins, receptor proteins, etc. Thus by almost a single step procedure, the isolation of a single protein from a protein mixture is potentially possible.

Amino Acid Composition in Proteins. The analysis of the amino acids in proteins is another step in understanding the informational expression. In the past, the determination of the amino acids present in the protein has involved the classical amino acid analyzer that was originally described by Speckman *et al.* (1958). The original instrument took often as long as 20 hours to obtain an analysis of a physiological fluid. Through the years considerable improvements have been made. The latest model of the amino acid analyzer uses a small metal high pressure column. Samples are loaded automatically by a micro processor that controls the instrument. The buffer system provides a stepwise sequence of the amino acids. Eluted amino acids in the flow tube pass through a 128°C water bath for rapid development of the ninhydrin color complex. The peak area and amino acid concentration is determined by the microprocessor. With protein hydrolysates, the time from injection to injection is 65 minutes. Whereas for physiological fluids, it takes 4.5 hours.

Amino acid analysis with gas chromatography involves making derivatives of the amino acid by silylation using hexamethyldisilazane (HMDS). Excellent results have also been obtained with trimethylsilyldiethylamine (TMSDEA). The separation is excellent and the time of separation is rapid (less than an hour normally). The main disadvantage to the GLC method is the need to make the derivate. Considerable data on the amino acid content of proteins have been obtained. It is also of interest that a number of unusual amino acids have been found in fungal proteins such as toxins.

Determining the Sequence of Amino Acids in Proteins. The ultimate step in the definition of the informational expression is the determination of the sequence of the amino acids in a purified protein. This process was originally a very unique feat when accomplished for the first protein, however, now it has become much more routine and commercial sequencers are now used. The approach generally involves a specific hydrolysis of the protein using specific enzymes which cleave the protein at specific sites. For example, a specific proteinase that cleaves at the carboxyl side of aspartic acid and glumatic acid residues. These specific cleavage agents are utilized to provide smaller peptides. Then the peptide fractions are isolated and the Edman degradation sequence is utilized. In the Edman degradation, one amino acid is removed at a time from the N terminal. The automatic sequencers analyze the amino acids in a sequential manner using high pressure liquid chromatography with the Edman degradation sequence. A number of fungal proteins have been analyzed and their amino acid sequence determined.

Proteases. Proteases play an important rôle in breaking down unwanted proteins and providing amino acids for further protein synthetic processes.

Induction of proteases in *N. crassa* required a protein substrate, and an activating protease (Drucker, 1975). The cells apparently need to be capable of protein synthesis, since cycloheximide inhibits the process but cell growth does not appear to be required. Betz (1976) found that cycloheximide stimulated intercellular protein degradation in a similar manner as starvation for the required amino acids. He suggested that stimulation of protein degradation in yeast was directly related to the inhibition of protein synthesis.

The coming years should see considerable advance in our understanding of the control factors of informational expression particularily in relation to the developmental process in fungi.

REFERENCES

Allen, R.C., and Maurer, H.P. (1972). "Electrophoresis and isoelectric focusing with polyacrylamid gels - Methods and Theories." Walter De Gruyter. New York, New York.

Anderson, L., and Anderson, N. G. (1977). High resolution two-dimensional electrophoresis of human plasma proteins. *Biochemistry* 74, 5321-5425.

Betz, H. (1976). Inhibition of protein synthesis stimulates intracellular protein degradation in growing yeast cells. *Biochem. Biophys. Res. Commun.* 72, 121-130.

Bloch, P. L., Phillips, T.A. and Neidhardt, F.C. (1980). Protein identification on O'Farrell Two-Dimensional Gels: Locations of 81 *Escherichia coli* Proteins. *J. Bacteriol.* 141, 1409-1420.

Brooks, C. J. and De Busk, A. G. (1973). Cellular compartmentation of aromatic amino acids in *Neurospora crassa*: I. Occupation of a protein synthesis pool by phenylalanine in Tyr-1 mutants. *Biochem. Genet.* 10, 91-103.

Crick, F. H. C. (1966). The genetic code. Cold Spring Harbor Symposium *Quant. Biol.* 31, 3-9.

Drucker, H. (1975). Regulation of exocellular proteases in *Neurospora crassa*: Metabolic requirements of the process. *J. Bacteriol.* 122, 1117-1125.

Elliott, S. G. and McLaughlin, C.S. (1978). Rate of macromolecular synthesis through the cell cycle of the yeast *Saccharomyces cerevisiae*. *Proc. Natl. Acad. Sci. U.S.A.* 75, 4384-4388.

Fraser, R. S. S. (1975). Turnover of polyadenylated messenger RNA in fission yeast: Evidence for the control of protein synthesis at the translational level. *Eur. J. Biochem.* 60, 477-486.

Gray, G. R. (1980). Affinity chromatography. *Analytical Chemistry* 52, 9-15R.

Gruber, K. A., Whitaker, J. M. and Morris, M. (1979). Molecular weight separation of proteins and peptides with a new high-pressure liquid chromatography column. *Analytical Biochemistry* 97, 176-183.

Heckman, J. E., Hecker, L.I., Schwartzbach, S.D., Barnett, W.E., Baumstark, B., and Rajbhandary, U.L. (1978). Structure and function of initiator methionine transfer RNA from the mitochondria of *Neurospora crassa*. *Cell* 13, 83-95.

Hemming, B. C., Gardner, J.S. Weber, D.J. and Hess, W.M. (1978). Isoelectric focusing of proteins from dormant teliospores of *Tilletia caries* and *T. controversa*. *Amer. Bot. Soc. (Abs)*.

Holley, R., Apgar, W.J., Everett, G.A., Madison, J.T., Marquisse, M., Merrill, S.H., Penswick, J.R., and Zamir, A. (1965). Structure of alanyl-tRNA. *Science* 147, 1462-1465.

Kellems, R. E., Allison, V.F. and Butow, R.A. (1974). Cytoplasmic type 80S ribosomes associated with yeast mitochondria: II. Evidence for the association of cytoplasmic ribosomes with the outer mitochondrial membrane *in situ*. *J. Biol. Chem.* 249, 3297-3303.

Kellems, R. E. and Butow, R.A. (1974). Cytoplasmic type 80S ribosomes associated with yeast mitochondria: III. Changes in the amount of bound ribosomes in response to changes in metabolic state. *J. Biol. Chem.* 249, 3304-3310.

Kim, S. H., Quigley, G.H., Suddath, F.L., McPherson, A. Sneden, D., Kim, J.J., Weinzierl, J., and Rich, A. (1973). X-ray crystallography of analyl-tRNA. *Science* 179, 285-288.

Kraft-Creech, J. and Lochmann, E.R. (1978). Protein synthesis on membrane-bound and free ribosomes in growing yeast. *Biochem. Biophys. Acta* 521, 426-434.

Mazumder, R., and Szer, W. (1977). Protein biosynthesis *In* "Biological information transfer," (Ed. M. Florkin, A. Neuberger, and L. VanDeenen). pp. 184-231. Elsevier Publishing Co. N.Y.

McNeil, R. G. and McLaughlin, C.S. (1974). Differential biological activity of three species of methionyl-tRNA in yeast. *Biochem. Biophys. Acta* 374, 176-186.

Nomura, M., Tissières, A. and Lengyel, P. (1974). "Ribosomes". Cold Spring Harbor Laboratory Press, Mass.

Ochoa, S., and Mazumder. R. 1974. Protein Synthesis in "Enzymes" Vol. X pp. 1-51. Academic Press, New York.

Ochoa, S. and DeHaro, C. (1979). Regulation of protein synthesis in eukaryotes. *Ann. Rev. Biochemistry* 48, 549-580.

O'Farrell, P. H. (1975). High resolution two-dimensional electrophoresis of proteins. *J. Biol. Chem.* 250, 4007-4012.

Petersen, J. G. L., Kielland-Brandt, M.C. and Nilsson-Tillgren, T. (1979). Protein patterns of yeast *Saccharomyces cerevisiae* during sporulation. *Carlsberg Res. Commun.* 44, 149-162.

Piperno, G., Huang, B. and Luck, D.J.L. (1977). Two-dimensional analyses of flagellar proteins from wild type and paralyzed mutants of *Chlamydomonas reihnardtii*. *Proc. Nat. Acad. Sci. USA* 74, 1600-1605.

Roberts, G. P., Leps, W.T., Silver, L.C. and Brill, W.J. (1980). Use of two-dimensional polyacrylamide gel electrophoresis to identify and classify *Rhizobium* strains. *Applied and Environmental Microbiology* 39, 41-42.

Rubin, L. B., Eremeeva, O.V., Fraikin, G.Y. and Shvinka, Y.E. (1973). A phytochrome regulation system and its role in the development of microorganisms. *Biol Nauki* 16, 49-55.

Rubinstein, M. (1979). Preparative high performance liquid partition chromatography of proteins. *Analytical Biochemistry* 98, 1-7.

Saravis, C. A., and Zamcheck, N. (1979). Isoelectric focusing in agarose. *J. Immunological Methods* 29, 91-96.

Shilo, B., Simchen, G. and Pardee, B. (1978). Regulation of cell-cycle initiation in yeast by nutrients and protein synthesis. *J. Cell Physiol.* 97, 177-188.

Soll, D., and Schimmel, P.R. (1974). Electrophoresis of proteins *In* "Enzymes" (Ed. P. D. Boyer), Vol. X. pp. 489-538, Academic Press, New York.

Speckman, D. A., Stein, W.H. and Moore, S. (1958). Chromatographic separation of amino acids. *Analytical Chemistry* 30, 1190-1194.

Ui, N. (1979). Rapid estimation of the molecular weights of protein polypeptide chains using high-pressure liquid chromatography in 6 M guanidine hydrochloride. *Analytical Biochemistry* 97, 65-71.

Ulaszewski, S. Theofanis, M., She, W., Rosenthal, P.J. Woodward, J.R., Cirillo, V.P., and Edmunds, Jr. L.N. (1979). Light effects in yeast: Evidence for participation of cytochromes in photoinhibition of growth and transport in *Saccharomyces cerevisiae* cultured at low temperatures. *J. Bacteriol.* 138, 523-529.

Walton, H. F. (1980). Ion exchange and liquid column chromatography. *Analytical Chemistry* 52, 15-27R.

Wool, I. G. (1979). The structure and function of eukaryotic ribosomes. *Ann. Rev. Biochem.* 49, 719-754.

PROTEIN SYNTHESIS IN AQUATIC FUNGI

J. S. Lovett and W. D. Sikkema

Department of Biological Sciences
Purdue University
West Lafayette, IN 47907, U.S.A.

INTRODUCTION

The role of protein synthesis and its regulation during sporulation and spore germination in aquatic fungi has received relatively little attention. Most of the work prior to 1975 has been reviewed in some detail (Lovett 1975, 1976) and needs no repetition here. We, therefore, propose to provide a brief review of the pertinent results published since 1975, and to then discuss some new work on sporulation in our laboratory. In doing this we would like to focus primarily on the three genera that have received the most experimental attention, *Achlya, Allomyces and Blastocladiella.* One thing that will be evident from this brief review of protein synthesis in aquatics is how little we really know about the basic physiology or biochemistry of development in the group. Those of us who work with these interesting organisms, while regretting their neglect, may, however, take some solace from the fact that those who work on sporulation and spore germination in higher fungi generally are not very much better off!

The induction of sporulation in water molds is usually accomplished by "stepping down" cultures from a growth medium to an inorganic starvation medium after a period of unrestricted growth. This is an effective procedure for obtaining reproducible and synchronous development of reproductive structures. The reason for its success in producing synchronous sporulation is probably that it avoids a gradual depletion of nutrients and the consequent appearance of the growth limiting conditions characteristic of batch cultures. The specific minimal requirements for inducing sporulation in aquatics have not generally been studied systematically to establish whether nitrogen depletion, carbon depletion, or both are essential. However, nitrogen appears to be important in *Allomyces* (Youatt *et al.*, 1971). It does seem clear that aquatics are competent to sporulate at almost any stage of growth; *Blastocladiella emersonii*, for example, can be induced within 30 min after zoospore inoculation (Hennessy and Cantino, 1972).

The process of sporulation in aquatics differs from conidiation by microfungi in two major ways. First, it is quite rapid, usually only requiring 4-5 hour from induction to spore release. Second, it typically involves the remodeling of a preexisting "cell" rather than the synthesis of new structures such as the formation of conidiophores and conidia. This difference may very likely be reflected by quite different underlying patterns of metabolism during sporulation. Spore production in water molds has, in addition, two well defined developmental stages. In the initial stage, the preexisting, typically multinucleate cell or hyphal tip is converted into a sporangium through the formation of a basal cross wall and a discharge papilla or papillae. In the second stage the actual differentiation of the zoospores takes place. The separation of sporulation into two stages seems justified by the common observation that development is often suspended at the papilla stage in aerial environments. These quiescent sporangia then proceed to cleave out spores when placed in an aquatic milieu. On the assumption that two discrete stages exist it would be of considerable interest to know whether the major zoospore components, or their precursors, were synthesized during the first stage and assembled during the second, produced throughout sporulation, or only in the second zoospore differentiation stage.

Germination of zoospores also involves the two discrete events of encystment and outgrowth. During encystment the flagella are retracted or discarded, preformed adhesive materials may be released (Sing and Bartnicki-Garcia, 1975), and a new cell wall is formed outside the plasma membrane. The second phase then involves the production of a rhizoid, a hyphal germ tube, or both, depending upon the species.

SPORES AND SPORE GERMINATION

There has been very little study of any aspect of metabolism in most aquatic zoospores, *Blastocladiella* perhaps being the exception (Cantino and Myers, 1972; Suberkropp and Cantino, 1973; Lovett, 1975; Selitrennikoff and Sonneborn, 1976; Gong and Lovett, 1977). It is clear that *B. emersonii* spores carry a complete but nonfunctioning system for protein synthesis (Lovett, 1975; Lovett *et al.*, 1977; Gong and Lovett, 1977) including stored messenger RNA (Schimmelpfeng *et al.*, 1976; Jaworski, 1976; Johnson *et al.*, 1977) that can be translated *in vitro* (Lovett, 1975; Lovett, unpublished). Less extensive studies with *Allomyces macrogynus* have indicated a similar status for the protein synthetic system in the zoospores of that species (Smith and Burke, 1979). The ribosomes in *Allomyces* also exist entirely within the nuclear cap, mainly in the form of monoribosomes but with a surprisingly high 25-35% in polysomal aggregates. Smith and Burke (1979) also found the poly (A) [putative poly(A)RNA or messenger RNA] distribution in *Allomyces* zoospores to be similar to *B. emersonii* with about two thirds in the nuclear cap organelle and the remainder elsewhere. At the present time the only direct evidence for factors that block protein synthesis in zoospores is our earlier report of a ribosome-associated inhibitor (Adelman and Lovett, 1974a). This has still not been identified, and thus remains open to verification. The highly organized compartmentalization of organelles in zoospores does, however, suggest that energy generation via the production and use of ATP may be tightly channeled. This would certainly help ensure an energy supply for motility while contributing to the suppression of biosynthesis.

In both *Blastocladiella* (Lovett, 1968; Leaver and Lovett, 1974) and *Allomyces* (Burke *et al.*, 1972; Olson, 1973; Olson and Fuller, 1971) protein synthesis is rapidly reactivated following zoospore or meiospore encystment; since synthesis and development can both occur for a considerable interval in the absence of new RNA synthesis (i.e., in the presence of actinomycin D), a significant fraction of the new proteins produced must have been translated from mRNA stored in the zoospore.

We are aware of no instance in aquatic fungi where the proteins synthesized in early germination have been identified, much less shown to be essential for a specific event in germination. This includes *Blastocladiella* zoospores (Lovett, 1968, 1975; Soll and Sonneborn, 1971; Silverman *et al.*, 1974), *Allomyces* zoospores (Burke *et al.*, 1972; Olson and Fuller, 1971), *Allomyces meiospores* (Olson, 1973), and *Allomyces* zygotes (Bruce and Mascarenhas, 1977a, 1977b). Some attempts to analyze this stage in *Blastocladiella* have been made in the respective laboratories of Dr. E. C. Cantino and Dr. D. R. Sonneborn, as well as our own. Cantino and Myers (1972) reported chitin synthase to preexist in zoospores and stressed its localization in the gamma particles that break down during encystment. Norrman *et al.* (1975) also demonstrated the presence of the first enzyme in the pathway leading to chitin, L-glutamine:D-fructose 6-phosphate aminotransferase. Selitrennikoff and Sonneborn (1976) have further shown that the remainder of the biosynthetic pathway leading to the immediate chitin precursor, UDP-GlcNAc, also preexists in the zoospores and is controlled by feedback inhibition. Thus, the system necessary for cell wall synthesis is already present prior to germination.

We have examined the microtubular protein pools on the assumption that new microtubules might be required for rhizoid initiation and/or development. However, examination of the soluble tubulin pools in the zoospore stage and during early development indicated that new tubulin synthesis was not a limiting factor (Abe and Lovett, unpublished). No new tubulin synthesis could be detected until 25-30 minutes into germination and a net pool increase only occurred after 50 minutes when the primary germ tube was already well established. The tubulin pool does rapidly expand by addition of subunits released during disassembly of the intracellular microtubular system of the zoospore. The pool expansion is, however, independent of the disassembly of the retracted flagellar axoneme; it occurs to the same extent when deflagellated zoospores germinate or when axoneme breakdown is blocked by cycloheximide. From this brief summary it is evident that the only specific candidates with known functions that have been examined so far preexist in zoospores at levels adequate to support early germination.

In another, more general approach we have tried to assess the number of proteins synthesized during zoospore germination and early development. Previous studies had already shown that *Blastocladiella* zoospores could encyst and form a short germ tube in the absence of new RNA synthesis, but this did require new protein synthesis (Lovett, 1968). Two approaches have been used to evaluate this requirement: pulse labeling followed by two-dimensional gel electrophoresis using the O'Farrell (1975) technique, and hybridization of polysomal mRNA from different stages to single copy tracer DNA. Radioautographs were made of two-dimensional gels prepared from extracts of cells labeled at 15 minute intervals during the first 75 minutes of germination in the presence or absence of 10 μg/ml actinomycin D

(Abe and Lovett, unpublished). Only two polypeptides out of 132 that could be detected by 30 minutes in untreated cells were absent from actinomycin D treated cells. Two additional spots that only appeared after 30 minutes in the absence of inhibitor were not blocked by actinomycin D, but the antibiotic did severely inhibit the appearance of additional spots after 30 minutes. The two actinomycin D-resistant proteins produced after 30 minutes suggest the possible post transcriptional control of their translation (or the rates of translation). The suppression of the synthesis of only two of the early proteins by actinomycin could be interpreted to mean that virtually all the earliest proteins normally made are translated from stored zoospore mRNA. However, the results obtained from hybridizing polysomal RNA from germinating cells in excess reactions with single copy tracer DNA indicate that the situation is not quite so simple (Johnson and Lovett, unpublished). The zoospores contain ca. 11,600 different mRNA sequences based on complexity measurements (14.5×10^6 nucleotides) and an average mRNA size of 1250 nucleotides (Johnson *et al.*, 1977). The polysomes from cells harvested at 30 minutes postinoculation were found to contain from 14,500-19,000 different mRNA sequences, depending on the germination medium used. If correct, these data mean that between 2900 and 7600 additional new transcripts over and above those already present in the zoospores may be expressed by translation on polysomes in the first 30 minutes. An additional and obvious conclusion must be that the number of labeled proteins we detected using 15 minute pulses was not really representative of the total number being made. This data suggests to us that the polypeptides that did appear on two-dimensional gels were very likely common metabolic and structural proteins representing the more abundant class of messengers. The original problem thus still remains: to identify the essential protein(s) that lead to developmental arrest when new mRNA synthesis is blocked in *Blastocladiella* zoospores, or the spores of other aquatics. Our hybridization data suggest that there may be a large number of candidates.

SPORULATION

In contrast to studies of development at the light and ultrastructural level, sporulation in aquatics has received little concentrated attention at the biochemical level. Griffin (1966) developed a system for inducing *Achlya bisexualis* development using a stepdown to 0.5 mM $CaCl_2$ and reported on RNA synthesis in such induced cultures (Griffin and Breuker, 1969). Techniques for the specific identification of mRNA were not then available, but some RNA synthesis was required throughout sporulation. Timberlake *et al.*, (1973) subsequently extended this study to an analysis of protein synthesis. This activity was also required thoughout the five hour period of sporangium and spore differentiation. They further found that the rates of protein synthesis doubled by one hour post induction and, perhaps most significant, that extensive protein turnover also occurred through at least 4.5 hours. The turnover rate was estimated at 75% per hour, or almost four complete turnovers of cellular protein in the five hours required for sporulation. At the same time the protein content decreased by 20% overall, while intracellular protease activity in crude extracts more than doubled in the first three hours. These results are quite dramatic and may represent an overestimate, but are generally consistent with the enzyme studies of Horgen and co-workers in the same system. The latter examined three specific enzymes, all presumably lysosomal in nature, for specific activity changes during induced sporulation in *Achlya*. The specific activities of acid phosphatase (O'Day and Horgen, 1974) and α-mannosidase (Horgen and O'Day, 1975) both increased several fold, reaching maximal levels at four hours post induction. In contrast ribonuclease activity (Sutherland *et al.*, 1976) dropped by 50% between two and three hours, returned to the normal level at four hours, and dropped by 50% again just prior to zoospore release. The RNase activity was found to be inversely proportional to the poly(A)-RNA content of the developing cultures. Inhibitor studies showed that the changes in all three enzymes required concomitant RNA and protein synthesis. The acid phosphatase and α-mannosidase both also decayed rapidly in the presence of either actinomycin D or cycloheximide, implying a requirement for continuous transcription and translation. The developmental role(s) of these enzymes, if any, is unknown. However, with the rapid rates of turnover reported by Timberlake *et al.* (1973) one must assume that the observed levels at each time represented a summation of the specific synthetic and degradative rates for these proteins. A change in either synthesis or degradation rates could cause a significant change in observed activity.

Studies of protein synthesis during zoosporangial and gametangial development in *Allomyces* species are not extensive. Most of the work on these stages of *Allomyces* has been directed toward RNA synthesis, particularly to establish whether it is required during the process, and the stage at which essential RNA is produced in induced cultures. Unfortunately there has been considerable disagreement in the interpretation of inhibitor experiments with *Allomyces* (Burke *et al.*, 1972; Burke and Germond, 1976; Fähnrich, 1974a, 1974b, 1976, 1977; Stumm and Croes, 1975; Youatt, 1976a). It seems clear from the report of Youatt (1976a) that the conflicts have most probably arisen from the use of different species and different cultural conditions. A key point made by Youatt is that the conditions used to grow and induce the cultures are very important. In particular it is necessary to ensure that the nitrogen source, or other critical variable, is not depleted or altered before deliberate experimental induction.

There appears to be a concensus that protein synthesis is required throughout both zoosporogenesis (Burke *et al.*, 1972) and gametogenesis (Fähnrich, 1974a, 1974b, 1976) based on the inhibition achieved with cycloheximide and puromycin, and the presence of polysomes (Stumm and Croes, 1975). There is no information yet available to indicate what kinds of proteins are being made, the qualitative and quantitative magnitude of synthesis, or whether extensive turnover occurs. It also seems probable that RNA transcription is required, at least during differentiation of gametangia (Stumm and Croes, 1975; Fähnrich, 1974a, 1974b; Youatt, 1976a) and sporangia (Fähnrich, 1976; Youatt, 1976a) despite the contrary report of Burke *et al.* (1972) that *A. arbuscula* mycelia produced sporangia in the presence of actinomycin D. Although the patterns of RNA synthesis have not been examined in detail, the developmental effects of inhibitors are consistent with mRNA as the developmentally essential product of transcription.

Both *Achlya* and *Allomyces* are filamentous organisms and it is difficult to know how much of a given activity in homogenates or extracts of sporulating mycelia is due to the developing sporangia at the hyphal tips and how much represents the supporting mycelia. Synchrony data on the cultures has also usually not been provided with the published experiments and this presents difficulty in the interpretation of the results. The isolation of gametangia to circumvent the mycelial contribution (Fähnrich, 1974b) represents an interesting but technically difficult solution which would, in any case, be restricted to the stage of gamete differentiation.

A rather long time ago we described the basic process for inducing zoosporogenesis in late log phase cultures of *Blastocladiella emersonii* (Murphy and Lovett, 1966; Lessie and Lovett, 1968). Except in details, the processes of sporulation in *B. emersonii* and both sporulation and gametogenesis in *Allomyces* hyphal tips seem remarkably similar. Thus, we may hope that much of the information obtained with each will apply to the other. In the case of *Blastocladiella*, the nonmycelial growth form does have the considerable advantage that the entire thallus becomes a zoosporangium which eliminates the uncertainity due to hyphal metabolism. Although sporogenesis in *Blastocladiella* and other fungi has been reviewed fairly recently (Lovett, 1975), it may be useful to mention a few specific points related to protein synthesis before taking up some recent studies.

First, sporulation can be induced at any stage by stepping cultures down from a growth sufficient to a starvation medium (Murphy and Lovett, 1966; Hennessy and Cantino, 1972; Lodi and Sonneborn, 1974; Barstow and Lovett, 1978). Our earlier work used late log phase cultures shifted over a 20 minute period to a dilute salts buffer solution. Pulse labelling experiments and polysome data obtained with these cultures yielded results quite consistent with active turnover of cellular proteins, i.e., significant rates of amino acid incorporation and rather high levels of polysomes without an increase in protein content (Murphy and Lovett, 1966; Adelman and Lovett, 1974a, 1974b). We also found that ribosomal RNA synthesis was virtually shut down within an hour after induction (Lovett, 1975). Using [^{3}H]-adenosine labelling, Jaworski and Thompson (1980) have reported that a very small amount of ribosomal RNA turnover continues even when the total RNA content decreases. More significantly they found evidence for active synthesis and turnover of poly(A)RNA throughout sporangium formation and most of zoospore differentiation. Of particular interest were their observations that: (a) some mRNA [poly(A)RNA] present in the growth phase was conserved in zoospores; and (b) that as much as one third of the zoospore stored mRNA was transcribed during the last 30 minutes of sporulation.

At the opposite exteme from the late log phase, Hennessy and Cantino (1972) were able to induce microcyclic sporogenesis in *Blastocladiella* as early as 30 minutes post inoculation. However, these cultures have not been examined for their biosynthetic capacities. In experiments to study the coordinate regulation of protein and RNA synthesis we have rapidly stepped down two hour-old cultures to a dilute salts medium (Lovett, unpublished). In such cultures the net accumulation of both RNA and protein ceases immediately after the shift. Then, after a 50 minute delay both the RNA and protein content of the cells begins to decrease; at first rapidly and then at a slower rate. The starved cells will sporulate but this was not followed to completion in the regulation experiments yielding this data. It should be noted, however, that the same 50 minute plateaus were also observed when other less stringent stepdown conditions were used to limit the amino acid supply.

Lodi and Sonneborn (1974) examined protein turnover in *Blastocladiella* cells induced to sporulate after six hours of growth. Using the release of [^{14}C]-leucine from uniformly labeled cells they found a turnover rate of 12% per hour during sporulation compared to less than 2% per hour in growing cells. They also identified two proteases that first appeared and increased in activity during sporulation. One, an alkaline protease, rose dramatically during the sporangial development stage and was subsequently released to the medium. The second, an acid protease, increased more slowly but was retained in the zoospores. The rate of protein turnover in zoospores was 1.2% per hour and increased to 5% per hour during germination. We have also noted a rapid increase in the proteolytic activity of cell extracts during germination and early development (Abe and Lovett, unpublished).

Finally, Correa *et al.* (1978) have found that the sporulation-specific alkaline protease is inhibited by antipain and phenylmethylsulfonylfluoride (a serine protease inhibitor). The same inhibitors completely block sporulation if added up to 30 minutes after induction, but not at all if added after 90 minutes. Antipain has no effect on growing cells when the alkaline protease is absent. Correa *et al.* (1978) suggested that the protease might function to modify or remove proteins in a step that was essential to the success of sporulation. The induction and secretion of proteases during *Blastocladiella* sporulation is quite similar to the activities of *Bacillus* species during endospore formation.

From this brief resumé of the data obtained with *Achlya*, *Allomyces* and *Blastocladiella* it seems evident that these and other aquatics probably share many characteristics in addition to their common aquatic habit. It also seems clear that from an experimental point of view they may share the problem of active protein turnover during sporulation; a situation already well known for the sporulation phases of bacteria, cellular slime molds and yeasts (See the paper by

Dr. Wolf in this symposium). As has been the case with other microbial systems it will be important to identify the short term metabolic events involved in adapting to the inducing medium, as contrasted to those specific to the sporulation process. This may be expected to be a problem restricted to the early stages of sporangium formation, i.e., while the cells are adjusting to nitrogen and/or carbon depletion and shifting to a new developmental program. The uniformity of the synthetic plateaus for RNA and protein observed in stepdown experiments suggests that in *Blastocladiella* at least the transition phase may require about 50-60 minutes. There is no pertinent data available but the delay could perhaps be related to the depletion of some finite protein fraction subject to rapid turnover upon N or C starvation. The initial metabolic response to starvation is certainly worthy of systematic study in any of these fungi, both in its own right and to provide a basis for further work on sporulation-specific events.

We have become convinced that the most fruitful approach for an analysis of sporulation is to focus on the biosynthesis of clearly defined structures or metabolic activities uniquely associated with the developmental process. An ideal strategy would be to study simultaneously the formation of components that are individually specific to the stages of sporangium and zoospore differentiation. In the case of *Blastocladiella* the most obvious candidates would be septation, papilla formation and proteases in sporangial development, and flagella axonemes (or microtubular precursors) and gamma particles, for zoospore differentiation. For practical purposes we have initially taken an atomistic approach and begun an in depth analysis of papilla differentiation. This project is far from complete but we think it is well enough along to be of some interest to this group.

We chose papilla development for a number of reasons. First, Skucas (1966) had shown that paillae of *Allomyces* could be separated from the sporangia by strong acid treatment. We subsequently found that *Blastocladiella* papilla development involved the localized deposition of material via secretory vesicles (Lessie and Lovett, 1968). A similar process of deposition has since been reported for *Allomyces* (Duff and Youatt, 1977) and *Entophlyctis* (Powell, 1976). Using a modification of the Skucas acid extraction technique, Youatt (1976b) isolated and partially characterized the papillae of *A. arbuscula* gametangia. She reported a complex composition with glycolipid, glucan and glycoprotein all present. The glycoprotein, however, appeared to be a major component of the papilla matrix. Finally, Lemos and Lodi (1975) demonstrated that fluorescein-labeled concanavalin A bound preferentially to *Blastocladiella* papillae and prevented papilla dissolution. Although these individual observations were fragmentary and derived from separate genera, taken together they provided the basis for a coherent, testable hypothesis concerning papilla development. Namely, that (a) the major component of *Blastocladiella* papillae is a glycoprotein or glycoproteins; (b) that papilla formation should involve the stage-specific induction of glycoprotein synthesis, including the protein and oligosaccharide moieties; (c) that the glycoprotein should be transported in secretory vesicles that migrate to a localized region of the cell; (d) that the papilla is formed by the deposition of the preformed glycoprotein via exocytosis; and (e) that the enzymes to degrade the chitinous cell wall in the same region must also be released or activated to make the pore that is filled by the papilla plug.

On the basis of the above, papilla differentiation appeared to provide a well defined vectorial system with which we could analyze each step in synchronously developing cell populations. *Blastocladiella* is a good organism for such a study both because it is non-mycelial, and because it is possible, using the induced minicycle system (Barstow and Lovett, 1978), to complete the entire cycle from zoospore to zoospore in 7 hours. Cells induced by a shift to dilute phosphate buffer at 3 hours complete papilla formation during the subsequent 3.5 hours. Such cells, containing only two nuclei, differentiate into sporangia and cleave out two zoospores in the same sequence and with nearly the same timing as multinucleate, late log phase cells. The smaller cells have a further advantage that they produce more papilla material relative to other cell constituents than larger cells; this at least in theory should aid in the isolation of papilla glycoprotein precursors and secretory vesicles.

Our initial approach has been to test the effects of a number of inhibitors on papilla formation and macromolecule synthesis is in order to establish the general properties of the system. Both actinomycin D and cycloheximide completely block papilla development and, on this basis, it is assumed that both RNA and protein synthesis are required. The process begins to escape from actinomycin D inhibition between 90 and 120 minutes after induction. It only becomes resistant to cycloheximide after 135 minutes when papillae first begin to appear in the culture. The point of half maximal inhibition by actinomycin precedes the point of maximal papilla formation in the controls by ca. 24 minutes, while half maximal inhibition of papillae by cycloheximide precedes the half maximal level of papilla formation by only 15 minutes. These results are consistent with a sequential transcription of the mRNA and its subsequent translation to produce the putative papilla glycoprotein precursor(s) just prior to papilla deposition. However, other processes that require RNA or protein synthesis would also be affected by these nondiscriminate inhibitors and these results merely place probable limits on the times of papilla glycoprotein synthesis.

To obtain more specific information about glycoprotein synthesis and transport we have tested the effects of several inhibitors reported to block glycosylation of proteins and exocytosis, respectively. For each inhibitor, the effects of concentration and time of inhibitor addition were first tested on papilla formation. Effective inhibitors were then further examined for their effects on protein synthesis via [^{3}H]-leucine incorporation, and glycosylation via [^{3}H]-mannose incorporation. Mannose incorporation was used as a test for effects on glycosylation because mannose is usually the major sugar found in fungal glycoproteins (Struck and Lennarz, 1980), because it was reported to be in papilla preparations from *Allomyces* (Youatt, 1976b), and, of course, because Con A binds to *B. emersonii* papillae (Lemos and Lodi, 1975). The glycosylation inhibitors bacitracin and tunicamycin, and the putative secretion inhibitors cytochalasin B and colchicine all blocked papilla development at appropriate concentrations. The mannose analogue, 2-deoxyglucose, also served as an effective inhibitor, but the microtubule inhibitor nocadazole had no effect. At 4 mM, 2-deoxyglucose (2-dGc) blocked papilla development and inhibited mannose incorporation but also inhibited leucine incorporation. This result suggested a non-specific effect of the sugar on cellular metabolism (2-dGlc can act as an analogue of both mannose and glucose and give rise to both UDP-2-dGlc and GDP-2-dGlc). Bacitracin at 2.5 mM prevented both mannose and leucine incorporation. In the absence of evidence that it acts in fungi as it does in bacteria, this result is considered equivocal and presumably non-specific. Tunicamycin, a lipophilic analogue of UDP-N-acetylglucosamine inhibited only mannose incorporation and papilla formation at 50 μg/ml, having no effect on [^{3}H]-leucine incorporation. This antibiotic is known to inhibit glycosylation in yeast (Kuo and Lampen, 1974) at the earliest stage of oligosaccharide formation, i.e., the attachment of N-acetylglucosamine to the lipid, dolichol phosphate. Its affect appears to be specific in *Blastocladiella* because it fails to inhibit general protein synthesis at effective concentrations (other sugars including N-acetylglucosamine have not yet been tested).

Colchicine only inhibited papilla development effectively at rather high concentrations (e.g., 30 mM), where it also inhibited both mannose and leucine incorporation. At a concentration of 8 mM where it failed to block leucine incorporation it also did not prevent papilla formation. This, again suggests a non-specific action in the cells. The microfilament-specific inhibitor, cytochalasin B at 82 mM inhibited papilla formation without any effect on mannose or leucine incorporation; this result indicates that cytochalasin B may have a direct and specific effect on exocytosis. We are now beginning a study of the intracellular effects of these inhibitors at the ultrastructural level with Dr. W. E. Barstow at the University of Georgia. The results of the inhibitor experiments just described seem entirely consistent with our working hypothesis that mannose-containing glycoproteins are synthesized and secreted during papilla development. Preliminary experiments have also shown that when cells that have developed papillae in media containing [^{3}H]-mannose are examined by whole-cell radioautography the majority of the isotope is found in the papilla region.

Using [^{3}H]-mannose as a way to label newly synthesized glycoproteins we have been able to identify at least one candidate for the papilla precursor in cell extracts. For this experiment two identical cultures were induced by a shift from PYG medium to 0.5 mM phosphate buffer after 3 hours of growth. The nonpapillate control culture was labeled with [^{3}H]-mannose between T_{20} minutes and T_{80} minutes poststep, and the experimental culture between T_{105} and T_{165} minutes; the latter pulse overlapped the appearance of visible papillae. Both cultures were harvested at the end of the pulse period and hot SDS extracts were prepared from homogenates of the cells. The extracts were then chromatographed on concanavalin A-Sepharose columns to bind glycoproteins. Subsequent elution with α-methylmannoside released the bound oligosaccharide(s). A significantly greater amount of labeled material was obtained from the papilla stage sample than from the nonpapillate control. The material bound by Con A was further fractionated by electrophoresis on 10% acrylamide-SDS gels. After slicing and counting the gels, a prominent peak with a molecular weight of ca. 70,000 was identified in the papilla stage sample. Some heterodisperse higher molecular weight material was also present. Neither component was evident in the Con A-bound fraction from the nonpapillate control cells. Following pronase digestion of the Con A-bound material to hydrolyze the protein, a heterodisperse elution profile was obtained upon Sephadex G-25 chromatography (after rebinding and elution of the oligosaccharide from Con A-Sepharose). Paper chromatography of a trifluoroacetic acid digest of the oligosaccharide revealed mannose as the only radioactive sugar present. Induced cells also incorporate radiolabeled glucose, galactose, and N-acetylglucosamine during the late phase of papilla formation. However, the specificity of this incorporation into glycoprotein or papillae has not yet been examined.

All of our results to date are consistent with the notion that the 70,000 dalton glycoprotein is a papilla precursor that is synthesized immediately preceding its deposition in the papilla. It is, of course, still essential to prove the identity of the *in vivo* labeled material with a papilla glycoprotein, and show a precursor-product relationship between the two. It is also possible that other glycoproteins may be synthesized in smaller amounts at the same stage based on the results of electrophoretic and chromatographic separations of the glycoproteins and oligosaccharides, respectively. It is of considerable interest in this regard that Jen and Haug (1979) have reported the binding of Con A-linked ferritin to the membranes and gamma particles of *Blastocladiella* zoospores.

In conclusion, we are encouraged by our results on papilla synthesis to believe that the system will be useful for model studies of differentiation in *B. emersonii*. Obviously much remains to be done, but we think the system will lend itself to a detailed analysis down to and including the expression of the specific gene products involved. We are convinced that it will only be possible to understand the control of proteins involved in development in *Blastocladiella* and other fungi when the details of the synthesis and the processing of the relevant polypeptides are known. We hope we have presented a convincing argument that the aquatic fungi provide a number of very interesting and useful, though under exploited, systems for developmental study.

ACKNOWLEDGEMENTS

Research of the authors discussed in this paper has been supported by the National Institute for Allergy and

Infectious Diseases (papilla development) and the National Science Foundation (spore germination protein synthesis).

REFERENCES

Adelman, T.G. and Lovett, J.S. (1974a). Evidence for a ribosome-associated translation inhibitor during differentiation of *Blastocladiella emersonii*, *Bichimica et Biophysica Acta* 335, 236-245.

Adelman, T.G. and Lovett, J.S. (1974b). Ribosome function *in vitro* and *in vivo* during the life cycle of *Blastocladiella emersonii*, *Biochimica et Biophysica Acta* 349, 240-249.

Barstow, W.E. and Lovett, J.S. (1978). Ultrastructure of a reduced developmental cycle (minicycle) in *Blastocladiella emersonii*, *Experimental Mycology* 2, 145-155.

Bruce, S.A. and Mascarenhas, J.P. (1977a). Gene activity during zygote germination and early sporophytic germling development in the aquatic fungus *Allomyces macrogynus*, *Experimental Mycology* 1, 194-210.

Bruce, S.A. and Mascarenhas, J.P. (1977b). The synthesis of RNA and protein by nongerminating zygotes of the aquatic fungus, *Allomyces macrogynus*, *Experimental Mycology* 1, 211-220.

Burke, D.J. and Germond, M. (1976). Effect of inhibitors on the morphology and growth of *Allomyces macrogynus*, *Journal of General Microbiology* 95, 213-219.

Burke, D.J., Seale, T.W. and McCarthy, B.J. (1972). Protein and ribonucleic acid synthesis during the diploid life cycle of *Allomyces arbuscula*, *Journal of Bacteriology* 110, 1065-1072.

Cantino, E.C. and Myers, R.B. (1972). Concurrent effect of visible light on γ-particles, chitin synthetase, and encystment capacity in zoospores of *Blastocladiella emersonii*, *Archives of Microbiology* 83, 203-215.

Correa, J.U., Lemos, E.M. and Lodi, W.R. (1978). Inhibition of sporulation in the water mold *Blastocladiella emersonii* by antipain, *Developmental Biology* 66, 470-479.

Duff, T. and Youatt, J. (1977). Discharge papilla development in the phycomycete *Allomyces*, *Archives of Microbiology* 112, 187-194.

Fähnrich, P. (1974a). Untersuchungen zur Entwicklung des Phycomyceten *Allomyces arbuscula*. I. Einfluss von Inhibitoren der Protein- und Nucleinsäuresynthese auf die Differenzierung Gametangien, *Archives of Microbiology* 98, 85-92.

Fähnrich, P. (1974b). Untersuchungen zur Entwicklung des Phycomyceten *Allomyces arbuscula*. II. Einfluss von Inhibitoren der Protein- und Nucleinsäursynthese auf die Gametogenese, *Archives of Microbiology* 99, 147-153.

Fähnrich, P. (1976). Inhibition of sporogenesis in *Allomyces arbuscula, Naturwissenschaften* 63, 247.

Fähnrich, P. (1977). Untersuchungen zur Entwicklung des Phycomyceten *Allomyces arbuscula*. III. Polysomenbildung und RNA-Metabolismus in den Gametangien während der Gametangiendifferenzierung, *Archives of Microbiology* 112, 255-261.

Gong, C.S. and Lovett, J.S. (1977). Regulation of protein synthesis in *Blastocladiella* zoospores: factors for synthesis in nonsynthetic spores, *Experimental Mycology* 1, 138-151.

Griffin, D.H. (1966). Effect of electrolytes on differentiation in *Achlya* sp., *Plant Physiology* 41, 1254-1256.

Griffin, D.H. and Breuker, C. (1969). Ribonucleic acid synthesis during the differentiation of sporangia in the water mold *Achlya, Journal of Bacteriology* 98, 689-696.

Hennessy, S.W. and Cantino, E.C. (1972). Lag-phase sporogenesis in *Blastocladiella emersonii*: induced formation of unispored plantlets, *Mycologia* 64, 1066-1087.

Horgen, P.A. and O'Day, D.H. (1975). The developmental patterns of lysosomal enzyme activities during Ca^{+2}-induced sporangium formation in *Achlya bisexualis*. II. α-Mannosidase, *Archives of Microbiology* 102, 9-12.

Jaworski, A.J. (1976). Synthesis of polyadenylic acid RNA during zoospore differentiation and germination in *Blastocladiella emersonii, Archives of Biochemistry and Biophysics* 173, 201-209.

Jaworski, A.J. and Thompson, K. (1980). A temporal analysis of the synthesis of the mRNA sequestered in zoospores of *Blastocladiella emersonii, Developmental Biology* 75, 343-357.

Jen, C.J. and Haug, A. (1979). Changes in surface properties of developing zoospores of *Blastocladiella emersonii:* binding of concanavalin A, *Journal of General Microbiology* 115, 145-151.

Johnson, S.A., Lovett, J.S. and Wilt, F.H. (1977). The polyadenylated RNA of zoospores and growth phase cells of the aquatic fungus, *Blastocladiella, Developmental Biology* 56, 329-342.

Kuo, S.C. and Lampen, J.D. (1974). Tunicamycin - an inhibitor of yeast glycoprotein synthesis, *Biochemical and Biophysical Research Communications* 58, 287-295.

Leaver, C.J. and Lovett, J.S. (1974). An analysis of protein and RNA synthesis during encystment and outgrowth (germination) of *Blastocladiella* zoospores, *Cell Differentiation* 3, 165-192.

Lemos, E.M. and Lodi, W.R. (1975). Effect of concanavalin A on sporulation of *Blastocladiella emersonii*, *Experimental Cell Research* 94, 437-440.

Lessie, P.E. and Lovett, J.S. (1968). Ultrastructural changes during sporangium formation and zoospore differentiation in *Blastocladiella emersonii*, *American Journal of Botany* 55, 220-236.

Lodi, W.R. and Sonneborn, D.R. (1974). Protein degradation and protease activity during the life cycle of *Blastocladiella emersonii*, *Journal of Bacteriology* 117, 1035-1042.

Lovett, J.S. (1968). Reactivation of ribonucleic acid and protein synthesis during germination of *Blastocladiella* zoospores and the role of the ribosomal nuclear cap, *Journal of Bacteriology* 96, 962-969.

Lovett, J.S. (1975). Growth and differentiation of the water mold *Blastocladiella emersonii*: cytodifferentiation and the role of ribonucleic acid and protein synthesis, *Bacteriological Reviews* 39, 345-404.

Lovett, J.S. (1976). Regulation of protein metabolism during spore germination. *In* "The Fungal Spore: Form and Function" (Ed. D.J. Weber and W.M. Hess). pp. 189-242. John Wiley and Sons, New York.

Lovett, J.S., Gong, C.S. and Johnson, S.A. (1977). Zoospore germination and early development of *Blastocladiella emersonii*. *In* "Eucaryotic Microbes as Model Developmental Systems" (Ed. D.H. O'Day and P.A. Horgen). pp. 402-424. Marcel Dekker, Inc., New York/Basel.

Murphy, Sr. M.N. and Lovett, J.S. (1966). RNA and protein synthesis during zoospore differentiation in synchronized cultures of *Blastocladiella*, *Developmental Biology* 14, 68-95.

Norrman, J., Giddings, T.H. and Cantino, E.C. (1975). Partial purification and properties of L-glutamine: D-fructose 6-phosphate aminotransferase from zoospores of *Blastocladiella emersonii*, *Phytochemistry* 14, 1271-1274.

O'Day, D.H. and Horgen, P.A. (1974). The developmental patterns of lysosomal enzyme activities during Ca^{2+}- induced sporangium formation in *Achlya bisexualis* I. Acid phosphatase, *Developmental Biology* 39, 116-124.

O'Farrell, P.H. (1975). High resolution two-dimensional electrophoresis of proteins, *Journal of Biological Chemistry* 250, 4007-4021.
Olson, L.W. (1973). Synchronized development of gametophyte germlings of the aquatic phycomycete, *Allomyces macrogynus*, *Protoplasma* 78, 129-144.
Olson, L.W. and Fuller, M.S. (1971). Leucine - lysine synchronization of *Allomyces* germlings, *Archiv fur Mikrobiologie* 78, 76-91.
Powell, M.J. (1976). Development of the discharge apparatus in the fungus *Entophlyctis*, *Archives of Microbiology* 111, 59-71.
Schimmelpfeng, L., Grierson, D. and Zetsche, K. (1976). Preformed poly(A)-containing RNA in zoospores of *Blastocladiella emersonii*, *Biochimica et Biophysica Acta* 435, 340-348.
Selitrennikoff, C.P. and Sonneborn, D.R. (1976). Post-translational control of *de novo* cell wall formation during *Blastocladiella emersonii* zoospore germination, *Developmental Biology* 54, 37-51.
Silverman, P.M., Huh, M.O.H. and Sun, L. (1974). Protein synthesis during zoospore germination in the aquatic phycomycete *Blastocladiella emersonii*, *Developmental Biology* 40, 59-70.
Sing, V.D. and Bartnicki-Garcia, S. (1975). Adhesion of *Phytophthora palmivora* zoospores: detection and visualization of concanavalin A-receptor sites appearing during encystment, *Journal of Cell Science* 19, 11-20.
Skucas, G.P. (1966). Structure and composition of zoosporangial discharge papillae in the fungus *Allomyces*, *American Journal of Botany* 53, 1006-1011.
Smith, B.A. and Burke, D.D. (1979). Evidence for the presence of messenger ribonucleic acid in *Allomyces macrogynus* mitospores, *Journal of Bacteriology* 138, 535-541.
Soll, D.R. and Sonneborn, D.R. (1971). Zoospore germination in *Blastocladiella emersonii*: cell differentiation without protein synthesis?, *Proceedings of the National Academy of Sciences (U.S.)* 68, 459-463.
Struck, D.K. and Lennarz, W.J. (1980). The function of saccharide-lipids in synthesis of glycoproteins. *In* "The Biochemistry of Glycoproteins and Proteoglycans" (Ed. W.J. Lennarz), pp. 35-83. Plenum Press, New York.
Stumm, C. and Croes, A.F. (1975). Polyribosomes in different stages of the life cycle of the water mold *Allomyces arbuscula*, *Archives of Microbiology* 102, 117-122.

Suberkropp, K.F. and Cantino, E.C. (1973). Utilization of endogenous reserves by swimming zoospores of *Blastocladiella emersonii*, *Archiv fur Mikrobiologie* 89, 205-221.

Sutherland R.B., Schuerch, B.M., Ball, S.F. and Horgen, P.A. (1976). The developmental patterns of lysosomal enzyme activities during Ca^{2+}-induced sporangium formation in *Achlya bisexualis*. III. Ribonucleases, *Archives of Microbiology* 109, 289-294.

Timberlake, W.E., McDowell, L., Cheney, J. and Griffin, D.H. (1973). Protein synthesis during the differentiation of sporangia in the water mold *Achlya*, *Journal of Bacteriology* 116, 67-73.

Youatt, J. (1976a). The effect of actinomycin D on the formation of gametangia and mitosporangia in *Allomyces*, *Archives of Microbiology* 109, 163-166.

Youatt, J. (1976b). Chemical nature of discharge papillae of *Allomyces*, *Transactions of the British Mycological Society* 66, 113-122.

Youatt, J., Fleming, R. and Jobling, B. (1971). Differentiation in *Allomyces*: the production of sporangia, *Australian Journal of Biological Sciences* 24, 1163-1167.

GENE ACTIVATION DURING DIFFERENTIATION OF THE RUSTS AND ANTHRACNOSE FUNGI

R.C. Staples and B.-F. Huang

Boyce Thompson Institute, Tower Road, Ithaca, NY 14853

INTRODUCTION

Rust fungi colonize living plants in nature, drawing on the productive capacity of the host for growth. Invasion from uredospores usually begins with development of infection structures on the germ tube, and it is completed when the haustorium penetrates a host cell. There are many reviews of this process, some quite recent. Several of these are by Allen (1976), Emmett and Parbery (1975), Littlefield and Heath (1979), Staples and Macko (1980), and Wynn and Staples (1980). We briefly describe differentiation in the bean rust fungus (*Uromyces phaseoli*) again in order to provide a morphological frame of reference and a time-scale for the biochemical discussion which follows.

For a discussion of gene activation in rusts, we review what is known of protein and nucleic acid synthesis in bean rust uredospores. We include aspects of the synthesis of DNA and messenger RNA (mRNA) in order to provide the available evidence about gene activation during the development of those infection structures which appear with high frequency on collodion membranes -- the appressorium and vesicle. In addition, we discuss related aspects of differentiation in *Colletotrichum truncatum* in order to compare the process in rusts with that in another fungal pathogen.

RUST FUNGI

Morphology of Uredospore Differentiation

Spore germination. With suitable moisture, uredospores of the bean rust fungus usually germinate within an hour to produce a single germ tube which is directed to elongate across the surface of its host by contact stimuli until a stomatal guard cell is encountered (for details see Littlefield and Heath, 1979; Staples and Macko, 1980). Germ tubes orient at right angles to the ridges and indentations of leaves as they grow out over the surface, but hyphal contact must be tight in order that there be an effective stimulus (Wynn and Staples, 1980). It has been shown that the germ tube is actually bound to the leaf surface apparently by a combination of fungal secretions and dissolved host wax (Onoe et al., 1972).

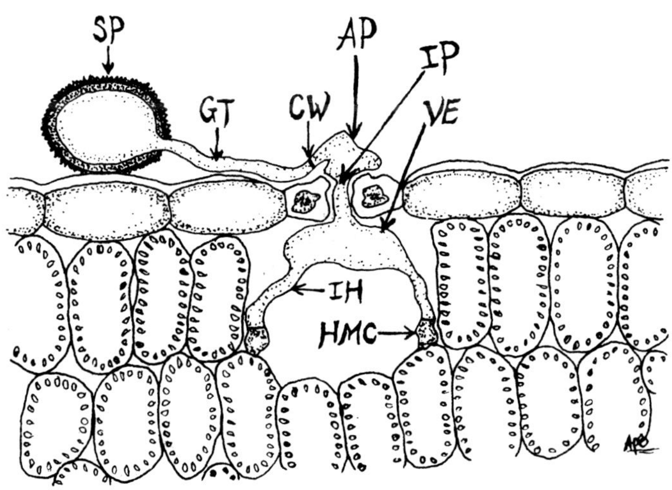

Fig. 1. Diagrammatic representation of infection structures produced by the bean rust fungus on the surface of a leaf. Penetration is between the stomatal guard cells into the substomatal cavity lined with mesophyll cells. Only those structures seen on collodion membranes are shown, and haustoria are omitted. Ap, appressorium; Gt, germ tube; CW, cross wall; Ih, infection hyphae; Ip, infection peg; HMC, haustorial mother cell; Sp, uredospore; Ve, vesicle.

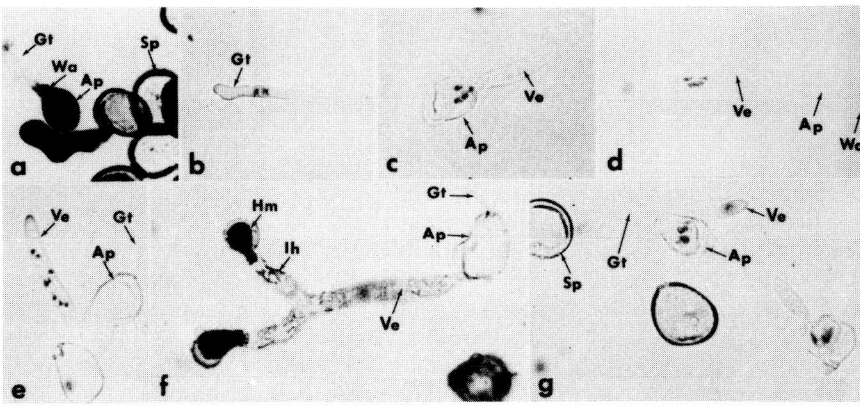

Fig. 2. Development of infection structures on collodion membranes containing 1% paraffin oil, a) cluster of bean rust uredospores germinated 4 h showing mature appressoria with cross walls. Spores stained with trypan blue in lactophenol, b) apex of germ tube showing dikaryotic state after staining by the Feulgen procedure, c) appressorium with young vesicle about five hours after germination showing four nuclei before migration into vesicle. Stained by Feulgen procedure, d) four appressorial nuclei after migration into the vesicle as visualized by the Feulgen stain, e) nuclei after division in the vesicle approximately six hours after germination. Stained by Feulgen procedure, f) vesicle, infection hyphae and haustorial mother cells which have appeared by eight hours after germination. Only the Hm have taken up trypan blue used as a stain. Haustorial mother cells are infrequent on membranes especially at eight hours but are more commonly observed (but less than 1%) by 16 hours. Infection hyphae can be singlet to triplet, g) appressoria and young vesicles showing two nuclei in each of the appressoria produced by spores germinated with Cordycepin. Vesicles are devoid of nuclei. Stained with Feulgen

stain. Separate studies showed that nuclear DNA had been synthesized (Staples et al., 1975). Ap, appressoria; Gt, germ tube; Hm, haustorial mother cell; Ih, infection hyphae; Ve, vesicle. The diameter of the spore in (f) is about 20 microns.

Development of infection structures. Infection structures required by rust fungi for penetration of the host are shown diagrammatically in Fig. 1. They begin to appear after the germ tube tip encounters a stomatal guard cell. In the rusts, the first structure is the appressorium. It is clearly defined and separated from the germ tube by a cross wall (Fig. 1, Fig. 2a). Once appressoria are induced, all infection structures develop until the haustorial mother cell has formed. On cellulose nitrate membranes (Maheshwari et al., 1967; Wynn and Gajdusek, 1968; Yaniv and Staples, 1974), the appressoria appear approximately three hours after germination (Fig. 3a), and mature by four hours. Division of the two germ tube nuclei (Fig. 2b) accompanies expansion of the germ tube tip as it develops to form the appressorium (Heath and Heath, 1978), and on membranes, the majority of the germ tubes have developed appressoria with four nuclei by the fourth hour (Fig. 2c). Next a peg appears, and on the host the cytoplasm migrates through it into the substomatal cavity below the stomatal guard cells (Fig. 1). On membranes, the peg is seen as a constriction between the appressorium and the vesicle (Fig. 2). After the nuclei have been carried into the expanding vesicle (Fig. 2d), they divide a second time (Fig. 2e). The majority of the vesicles have formed by about the sixth hour, and infection structure development usually ceases with appearance of the infection hyphae approximately by eight hours (Fig. 2f).

Haustorial mother cells occasionally develop on infection hyphae when the spores are germinated on membranes (Fig. 2f), but a third recognition signal may normally be required to develop haustorial mother cells with high frequency. Formation of haustoria appears to involve specific recognition of the mesophyll wall (Staples and Macko, 1980).

Protein Synthesis During Germination and Differentiation

DNA synthesis. One of the earliest biosynthetic events to occur in bean rust uredospores induced to form infection structures is the synthesis of nuclear DNA. Studies of DNA synthesis were carried out by allowing the spores to germinate on membranes in the presence of radioactive adenosine

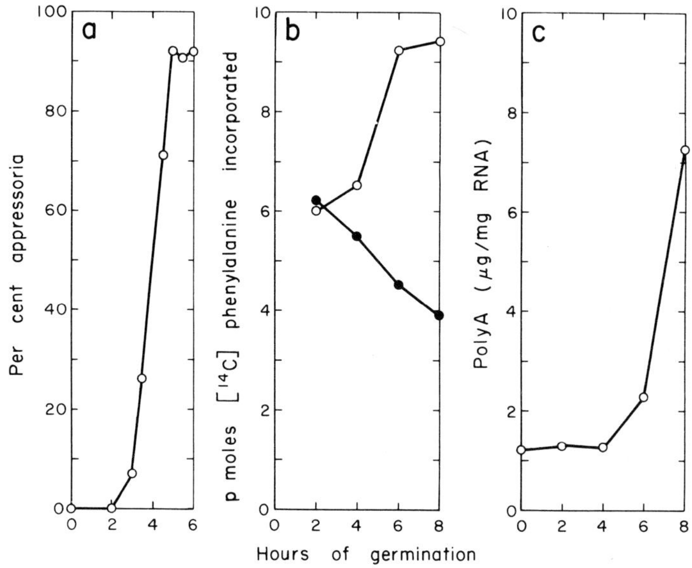

Fig. 3. Time-course studies showing changes with germination of a) appressoria formed as percentage of total, b) amount of template RNA synthesized as determined using a cell-free assay system prepared from Escherichia coli depleted of mRNA, procedure described by Ramakrishnan and Staples, 1970, c) amount of polyadenylic acid (polyA) in spores expressed in terms of RNA content of spores and determined as described by Staples et al. (1975).

(Staples, 1974). After periods of germination, DNA was extracted from the spores and centrifuged to equilibrium on gradients of CsCl. This process separated the mitochondrial DNA from nuclear DNA and also provided a rigorous identification because the gradient allowed a precise determination of their buoyant densities (Fig. 4). It was found that uredospores begin the synthesis of nuclear DNA sometime after the second hour of germination, about the time when the nuclei begin to divide (Fig. 4b). Until then, DNA synthesis is entirely confined to the mitochondria (Fig. 4a), and replication of nuclear DNA does not occur (Staples, 1974).

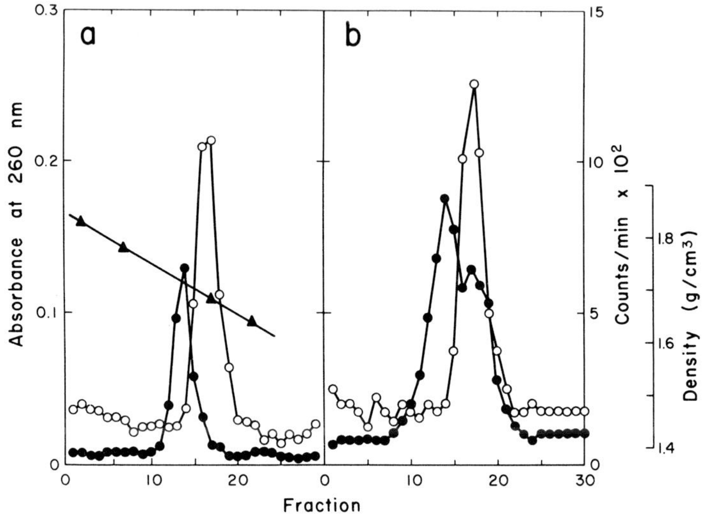

Fig. 4. Synthesis of DNA during germination of bean rust uredospores on collodion membranes sprayed with (^{3}H)Adenosine. DNA was extracted, treated with RNase-free DNase and centrifuged to equilibrium on gradients of CsCl. a) germination with isotope for 2h, b) germination for 4h but isotope present only from the second to fourth hour when the spores were harvested. Closed circles, counts/min; open circles, absorbance; closed triangles, density of CsCl gradient. Mitochondrial DNA is heavier (1.718 g/cm^3) than nuclear DNA (1.695 g/cm^3). Adapted from Staples, 1974.

Controls showed that synthesis of nuclear DNA did not occur in noninduced spores which lacked appressoria on the germ tubes.

If viable nuclei are required for formation of appressoria, irradiation of germinating uredospores with gamma rays should inhibit differentiation but not germ tube elongation. The nucleus is the sensitive target of gamma radiation. Death of such fungal cells as *Neurospora crassa* occurs from nuclear damage not from a general cellular effect, and cyto-

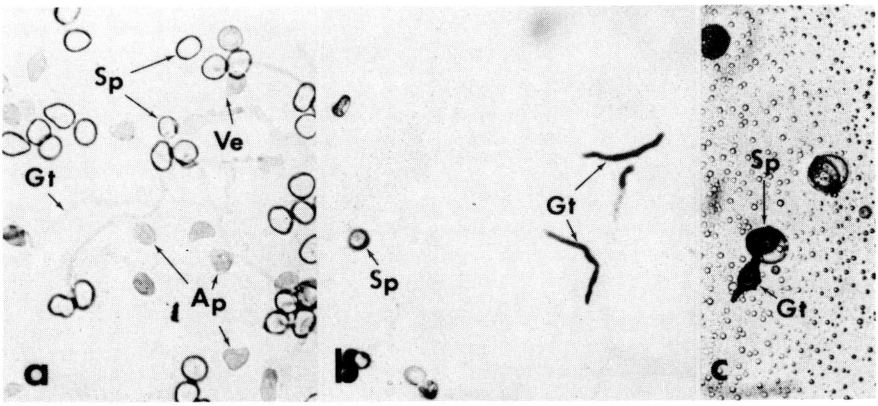

Fig. 5. A comparison of the effect of gamma radiation on germinating bean rust uredospores. a) nonirradiated spores germinated 5 h, b) spores germinated 2 h, then irradiated with 460 rads, and germinated an additional 3 h, c) spores germinated 2 h, then irradiated with 1840 rads, and germinated an additional 3 h. Ap, appressorium; Gt, germ tube; Sp, spore; Ve, vesicle. Spores were stained with Trypan blue in lactophenol. Spore diameter is about 20 microns.

plasmic functions like synthesis of protein are not inhibited except at extremely high doses (Davies and Evans, 1966). Consequently we exposed spores germinating on collodion membranes to gamma radiation from a ^{137}Ce source for various periods of time (46 rads/min) to see if appressoria would appear. As shown in Fig. 5, spores irradiated with 460 rads produced only dikaryotic germ tubes (Fig.5b) while 83% of the non-irradiated spores produced infection structures (Fig. 5a). Less than one percent of the spores irradiated with 230 rads produced infection structures. Despite the radiation effects, however, all germ tubes elongated normally regardless of the level of radiation applied. At the very high level of 1840 rads, however, germ tube elongation was inhibited (Fig. 5c). Staining the spores by the Feulgen procedure showed that all germ tubes were dikaryotic and that nuclear division had occurred only when infection structures were formed. Thus functional nuclei are required for differentiation of infection structures but not germ tube elongation.

The relationship between DNA synthesis, nuclear viability and differentiation was examined by using several metabolic inhibitors (Staples et al., 1975). In these studies, actinomycin D blocked both synthesis of DNA and nuclear division so that only dikaryotic germ tubes appeared as occurred after irradiation (similar to Fig. 2b and Fig. 5b). Terminal swellings do eventually develop at the tip of the germ tubes; however, cross walls between the germ tube and the appressorium do not form and nuclear division is never observed (R.C. Staples, unpublished information). Actinomycin D inhibits synthesis of DNA (Staples et al., 1975), and no earlier biosynthetic event has been found which is associated with formation of infection structures.

In agreement with this, it was found that Cordycepin also inhibits nuclear activity but blocks only vesicle development, not appearance of appressoria. This drug inhibits nuclear division but not synthesis of DNA (Staples et al., 1975), and differentiating spores exposed to it develop apparently normal appressoria and vesicle initials. But the vesicles do not elongate and neither nuclear division nor nuclear migration occurs (Fig. 2g). We concluded that while nuclear division need not precede formation of appressoria, replication of DNA is required.

DNA polymerase. Bean rust uredospores contain mitochondrial and nuclear DNA polymerases, and these were purified from bean rust uredospores (Yaniv and Staples, 1978; Staples and Yaniv, 1978). The enzymes were readily soluble when extracted from resting spores; however, only small quantities of mitochondrial DNA polymerase have been obtained from differentiated spores apparently because the nuclear enzyme becomes bound with the onset of replication (R.C. Staples, unpublished information).

The molecular weight of the nuclear DNA polymerase is about 175,000 daltons. The optimum pH for enzyme activity was found to be 7.5, and the reaction requires the simultaneous presence of dATP, dCTP, dTTP, dGTP, a DNA templet and magnesium ions. The enzyme is sensitive to N-ethylmaleimide. All of these properties suggest that the enzyme is a DNA α-polymerase although the classification may be irrelevant because fungi lack a β-polymerase (Plevani and Chang, 1977). The other enzyme is located in the mitochondria (Staples and Yaniv, 1978). Its molecular weight is 150,000 daltons, and it has an isoelectric point of 5.4. Good activity is obtained when the enzyme is primed with poly(dA)$(dT)_{10}$. This enzyme is also inhibited by N-ethylmaleimide.

The presence of these enzymes in uredospores demonstrates that synthesis of DNA, one of the earliest events required for appressorium formation, can be carried out by spores which already contain a full complement of DNA polymerases. Knowledge of how nuclear DNA polymerase is activated when the spores are induced to differentiate remains one of the important voids in our information about infection structure development.

Protein synthesis. The synthesis of proteins and the protein synthetic apparatus has been extensively studied in rust fungi especially in germinating uredospores (Staples and Yaniv, 1976); however, there have been few comparable studies during formation of infection structures (Staples et al., 1975). It is known that protein content of bean rust uredospores increases 20% to 30% during the first three hours of germination (Trocha and Daly, 1970), and G. Wolf (personal communication) has demonstrated that the spores readily synthesize a number of soluble proteins during germination. But a similar exploration has not been extended to include the period of infection structure development when ribosomal activity increases 3-fold (Yaniv and Staples, 1974).

During studies of gene activity by uredospores during differentiation, Huang and Staples (1980) have examined the synthesis of proteins by bean rust uredospores. Spores were pulse-labeled with (^{35}S)-methionine for consecutive intervals of 1.5 h (Fig. 6). Spores induced to differentiate develop appressoria beginning about 3.0 h while vesicles develop by 6.0 h. Of the many proteins synthesized, at least three (P14, P23, P26) are produced by all spores during the first three hours of germination (Fig. 6 a,b). Some of these germination related proteins are not observed after the infection structures begin to appear (Fig. 6 c,d), and many others continue to be synthesized by both dif-

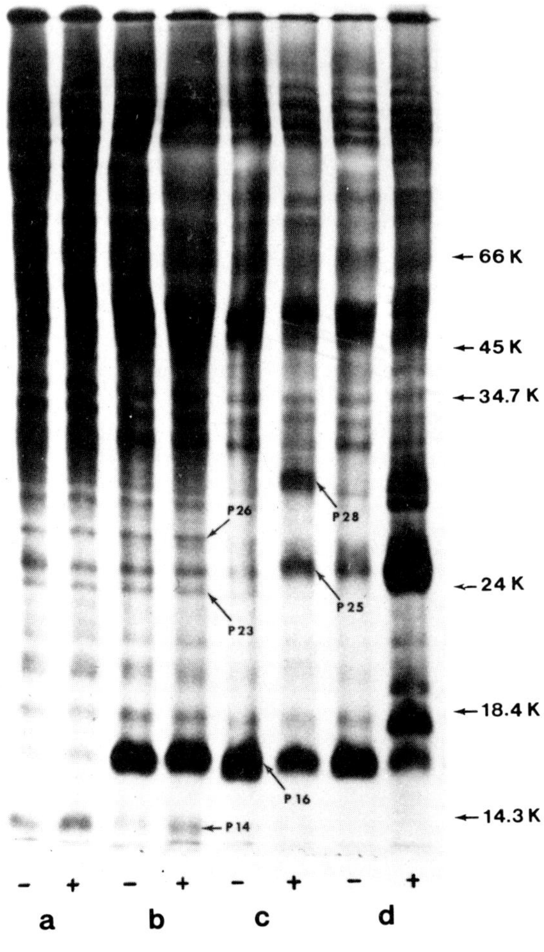

Fig. 6. Autoradiogram of proteins extracted from spores at the end of various time-periods after incubation with [^{35}S] methionine during germination on collodion membranes. Proteins were treated with SDS and separated on a polyacrylamide slab gel. Isotope sprayed onto membranes after germination for a) 0-1.5 h; b) 1.5-3.0 h; c) 3.0-4.5 h; d) 4.5-6.0 h. (+), paraffin oil added to membranes to induce differentiation; (-), oil omitted from membrane and only germ tubes developed on spores. Numbers at the right indicate position of molecular weight markers in daltons. Adapted from Huang and Staples (1980).

ferentiated and nondifferentiated spores. At least one protein (P16) apparently is related to elongation of the germ tube, since its radioactivity intensified after 1.5 h when germ tubes are elongating. It is synthesized throughout the differentiation period.

Beginning with appressorium formation (Fig. 6c+), two groups of proteins (P25, P28) are formed only by spores destined to develop infection structures. Labeling of additional proteins becomes apparent by 4.5 h when formation of infection structures is most vigorous, and none of these proteins is synthesized by spores not induced to form infection structures (Fig. 6d-).

To determine if some of the lighter new proteins which appeared in differentiated spores between 3 h and 4.5 h (Fig. 6c+) are formed by processing of some of the heavier proteins to lighter forms, we carried out pulse-chase trials (Huang and Staples, 1980). Most of the radioactivity that is lost disappeared from the lighter proteins while labelling in the heavier proteins remains, and there did not appear to be processing from the heavier proteins. In agreement was the fact that we were studying proteins labeled for equal lengths of time during the period of germination and differentiation.

It is clear from these studies that uredospores synthesize a wide range of proteins when induced to germinate and differentiate. Spores committed to develop infection structures shift their synthetic program, and a wide range of proteins are produced which are different from those present in germinating spores.

The special proteins like P25 and P28 that are associated with appressorium formation begin to be produced during the time when DNA replication is beginning in the nuclei, and it was of interest to determine if any proteins bound DNA. While a number of DNA-binding proteins are located on the gels (Fig. 7), a change in the pattern of DNA-binding proteins during differentiation of the spores was not detected.

Messenger RNA. We have known for some time that uredospores contain a complete system for synthesis of proteins (Yaniv and Staples, 1974). These components include the ribosomes, mRNA, tRNA, and the enzymes required for translation. Of the components studied quantitatively, we found that tRNA and ribosomal RNA are synthesized in small amounts through-

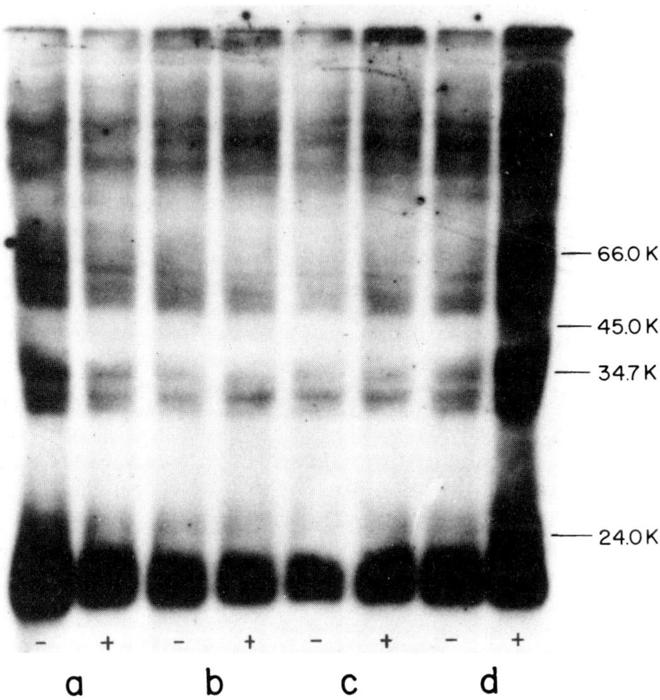

Fig. 7. Unlabelled proteins were extracted from spores at different stages of germination and separated by SDS-polyacrylamide gel electrophoresis. The proteins were transferred from the gel to two sheets of cellulose nitrate paper as described by Bowen et al. (1980). The cellulose nitrate sheet was then flooded with ^{32}P-DNA prepared from yeast and the DNA-binding proteins were visualized by autoradiography. Symbols are used as in Fig. 6.

out germination and differentiation. Furthermore, polyribosomes are present which actively incorporate amino acids in a cell-free system for protein synthesis, and the activity of this program of mRNA translation increases sharply between the fourth to the sixth hour in spores which form infection structures (Fig. 3b). These activities correlate nicely with the synthesis of proteins shown in Fig. 6, and the induction of uredospores to form appressoria must also induce spores to make mRNA available for translation to support construction of the infection structures.

In an effort to determine whether mRNA is synthesized or activated by the spore from endogenous reserves, we assayed the content of polyadenylic acid (polyA) in uredospores (Staples et al., 1975). Although mindful of the fact that not all messenger RNA contains polyA, particularly the histone messengers; assays for polyA are reliable and offer an alternative approach for estimating messenger activity to that provided by cell-free incorporation assays. PolyA analyses (Fig. 3c) show that while uredospores contain reasonable levels of polyA (1.2 μg/mg RNA), synthesis of additional polyA did not begin until after six hours when the vesicle develops.

Studies of nuclear division reviewed above have revealed that Cordycepin inhibits nuclear division and completion of the vesicle but not the appressorium (Fig. 2g). Cordycepin is a potent inhibitor of polyA synthesis, and its effects on development correlate closely in time with the synthesis of polyA.

Taken together, the studies on polyA synthesis and the effects of Cordycepin suggest that synthesis of mRNA occurs relatively late in the sequence of spore differentiation. If so, the shift in protein synthesis illustrated in Fig. 6 must be controlled at the level of translation from mRNA stored in the spore. In contrast, gene activation apparently does not occur until about the time of the second round of nuclear division after nuclear migration to the vesicle. Thus all of the first steps of infection structure development -- nuclear division, appressorium formation and initiation of the vesicle -- appear to occur by activation of translation of endogenous mRNA before the synthesis of additional mRNA is started. This shift in messenger synthesis, which coincides with occurrence of the second round of nuclear division in the vesicle, is a significant change in spore metabolism as it undergoes development of the infection structures. The shift occurs spontaneously because once induced all infection structures develop up to the haustorial mother cell without additional stimuli (Staples and Macko, 1980; Wynn and Staples, 1980).

APPRESSORIUM DEVELOPMENT IN AN ANTHRACNOSE FUNGUS

Morphology of Development

Conidia of the anthracnose fungi germinate to form appressoria when the germ tube contacts a solid surface

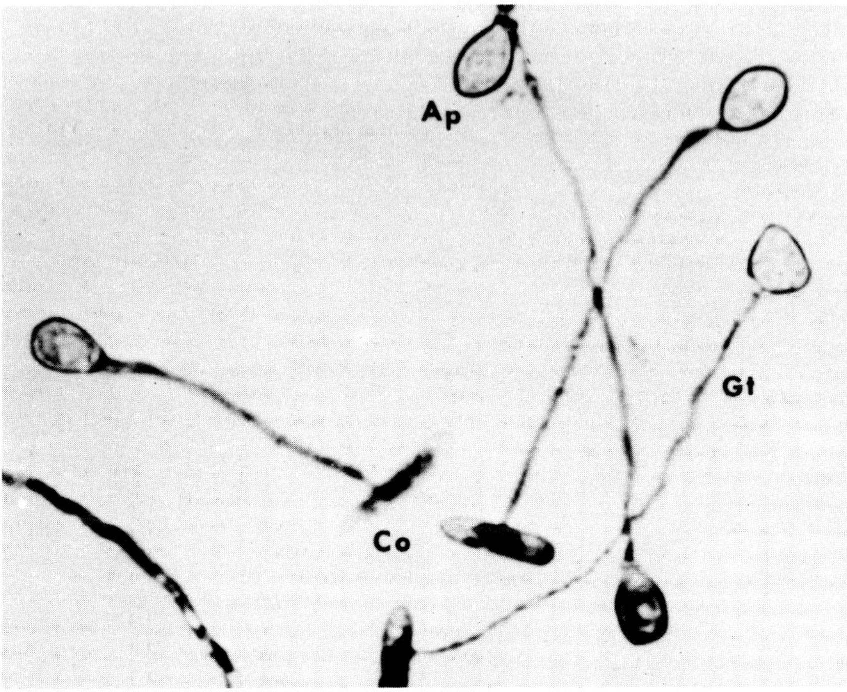

Fig. 8. Conidia of Colletotrichum lindemuthianum germinated 6 h on glass in a thin film of bean decoction. The germ tubes are unusually elongated because the film of bean extract was thicker than usual. The appressoria are thick walled and opaque. Ap, appressorium; Co, conidia; Gt, germ tube. Conidia are about 22 microns long and were stained with trypan blue in lactophenol.

(Fig. 8). After a first nuclear division, the conidia of *Colletotrichum truncatum* divide into two cells (Fig. 9), then produce a germ tube in which a second division of the nuclei occurs followed by nuclear migration into the developing appressorium. At maturity the appressorium is a distinctively dark-walled structure separated from the germ tube by a cross wall and containing two nuclei. Growth of the conidium resumes when a penetration peg protrudes from the underside of the appressorium either to rupture the host epidermal cell (Mercer et al., 1975) or to produce mycelia (Staples et al., 1976).

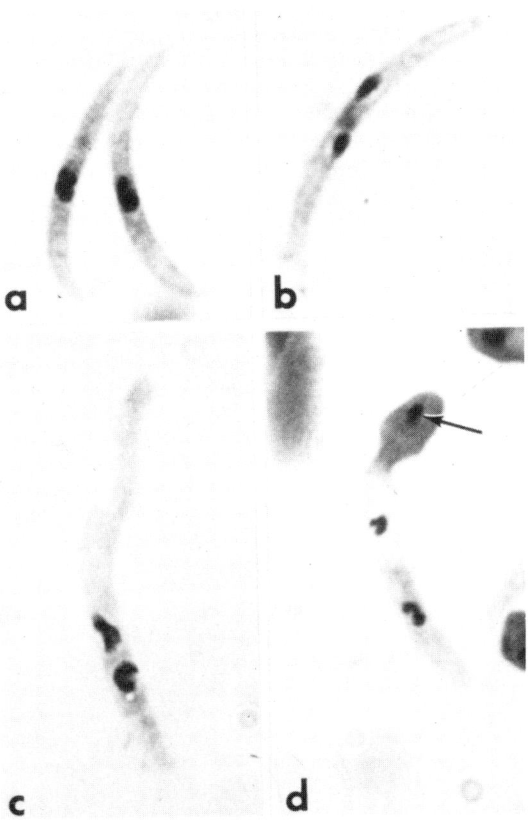

Fig. 9. Conidia of C. truncatum stained by the Feulgen procedure showing the nuclear condition during germination and development of appressoria on glass in a thin film of bean decoction. a) conidia before germination, b) conidium after first cell division has occurred. Each nucleus is in a different cell, but cell wall is not visible, c) germ tube has been emitted from one cell, d) second nuclear division has occurred and one nucleus has migrated into developing appressorium (arrow). A further nuclear division will occur to provide the mature appressorium with two nuclei as shown in Staples et al. (1976). Length of conidium is about 25 microns.

Nuclear Division and Appressoria

Conidia of C. *truncatum* translate the contact stimulus to form the appressorium only during the second round of nuclear division that takes place in the germ tube (Staples et al., 1976). When the second nuclear division is blocked by an inhibitor such as fluorodeoxyuridine, appressoria fail to appear. Upon removal of the drug, nuclear division resumes but only mycelia appear (Fig. 10). These and other studies have suggested that transcription of genes which guide production of appressoria can occur only between the first and second nuclear division in a time-dependent fashion.

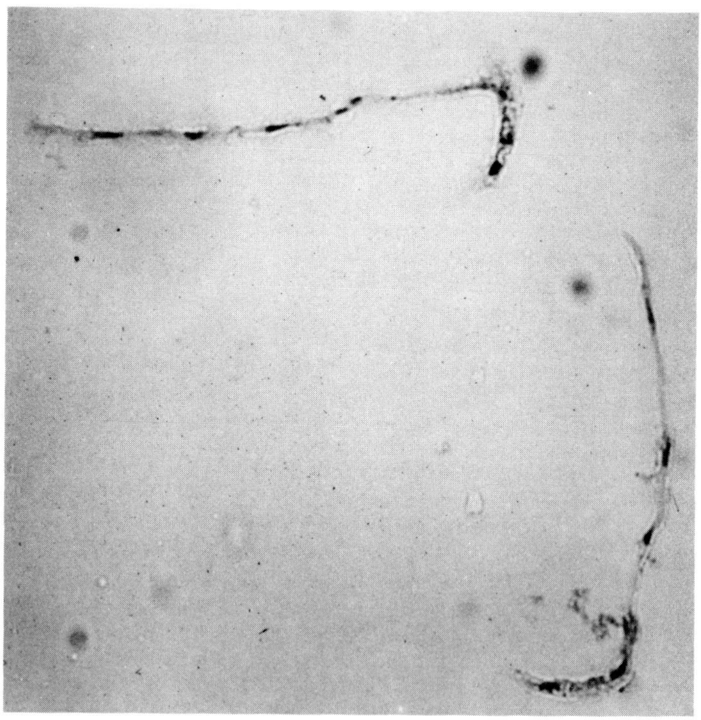

Fig. 10. Conidia of C. truncatum germinated in the presence of 5-fluorodeoxyuridine for 6 h. By this time one nuclear division had occurred, the conidium had divided into two cells, and germ tubes had been produced; further development was inhibited. The medium was decanted and replaced by fresh bean decoction, after which nuclear division resumed, but appressoria never appeared. Spores stained by the Feulgen procedure, conidia 25 microns in length.

That activation of specific genes is involved in production of appressoria is suggested not only by the development of the morphologically distinct appressorium, but by the fact that Actinomycin D blocks the development of appressoria but not mycelia. Neither mycelial proliferation, nuclear division, nor DNA synthesis is affected by the drug, and appressoria appear when the inhibition is relieved (Staples et al., 1976). Messenger RNA specific for appressorium development must be synthesized by the spore in response to germ tube contact with a hard surface, and in agreement with this, it was found that polyA(+)RNA is synthesized beginning with the second nuclear division (Staples et al., 1976). As was the case with bean rust uredospores, development of conidial stages earlier than the second nuclear division apparently is dependent upon translation of mRNA stored in the spore (Staples et al., 1976), while later stages appear when mRNA is synthesized.

SOME ADDITIONAL CONCLUSIONS

With the anthracnose fungi, the control point being exercised at the time of the second nuclear division apparently enables the conidium to proceed rapidly with development of mycelia should a host surface not be encountered. This decision is an unnecessary one for the rust fungi which colonize only living hosts. At the extreme end of the adaptive scale, the rusts nevertheless retain this control which must function or development of the infection hyphae ceases (Fig. 2g). However, the control is passive in the rusts because once appressoria are induced all infection structures appear in sequence until the haustorial stage is reached (Wynn and Gajdusek, 1968).

Despite the differences, it is obvious that both uredospores and conidia have developed the capacity to respond to environmental cues by reprogramming their nuclei to produce the specialized structures required for invasion of their hosts. That uredospores do this from limited endogenous reserves, producing a wide range of proteins in the process, is a truly remarkable adaptation about which we still know very little.

ACKNOWLEDGEMENTS

Research of the authors supported in part by a grant from the US Department of Agriculture. We especially thank Dr. Fred Abeles of the US Department of Agriculture for his suggestion that nuclei can be inactivated by gamma radiation.

REFERENCES

Allen, P.J. (1976). Control of spore germination and infection structure formation in the fungi. *In* "Encyclopedia of Plant Physiology" (R. Heitefuss and P.H. Williams, Eds.), Vol. 4, pp.51-85. Springer-Verlag, New York.

Bowen, B., Steinbert, J., Laemmli, U.K., and Weintraub, H. (1980). The detection of DNA-binding proteins by protein blotting, *Nucleic Acids Res.* 8,1-20

Davies, D.R., and Evans, H.J. (1966). The role of genetic damage in radiation-induced cell lethality, *Adv. Radiat. Biol.* 2, 243-353.

Emmett, R.W., and Parbery, D.G. (1975). Appressoria, *Annu. Rev. Phytopathol.* 13, 147-167.

Heath, M.C., and Heath, I.B. (1978). Structural studies of the development of infection structures of cowpea rust, *Uromyces phaseoli* var. *vignae*. I. Nucleoli and nuclei, *Can. J. Bot.* 56, 648-661.

Huang, B.-F., and Staples, R.C. (1980). Synthesis of proteins during differentiation of the bean rust fungus, *Exp. Mycol.* (submitted for publication).

Littlefield, L.J. and Heath, M.C. (1979). "Ultrastructure of rust fungi" Academic Press, New York.

Maheshwari, R., Hildebrandt, A.C., and Allen, P.J. (1967). The cytology of infection structure development in urediospore germ tubes of *Uromyces phaseoli* var. *typica* (Pers.) Wint., *Can. J. Bot.* 45, 447-450.

Mercer, P.C., Wood, R.K.S., and Greenwood, A.D. (1975). Ultrastructure of the parasitism of *Phaseolus vulgaris* by *Colletotrichum lindemuthianum*, *Physiol. Plant Pathol.* 5, 203-214.

Onoe, T., Tani, T., and Naito, N. (1972). Scanning electron microscopy of crown rust appressorium produced on oat leaf surface, *Kagawa Daigaku Nogakubu Gakujutsu Hokoku* 24, 42-47.

Plevani, P. and Chang, L.M.S. (1977). Enzymatic initiation of DNA synthesis by yeast DNA polymerases. *Proc. Natl. Acad. Sci. USA* 74, 1937-1941.

Ramakrishnan, L. and Staples, R.C. (1970). Changes in ribonucleic acids during uredospore differentiation, *Phytopathology* 60, 1087-1091.

Staples, R.C. (1974). Synthesis of DNA during differenation of bean rust uredospores, *Physiol. Plant Pathol.* 4, 415-424.

Staples, R.C., App, A.A., and Ricci, P. (1975). DNA synthesis and nuclear division during formation of infection structures by bean rust uredospore germlings, *Arch. Microbiol.* 104, 123-127.

Staples, R.C. Laccetti, L., and Yaniv, Z. (1976). Appressorium formation and nuclear division in *Colletotrichum truncatum*, *Arch. Microbiol.* 109, 75-84.

Staples, R.C., and Macko, V. (1980). Formation of infection structures as a recognition response in fungi, *Exp. Mycol.* 4, 2-16.

Staples, R.C., and Yaniv, Z. (1976). Protein and nucleic acid metabolism during germination. *In* "Encyclopedia of Plant Physiology" (R. Heitefuss and P.H. Williams, Eds.) Vol. 4, pp.86-103. Springer-Verlag, New York.

Staples, R.C., and Yaniv. Z. (1978). Mitochondrial DNA polymerase from germinated bean rust uredospores, *Exp. Mycol.* 2, 290-294.

Trocha, P., and Daly, J.M. (1970). Protein and ribonucleic acid synthesis during germination of uredospores, *Plant Physiol.* 46, 520-526.

Wynn, W.K., and Gajdusek, C. (1968). Metabolism of glucomannan-protein during germination of bean rust spores, *Contrib. Boyce Thompson Inst.* 24, 123-138.

Wynn, W.K., and Staples, R.C. (1980). Tropisms of fungi in host recognition. *In* "Plant Disease Control: Resistance and Susceptibility" (R.C. Staples and G.H. Toenniessen, Eds.) (In press). Wiley Interscience, New York.

Yaniv, Z., and Staples, R.C. (1974). Ribosomal activity in uredospores germinated on membranes, *Phytopathology* 64, 1111-1114.

Yaniv, Z., and Staples, R.C. (1978). DNA polymerase from nongerminated bean rust uredospores, *Exp. Mycol.* 2, 279-289.

PROTEINASES AND SPORULATION IN YEAST

D. H. Wolf

Biochemisches Institut der Universität Freiburg
D-7800 Freiburg im Breisgau, F.R.G.

INTRODUCTION

The differentiation process of sporulation in yeast is a complex process, which involves meiosis and subsequent ascospore formation. Diploid cells of *Saccharomyces cerevisiae* can be induced to undergo meiosis and ascospore formation by starvation on nitrogen free acetate medium (Croes 1967 a, b). Since sporulation occurs in the absence of an exogenous nitrogen source, the internal amino acid pool and protein degradation are the only means to provide the sporulating cell with amino acids for the synthesis of new proteins required for the differentiating cell. Furthermore, some enzymes of the vegetative cell, which may be harmful for the onset of differentiation may be rapidly inactivated. One inactivation mechanism might be proteolysis.

TABLE 1

Proteinase activities found in Saccharomyces cerevisiae before the detection of additional activities by the aid of proteinase mutants

Enzyme	Characteristics	Molecular weight
Proteinase A	Acid endo-peptidase	60,000
Proteinase B	Serine endo-peptidase	32,000-44,000
Carboxypeptidase Y	Serine exo-peptidase	61,000
Aminopeptidase I (vacuolar Aminopeptidase)	Exo-peptidase (Zn^{2+})	640,000
Aminopeptidase II	Exo-peptidase	120,000
Aminopeptidase III	Exo-peptidase	30,000
Dipeptidase	Mg^{2+}	120,000

Seven proteolytic enzymes had been found and characterized

in *Saccharomyces cerevisiae* before proteinase mutants led to the detection of a variety of additional proteolytic activities (Table 1): Two proteinases cleaving proteins endoproteolytically - proteinases A and B -, one peptidase cleaving peptides from the carboxy terminus - carboxypeptidase Y -, three peptidases starting peptide hydrolysis at the amino terminus - aminopeptidases I, II and III and one dipeptidase (for reviews see Wolf (1980); Wolf and Holzer (1980)).

RESULTS AND DISCUSSION

First experiments to get insight into involvement of proteolysis in sporulation were done by measuring overall protein degradation during this differentiation process. Cell proteins were radioactively labelled with |^{3}H|leucine during vegetative growth and then transferred to sporulation medium containing unlabelled leucine as chase. Degradation of the pre-existing vegetative proteins was measured by following the decrease of trichloroacetic-acid-precipitable radioactivity from the prelabelled cellular proteins.

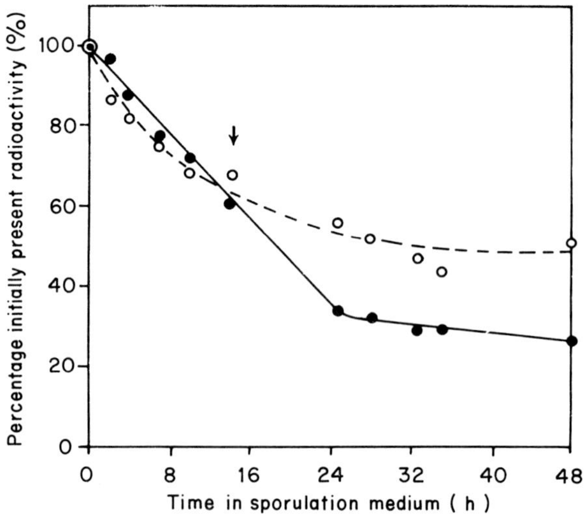

Fig. 1 Protein degradation after transfer to sporulation medium in sporulating cells of strain 740 (●—●) and non-sporulating cells of haploid H 4 (○—○). (From Betz and Weiser, 1976a)

Fig. 1 shows a continuous degradation of more than 60% of the vegetative cell protein in diploid sporulating cells during the first 24 hours, the time until first asci appear. In contrast, haploid cells, not able to sporulate, show a slow down of protein degradation and only about 40% of the protein is degraded during the same time (Betz and Weiser, 1976a). The rather comparable initial rates of protein degradation in sporulating diploid cells and non-sporulating haploid cells may be due to a similar adaption process to nitrogen starvation for both cell types on sporulation medium. Surprisingly, vegetative proteins and proteins synthesized during early (1.5 h to 3 h) and late (10 h to 12 h) phase of sporulation were degraded at similar rates (Fig. 2) (Betz and Weiser, 1976a).

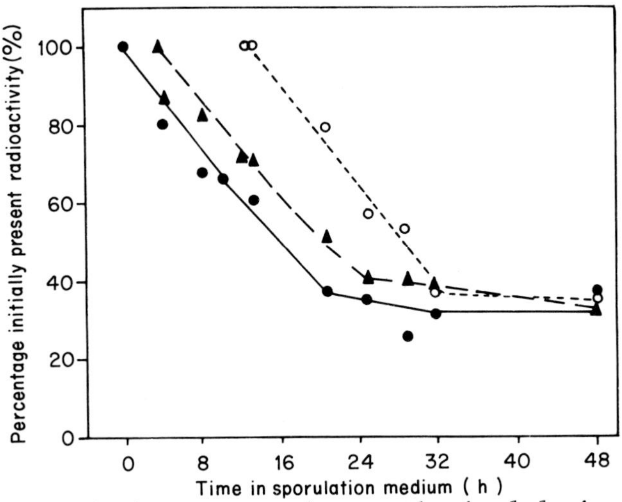

Fig. 2 Degradation of proteins synthesized during vegetative growth and during sporulation. Proteins of sporulating strain 740 were labelled with |³H|leucine between 7 h and 8 h of vegetative growth and during sporulation as indicated in the text. - The breakdown of labelled proteins was followed under chase conditions. (●—●) *|³H|radioactivity in protein labelled between 7 h and 8 h of vegetative growth;* (▲—▲) *|³H|radioactivity in protein labelled between 1.5 h and 3 h after transfer to sporulation medium;* (○—○) *|³H|radioactivity in protein labelled between 10 h and 12 h after transfer to sporulation medium* (From Betz and Weiser, 1976a).

The degradation process seems to concern proteins in all cellular compartments as can be found when differential centrifugation is undertaken with homogenates of sporulating cells (Betz and Weiser, 1976a).

Is protein degradation during sporulation dependent on energy production and protein synthesis?

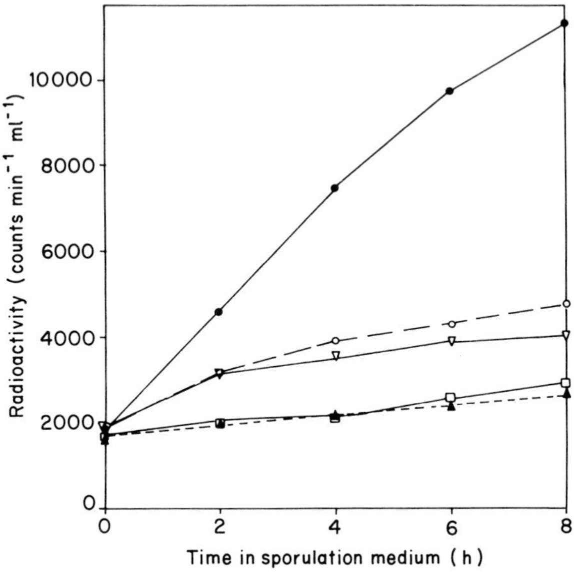

Fig. 3 Degradation of vegetative proteins after transfer of cells to sporulation medium in the presence of inhibitors of energy metabolism and of protein synthesis. Degradation of vegetative proteins in strain 740 after transfer to sporulation medium was followed by determination of the trichloroacetic-acid suluble radioactivity released from the $|^3H|$ *labelled vegetative proteins. (●—●) Control, without any inhibitor addition; (○—○) + 100 µg cycloheximide/ml culture; (▲—▲) + NaCN and NaF (10 mM each); (▽–▽) + 10 µg valinomycin/ml culture; (□-□) + 0.01 mM carbonylcyanide chlorophenylhydrazone.* (From Betz and Weiser, 1976a).

As can be seen from Fig. 3, there is strong inhibition of protein degradation visible when sporulating cells were treated with the energy uncouplers carbonylcyanide chlorophenylhydrazone and valinomycin or a mixture of cyanide and fluoride upon onset of sporulation. Thus, protein degradation during the differentiation process of sporulation seems to depend on the intracellular production of ATP (Betz and Weiser, 1976a). Inhibition of protein synthesis by cycloheximide led to reduction of the degradation rate of about 60 to 70% (Fig. 3), (Betz and Weiser, 1976a). This reduction of the degradation rate by cycloheximide is probably due to inhibition of synthesis of proteinases or other proteins required for the degradation process.

Studies were undertaken to find out, which cell proteins disappear during ascopore formation, a process most likely due to proteolysis.

TABLE 2

Specific enzyme activities in vegetative cells and cells suspended in sporulation medium of sporulating strain 740 and non-sporulating haploid H 4 (From Betz and Weiser, 1976b).

Enzyme	Specific activity after suspension in sporulation medium of			
	strain 740 for		haploid H4 for	
	0 h	24 h	0 h	24 h
	U/mg protein			
Malate dehydrogenase	2.6	1.0	1.5	0.7
Isocitrate lyase	0.176	0.035	0.071	0.034
NAD-dependent glutamate dehydrogenase	0.36	0.02	0.72	0.06
Fructose bisphosphatase	0.127	0.030	0.099	0.030
Alcohol dehydrogenase	3.06	1.07	2.06	0.72
Hexokinase	1.77	0.76	1.40	0.68
Glutamate oxalacetate transaminase	0.295	0.125	0.275	0.137
Aminopeptidase I	0.0032	0.0157	0.0071	0.0137
	A_{260} x min^{-1} x mg protein^{-1}			
Ribonuclease	0.017	0.086	0.039	0.059

Table 2 shows a variety of enzymes measured. Enzyme activities of malate dehydrogenase, isocitrate lyase, NAD-dependent glutamate dehydrogenase, fructose bisphosphatase, alcohol dehydrogenase, hexokinase and glutamate oxalacetate transaminase disappear to a considerable extent during the sporulation process. However, decrease of these enzyme activity is not sporulation specific as these activities decrease also in haploid cells transferred to sporulation medium (Betz and Weiser, 1976b). Investigation, whether loss of enzyme activities could be due to proteolytic degradation, NAD-dependent glutamate dehydrogenase, isocitrate lyase, malate dehydrogenase and fructose bisphosphatase were measured under conditions which had been shown to inhibit protein degradation in sporulating cells, namely after addition of cycloheximide or carbonylcyanide chlorophenylhydrazone. Both drugs inhibited inactivation of the above enzymes indicating that metabolic energy and protein synthesis is necessary for the inactivation events as is the case for protein degradation during early stages of sporulation (Betz and Weiser, 1976b).

As can be seen from Table 2 hydrolytic activities of aminopeptidase I and also ribonuclease increase in haploid and diploid cells in sporulation medium. However, increase in sporulating diploids is considerably larger than in haploids. The increase of hydrolytic activities will be discussed in more detail later on.

During the development of metabolic inactive asci, a continuous decrease in the rate of oxygen uptake was observed (Küenzi *et al.* 1974). When whole diploid cells were analyzed spectrophotometrically for cytochrome content during the course of sporulation, a reduction of cytochromes c, b and aa_3 levels was visible (Fig. 4). Mature asci contained 20% or less of these cytochromes compared to vegetative cells. In haploid cells a reduction of about 60% of the cytochrome content and of cytochrome c oxidase activity was found under nitrogen starvation indicating that the observed losses are not strictly sporulation specific (Betz and Weiser, 1976b). However, as reduction of respiratory activity seems to be a prerequisite for the acquisition of the semidormant state, experiments were undertaken to discriminate whether cellular cytochrome content during sporulation may be due to selective degradation of the hemoproteins or due to non-selective breakdown of whole mitochondrial membranes. One might expect changes in the ratio of cytochrome c oxidase and another protein, i. e. F_1-ATPase during ascospore formation if cytochrome c oxidase were selectively degraded. Submitochondrial particles from vegetative cells and from purified asci were isolated and analyzed for their cytochrome c oxidase and F_1-ATPase content and the activity ratio was compared (Table 3).

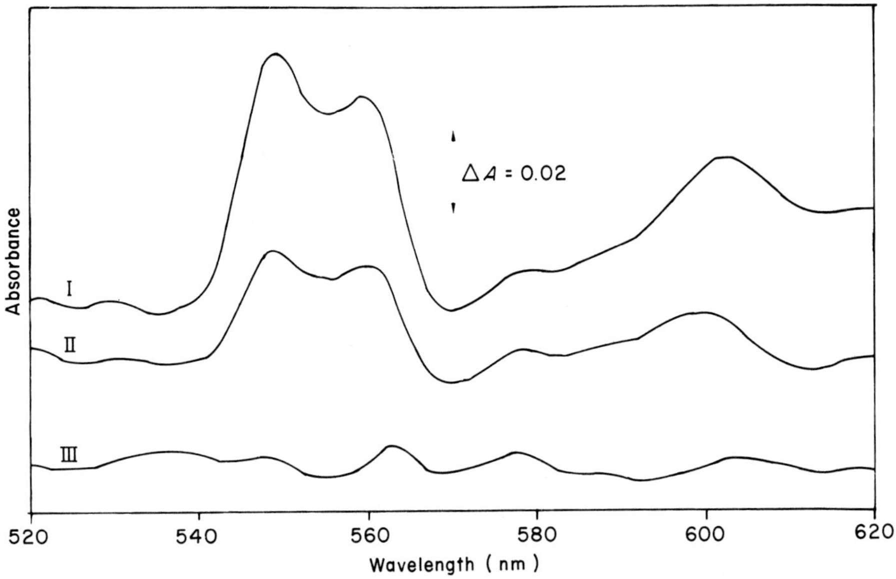

Fig. 4. Cytochrome spectra of whole vegetative cells (I), cells sporulated for 18 h (II) and purified asci (III) of strain 740. Cytochromes c (549 nm), b (559 nm) and aa_3 (603 nm). (From Betz and Weiser, 1976b).

TABLE 3

F_1-ATPase and cytochrome c oxidase activities in submitochondrial particles from vegetative cells and from asci of strain 740 (From Betz and Weiser, 1976b).

Submitochondrial particles from	Activity of		Activity ratio (a/b)
	cytochrome c oxidase (a)	F_1-ATPase (b)	
	units/mg	U/mg protein	
Vegetative cells	0.0072	0.98	0.0073
Asci	0.0030	0.45	0.0067

No significant alteration of the activity ratio of cytochrome c oxidase to F_1-ATPase was visible during the sporulation process (Betz and Weiser, 1976b). It was concluded that the loss of cellular cytochromes during sporulation is due to a non-specific degradation of mitochondrial membrane proteins. (Betz and Weiser, 1976b).

Which proteinases are involved in degradation processes during sporulation? First indication was thought to come from studies on proteinase regulation during the differentiation process. As first enzymes, the vacuolar proteinases A, B and carboxypeptidase Y were investigated in sporulating diploid cells.

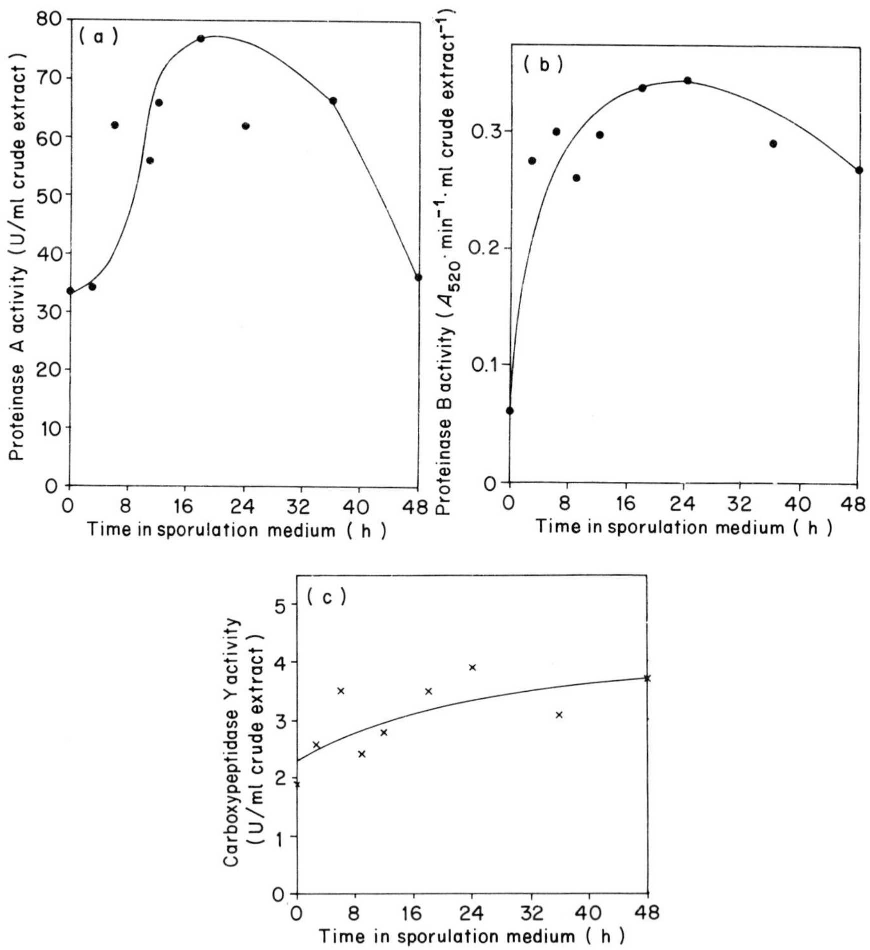

Fig. 5. Proteinase activities during sporulation (From Betz and Weiser, 1976a)

Fig. 5 shows a more than two fold increase of proteinase A and a more than 6 fold increase of proteinase B activities during the first 24 hours on sporulating medium. Carboxypeptidase Y did not increase significantly (Betz and Weiser 1976a). However, Klar and Halvorson (1975) found an about 8 fold increase in carboxypeptidase Y activity. From these results one may deduce a vital function for proteinase A and especially for proteinase B during the sporulation process and possibly also a necessary function for carboxypeptidase Y.

However, final answers concerning the question which proteinases are involved in the sporulation process and which functions are performed by these proteinases as well as the question whether there exist other as yet unknown proteolytic enzymes, which might be involved in the sporulation process, should come from proteinase mutants and their biochemical analysis. Lack of one or more of the known proteolytic activities should uncover their function during the differentiation process and in addition lead to the detection of unknown proteolytic enzymes with overlapping specificities with respect to the measured substrates.

We isolated carboxypeptidase Y mutants with the aid of the histochemical stain N-acetyl-DL-phenylalanine-ß-naphtylester (Wolf and Fink, 1975): Mutants not able to cleave this carboxypeptidase Y substrate were collected on the basis that such mutagenized colonies could not liberate ß-naphthol which otherwise can react with the diazonium salt Fast Garnet GBC to form a red color, and thus did not stain (Fig. 6). Analysis of the mutation revealed a lesion in the structural gene of carboxypeptidase Y (Wolf and Weiser, 1977). When diploid cells homozygous for the carboxypeptidase Y lesion were sporulated, no alteration in sporulation frequency as compared to wild type was observed (Wolf and Fink, 1975). There exist two possibilities for this finding:

1) Carboxypeptidase Y is not involved in any biochemical process during sporulation
2) Other enzymes exist, which can fill in for carboxypeptidase Y when the enzyme is absent.

Growth experiments using a carboxypeptidase Y substrate, namely N-carbobenzoxy-glycyl-L-leucine as sole nitrogen source, led to the finding that the carboxypeptidase Y mutant could still grow on this peptide. This indicated that another carboxypeptidase with overlapping specificity existed, which was able to liberate leucine from the dipeptide for growth. The new carboxypeptidase, called carboxypeptidase S, in contrast to carboxypeptidase Y, was shown to be a metal ion dependent enzyme (Wolf and Weiser, 1977). Double mutants lacking carboxypeptidase Y and carboxypeptidase S were isolated on the basis that a cell lacking both enzymes might not be able to grow on the carboxypeptidase Y - carboxypeptidase S substrate N-carbobenzoxy-glycyl-L-leucine as sole nitrogen source. Mutants were isolated and the carboxypeptidase S lesion in all mutants was shown to segregate 2:2 in meiotic tetrads, indicating a single gene mutation (Fig. 7).

Diploids homozygous for carboxypeptidase Y and carboxypeptidase S absence were constructed and their phenotype was compared with wild type diploids. Protein turnover in the double mutant was 3% per hour and not altered as compared to wild type under sporulation conditions. Comparison of sporutation frequency between homozygous carboxypeptidase Y - carboxypeptidase S - double mutants and wild type cells revealed an about 16% reduction in case of the double mutant. From this small reduction, we do not want to draw any conclusion about involvment of carboxypeptidases Y and S in sporutation. It seems clear, however, that both carboxypeptidases do not catalyze any vital event that is strictly dependent on the presence of any one of the two enzymes. However, if for instance both enzymes are involved in a more general process like generation of amino acids from intracellular peptides necessary for the synthesis of new proteins, one may again assume other enzymes present in the cell, which can fill in for both carboxypeptidases, when they are absent. For instance, the cell contains at least three aminopeptidases which might also be involved in intracellular peptide hydrolysis (for reviews see Wolf and Holzer, 1980; Wolf 1980).

We checked possible hydrolysis of 39 aminoterminal blocked dipeptides in order to find new carboxypeptidase activities in the carboxypeptidase Y - carboxypeptidase S double mutant. Separation of mutant crude extract proteins by molecular weight using Sephadex chromatography revealed four new activities (Fig. 8). Three activities split N-carbobenzoxy-L-alanyl-L-phenylalanine (Peak I, II, IV) and one activity hydrolyzes N-carbobenzoxy-L-seryl-L-phenylalanine (Peak III). Activities in peak I and II are partly inhibited by phenylmethylsulfonylfluoride, antipain and leupeptin. Activity in peak III is inhibited by EDTA and activity in peak IV is inhibited by phenylmethylsulfonylfluoride (D.H. Wolfand C. Ehmann, submitted). Multiple deficient carboxypeptidase mutants will give us the answers about carboxypeptidase function in sporulation.

The considerable increase of the proteinase B activity during sporulation (Betz and Weiser, 1976a) had led to the proposal that this enzyme plays a vital role during this differentiation process. We isolated mutants of the enzyme.

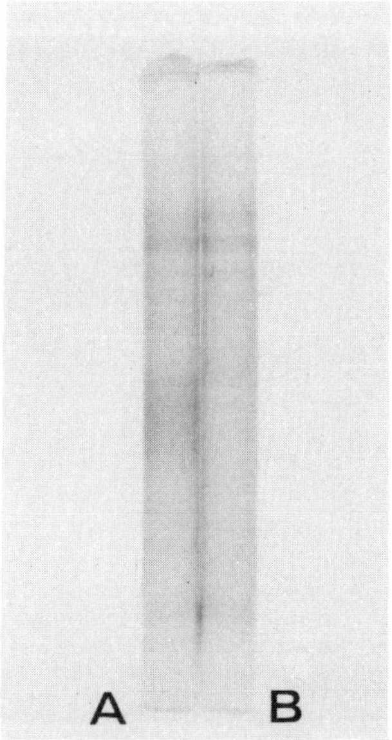

Fig. 6. Disc gel electrophoresis of carboxypeptidase Y mutant (B) and wild type S288c (A) crude extracts followed by activity staining of carboxypeptidase Y (For details see Wolf and Fink, 1975)

TABLE 4

Strain	Mutation	Source	Specific activity (units/mg) Proteinase A	Proteinase B	Carboxypeptidase Y
233	Wild type	Whole cell extract	2.1	39×10^{-3}	6.1×10^{-3}
HP 232-2B	prb1	Whole cell extract	1.9	0.2×10^{-3}	6.1×10^{-3}
HP 232-2B	prb1	Vacuoles	12.6	0.76×10^{-3}	25×10^{-3}

Proteinase activities in wild type and proteinase B mutant extract and purified mutant vacuoles. (For details see Wolf and Ehmann, 1979)

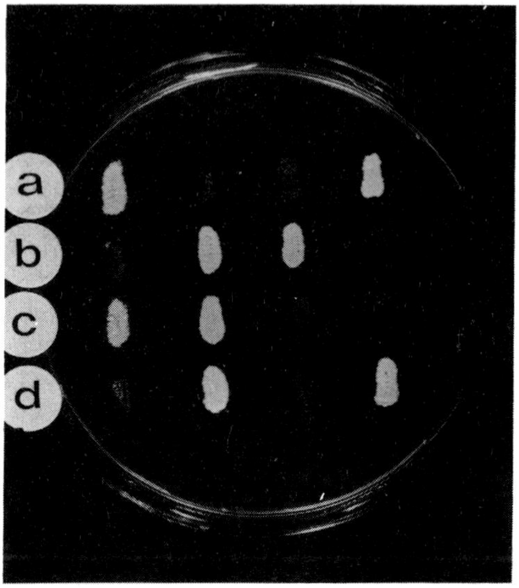

Fig. 7. Segregation pattern of four carboxypeptidase S mutants (a) cps 14; (b) cps 26; (c) cps 31; (d) cps 33. Tetrads grown on agar complete medium plates were replica plated onto agar plates containing mineral medium with 0.5% N-carbobenzoxy-glycyl-L-leucine as sole nitrogen source

Using one of the temperature sensitive cell lysis strains of Hartwell (strain 233), we screened for mutants which were not able to cleave the specific proteinase B substrate Hide Pow-

TABLE 5

Sporulation behaviour of proteinase B mutant
(From Wolf and Ehmann, 1979)

Diploid strain	Approximate time needed for appearance of first asci	Ascospores
	h	%
prb^{+}/prb^{+}	18	25.5
prb^{+}/prb^{-}	18	26.7
prb^{-}/prb^{-}	26	10.7

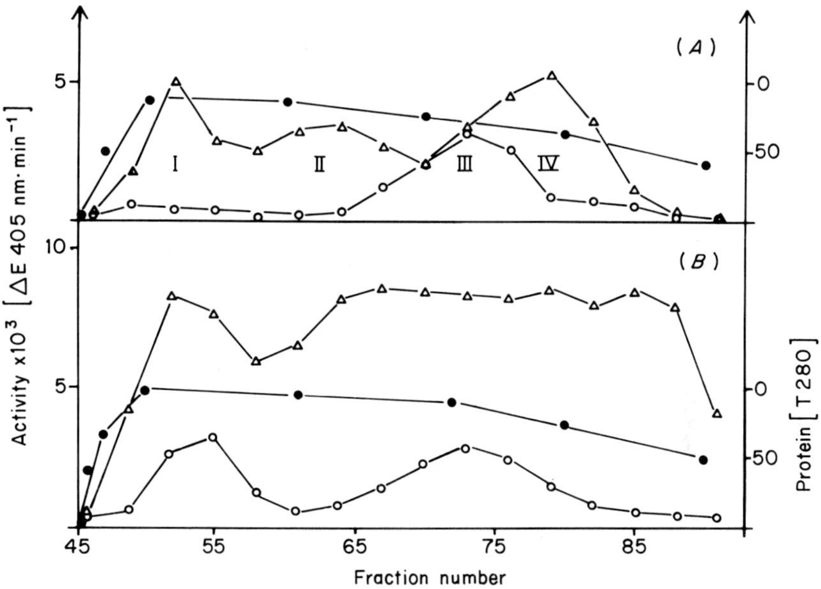

Fig. 8. New carboxypeptidase activities in a carboxypeptidase Y - carboxypeptidase S double mutant. (A) mutant; (B) wild type (control). Cell extracts were chromatographed on Sephadex G-150 to separate the activities. Activity against N-carbobenzoxy-L-alanyl-L-phenylalanine (△—△); N-carbobenzoxy-L-seryl-L-phenylalanine (○—○). (●—●) Protein. (From D. H. Wolf and C. Ehmann, submitted)

der Azure on petri plates after lysis of the mutagenized colonies (Fig. 9).

One of the isolated mutants (HP233) was analyzed in detail (Wolf and Ehmann , 1978; Wolf and Ehmann, 1979). The mutation which segregates 2:2 in meiotic tetrads, indicating a single gene mutation, was shown to lead to absence of proteinase B from the vacuolar compartment of the cell, where location of the enzyme had been established in wild type cells (Table 4).

Diploids homozygous for the absence of proteinase B, heterozygous for this mutation and wild type were constructed and analyzed for their protein turnover capacity under sporulation conditions and their sporulation behaviour. As can be seen from Fig. 10, proteinase B mutant diploids (prb^-/prb^-) exhibit a more than 40% reduced protein degradation as compared to wild type diploids (prb^+/prb^+) or strains heterozygous for the prb mutant gene (prb^+/prb^-). Sporulating homozygous diploids lacking proteinase B activity (prb^-/prb^-)

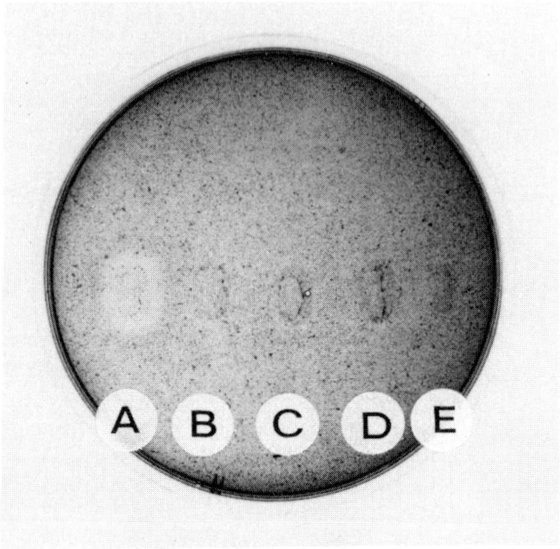

Fig. 9. Proteinase B mutant screening on Hide Powder Azure
(A) Cell lysis strain 233, wild type;
(B) HP58, mutant (prb⁻)
(C) HP81, mutant (prb⁻)
(D) HP163, mutant (prb⁻)
(E) HP 232, mutant (prb⁻)
(From D. H. Wolf and C. Ehmann, 1978)

exhibit a time interval prolonged by about 50% until first asci appear as compared to wild type (prb^+/prb^+) diploids or heterozygous prb^+/prb^- cells (Table 5). Sporulation frequency of mutant diploids is also reduced to about 50% (Table 5).

The results indicate a necessary function of the vacuolar proteinase B in protein turnover providing sporulating cells with peptides, which are finally cleaved into amino acids required for new protein synthesis during this differentiation process.

To test the influence on sporulation of a simultaneous lack of the three proteolytic enzymes - proteinase B, carboxypeptidase Y and carboxypeptidase S -, we constructed triple mutants. Diploid triple mutants (prb^- prc^- cps^-/prb^- prc^- cps^-) lacking proteinase B, carboxypeptidase Y and carboxypeptidase S, had nearly lost their sporulation ability (sporulation frequency < 3%). This underlines the necessary function of proteinase B during this process and indica-

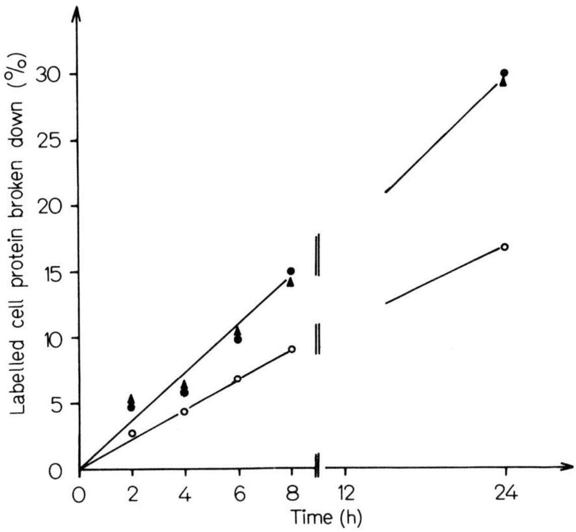

Fig. 10. Protein degradation during sporulation in proteinase B mutant. Wild type diploids (prb^+/prb^+) *(●—●), heterozygous diploids* (prb^+/prb^-) *(▲—▲), and homozygous mutant diploids* (prb^-/prb^-) *(o—o) were grown on complete medium in the presence of* $|^3H|$*-L-leucine and transferred onto 1% potassium acetate sporulation medium containing non-radioactive leucine as chase. Samples were taken at times indicated and release of trichloroacetic acid soluble radioactivity from the prelabelled protein was determined* (From D. H. Wolf and C. Ehmann, 1979)

tes a necessary function for one or both carboxypeptidases in concert with proteinase B during this differentiation process. This function most likely lies in the generation of amino acids from intracellular protein for new protein synthesis.

We had found four new carboxypeptidase activities. The question was open whether the yeast cell contains also additional endoproteinase activities. To test this, proteinase mutants are also highly desirable as the missing known activities do not interfere any more with the unknown enzymes. We searched for new activities in the triple mutant lacking proteinase B, carboxypeptidase Y and carboxypeptidase S (T. Achstetter, C. Ehmann and D. H. Wolf, submitted). New endoproteolytic activities were searched with substrates of the relatively specific serum proteinases, p-nitroanilide derivatives of peptides blocked at the amino terminus. None of the known proteinases remaining - proteinase A and the amino-

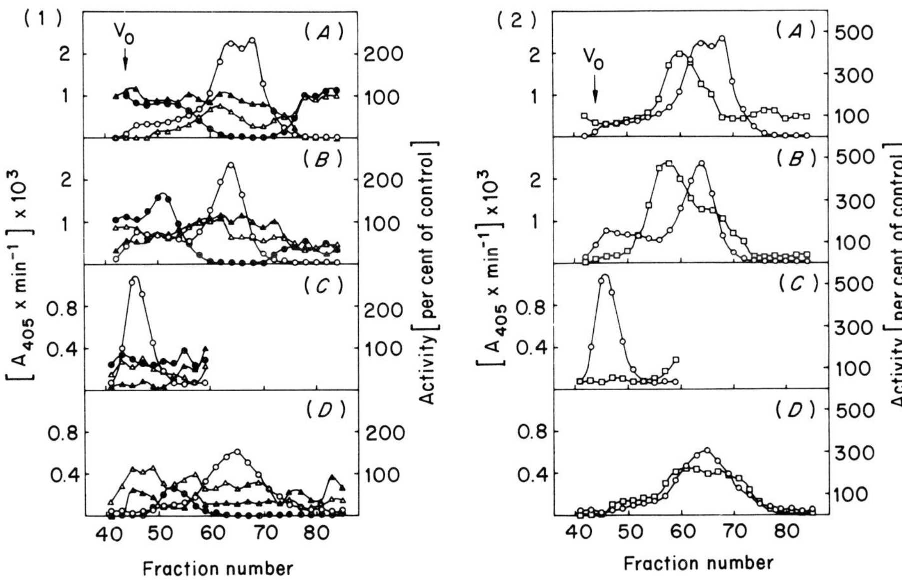

Fig. 11. *New proteolytic activities separated by Sephadex G-150 chromatography in a mutant extract lacking carboxypeptidases Y and S and proteinase B. Proteolytic activity (○—○)*
1) Response of the proteolytic activities (in %) to the proteinase inhibitors phenylmethylsulfonylfluoride (saturated) (△—△), chymostatin, 250 µg/ml (▲—▲), ethylenediamine tetraacetic acid (●—●)
2) Response of the proteolytic activities (in %) to Co^{2+} (1 mM) (□—□).
(A), activity against benzoyl-L-prolyl-L-phenylalanyl-L-arginyl-p-nitroanilide
(B), activity against carbobenzoxy-L-isoleucyl-L-glutamyl-glycyl-L-arginyl-p-nitroanilide
(C), activity against carbobenzoxy-glycyl-glycyl-L-leucyl-

(D), activity against succinyl-L-alanyl-L-alanyl-L-prolyl-L phenylalanyl-p-nitroaninilide p-nitroanilide
Cells were grown at 30°C for 22 hours in complete medium. Extract was centrifuged at 100,000 x g. Proteolytic activity was measured spectrophotometrically at 405 nm, 30°C and pH 7.5 as liberation of p-nitroaniline from the p-nitroanilide peptide derivatives (0.5 mM).
(From T. Achstetter, C. Ehmann and D. H. Wolf, submitted)

TABLE 6

New proteolytic activities in an extract of a triple mutant devoid of carboxypeptidases Y and S and proteinase B. Conditions were as indicated in legend to Fig. 11.
(From T. Achstetter, C. Ehmann and D. H. Wolf, submitted)
Bz, benzoyl; Cbz, carbobenzoxy; Tos, tosyl; Suc, succinyl; Nan, p-nitroanilide

Peptide Substrate	Specific activity nmol x min^{-1} x mg^{-1}
Bz - Pro - Phe - Arg - Nan	0.69
Tos - Gly - Pro - Lys - Nan	0.1
Cbz - Gly - Pro - Arg - Nan	0.1
Bz - Ile - Glu - Gly - Arg - Nan	0.81
Bz - Phe - Val - Arg - Nan	0.33
H - D - Phe - Pip - Arg - Nan	not detectable
H - D - Val - Leu - Lys - Nan	0.25
H - D- Val - Leu - Arg - Nan	0.32
Cbz - Val - Gly - Arg - Nan	0.19
Bz - Asp - Trp - Arg - Nan	0.12
Suc - Ala - Ala - Ala - Nan	not detectable
Cbz - Gly - Gly - Leu - Nan	0.37
Suc - Ala - Ala - Pro - Phe - Nan	0.23
Suc - Gly - Gly - Phe - Nan	0.12

peptidases - is able to cleave the substrates under the conditions used. Table 6 shows proteolytic activities in the triple mutant present against a variety of different proteinase substrates. When the proteins of a 100,000 x g supernate of a crude extract of the triple mutant are separated according to molecular weight by Sephadex chromatography and the eluted fractions are tested against four different substrates, several main proteinase activity peaks occur. (Fig. 11 (1), (2) A-D). If one analyzes these activties in

TABLE 7

New proteolytic activities in the soluble fraction of a triple mutant devoid of carboxypeptidases Y and S and proteinase B identified by action of proteinase inhibitors and $CoCl_2$ after Sephadex gel filtration. Activity in fraction 42, 45, 46, 49, 51 and 53 is inhibited by $CoCl_2$. For conditions see Fig. 11.

- *no inhibition*

(+) *weak inhibition, possibly caused by overlapping specificities*

\+ *0 - 30% inhibition*

++ *30 - 60% inhibition*

+++ *60 - 100% inhibition*

(Fom T. Achstetter, C. Ehmann and D. H. Wolf, submitted)

Column fraction	Proteolytic activity	Substrate cleaved	EDTA	$PhMeSo_3F$	Chymostatin	Activated by
42	1	D	+++	++	+++	
45	2	A	-	+++	-	
45	3	B	-	(+)	++	
46	4	C	(+)	(+)	+++	
49	5	D	+++	-	++	
51	6	B	-	(+)	(+)	EDTA
53	7	D	+	+	+++	
57	8	A,B,D	++	-	-	$CoCl_2$
60	9	A	+++	+	+	$CoCl_2$
63	10	D	+++	+	+++	$CoCl_2$
68	11	A	+++	+++	(+)	

more detail using proteinase inhibitors of different action spectrum or metal ions (Fig. 11 (1), (2) A-D) one finds different proteolytic activities hidden under these main activity peaks. A preliminary analysis resulted in 11 different activities in the soluble fraction of desintegrated triple mutant cells (Table 7).

At least two proteolytic activities were found in the membranous fraction of desintegrated triple mutant cells: One activity, called proteinase M, which is able to split p-nitroaniline directly from a variety of p-nitroanilide derivatives of peptides (first column of Table 8) and one activity

TABLE 8

New proteolytic activities in the precipitate of desintegrated cells of a triple mutant devoid of carboxypeptidases Y and S and proteinase B. For test conditions see legend to Fig. 11 (From T. Achstetter, C. Ehmann and D. H. Wolf, submitted)

	Specific activity $\lvert nmol \times min^{-1} \times mg^{-1} \rvert$	
Peptide Substrate	without addition	with aminopeptidase
Bz-Pro-Phe-Arg-Nan	0.4	not determined
Tos-Gly-Pro-Lys-Nan	< 0.1	not determined
Cbz-Gly-Pro-Arg-Nan	< 0.1	not determined
Bz-Ile-Glu-Gly-Arg-Nan	0.68	not determined
Bz-Phe-Val-Arg-Nan	< 0.1	0.45
H-D-Phe-Pip-Arg-Nan	< 0.1	not detectable
H-D-Val-Leu-Lys-Nan	< 0.1	0.21
H-D-Val-Leu-Arg-Nan	< 0.1	0.28
Cbz-Val-Gly-Arg-Nan	0.36	not determined
Bz-Asp-Trp-Arg-Nan	0.13	not determined
Suc-Ala-Ala-Ala-Nan	< 0.1	not determined
Cbz-Gly-Gly-Leu-Nan	< 0.1	not determined
Suc-Ala-Ala-Pro-Phe-Nan	< 0.1	not determined
Suc-Gly-Gly-Phe-Nan	< 0.1	not determined

called proteinase P, which can only cleave the peptide bond between two amino acids of the p-nitroanilide derivatives of the peptide measured. This makes further aminopeptidase cleavage necessary to liberate p-nitroaniline from the cleaved peptide and to detect the proteinase (second column of Table 8). Action of the newly detected proteinases in the differentiation process of sporulation has to be elucidated in future experiments.

ACKNOWLEDGEMENTS

This work was supported by the Deutsche Forschungsgemeinschaft, Bonn - Bad Godesberg and by the Fonds der Chemie. The expert assistance of Mrs. H. Gottschalk during preparation of this manuscript is gratefully acknowledged.

REFERENCES

Betz, H. and Weiser, U. (1976a), Protein degradation and proteinases during yeast sporulation, *European Journal of Biochemistry* 62, 65-76

Betz, H. and Weiser, U. (1976b), Protein degradation during yeast sporulation. Enzyme and cytochrome patterns, *European Journal of Biochemistry* 70, 385-395

Croes, A. F. (1967a), Induction of meiosis in yeast. I. Timing of cytological and biochemical events, *Planta* 76, 209-226

Croes, A. F. (1967b), Induction of meiosis in yeast. II. Metabolic factors leading to meiosis, *Planta* 76, 227-237

Klar, A. J. S. and Halvorson, H. O. (1975), Proteinase Activities of *Saccharomyces cerevisiae* during sporulation, *Journal of Bacteriology* 124, 863-869

Küenzi, M. T., Tingle, M. A. and Halvorson, H. O. (1974), Sporulation of *Saccharomyces cerevisiae* in the absence of a functional mitochondrial genome, *Journal of Bacteriology* 117, 80-88

Wolf, D. H. (1980), Control of metabolism in yeast and other lower eukaryotes through action of proteinases. *In* "Advances in Microbial Physiology" (Eds. A. H. Rose and J. G. Morris) pp. 267-337. Academic Press, London - New York

Wolf, D. H. and Ehmann, C. (1978), Isolation of yeast mutants lacking proteinase B activity, *Federation of the European Biochemical Societies Letters* 92, 121-124

Wolf, D. H. and Ehmann, C. (1979), Studies on a proteinase B mutant of yeast, *European Journal of Biochemistry* 98, 375-384

Wolf, D. H. and Fink, G. R. (1975), Proteinase C (carboxypeptidase Y) mutant of yeast, *Journal of Bacteriology* 123, 1150-1156

Wolf, D. H. and Holzer, H. (1980), Proteolysis in yeast. *In* "Transport and utilization of amino acids, peptides and proteins by microorganisms" (Ed. J. W. Payne) pp. 431-458. John Wiley, Chichester

Wolf, D. H. and Weiser, U. (1977), Studies on a carboxypeptidase Y mutant of yeast and evidence for a second carboxypeptidase activity, *European Journal of Biochemistry* 73, 553-556

V. ENVIRONMENTAL CONTROLS

ENVIRONMENTAL CONTROLS: ECOLOGICAL ASPECTS
DISCUSSANT'S INTRODUCTION

A.S. Sussman

Division of Biology (Botany),
University of Michigan, Ann Arbor MI 48109 U.S.A.

It is my intention to set an ecological context for the presentations at this session. After all, the controls with which we deal in the laboratory should relate to those in ecological settings through which the selection pressures that are the motive power for evolution operate.

PHYSICAL CONTROLS

High temperatures, and low, have been used since the turn of this century to break the dormancy of certain spores. No more will be said about cold-activation other than that most organisms whose spores require such treatment to germinate are pathogens whose development would be expected to be coordinated with that of their host. However, the disproportionately large amount of work done with pathogens must be kept in mind as we evaluate such generalizations.

What do we know of the role of heat-activation in nature? It might be expected that the spores of thermophiles would require high temperatures for germination and, indeed, this is the case but, as Jack and Tansey (1977) have shown, not all such organisms have such a requirement. Pyrophiles, by definition, should need fire at some stage in development but the questions for us is, "Do their *spores* have such a need?" The effects of forest fires on the soil mycoflora of a *Pinus contorta* association were studied by Widden and Parkinson (1975) who found that *Gelasinospora* sp. occurred *only* in the burned plot. Extracts of burned litter were inhibitory to the species studied except for *Cylindrocarpon destructans*. Unfortunately, the reasons for the survival of *Gelasinospora* sp. are not known in this case but Wicklow (1975) has found *G. cerealis* and other pyrophiles to require heat activation. On the other hand, Wicklow and Zak (1979) have shown that members of the "carbonicolous/coprophilous ascomycete community" they studied in Illinois require mineral factors from burnt-over soil as well as heat-activation in order to produce sporocarps, that is, to survive optimally. Furthermore, fire serves to reverse the "widespread soil fungistasis" so factors beyond activation of spores are involved in this natural situation as well. To add to the complexity, the spores of pyrophiles appear to have great tolerance of the reduced moisture holding capacity of soils that have been burned over, while benefitting as well from the reduced competition from organisms that do not survive heating in this habitat (El-Abyad and Webster, 1968a,b). Of course, pyrophiles need considerable heat resistance in order to survive fire in the first place. Consequently, the ecological role of heat-activation of fungal spores cannot be viewed in isolation from other stages in the life history or from other environmental variables. Nor can such a role be assured because the elimination of competition and other selective advantages may derive from exposure to heat.

This conclusion is borne out in studies of compost growers and coprophiles, which have many characteristics in common, especially the succession of organisms represented in their habitat (Hudson, 1968). In these environments, no evidence has been adduced yet to implicate heat-activation as conferring selective advantage upon the organisms involved. In fact, Harper and Webster's (1964) detailed studies of fungal succession on dung, and Chang and Hudson's (1967) of straw composts lead to the conclusion that the high temperatures developed in the gut of mammals and in composts, respectively, serve to restrict competition by killing mesophilic microorganisms, thereby placing a premium upon the ability of coprophiles and compost growers to survive in their habitats (Sussman and Douthit, 1981). Where activation may be involved in these cases is in the alteration of the surface of spores by enyzmes in the gut as Janczewski (1871) first proposed. Since that time, other evidence exists for enzymatic effects of this kind in a variety of organisms (Sussman and Douthit, 1981).

Conditions in composts probably are more rigorous than in dung alone because of heating caused by their fermentation, so spores of fungi whose normal habitat is the latter usually do not survive in the former (Renard and Cailleux, 1973). Therefore, it may be that, contrary to the situation in dung, the high temperatures in composts may activate some fungal spores but surprisingly little work has been reported on this subject. A likely possibility is spores of *Neurospora*, and of organisms from similar habitats, which are activated by heat and chemicals together, as proposed by Sussman *et al.* (1959), a combination similar to that which is effective with basidiospores of *Coprinus radiatus* (Mills and Eilers, 1973), another coprophile. This suggestion stems from the observation that the presence of furfural reduces the heat needed to activate ascospores of *Neurospora*. Thevelein and Van Assche (1980) have made similar observations with spores of *Phycomyces*, which respond to lower temperatures in the presence of alcohols and other solvents. There are two reasons for proposing that such combined treatments are of significance in nature: the temperatures required in the presence of chemicals are low enough to make them more likely in dung, composts and other habitats where some heat is available; and, in the case of *Neurospora*, at least, sub-activation temperatures restore older spores' sensitivity to chemical activators (Sussman *et al.*, 1959). Nevertheless, these are only speculations as yet and have not been tested in appropriate natural habitats.

How widely distributed among fungi is the need for heat-activation? Aside from the coprophilous organisms and the pyrophiles discussed above and those listed in earlier reports (Sussman and Halvorson, 1966) few others have been found to have this requirement. However, the combined need for heat and chemicals may be of wider prevalence than among the few cases mentioned earlier. Furthermore, a type of dormancy may be imposed upon some spores that induces the need for heat-activation. Thus, Bromfield (1964) discovered that the chilling of uredospores of *Puccinia graminis* f. sp. *tritici* induces a "cold dormancy" that can be reversed by temperatures above the optimum for growth so it is possible that other cases of this kind will be found. It this is so, the need for vapor-phase hydration may also be induced so care must be taken in the way spores are rewetted (Maheshwari and Sussman, 1971).

As in the case of fungi whose spores require cold-activation, those that require light are mainly pathogens (Sussman and Halvorson, 1966). There is some evidence that the germination of uredospores of *Puccinia graminis* f. sp. *tritici* is mediated by a phytochrome-like system, based upon comparisons of action spectra for promotion and inhibition (Lucas *et al.*, 1975; Schneider and Murray, 1979). Consequently, it may be that the development of the pathogens is coordinated with that of its host by this means. However, most of the spores that have been reported to require light for germination respond to wavelengths in the blue region (Sussman and Halvorson, 1966). In these cases, it is possible that blue light is a cue that suggests the presence of moisture, in that plant leaves filter out most other wavelengths so that germination would take place in the shade provided by them. That light may signal the presence of water in the environment is suggested by the negation of the light sensitivity of uredospores of *P. graminis* f. sp. *tritici* by soaking in water (Chang, *et al.*, 1974). Further attesting to the diversity of light receptors in the fungi are the conidia of *Trichometasphaeria turcica*, whose primary hyphae require green light to produce appressoria, perhaps an adaptation to life of the surface of leaves of maize, from which light of this quality is reflected (Benedict, 1976).

CHEMICAL CONTROLS

A quotation of Lucas (1947) makes the importance of chemical controls manifest "...the links between contemporary and successive members of the community may depend as much upon the physiological processes of the members and their external by-products as upon the more readily understood prey-predator relationship." In respect to spores, chemical activators may be viewed as environmental signals that presage conditions that are favorable to growth and development and these "ecological ectocrines" (Lucas, 1947) are the products of organisms that interact in natural situations. A fuller discussion of these substances is provided by Sussman (1981), and the paper by Macko in this volume covers the role of self-inhibitors and self-stimulators of germination.

As would be expected, these chemical signals may be stimulatory or inhibitory. Among the former are the antibiotics, although it cannot be assumed that each has a ecological role. Nevertheless, it is known that toxic bacterial metabolites reduce the germination of spores of *Tilletia caries* and of *Hypoxylon pruinatum* (Kollmorgen and Jones, 1975; Wood and French, 1965), and other examples exist as well. Mycotoxins like diacetoscirpenol and rubratoxin B also are known to inhibit fungal spore germination (Reiss, 1973). But most of the inhibitory effects that have been recorded are from substances produced by higher plants. In addition to a large array of substances released from swelling seeds and all parts of plants, phyloalexins are known to inhibit fungal spores (Mineo and Uritani, 1978; Shirata, 1978). Self inhibitors are a special group of substances that usually are not ectocrines but which may have an important role in suppressing germination until appropriate environmental conditions arise. The recent elucidation of the chemical structure of discadenine, the self-inhibitor of spores of *Dictyostelium discoideum* (Nomura *et al.*, 1977), supercedes the work of Bacon *et al.* (1973) on this compound, and suggests that it has cytokinin-like activity. This finding suggests some linkage with higher plants but the details remain to be worked out.

Stimulators of germination may be specific activators of the germination process or nutrients. It may be difficult to distinguish between these two classes of activators but the amount needed and whether they have a role in energy metabolism must be considered. In any event, microorganisms and others, including animals (Pugh, 1972), produce such compounds and many fungal spores are affected. Chemicals released by host plant roots in the rhizosphere appear to induce the development of pathogens and other selective advantages may derive from the release of such stimulators. For example, the production of cumin seed exudates appears to be coordinated with the stages in the germination of the spores of the pathogen (Chaturvedi, *et al.*, 1974). Self-stimulators, like self-inhibitors, may be "ecological ectocrines" as well for usually they may be produced by other organisms. An interesting exception is *Leccinum* sp., whose mycelium is the only source of the self stimulator of spore germination yet found (Fries, 1979).

NUTRIENT-DEPRIVATION VERSUS INHIBITORS

Interpretation of data on "ecological ectocrines" sometimes is obfuscated by the difficulty of distinguishing between the inhibitory effects of toxic substances and nutrient deprivation. We owe to Lockwood (1977) the reminder that nutrient deprivation can explain many of the data on the "widespread soil fungistasis," although there is evidence for the presence of inhibitors as well (Pavlica *et al.*, 1977). That the interactions of the phylloplane microorganisms can be explained similarly has been shown by Blakeman and Brodie (1977) and others so other natural associations may be found to illustrate this effect as well.

Much remains to be learned about this hypothesis in relation to dormant systems. Clearly, it can explain a form of exogenous dormancy like that imposed by the "widespread soil fungistasis," as reported recently for sclerotia of *Stromatinia gladioli* by Jeves and Coley-Smith (1980). In this case, the host-specific germination stimulant produced by *Gladiolus* sp. overcomes the "widespread soil fungistasis" to which intact sclerotia are subject. Therefore, the natural environment of a propagule may have to be examined before even the physiology of activators can be fully understood.

REFERENCES

Bacon, C.W., Sussman, A.S. and Paul, A.G. (1973), Identification of a self-inhibitor from spores of *Dictyostelium discoideum*, *J. Bacteriol.* 113, 1061-1063.

Benedict, W.G. (1976), Light-dependent morphogenesis of conidia of *Trichometasphaeria turcica* in vitro, *Can. J. Bot.* 54, 552-555.

Blakeman, J.P., Brodie, I.D.S. (1977), Competition for nutrients between epiphytic micro-organisms and germination of spores of plant pathogens on beetroot leaves, *Physiol. Plant Pathol.* 10, 29-42.

Bromfield, K.R. (1964), Cold-induced dormancy and its reversal in uredospores of *Puccinia graminis* var. *tritici*, *Phytopathology* 54, 68-74.

Chang, H.-S., Wilcoxson, R.D. and Calpouzos, L. (1974), Inhibition of uredospore germination by light partially relieved by soaking spores in water, *Phytopathology* 64, 158.

Chang, Y. and Hudson, H.J. (1967), The fungi of wheat straw compost I. Ecological studies, *Trans. Br. Mycol. Soc.* 50, 649-666.

Chaturvedi, A.N., Murralia, R.N. and Siradhana, B. (1974), The influence of cumin seed exudates on fungal spore germination, *Plant Soil* 40, 49-56.

El-Abyad, M.S.H. and Webster, J. (1968a), Studies on pyrophilous discomycetes I. comparative physiological studies, *Trans. Br. Mycol. Soc.* 51, 353-367.

El-Abyad, M.S.H. and Webster, J. (1968b), Studies on pyrophilous discomycetes II. competition, *Trans. Br. Mycol. Soc.* 51, 369-375.

Fries, N. (1979), The taxon-specific spore germination reaction in *Leccinum*, *Tran. Br. Mycol. Soc.* 73, 337-341.

Harper, J.E. and Webster, J. (1964), An experimental analysis of the coprophilous fungus succession, *Trans. Br. Mycol. Soc.* 47, 511-530.

Hudson, H.J. (1968), The ecology of fungi on plant remains above the soil, *New Phytol.* 67, 837-874.

Jack, M.A., Tansey, M.R. (1977), Growth, sporulation and germination of thermophilic fungi incubated in sun-heated soil, *Mycologia* 69, 109-117.

Janczewski, E.V.G. (1871), Morphologische Untersuchungen über *Ascobolus furfuraceus*, *Bot. Ztg.* 29, 255-262.

Jeves, T.M. and Coley-Smith, J.R. (1980), Germination of sclerotia of *Stromatinia gladioli*, *Trans. Br. Mycol. Soc.* 74, 13-18.

Johnson, C.L. and Preece, T.F. (1979), Natural history of *Pilobolus kleinii*: Experiments with sporangia in the cow digestive tract and the early stages of growth, *Trans. Br. Mycol. Soc.* 72, 453-457.

Kollmorgen, J.F. and Jones, L.C. (1975), The effects of soil-borne micro-organisms on the germination of chlamydospores of *Tilletia caries* and *T. foetida*, *Soil Biol. Biochem.* 7, 407-410.

Lockwood, J.L. (1977), Fungistasis in soils, *Biol. Rev. Cambridge Phil. Soc.* 52, 1-43.

Lucas, C.E. (1947), The ecological effects of external metabolites, *Biol. Rev. Cambridge Phil. Soc.* 22, 270-295.

Lucas, J.A., Kendrick, R.E. and Givan, C. (1975), Photocontrol of fungal spore germination, *Plant Physiol.* 56, 847-849.

Maheshwari, R. and Sussman, A.S. (1971), The nature of cold-induced dormancy in urediospores of *Puccinia graminis* var. *tritici*, *Plant Physiol.* 47, 389-395.

Mills, G.L. and Eilers, F.I. (1973), Factors influencing the germination of basidiospores of *Coprinus radiatus*, *J. Gen. Microbiol.* 77, 393-401

Mineo, K. and Uritani, I. (1978), Isolation and characterization of factors in sweet potato root which agglutinate germinated spores of *Ceratocystis fimbriata*, black rot fungus, *Plant Physiol.* 62, 751-753.

Nomura, T., Tanaka, T., Abe, H. and Uchiyama M. (1977), Cytokinin activity of discadenine: a spore germination inhibitor of *Dictyostelium discoideum*, *Phytochemistry* 16, 1819-1820.

Pavlica, D.A., Hora, T.S., Bradshaw, J.J., Skogerboe, R.K. and Baker, R. (1977), Volatiles from soil influencing activities of soil fungi, *Phytopathology* 68, 758-765.

Pugh, G.C.F. (1972), The contamination of birds' feathers by fungi, *Ibis* 114, 172-177.

Reiss, J. (1973), Der Einfluss von Mycotoxinen auf Pflanzen, *Z. Lebensm. Unters.-Forsch* 151, 256-259.

Renard, Y. and Cailleux, R. (1973), Contribution à l'étude des microorganismes du compost destiné à la culture de champignon de couche, *Rev. Mycol. (Paris)* 37, 36-47.

Schneider, M.J. and Murray, B.J. (1979), Photochrome mediated germination in the fungus *Puccinia graminis*, *Photochem. Photobiol.* 29, 1051-1052.

Shirata, A. (1978), Production of phytoalexin in cortex tissue of mulberry shoot, *Ann. Phytopathol. Soc. Jpn.* 44, 485-492.

Sussman, A.S. and Douthit, H.A. Jr. (1981) Environmental aspects of the germination of spores. *In* "The Encyclopedia of Plant Physiology, New Series" (Ed. O.L. Lange, P.S. Nobel, C.B. Osmond and H. Ziegler) In Press, Springer, Berlin.

Sussman, A.S. and Halvorson, H.O. (1966). "Spores: Their Dormancy and Germination" Harper & Row, New York and London.

Sussman, A.S., Lowry, R.J. and Tyrrell, E. (1959), Activation of *Neurospora* ascospores by organic solvents and furans, *Mycologia* 51, 237-247.

Thevelein, J.M. and Van Assche, J.A. (1980), Unpublished data.

Wicklow, D.T. (1975), Fire as an environmental cue triggering ascomycete development in prairie soils, *Mycologia* 67, 852-862.

Wicklow, D.T. and Zak, J.C. (1979), Ascospore germination of carbonicolous ascomycetes in fungistatic soils: an ecological interpretation, *Mycologia* 71, 238-242.

Widden, P. and Parkinson, D. (1975), The effects of a forest fire on soil microfungi, *Soil Biol. Biochem.* 7, 125-138.

SPORE ACTIVATION

D.A. Cotter

Department of Biology
University of Windsor
Windsor, Ontario, Canada

INTRODUCTION

In discussing the phenomenon of spore activation I will rely heavily on work which has been done with the cellular slime mould *Dictyostelium discoideum* Raper. A reading of other chapters in this book as well as the preceeding volumes in this series would place the present manuscript in proper perspective. See "The Fungus Spore" (Madelin, 1966) and "The Fungal Spore Form & Function" (Weber and Hess, 1976).

The point I would like to emphasize from the diagram in Fig. 1 is that the mature fruiting body consists of an upright column of dead stalk cells which supports an oval mass of spores at the top of the fruiting body (Raper, 1935; Whittingham and Raper, 1960). The mass of spores is contained in a viscous material consisting of extracellular proteins, some of which aided in the formation of the spore and stalk cells (Hohl and Hamamoto, 1969). Other components of the matrix are complex carbohydrates as well as trace quantities of trehalose and glucose (Ceccarini and Filosa, 1965); N,N-dimethylguanosine (Bacon and Sussman, 1973) and discadenine (Abe *et al.*, 1976) are also components of the matrix.

I shall use the following definitions suggested by Sussman (1976) in discussing the topic of spore activation:

1. Dormancy. Any rest period or reversible interruption of the phenotypic development of an organism.
2. Activation. A treatment that leads to an increase in germination.
3. Germination. The first irreversible stage which is recognizably different from the dormant organism, as judged by morphological, cytological, physiological, or biochemical criteria.

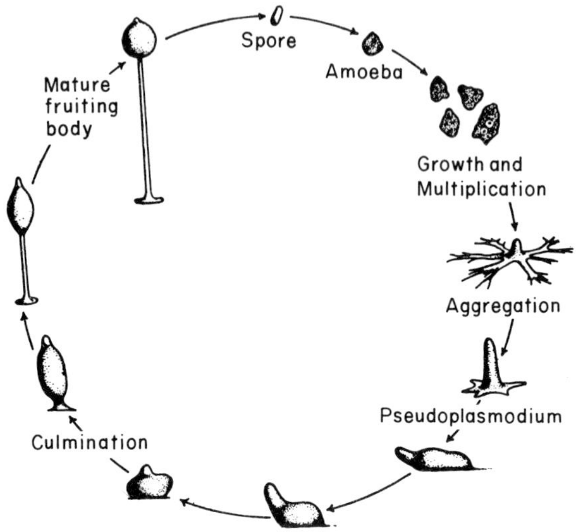

Fig. 1. Life cycle of Dictyostelium discoideum (after Wright, 1963).

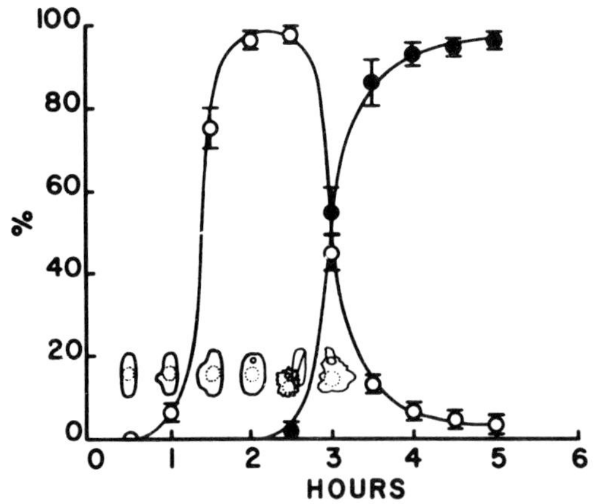

Fig. 2. Kinetics of spore germination in Dictyostelium discoideum, showing the mean and standard deviation for eight experiments with one day old spores. Symbols: open circles, swollen spores; closed circles, myxamoebae. The semidiagrammatic sequence of germination includes lag, early swelling, middle swelling, late swelling, emergence, and post emergence (Cotter and Raper, 1968b).

Resistant structures such as *D. discoideum* spores, *Neurospora* sp. ascospores, and *Phycomyces blakesleeanus* sporangiospores are said to possess constitutive dormancy since they do not germinate unless activated by chemical or physical treatments (Sussman, 1976). In the case of *D. discoideum* the process of spore germination, after activation, may be divided into three major stages: post-activation lag, spore swelling, and myxamoeba emergence (Fig. 2). The spore contains its own sources of carbon and energy; oxygen, pH maintenance, and a suitable temperature are the only external requirements for rapid germination (Cotter and Raper, 1968a,b).

As can be noted from Fig. 3, oxygen uptake is the first physiological sign that activation has occurred. Precursor uptake into RNA and protein appears to begin during the transition between the late lag phase and the early swelling phase of germination (Bacon and Sussman, 1973; Yagura and Iwabuchi, 1976; Giri and Ennis, 1977). Polysomes can first be detected during early spore swelling (Feit *et al.*, 1971; Yagura and Iwabuchi, 1976; Giri and Ennis, 1977). Spores of *D. discoideum* are similar to spores of *Botryodiplodia theobromae* (Brambl and Van Etten, 1970) in that they contain preformed mRNA which may be translated along with newly made mRNA during the germination process (Bacon and Sussman, 1973; Giri and Ennis, 1978; MacLeod *et al.*, 1980; Dowbenko and Ennis, 1980). A total of 300 stage specific proteins including actin are synthesized during the swelling, emergence, and post-emergence stages of germination (Giri and Ennis, 1977, 1978; Dowbenko and Ennis, 1980; MacLeod *et al.*, 1980). One of these proteins may be a spore swelling enzyme produced at the end of the post-activation lag stage of germination (Hohl, 1976). The developmental kinetics of three proteins synthesized during germination are shown in Fig. 4 (Cotter *et al.*, 1979; Tisa and Cotter, 1979, 1980; Chan and Cotter, *unpublished results*).

ACTIVATION WITH PHYSICAL AND CHEMICAL TREATMENTS

Once a spore has begun to swell it is, by definition, germinated. Thus we will be mainly concerned in this review with the activation treatments themselves as well as the events which occur during the lag phase before spore swelling begins.

In the earliest studies with *D. discoideum* it was noted that the spores were constitutively dormant and did not release myxamoebae if merely suspended in buffer; however, each spore released a single myxamoeba if peptone were added

3.

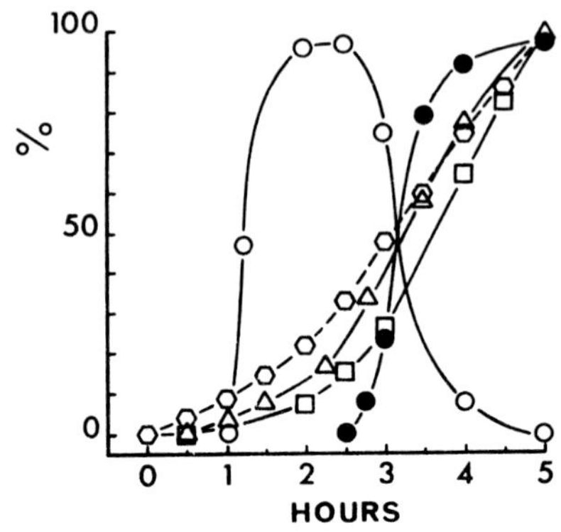

4.

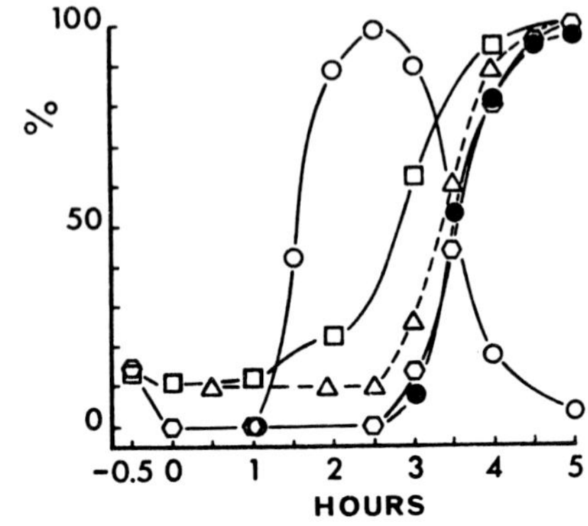

Fig. 3. Physiological events during germination of two day old spores. The values of oxygen uptake and incorporation are expressed as a percentage of the 5 hour values; the data are replotted from Cotter *et al.,* (1979) *and* Yagura *et al.,* (1976). *Symbols: open circles, percent spore swelling; closed circles, percent myxamoebae; hexagons, oxygen uptake; triangles, uridine incorporation; squares, amino acid incorporation.*

Fig. 4. The developmental kinetics of three enzymes during spore germination of two day old spores. Enzyme specific activity is expressed as a percentage of the 5 hour values. Symbols: open circles, percent spore swelling; closed circles, percent myxamoebae; squares, alkaline phosphatase; triangles, trehalase; hexagons, beta-glucosidase.

to the buffer (Cotter and Raper, 1966). When the component parts of peptones were tested (exclusive of peptides), it was the amino acid fraction which was mainly responsible for germination; vitamins, salts, and glucose did not induce germination. A mixture of the hydrophobic amino acids tryptophan, phenylalanine, and methionine partially substituted for the peptone. Once the spore swelling process was underway, the amino acid mixtures could be removed without retarding release of the myxamoebae (Cotter and Raper, 1966). Mild heat treatment of spores before plating them on agar containing amino acid mixtures resulted in an increased rate and percentage of germination. In fact, washed dormant spores would germinate in the absence of amino acids if first heated in phosphate buffer at 45 C for 30 min before being incubated at 23.5 C in phosphate buffer. This phenomenon is not unique to slime molds since heat treatment will activate a number of fungal spores from widely different taxonomic groups (Table 1). When spores of *Dictyostelium discoideum* are heated at 45 C for longer than 30 min there is some inactivation of the population (Cotter and Raper, 1966, 1968a, 1968b; Cotter and George, 1975). Heat-activation may be directly linked to amino acid activation, since dormant spores are devoid of pool levels of tryptophan, phenylalanine, and methionine; heat-shock induces the accumulation of these amino acids in the activated spores (Bacon and Sussman, 1973). The cardinal temperatures for *D. discoideum* spore swelling are a minimum at about 6 C, an optimum near 29 C, and a maximum near 35 C. The temperatures for

TABLE 1

*Optimal Temperature and Duration of Treatment Required for the Heat-activation of Selected Fungal Spores (*Sussman, 1976*).*

Organism	Temperature	Duration
Dictyostelium discoideum	45 C	30 min
Puccinia graminis	40 C	5 min
Coprinus radiatus	44 - 46 C	4 hr
Mucor miehi	45 C	5 hr
Phycomyces blakesleeanus	50 C	3 min
Neurospora tetrasperma	50 - 65 C	10 - 20 min
Byssochlamys fulva	75 C	5 - 10 min

myxamoebae emergence are similar to those for vegetative growth, the optimal being near 22 C and maximal at 27 C. Dormant *D. discoideum* spores are resistant to freezing, drying, sonic treatment, and osmotic pressure changes; resistance is lost upon initiation of spore swelling (Cotter and Raper, 1968b). Spores which are heat-activated remain activated even if dried or quickly frozen. These spores swell and complete emergence when returned to adequate moisture levels at room temperature (Cotter and Raper 1968b; Bacon and Sussman, 1973).

Neither light nor transmission electron microscopy revealed morphological differences between activated spores and dormant spores prior to swelling (Cotter and Raper, 1968b; Cotter *et al.*, 1969; Cotter and George, 1975). The first sign of incipient swelling is a slight phase darkening of the spore; transmission electron microscopy reveals that the highly crenated mitochondria begin to lose their peripheral layer of ribosomes at about this time (Cotter *et al.*, 1969). Since electron microscopy initially failed to provide information on the primary site of activation, a thermodynamic approach was developed to study reversible activation of *D. discoideum* spores (Cotter, 1973).

Reversible spore activation of *D. discoideum* is characterized by a narrow range of temperatures centered near 43 C (Fig. 5). Cotter (1973) argued that the sharpness and

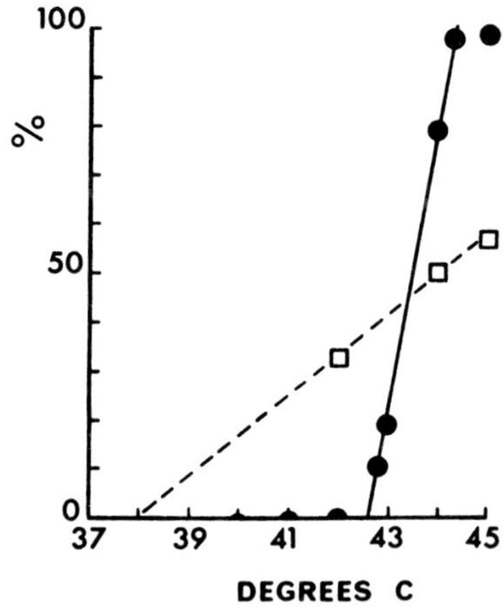

Fig. 5. Trypsin denaturation and D. discoideum spore activation. The ordinate refers to the percentage of denatured trypsin and/or the percentage of myxamoebae released from spores heated at various temperatures for 30 min. Heated spores were incubated at 23.5 C on buffered agar for 7 hr; each plotted point (circles) represents a determination of 1,200 objects. The data from Anson and Mirsky (1934) *for trypsin denaturation are plotted as open squares.*

large slope of the usual heat-activation curve (Fig. 5) suggested a process which required a high degree of cooperativity and that the thermal energy effected a conformational change in some molecules or aggregate of molecules. It was hypothesized that the sharpness of the transition point and the large activation energy ruled out either small molecules or those of varied molecular weight. It was concluded that the key substances which should be considered as the major candidates for heat-mediated transitions were macromolecules and macromolecular assemblies containing RNA, DNA, lipids, and proteins. The change in entropy (using the Van't Hoff equation) for the heat-activation of spores of *D. discoideum* and other microorganisms was compared to values for the denaturation of proteins (Table 2). Values such as these are obtained by the Arrhenius method whereby the logarithm of a parameter is plotted against the reciprocal of the temperature. The resulting values for spore activation of *D. discoideum*, *N. tetrasperma*, and *P. blakesleeanus* are much larger than for globular protein denaturation or endospore activation (Table 2).

TABLE 2

The Change in Standard Entropy with Protein Denaturation or Spore Activation (modified from Cotter, 1973)

Protein or spore	Temperature used for entropy calculation	Standard entropy J/^{o}K mole
Trypsin (pH 2)	44°C	891
Ribonuclease (pH 2.5)	50°C	1423
Chymotrypsinogen (pH 2.5)	50°C	1506
Chymotrypsinogen (pH 2.5)	75°C	2845
Bacillus subtilis	5 - 95°C	21 - 33
Bacillus megaterium	50 - 75°C	615
Neurospora tetrasperma	51°C	6276
Dictyostelium discoideum	44°C	7113
Phycomyces blakesleeanus[1]	41.9^{o}C	3644

References
1) Thevelein *et al.*, 1979

Direct activation of the genome would not appear to be a result of spore activation treatments since transcription does not begin until swelling is initiated (see Fig. 3). Melting curves of RNA are not sharp nor is a large slope found for the transition from the native to the denatured state (Cox and Littauer, 1960). The nuclear DNA of *D. discoideum* has a Tm of 78.8 C while the mitochondrial DNA possesses a Tm of 80.0 C (Sussman and Rayner, 1971). A similar argument can be made for *Neurospora* ascospores which are activated at between 50 - 60 C; the Tm of the DNA is above 85 C (Sussman, 1976). The genome of *D. discoideum* contains oligo dT sequences (Firtel and Lodish, 1973; Jacobson *et al.*, 1974) which should be capable of denaturation at lower temperatures than the bulk of the DNA. However these same sequences should be more susceptible to thymine dimer formation during ultraviolet light treatment than the bulk of the DNA. The slime mold *D. discoideum* is unlike many organisms in that the spores are more susceptible to ultraviolet light inactivation than are the vegetative cells (Cotter, 1973). Indeed, the DNA of dormant spores can be functionally destroyed by U.V. light at 260 nm; the U.V. killed spores can still be heat-activated and only the emergence stage of germination is prevented by this treatment (Cotter, 1973). Inactivated spores retain normal oxygen uptake but do not synthesize RNA, alkaline phosphatase, trehalase, and beta-glucosidase during spore swelling (Demsar, Hamer and Cotter, *unpublished results*). The large majority of inhibitors of protein and RNA synthesis added during spore activation prevent precursor incorporation but allow spores to proceed to the end of the swelling stage of germination (Table 3). Thiolutin is the only RNA synthesis inhibitor which blocks spore swelling (Giri and Ennis, 1977); it is not known if thiolutin is specific to only inhibition of RNA synthesis. It might be argued that inhibitors of macromolecular synthesis do not enter spores during activation. In opposition to this idea it should be noted that spores activated in the presence of edeine and then washed free of the drug after activation proceed to swell but do not release myxamoebae. This drug is unlike cycloheximide which reversibly inhibits myxamoeba emergence (Cotter and Raper, 1966, 1970). In summary, the target of heat-induced activation does not appear to involve the direct stimulation of macromolecular synthesis since spore swelling may occur in the absence of macromolecular synthesis. Bacterial endospore "germination" also does not require macromolecular synthesis (see Rossignol and Vary, 1979a).

Transmission electron microscopic examination reveals that prespore cells of *D. discoideum* contain prespore vesicles which fuse with the plasma membrane during the sporulation process (Hohl and Hamamoto, 1969). Thus, the plasma membrane of the nascent spore may be partially or

TABLE 3

Inhibitors of RNA and Protein Synthesis Block Precursor Incorporation but the Majority of Drugs do not Deactivate Spores

Inhibitor	Concentration (μg/ml)	Percent Incorporation	Swelling Blocked	Emergence Blocked	Reference
Control (Dormant Spores)	- -	5.6 % leucine	- -	- -	1
Control (Germinating Spores)	- -	100.0 % leucine	No	No	1
Edeine	100	1.0 % leucine	No	Yes	1
Cycloheximide	200	6.5 % leucine	No	Yes	1
Puromycin	1000	0.4 % leucine	No	Yes	1
Chloramphenicol	4000	- -	No	Yes	2
Daunomycin	250	0.4 % uracil	No	Yes	3
Lomofungin	100	1.7 % uracil	No	Yes	3
Thiolutin	100	2.9 % uracil	Yes	Yes	3

References: 1) Cotter *et al.*, 1979 2) Kobilinsky and Beattie, 1977 3) Giri and Ennis, 1977 completely replaced by the prespore vesicle membrane. This switch in plasma membrane components could be involved in the maintenance and later breaking of spore dormancy. It had been speculated by Cotter and Raper (1968b) that a change in permeability accompanied activation which suggested that the plasma membrane was involved in both activation and deactivation processes. Recently Hohl *et al.* (1978) found that the *D. discoideum* spore plasma membrane changed as a result of heat-induced spore activation. At the end of heat activation the distribution of spore plasma membrane particles between the two fracture faces (protoplasmic face and exoplasmic face) is drastically changed. While in dormant spores the particle ratio of PF/EF was 1:1, it increased up to 9:1 in heat activated spores, indicating a subtle change in plasma membrane properties (Hohl *et al.*, 1978). Less drastic effects have been observed with another fungus, *Phycomyces*, in which some of the spore plasma membrane particles form tiny aggregates during heat activation (Malhotra and Tewari, 1973). The permeability of *D. discoideum* spores increased within 30 min following heat activation as determined by efflux measurements of radioactively labelled spores. At the onset of swelling this efflux was accelerated. During the swelling stage of germination the osmotically active material within the spores increased, part of which could be recovered from the supernatant. The combined experiments pointed to the plasma membrane as the target site of heat-induced activation of *D. discoideum* spores (Hohl *et al.*, 1978). Sussman (1976) previously hypothesized that a permeability change is involved in the activation of *Neurospora* ascospores.

Could the phospholipid portion of the spore plasma membrane be the target of thermal activation?

Lipid melting in cellular biological membranes usually begins at temperatures below the normal growth optimum and is spread over a 25 C range for many organisms (Melchior and Steim, 1976). From a theoretical point of view bulk lipid transitions of the plasma membrane would not appear to be sharp enough to account for *D. discoideum* spore activation nor would this type of phase transition occur at temperatures of 42 - 45 C. One reason for this is that membranes containing cholesterol exist over a wide temperature range in a semifluid state without undergoing a phase transition (Melchior and Steim, 1976). The actual level of sterol ester increases during differentiation of *D. discoideum* cells. The level of sterol ester is 1% (w/w) of the total lipid in vegetative amoebae and increases to 15% (w/w) in mature fruiting bodies (Long and Coe, 1974). In studies with *D. discoideum* vegetative cells Herring and Weeks (1979) have concluded that the plasma membrane has the usual amount of sterol which should broaden any phospholipid phase transition. However, these workers were nevertheless surprised by their inability to detect any significant phase change over the temperature range of 0 - 40 C using ESR as well as other techniques. The above studies were done using vegetative cells and so the critical experiments with *D. discoideum* spores have not been done.

As pointed out by Sussman (1976) clear distinctions between proteins and lipids may be hard to make when membrane complexes and lipoproteins are considered.

Could proteins of the plasma membrane be the target of thermal activation?

As discussed above, the data in Table 2 indicate that the positive entropy change for fungal spore activation is much higher than that for complete protein denaturation. However, accompanying the order → disorder transition due to the unfolding of proteins during denaturation is a process of disorder → order as hydrophobic amino acid side-chains are exposed to water (Tanford, 1968). Thus, it was argued that if protein denaturation were partial and involved only the hydrophilic amino acid residues, then a relatively high positive entropy would result during a helix → random coil transition (Cotter, 1973). In fact, Bresler (1971) calculated that this type of transition for a protein with a molecular weight of 50,000 would have a positive entropy change of 8,368 J/^{o}K mole which is in the range of fungal spore activation entropies (Table 2). The actual entropy change for the helix → coil transition is directly dependent on the molecular weight of the protein and increases with increasing molecular weight (Bresler, 1971). It is possible then that under certain conditions denaturation may be partial and only involve the hydrophilic residues of proteins. A class of proteins which could fulfill the requirement that only hydrophilic residues be involved in reversible partial denaturation (i.e. conformational changes) during *D. discoideum* activation might be proteins located in cellular membranes such as the plasma membrane and/or the mitochondrial inner membrane. While membrane proteins generally have the same ratio of hydrophilic to hydrophobic amino acids as water soluble proteins, the former are amphiphilic (Vanderkooi and Green, 1971; Singer and Nicolson, 1972).

That is, the difference between the two is that certain membrane proteins consist of separate hydrophilic and hydrophobic domains (see Tanford, 1978). The hydrophobic regions of these membrane proteins might be resistant to heat-induced denaturation because they would be surrounded by a hydrophobic lipid matrix; since only hydrophilic regions of these proteins might be available for denaturation under aqueous conditions, then a relatively high positive entropy change would occur upon partial denaturation (conformational or allosteric change).

The possible hydrophilic nature of the activatable proteins in *D. discoideum* links the availability of water to the degree of possible protein denaturation and allows the prediction that water is required for spore activation. It has been confirmed that the activity of water is critical for activation, deactivation and maintenance of either state (Cotter, 1977). Since oxygen consumption may be proportional to the degree of hydration of many types of dormant organisms (see Sussman in Gould and Hurst, 1969) a link was suggested between the two by Cotter (1973). He thus proposed that the inner mitochondrial membrane might be the site of heat-induced activation of *D. discoideum* spores. In particular it was hypothesized that a protein in the inner mitochondrial membrane might regulate mitochondrial activity through conformational changes (Cotter, 1973; Cotter *et al.*, 1976).

Hohl *et al.*, (1978) have suggested that the particles of the spore plasma membrane observed by freeze-fracture electron microscopy are quite homogenous in size. Thus, the concept of an activation treatment causing a conformational change in a class of proteins might apply to these membrane particles. Hohl *et al.*, (1978) also considered it possible that activation might involve more than one "denaturable" target site. We have recently demonstrated that beta-glucosidase is almost totally denatured in spores during heat-shocking while the activity of alkaline phosphatase is reduced by only 20 percent (Fig. 4). Other enzymes denatured more than 70 percent during heat shocking include alpha-glucosidase, beta-galactosidase and alpha-mannosidase (Tisa and Cotter, 1980). Surprisingly, acid phosphatase, unlike the other lysosomal enzymes, is not denatured during heat induced activation (Tisa and Cotter, 1979). *Phycomyces* spore trehalase also appears to undergo heat mediated transitions during activation which may not be involved in germination (Van Assche *et al.*, 1972; Van Assche *et al.*, 1978). Recently, spores of *Phycomyces* were scanned with a differential scanning calorimeter; the spectrum obtained was clearly influenced by the previous activation of the spores by heat or by acetate (Van Cauwelaert and Verbeke, 1979). When *Phycomyces* spores were allowed to return to dormancy the original spectrum of dormant spores was obtained. The high temperature at which the difference in the spectrum between activated and dormant spores was found, suggests that activation of spores is obtained through a conformational change of a protein (Van Cauwelaert and Verbeke, 1979). Using high pressure to vary the heat activation temperature of *Phycomyces* spores revealed once again that a protein conformational change may be the basis for fungal spore activation instead of a membrane phase transition (Thevelein *et al.*, 1979).

It seems probable that a combination of conformational or allosteric changes in specific "membrane activation proteins" along with denaturation of other spore proteins could account for the relatively high apparent activation entropies of fungal spores undergoing thermal stimulation. An extension of this reasoning could rationalize the different apparent activation entropies among fungal spores as well as bacterial endospores. Heat induced activation of bacterial endospores might primarily utilize denaturation of membrane proteins involving both hydrophilic and hydrophobic domains which results in low apparent activation entropies (Table 2). The function of germinants such as L-alanine, L-proline, adenosine and D-glucose might be to allosterically induce a counter conformational change in the denatured hydrophobic domain of the "activated proteins". This corrected conformational state might then be equivalent to the higher entropy state calculated for fungal spores. Experimental data exist showing that the target or trigger sites of bacterial endospores are "plasma membrane" proteins (Rossignol and Vary, 1979; Janoff *et al.*, 1979). In the case of *Bacillus megaterium* it has been possible to reversibly inactivate the proline target sites of the plasma membrane without inactivating the glucose target sites (Rossignol and Vary, 1979a,b; Racine *et al.*, 1979; Janoff *et al.*, 1979). It may be significant to our thesis that the L-alanine binding sites of bacterial endospores are hydrophobic in nature and can be reversibly inhibited by various hydrophobic compounds (Yasuda-Yasaki *et al.*, 1978).

While we now believe that the evidence favors proteins in the plasma membrane as the primary target of heat-induced activation, we feel that the mitochondria of *D. discoideum* have a crucial secondary role to play in the regulation of the dormant and activated states.

Further support for the hypothesis that protein

conformational changes may be involved in spore activation is obtained from a study of additional treatments and agents which induce *D. discoideum* spore germination at room temperature (Table 4). Many of these agents are known to protein chemists as typical protein perturbants or denaturants. In the case of *Neurospora* ascospores and several other spores, lipid soluble agents appear to dominate the lists of activating agents (Sussman, 1976). It is possible that lipid soluble compounds such as furfural and alcohols may interact with hydrophobic proteins of *Phycomyces* spores (Thevelein *et al.*, 1979). The treatments listed in Table 4 are some of the activation methods which are practicable. I shall exclude gamma irradiation from further comment since I have no direct experience with the activation technique; I have noted however that gamma irradiation will also activate *Phycomyces* spores (Van Assche *et al.*, 1977).

The activating treatments have many characteristics in common. For example, spore populations subjected to suboptimal activation show a short lag and then the activated portion of the population quickly swells. The optimal conditions for activation can be obtained by variation of either the duration of the application or the "concentration" of the treatment. Whenever the optimal duration or "concentration" are slightly exceeded, then the post-activation lag is extended before swelling begins. Activation does not occur with chemical treatment at 0 C for any of the agents listed in Table 4. Whenever spores are activated with two treatments simultaneously, the resulting lag phase is never shorter than the lag phase of the single treatment producing the longest lag; that is, the treatment producing the longest lag overrides the treatment producing the shortest lag. The addition of the chemical agents to dormant *D. discoideum* spores causes plasmolysis. Upon deplasmolysis the spores are still not activated. Activation only occurs when the partially deplasmolyzed spores are subjected to a saltatory reduction in the concentration of the activating agent. Slow addition of buffer to avoid the saltatory effect results in very little spore activation. Chemical compounds which do not penetrate spores do not cause activation. An example of this phenomenon can be observed when the polyol series of compounds are studied. Methanol and ethylene glycol initially cause plasmolysis which is quickly relieved; glycerol penetrates spores very slowly and thus activation by rapid concentration reduction occurs only after 24 hr. Erythritol causes plasmolysis which is not relieved in 24 hr and no germination occurs when the compound is removed from spores. These experiments

TABLE 4

Dictyostelium discoideum Activation Treatments

Treatment	Duration of Post-activation lag	Reference
1 % peptone	30 min	1
Heat at 45°C for 30; 60; 90 min	1; 1.5; 2 hr	2, 3
Gamma irradiation 270 Krads	-	4
20 % dimethyl sulfoxide 30; 60 min	30; 45 min	5, 6, 7
8 M urea for 30; 60 min	35; 50 min	5, 8
6 M guanidine HCl for 60 min	30 min (only 64 % germination)	5, 9
2 M dimethylurea for 60 min	30 min	5, 9
2 M tetramethyl urea for 30 min	2.5 hr (only 71 % germination)	9
3 M ethylene glycol for 60 min	30 min	10
3 M glycerol for 24 hr	30 min (only 66 % germination)	10
30 % v/v methanol for 60 min	2 hr	*unpublished*

References:
1) Cotter and Raper, 1966
2) Cotter and Raper, 1968b
3) Cotter and George, 1975
4) Khoury *et al.*, 1970
5) Cotter, 1975
6) Ennis and Sussman, 1975
7) Cotter *et al.*, 1976
8) Cotter and O'Connell, 1976
9) Cotter, 1979
10) Cotter, 1977

make it unlikely that the spore wall or external face of the plasma membrane are the site of chemically induced activation.

Air-dried *D. discoideum* spores or those incubated in 2 M sucrose are not activated with the usual heat treatment of 45 C for 30 min. Dry *Phycomyces* spores also cannot be activated (Van Cauwelaert and Verbeke, 1979). The temperature for maximum activation and the temperature for thermal inactivation of *D. discoideum* spores are raised 7 - 10 C in 2 M sucrose (Cotter, 1977). This is the type of response expected for proteins protected from denaturation by low water activity (see Sussman and Halvorson, 1966).

DEACTIVATION TREATMENTS

Virtually all of the activation treatments discussed above are reversible when the spores are incubated under nonpermissive germinating conditions. The treatments which deactivate spores of *D. discoideum* all appear to be related to environmental perturbations (Table 5). Thus moderate osmotic pressure (0.25 M sucrose), low oxygen tension, low or high pH, and low or high temperature cause deactivation of spores. Deactivation will occur only if the restrictive environmental condition is imposed before the spores enter the swelling stage of germination. Deactivating spores do not incorporate precursors or synthesize enzymes such as trehalase (Table 6).

To the uninitiated, the typical data for deactivation experiments may seem complicated (Fig. 6). The data simply show that under restrictive environmental conditions activated spores still proceed to the end of the post-activation lag stage of germination and then halt. The longer the spores remain at the end of the post-activation lag stage of germination (fail-safe period) the greater the deactivation of the spore population (Cotter, 1975, 1977). Naturally, spores which are close to swelling (before being placed under deactivating conditions) will spend more time in the "fail-safe" period than freshly activated spores. Thus spores near the swelling stage deactivate faster than freshly activated spores (Fig. 6A, closed circles). Note that deactivated spores may be reactivated by any of the techniques of Table 4. Spores which do not halt at the end of the lag phase are killed by many of the harsh environmental conditions, which emphasizes the possible utility of the deactivation mechanism. There are two partial exceptions to the above statement; moderate osmotic pressure (partial dehydration) and 0°C incubation result in

TABLE 5

Restriction of Electron Transport Deactivates Dictyostelium discoideum Spores

Deactivating conditions	Concentration of agent	% reduction in O_2 uptake over 5 hr	Time for deactivation
Activated (control)[a]	-	0	-
Dormant (control)	-	95	-
Anaerobiosis[b]	O_2 less than 2%	98	95 % by 6 hr
Azide[a]	2×10^{-3} M	81	95 % by 4 hr
Cyanide[a]	2×10^{-3} M	76	99 % by 5 hr
SHAM[b,c]	5×10^{-3} M	18	0 % by 5 hr
37°C[a]	-	63	95 % by 5 hr
0°C[a]	-	-	95 % by 48 hr
Sucrose[b]	0.25 M	72	95 % by 6 hr
pH 2[a]	-	93	76 % by 5 hr
pH 12[a]	-	87	77 % by 5 hr

[a]Spore activation with DMSO treatment.

[b]Spore activation with heat treatment.

[c]SHAM, salicylhydroxamic acid.

Taken from Cotter *et al.*, 1979

TABLE 6

Unfavorable Environmental Conditions Inhibit Precursor Uptake, Block Trehalase Synthesis, and Deactivate Dictyostelium discoideum Spores

Deactivating agent or treatment	% inhibition of uridine incorporation	% inhibition of amino acid incorporation	Trehalase specific activity	Trehalase sp. act. (ratio of treated to dormant)
Control (activated)	0	0	0.414	6.12
Control (dormant)	94.4	94.4	0.067	1.00
$0^{o}C$[a]	99.0	97.3	0.077	1.15
$37^{o}C$	98.4	97.6	0.058	0.87
pH 2	99.0	96.0	0.084	1.25
pH 12	98.6	99.0	0.082	1.22
0.25 M sucrose[b]	96.3	88.0	0.086	1.28
2 mM azide	96.9	95.7	0.069	1.03
2 mM cyanide	99.0	99.4	0.065	0.97

[a]All activated spores were deactivated by exposure to unfavorable environmental conditions for 5 hr, except for spores deactivated at $0^{o}C$ for 48 hr.

[b]Spores were activated with DMSO treatment except for heat-activated spores which were deactivated with 0.25 M sucrose treatment.

Taken from Cotter *et al.*, 1979.

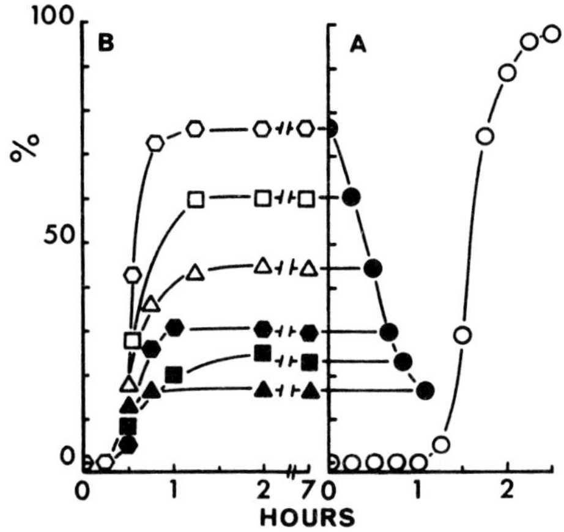

Fig. 6. Spore deactivation during the lag phase caused by 0.25 M sucrose. Spores were heat-activated at 45 C for 30 min in buffer and allowed to germinate at 23.5 C (panel A, open circles). During the first hour of germination spore samples were removed from the germination tube every 10 to 15 min and placed in 0.25 M sucrose at 23.5 C for 3 hr. At exactly 3 hr after the addition of sucrose, the carbohydrate was removed and the spore samples were allowed up to 7 hr of incubation in fresh buffer at 23.5 C to complete germination (panel B). The final percentage of germination was replotted (panel A, closed circles) against the time the spore samples were originally placed in sucrose. Taken from Cotter *et al.*, 1979.

effective deactivation but do not kill swollen spores or vegetative amoebae over short-time periods. The latter two environmental variables however, can be slightly altered so that only nonswollen spores are resistant to complete drying and freezing at 0 C (Cotter and Raper, 1968b; Bacon and Sussman, 1973). If spore populations undergoing deactivation are returned to permissive conditions, then the non-deactivated spores swell with a lag of only 15 - 20 min indicating that they have previously reached the "fail-safe" point or period (Fig. 6 B). The actual time required for spores to deactivate is determined by the following: the method of activation, the method of deactivation, the duration of the deactivation treatment, and the temporal position of the spores in post-activation lag when the deactivation treatment is applied.

Spores which are within 7 - 10 min of swelling are committed to swell even when placed under deactivating conditions.

Mutants of *D. discoideum* have been isolated which have altered responses to deactivation conditions; for example, one mutant, SG1, will swell and release myxamoebae in 0.25 M sucrose (Cotter and Dahlberg, 1977). Such results imply that harsh environmental shifts do not physically inhibit the spores but instead the environmental shifts serve as signals that the environment is not conducive to future optimal growth (Cotter, 1977).

Spores of *D. discoideum* possess a complete mitochondrial pathway which does not appear to function in the dormant state (Kobilinsky and Beattie, 1977). Oxygen uptake in vegetative cells is 87 % inhibited by cyanide. Our studies also imply that germinating spores contain at least two major pathways for oxygen consumption. The major pathway which represents 76 % of oxygen uptake is inhibited by cyanide. Addition of cyanide to dormant as well as to germinating spores reveals the presence of a second pathway for oxygen uptake which is inhibited by SHAM. Inhibition of the SHAM-sensitive pathway does not result in spore deactivation; inhibition of the cyanide sensitive pathway causes deactivation (Table 5). Remembering that macromolecular synthesis inhibitors do not cause deactivation (Table 3) we suggest that mitochondrial control may be an associated part of the activation mechanism (Cotter *et al.*, 1979). It is possible that like *Neurospora* and *Phycomyces* the initial metabolism after activation of *D. discoideum* spores may be primarily anaerobic (Eilers *et al.*, 1970; Rudolph and Furch, 1970; Furch, 1972). A highly aerobic metabolism may be required to allow *D. discoideum* spores to swell and synthesize macromolecules (Cotter *et al.*, 1979). Heat-activated *Neurospora* ascospores are also known to deactivate when incubated under anaerobic conditions (Sussman and Halvorson, 1966).

AUTOINHIBITORS AND AUTOACTIVATORS

The germination process of *D. discoideum* may also be regulated by the presence of autoinhibitors which are located in the matrix of the fruiting body (Snyder and Ceccarini, 1966; Cotter and Raper, 1968b, Bacon and Sussman, 1973). One such autoinhibitor called discadenine by Abe *et al.* (1976) appears to duplicate the effects of environmental shifts; that is, in low concentrations the compound deactivates spores if present during the post-activation lag stage of germination (Cotter and Raper, 1968b). Under these conditions protein synthesis is inhibited and oxygen uptake is minimal (Bacon and Sussman, 1973). Tanaka *et al.* (1975) have hypothesized that discadenine has binding sites on the plasma membrane. The addition of cAMP to activated spores will partially overcome the inhibition caused by low concentrations of the autoinhibitor (Katilus and Ceccarini, 1975).

Recently it was found that aged wild type *D. discoideum* spores will spontaneously germinate and release autoactivators upon spore swelling (Dahlberg and Cotter, 1978a). As mentioned previously it is during spore swelling that an accelerating efflux of osmotically active materials is observed (Hohl *et al.*, 1978). The autoactivators have been separated into three fractions which are similar but not identical to spore activators released by bacteria (Hashimoto *et al.*, 1976; Dahlberg and Cotter, 1978b).

CONCLUSIONS

The weight of evidence would seem to favor proteins of the plasma membrane as the prime target of heat-induced activation, at least, for spores of *D. discoideum* and perhaps *P. blakesleeanus*. It is probable that heat activation also causes conformational changes in proteins not involved in germination of *D. discoideum*. The fact that the above spores can be activated with different classes of agents, including autoactivators, might mean that the proposed "membrane activation proteins" (trigger sites) are somewhat heterogeneous.

Our present working model for spore activation and subsequent germination conceives of the entire process as a cascade phenomenon which can be regulated by activators, autoactivators, autoinhibitors, and perhaps cAMP. The "fail-safe" point during post-activation lag may be a metabolic period during which mitochondria control whether spores proceed to swell under favorable conditions or deactivate under unfavorable conditions.

REFERENCES

Abe, H., Uchiyama, M., Tanaka, Y., and Saito, H. (1976). Structure of discadenine, a spore germination inhibitor from the cellular slime mold, *Dictyostelium discoideum. Tetrahedron Letters* 42, 3807-3810.

Anson, M.L. and Mirsky, A.E. (1934). The equilibrium between active native and inactive denatured trypsin. *J. Gen. Physiol.* 17, 393-398.

Bacon, C.W. and Sussman, A.S. (1973). Effects of the self-inhibitor of *Dictyostelium discoideum* on spore metabolism. *J. Gen. Microbiology* 76, 331-344.

Brambl, R.M. and Van Etten, J.L. (1970). Protein synthesis during fungal spore germination. V. Evidence that the ungerminated conidospores of *Botryodiplodia theobromae* contain messenger ribonucleic acid. *Arch. Biochem. Biophys.* 137, 442-452.

Bresler, S.E. (1971). "Introduction to Molecular Biology" Academic Press, New York.

Ceccarini, C. and Filosa, M. (1965). Carbohydrate content during development of the slime mold, *Dictyostelium discoideum. J. Cell. Comp. Physiol.* 66, 135-140.

Cotter, D.A. (1973). Spore germination in *Dictyostelium discoideum* I. The thermodynamics of reversible activation. *J. Theoretical Biology* 41, 41-51.

Cotter, D.A. (1975). Spores of the cellular slime mold *Dictyostelium discoideum. In* "Spores VI" (Ed. P. Gerhardt, R. Costilow and H. Sadoff). pp. 61-72, Amer. Soc. Microbiol., Washington.

Cotter, D.A. (1977). The effects of osmotic pressure changes on the germination of *Dictyostelium discoideum* spores. *Canadian J. Microbiology* 23, 1170-1177.

Cotter, D.A. (1979). Activation of *Dictyostelium discoideum* spores with guanidine and methyl derivatives of urea. *Current Microbiology* 2, 27-30.

Cotter, D.A. and Raper, K.B. (1966). Spore germination in *Dictyostelium discoideum. Proceedings of the National Academy of Science U.S.* 56, 880-887.

Cotter, D.A. and Raper, K.B. (1968a). Factors affecting the rate of heat-induced spore germination in *Dictyostelium discoideum*. *J. Bacteriol.* 96, 86-92.

Cotter, D.A. and Raper, K.B. (1968b). Properties of germinating spores of *Dictyostelium discoideum*. *J. Bacteriol.* 96, 1680-1689.

Cotter, D.A., Miura-Santo, L.Y. and Hohl, H.R. (1969). Ultrastructural changes during germination of *Dictyostelium discoideum* spores. *J. Bacteriol.* 100, 1020-1026.

Cotter, D.A. and Raper, K.B. (1970). Spore germination in *Dictyostelium discoideum:* Trehalase and the requirement for protein synthesis. *Dev. Biology* 22, 112-128.

Cotter, D.A. and George, R.P. (1975). Germination and mitochondrial damage in spores of *Dictyostelium discoideum* following supraoptimal heating. *Archives for Microbiology* 103, 163-168.

Cotter, D.A., Morin, J.W. and O'Connell, R.W. (1976). Spore germination in *Dictyostelium discoideum* II. Effects of dimethyl sulfoxide on post-activation lag as evidence for the multistate model of activation. *Archiv. Microbiol.* 108, 93-98.

Cotter, D.A. and O'Connell, R.W. (1976). Activation and killing of *Dictyostelium discoideum* spores with urea. *Canadian J. Microbiol.* 22, 1751-1755.

Cotter, D.A. and Dahlberg, K.R. (1977). Isolation and characterization of *Dictyostelium discoideum* spore mutants with altered activation requirements. *Experimental Mycology* 1, 107-115.

Cotter, D.A., Garnish, F.J., and Tisa, L.S. (1979). The physiological effects of restrictive environmental conditions on *Dictyostelium discoideum* spore germination. *Canadian J. Microbiology* 25, 24-31.

Cox, R.A. and Littauer, U.Z. (1960). Properties of low molecular weight RNA from *E. coli*. *J. Mol. Biol.* 2, 166-168.

Dahlberg, K.R. and Cotter, D.A. (1978a). Autoactivation of spore germination in mutant and wild type strains of *Dictyostelium discoideum*. *Microbios* 23, 153-166.

Dahlberg, K.R. and Cotter, D.A. (1978b). Activators of *Dictyostelium discoideum* spore germination released by bacteria. *Microbios Letters* 9, 139-146.

Dowbenko, D.J. and Ennis, H.L. (1980). Regulation of protein synthesis during spore germination in *Dictyostelium discoideum*. *Proc. Natl. Acad. Sci. U.S.* 77, 1791-1795.

Eilers, F.I., Ikuma, H. and Sussman, A.S. (1970). Changes in metabolic intermediates during activation of *Neurospora* ascospores. *Can. J. Microbiol.* 12, 1351-1356.

Ennis, H.L. and Sussman, M. (1975). Mutants of *Dictyostelium discoideum* defective in spore germination. *J. Bacteriol.* 124, 62-64.

Feit, I.N., Chu, L.K. and Iverson, R.M. (1971). Appearance of polyribosomes during germination of spores in cellular slime mold *Dictyostelium purpureum*. *Experimental Cell Research* 65, 439-444.

Firtel, R.A. and Lodish, H.F. (1973). A small nuclear precursor of messenger RNA in the cellular slime mold *Dictyostelium discoideum*. *J. Mol. Biol.* 79, 295-314.

Furch, B. (1972). Zur warmeaktivierung der sporen von *Phycomyces blakesleeanus*. Das auftreten von gärungen unter aeroben bedingungen. *Protoplasma* 75, 371-379.

Giri, J.G. and Ennis, H.L. (1977). Protein and RNA synthesis during spore germination in the cellular slime mold, *Dictyostelium discoideum*. *Biochem. Biophys. Res. Commun.* 77, 282-289.

Giri, J.G. and Ennis, H.L. (1978). Developmental changes in RNA and protein synthesis during germination of *Dictyostelium discoideum* spores. *Dev. Biol.* 67, 187-201.

Hashimoto, Y., Tanaka, Y. and Yamada, T. (1976). Spore germination promoter of *Dictyostelium discoideum* excreted by Aerobacter aerogenes. *J. Cell Sci.* 21, 261-271.

Herring, F.G. and Weeks, G. (1979). Analysis of *Dictyostelium discoideum* plasma membrane fluidity by electron spin resonance. *Biochim. Biophys. Acta* 552, 66-77.

Hohl, H.R. (1976). Myxomycetes. *In* "The Fungal Spore Form and Function" (Ed. D. Weber and W. Hess). pp. 463-500. John Wiley and Sons, New York.

Hohl, H.R. and Hamamoto, S.T. (1969). Ultrastructure of spore differentiation in *Dictyostelium:* The prespore vacuole. *J. Ultrastructural Research* 26, 442-453.

Hohl, H.R., Bühlmann, M. and Wehrli, E. (1978). Plasma membrane alterations as a result of heat activation in *Dictyostelium* spores. *Arch. Microbiol.* 116, 239-244.

Jacobson, A., Firtel, R.A. and Lodish, H.F. (1974). Transcription of poly(dt) sequences in the genome of the cellular slime mold, *Dictyostelium discoideum*. *Proc. Natl. Acad. Sci. U.S.* 71, 1607-1611.

Janoff, A.S., Coughlin, R.T., Racine, F.M., McGroarty, E.J. and Vary, J.C. (1979). Use of elctron spin resonance to study *Bacillus megaterium* spore membranes. *Biochem. Biophys. Res. Comm.* 89, 565-570.

Katilus, J. and Ceccarini, C. (1975). Purification and new biological properties of the slime mold germination inhibitor. *Dev. Biol.* 42, 13-18.

Khoury, A.T., Deering, R.A., Levin, G. and Altman, G. (1970). Gamma-ray-induced spore germination of *Dictyostelium discoideum. J. Bacteriol.* 104, 1022-1023.

Kobilinsky, L. and Beattie, D.S. (1977). Respiratory competence of *Dictyostelium discoideum* spores. *J. Bacteriol.* 132, 113-117.

Long, B.H. and Coe, E.L. (1974). Changes in neutral lipid constituents during differentiation of the cellular slime mold, *Dictyostelium discoideum.* J. Biological. Chem. 249, 521-529.

MacLeod, C., Firtel, R.A. and Papkoff, J. (1980). Regulation of actin gene expression during spore germination in *Dictyostelium discoideum. Developmental Biol.* 76, 263-274.

Madelin, M.F. (Ed.) (1966). "The Fungus Spore" Butterworths, London.

Malhotra, S.K. and Tewari, J.P. (1973). Molecular alterations in the plasma membrane of sporangiospores of *Phycomyces* related to germination. *Proc. Roy. Soc. (Lond.) B,* 184, 207-216.

Melchior, D.L. and Steim, J.M. (1976). Thermotropic transitions in biomembranes. *Ann. Rev. Biophys. and Bioeng.* 5, 205-238.

Racine, R.M., Dills, S.S. and Vary, J.C. (1979). Glucose-triggered germination of *Bacillus megaterium* spores. *J. Bacteriol.* 138, 442-445.

Raper, K.B. (1935). *Dictyostelium discoideum,* a new species of slime mold from decaying forest leaves. *J. Agricultural Research* 50, 135-147.

Rossignol, D.P. and Vary, J.C. (1979a). Biochemistry of L-proline-triggered germination of *Bacillus megaterium* spores. *J. Bacteriol.* 138, 431-441.

Rossignol, D.P. and Vary, J.C. (1979b). L-proline site for triggering *Bacillus megaterium* spore germination. *Biochem. Biophys. Res. Comm.* 89, 547-551.

Rudolph, H. and Furch, B. (1970). Untersuchungen zur aktivierung von sporenhomogenaten durch warmebehandlung. *Arch. Mikrobiol.* 72, 175-181.

Singer, S.L. and Nicolson, G.L. (1972). The fluid mosaic model of the structure of cell membranes. *Science* 175, 720-731.

Snyder, H.M. and Ceccarini, C. (1966). Interspecific spore inhibitor in the cellular slime moulds. *Nature* 209, 1152.

Sussman, A.S. (1969). The prevalence and role of dormancy. *In* "The Bacterial Spore" (Ed. G.W. Gould and A. Hurst). pp. 1-38. Academic Press, New York.

Sussman, A.S. (1976). Activators of fungal spore germination. *In* "The Fungal Spore Form and Function" (Ed. D. Weber and W. Hess). pp. 101-139. John Wiley and Sons, New York.

Sussman, A.S. and Halvorson, H.O. (1966). "Spores their Dormancy and Germination" Harper and Row, New York.

Sussman, M.S. and Rayner, E.P. (1971). Physical characterization of deoxyribonucleic acids in *Dictyostelium discoideum.Arch. Biochem. Biophys.* 144, 127-137.

Tanaka, Y.M., Hashimoto, Y., Yanagisawa, K., Abe, H. and Uchiyama, M. (1975). Partial structure of a spore germination inhibitor from a cellular slime mold, *Dictyostelium discoideum. Agr. Biol. Chem.* 39, 1929-1932.

Tanford, C. (1968). Protein denaturation. *Adv. Protein Chem.* 23, 121-282.

Tanford, C. (1978). The hydrophobic effect and the organization of living matter. *Science* 200, 1012-1018.

Thevelein, J.M., Van Assche, J.A., Carlier, A.R. and Heremans, K. (1979). Heat activation of *Phycomyces blakesleeanus* spores: Thermodynamics and effect of alcohols, furfural, and high pressure. *J. Bacteriol.* 139, 478-485.

Tisa, L.S. and Cotter, D.A. (1979). The expression of acid and alkaline phosphatase activities during the germination of *Dictyostelium discoideum* spores. *Microbios Letters* 11, 85-94.

Tisa, L.S. and Cotter, D.A. (1980). Expression of glucosidase activities during germination of *Dictyostelium discoideum* spores. *J. Bacteriol.* 141, 436-442.

Van Assche, J.A., Carlier, A.R. and Dekeersmaeker, H.I. (1972). Trehalase activity in dormant and activated spores of *Phycomyces blakesleeanus*. *Planta* 103, 327-333.

Van Assche, J., Carlier, A.R. and Van Tieghem, L.L.C. (1977). The effect of gamma radiation on breaking of dormancy in *Phycomyces* spores. *Arch. Microbiol.* 113, 95-97.

Van Assche, J.A., Van Laere, A.J. and Carlier, A.R. (1978). Trehalase metabolism in dormant and activated spores of *Phycomyces blakesleeanus*. *Planta* 139, 171-176.

Van Cauwelaert, F.H. and Verbeke, M.N. (1979). Differential scanning calorimetric observations concerning the activation mechanism of spores of *Phycomyces blakesleeanus. Biochem. Biophys. Research Communications* 89, 414-419.

Vanderkooi, G. and Green, D.E. (1971). New insights into biological membrane structure. *Bioscience* 21, 409-415.

Weber, D.J. and Hess, W.M. (ed.) (1976). "The Fungal Spore Form and Function" John Wiley and Sons, New York.

Whittingham, W.F. and Raper, K.B. (1960). Non-viability of stalk cells in *Dictyostelium*. *Proceedings of the National Academy of Science U.S.* 46, 642-649.

Wright, B.E. (1963). Endogenous activity and sporulation in slime molds. *Ann. New York Acad. Sci.* 102, 740-754.

Yagura, T. and Iwabuchi, M. (1976). DNA, RNA, and protein synthesis during germination of spores in the cellular slime mold, *Dictyostelium discoideum*. *Exp. Cell Research* 100, 78-87,

Yasuda-Yasaki, Y., Namiki-Kanie, S. and Hachisuka, Y. (1978). Inhibition of *Bacillus subtilis* spore germination by various hydrophobic compounds: Demonstration of hydrophobic character of the L-alanine receptor site. *J. Bacteriol.* 136, 484-490.

SPORE GERMINATION: HEAT MEDIATED EVENTS

B. Furch

Botanisches Institut der Universität
Biologiezentrum, D-2300 Kiel, (FRG)

INTRODUCTION AND GENERAL CONSIDERATIONS

The much investigated dormancy and its cessation by activation is a widespread phenomenon in the plant kingdom. Intensive research has been carried out on the activation of dormant seeds, but there is considered a great lack of knowledge e.g. Khan (1977).

As for bacterial and fungal spores at least the concept seems to be outlined more clearly: the terms dormancy, activation, and germination have been validly defined for both classes and these definitions are widely accepted: Sussman (1966). Thus we discern between exogen dormancy (further development is delayed by chemical or physical influences of the environment) and constitutive dormancy (caused by internal factors such as self-inhibitors, metabolic blocks, penetration barrieers).

Characteristically constitutive dormancy can be overcome by physical or chemical agents without their being essential for further development. Many activating chemicals can even prevent germination when left to act upon the objects for too long e. g. Halbsguth and Rudolph (1958); Lingappa *et al.* (1970).

To break dormancy, to activate, means to change the cryptobiotic spore into an actively metabolizing one and thus to start the germination. In contrast to the view of Manners (1966), who declares that a spore has germinated only 'when the germ tube is protruded', today most investigators agree to speak of germination when 'the first irreversible stage... is recognizably different from the dormant organism, as judged by physiological or morphological criteria':Sussman (1966).

By investigating germination started off by an activation it is hoped that the generally valid conclusions can be drawn about the basic biochemical and ultrastructural events which control cellular metabolism and differentiation.

This review mainly presents a comprehensive survey of those biochemical and ultrastructural heat mediated events which became known after the Second International Symposium on Fungal Spores in 1974. Hence, in order to understand the nature of the problem and the significance of the data we must not forget the excellent reviews published until then e.g. Madelin (1966); Sussman (1966); Bergman *et al.* (1969); Sussman and Douthit (1973).

Even given these limits, selection is inevitable because of the abundance of material. The author therefore refers mainly to the conditions found in the sporangiospores of *Phycomyces blakesleeanus*.

Summarizing the present state of investigations on the activation of spores we can say that for no object a definite conclusion can be drawn as to how activating stimuli trigger germination in detail e.g. Gould and Hurst (1969); Sussman (1976); Anderson (1978). Lately,however, the first site of activation seems to be within reach owing to newer investigations and reflections mainly made by Cotter (1973); Cotter and George (1975); Sussman (1976); Hohl *et al.* (1978); Van Thevelein *et al.* (1979); Furch (1979). The particular data, however, sometimes are controversially emphasized, e.g. Sussman (1976) versus Cotter (1973).

Neither can the more general question, whether there is one process of activation for all objects be answered up to now. On the one hand, different activating systems are probable because of the great diversity among the fungi; on the other hand, similar activation effects on the sporangiospores of *Phycomyces blakesleeanus*, on the ascospores of *Neurospora tetrasperma*, and of *N. crassa*, and on the basidiospores of *Coprinus radiatus* make it likely that there is 'a common mechanism of activation at least in the spores of these molds': Van Thevelein *et al.* (1979).

When studying activation or germination phenomena it is widely agreed that it is unimportant whether germination is started chemically or physically (heat). Both starter mechanisms apparently can be used for the purpose e.g. Sussman (1976). There is, however, one advantage to heat activation:the starter, risen temperature applied for a short time can easily be switched off by cooling down the spore suspension to normal growth temperature. Acetate on the contrary, and other chemical activators probably too, cannot be totally removed. This could already be shown by Borchert (1962) when he autoradiographically investigated the spores of *Phycomyces blakesleeanus*, which had been activated by acetate. Possibly this mode of activation is irreversible because of the rest amounts of acetate in the spore:even if the medium does not allow germination (bidistilled water) after the spore has been activated, *Phycomyces* remains in the state of activation. If activated by heat, however, activation ceases given the conditions described above. We do not yet know what happens to chemical activators and whether they are metabolized after all. This makes it difficult to analyse the mechanism of activation. Hence it surely is no accident that lately heat activation has been the method of choice.

MORPHOLOGICAL AND ULTRASTRUCTURAL CHANGES

Fig. 1 is the key to understanding of fig. 2 to 4 and thus gives us an idea about the experimental conditions used to investigate heat mediated events in *Phycomyces blakesleeanus*. Time is represented on the abscissa. Note that it is spread in the area of the scanning pattern. The hint 'dry spores' with the arrow pointing at 'air' means that at this particular point the spores are found in dry condition i.e. in their storage form. In experiment they are kept incubated in a dilute suspension (1-10 mg/ml) in a minimal medium to avoid clotting and selfactivation:Furch (1971;1974a). Aerobic conditions are provided for by stirring the suspension. In favour of clearness the preincubation phase always preceding an experiment has not been noted. The first four hours of experiment represent the phase of dormancy=control:at a temperature of 18°C the spores are dormant and show only a minor germination rate of about 3%, the 3 minute period of heat activation at 50°C is shown by the grey scanning pattern in all figures. Already during activation and up to 6 hours afterwards the heat mediated events are investigated by taking regular measurements. The events are represented in their relative intensity. The dotted line means that no effect was measurable at that moment.

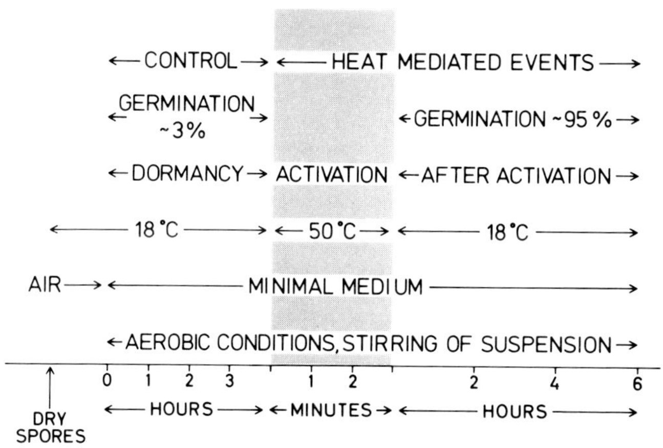

Fig.1. Key of understanding of fig. 2 to 4 and explanation of the experimental conditions used to investigate the heat mediated events in Phycomyces blakesleeanus spores.

Spore shape

Fig. 2 shows dry spores in the lower left corner. Released from the sporangium they have dried out and thus got invaginated because of the shrinking of the protoplast. Because spores, which are dried out to this extent have a very low water potential of -70 to -90 bar they take up water in a split second when suspended: Furch (1978). The dormant spore now forms an ellipsoid of revolution measuring 14 x 7 μm. After activation its halo having been visible in phase contrast disappears, it starts to swell and to round off. Finally after 2 to 4 hours its volume has increased three fold: Galle (1964). Inside the spore the light microscope

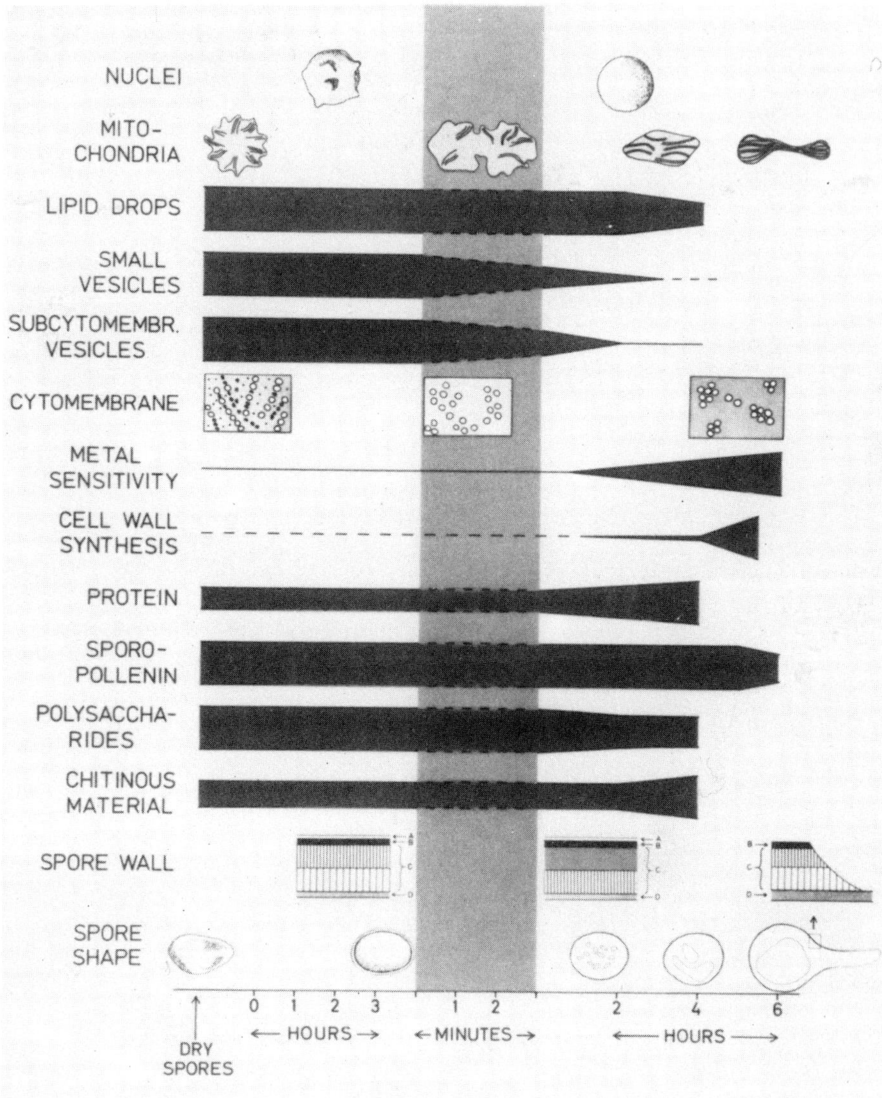

Fig. 2. Morphological, ultrastructural, and related heat mediated events in Phycomyces blakesleeanus spores. For further explanations see text and fig. 1.

shows small vacuoles which finally join to form a central vacuole. About 6 hours after activation the germ tube protrudes: Sommer and Halbsguth (1957).

Spore wall

Hess (1978) writes in a review: most dormant spores with thick walls are impossible to properly fix and embed. Looking at *Phycomyces* and on *Neurospora* with the electron microscope the resolution of wall structures also is partly unsatisfactory and the interpretation of TEM-pictures sometimes is controversial e.g. Byrne (1975) versus Lowry and Sussman (1968). Contrary to the results of Borchert (1962) and Tu and Malhotra (1976), according to Furch and Pambor (1978a,b) the wall of dormant spores like the spore wall of *Neurospora tetrasperma* consists of 5 elements: A=layer of mucus on the perispore; B=perispore; C=epispore consisting of 2 layers which show different contrast (OsO_4) and D=endospore. Shortly after heat activation the wall changes characteristically: the layer of mucus on the perispore disappears, epispore and endospore slacken until finally about 6 hours after activation the endospore alone, which thickened by then, forms the wall of the germ tube.

Chitinous material; polysaccharides; sporopollenin; protein

Already one hour after heat activation the structure of the spore wall changes. The wall's contents of polysaccharides decreases from 50 to 35%, whereas proteins increase from 11 to 22% up to 4 hours after activation. Also the contents of chitinous material doubles from 3 to 6% within 4 hours from activation whereas the contents of sporopollenin within the spore wall only starts to decrease after 4 hours from activation. The wall of the germ tube does not contain sporopollenin: Furch and Gooday (1978).

Cell wall synthesis

The process of wall reorganization also can be proved by increasing activity of the enzymes involved: Cohen (1974); Van Laere *et al.* (1976);

Van Laere and Carlier (1978); Malhotra (1978).

Metal sensitivity

Further on wall reorganization is indicated directly by changes in the permeability for cations. Like the dormant spores of *Neurospora* those of *Phycomyces* are relatively impermeable to cations measured by the spore's susceptibility to poisoning heavy metals like copper, lead and silver: Furch (unpublished). In accordance with these results only the wall of the mycel of *Phycomyces* can be dissolved by unspecific carbohydrases, lipases and proteases; the walls of dormant spores remain untouched by these enzymes: Furch (1974;1979).

Since the biosynthesis of newly built cell walls of fungi has been investigated intensively by various authors (for literature see Farkas (1979)) it seems to be clear that for most fungal spores synthesis of wall material occurs already during the swelling phase of the spores: Smith *et al.* (1976).

Cytomembrane

According to Malhotra and Tewari (1973) and Furch and Pambor (unpublished) by means of freeze fracture and freeze etching in dormant spores of *Phycomyces blakesleeanus* especially in the EF-face 3 different sorts of particles can be shown: homogeneous particles (about 35 nm in diameter, indicated by open circles) compound particles (30 nm, indicated by closed circles), and small particles (5 nm, dots). Already few minutes after heat activation, the small and compound particles cannot be detected any more. The homogeneous particles, which before had a pearl string like distribution in the EF-face now begin to aggregate. Since all three kinds of particles can also be detected in the PF-face, they should be intramembranous particles. Tu and Malhotra (1977) rather think them to be indentions caused by the cell wall. But we never saw any indentions of the cell wall in our strain K1+. Neither the theoretical background of their aggregation, nor their nature and functional aspects of their aggregation are known. It might be caused by a temperature induced phase transition of the phospholipids caused by heat activation. Probably at least part of them are related to controlled dormancy and germination:Malhotra and Tewari (1973); similarly as for *Dictyostelium discoideum*:Hohl *et al.* (1978).

Subcytomembraneous vesicles; small vesicles; lipid drops

Increases and decreases in the degree of vacuolation are commonly observed during the germination of fungal spores. Also density changes of the spore cytosol are reported upon as a result of the beginning germination. Jervis (1978), for *Neurospora crassa*, and Galle (1964) and Van Laere *et al.* (1980), for *Phycomyces blakesleeanus*, think this phenomenon to be the first clearcut morphological indication of the spore having overcome dormancy.

Lipid drops are observed to disappear in activated spores of *Phycomyces* from two hours after activation. It seems, however, to be an important ultrastructural feature that a layer of subcytomembranous vesicles of 10 nm in diameter and, somewhat closer to the inner of the spore, a cluster of small vesicles also disappear during activation:Pambor (unpublished). The nature and function of these vesicles are unknown.

Mitochondria

Large, irregularly lobed mitochondria have been found more frequently in thick-walled dormant spores e.g. Hawker and Madelin (1976). Like the mitochondria in dormant ascospores of *Neurospora tetrasperma* those in *Phycomyces blakesleeanus* are circular in cross section with a regularly lobate circumference:Lowry and Sussman (1968); Malhotra and Tewari (1973); Pambor (1978). The cristae point to the center so that the average of the 20 to 30 cristae of each section appear star-shaped. During heat activation the regular invaginations become smoothed and the mitochondria begin to increase markedly in size. Three hours after activation, they are thin and elongated with cristae now parallel to the outer membrane. because the area of sectioned mitochondria diminishes markedly, and, comparing the number of sectioned cristae,

Pambor (1978) thinks the mitochondria divide by fractionation (division rate of 1 in dormant spores to 4 in spores three hours after activation). Shortly before the germ tube is protruded, they appear roundly to elliptically shaped in section. The membranes, having shown negative contrast before, now show normal contrast. The multiplication of mitochondria due to heat activation coincides with the findings of Halbsguth and Rudolph (1958), who observed a steady increase of spores' oxygen consumption up to 6 hours after activation.

Nuclei

A clearly documented event during spore germination is the rise in nuclear volume of heat activated spores e.g. Lowry and Sussman (1968) for *Neurospora*. This is also the case with *Phycomyces* spores. Soon after activation the lobate nuclei of the dormant spores round up and swell. Nuclear division occurs shortly prior to germ tube protrusion.

On the whole the findings on heat mediated morphological and ultrastructural events show that the spores react very quickly, often already when the treatment commences. How far these partly very drastic changes at the organelle level directly correlate with the trigger mechanism cannot be definitely decided. They may simply be a consequence of germination induction.

PHYSIOLOGICAL CHANGES

The heat mediated events have been especially intensively studied in biochemical experiments with *Phycomyces blakesleeanus* spores, see fig 3 and 4. Gas exchange measurements of Halbsguth and Rudolph (1958) show the strongly increasing output of CO_2 already during heat treatment. It leads to a short-termed RQ of 2. Partly this is due to a push-up of fermentation processes, which is also shown in a raise in activity of catabolic enzymes: Furch (1972). The activity of key enzymes of glycolysis (HK, GAPDH, PFK, and PK) is increased, this even leads to an increased throughput in this sequence in cell free extracts of spores: Rudolph and Furch (1970); Rudolph (1972); Furch (1979).

Especially drastic is the raise in trehalase activity which occurs in two different forms: Van Assche *et al.* (1972); Van Assche and Carlier (1975); see fig. 5. The cyclic event of activation and deactivation shows, as with trehalase, also in intact spores: Halbsguth and Rudolph (1958); see fig. 6. Since in ascospores of *Neurospora tetrasperma*, which also can be heat activated, storage carbohydrate breakdown is confined by this same enzyme trehalase, and thus the spores are kept dormant: Hill and Sussman (1964), a similar mechanism might

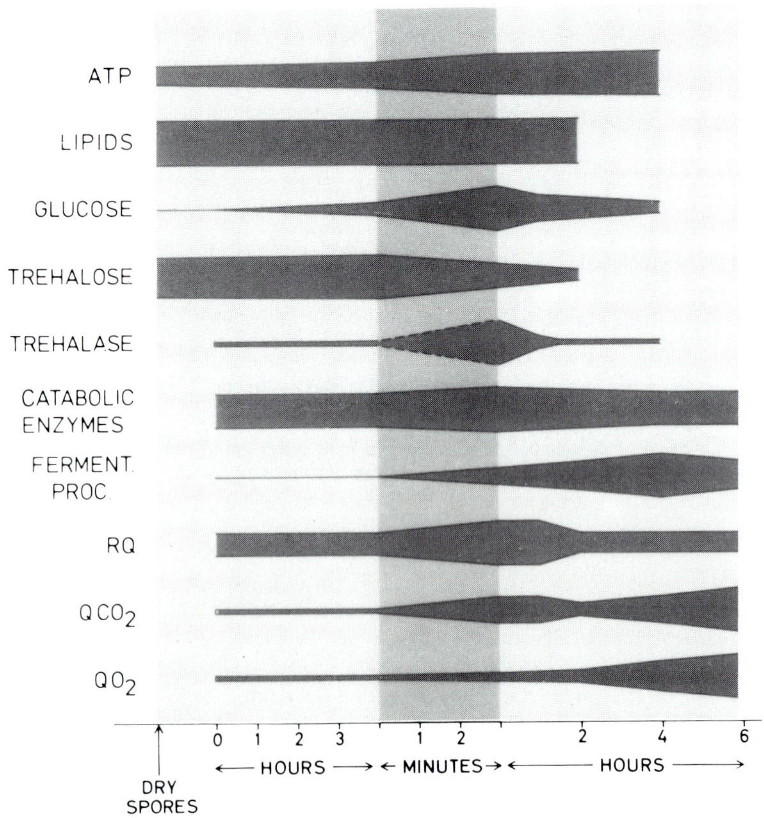

Fig. 3. Physiological heat mediated events in Phycomyces blakesleeanus spores. For further explanations see text and fig. 1.

be assumed for sporangiospores of *Phycomyces blakesleeanus*. Although the storage carbohydrate of *Phycomyces*, trehalose, is broken down to glucose after heat activation, the fact that glucose, applied to dormant spores is taken up and even leads to a tenfold increase in respiration, however, does *not* activate the spores is not in support of trehalase playing a key role in breaking dormancy: Rudolph and Ochsen (1969). Also do gamma-rays

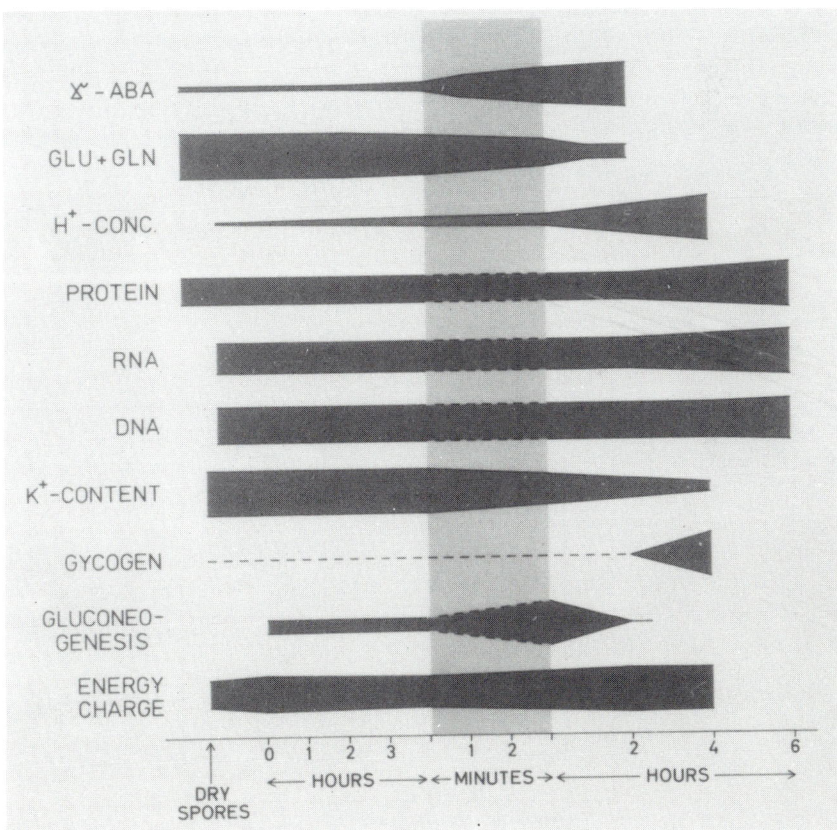

Fig. 4. Physiological heat mediated events in Phycomyces blakesleeanus spores. For further explanations see text and fig. 1.

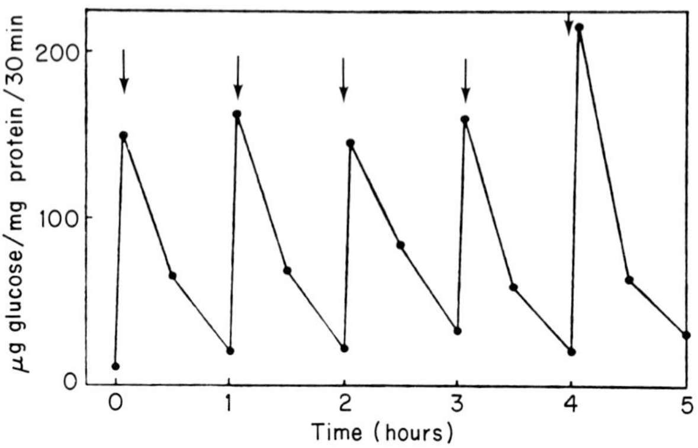

Fig.5. Increase and decay in trehalase activity in spores of Phycomyces blakesleeanus (strain K1+) following repeated treatment at 50°C for 3 minutes (arrows) interrupted by periods of 1 hour at 18°C (treatment for the deactivation of spores). From Van Assche et al. (1972).

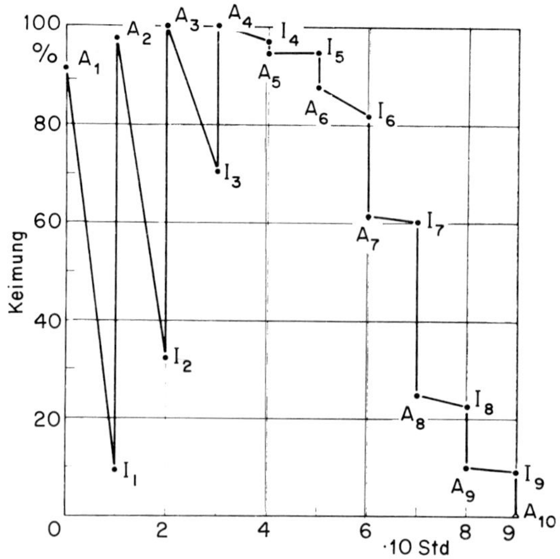

Fig.6. Repeated activation and deactivation of spores of Phycomyces blakesleeanus (strain K1+). A_1-A_{10}=activation; I_1-I_{10}=deactivation. From Halbsguth and Rudolph (1958).

(300 krad), acid (0,1 m HCl), or reducing substances ($Na_2S_2O_4$, $Na_2S_2O_5$, and $NaHSO_3$) lead to a percentage of germination similar to that of heat activation, *without* trehalase activity increasing: Van Assche *et al.* (1977;1978).

Lipids are not used to gain energy in activated spores, the rise in energy charge from 0.5 in dormant spores to 0.8 in activated ones should thus be an effect of the breakdown of storage carbohydrates and the amino acid metabolism mainly of glutamine and glutamate:Furch (1974;1979); Furch *et al.* (1976).

It might be important in understanding the heat activation mechanism that Tu and Malhotra (1973) could show that adenyl cyclase is part of the membrane system of dormant and activated spores. According to the authors it is conceivable that its activity is mediated by heat, too.

At present it is not clear, why the spores exsude K^+ ions into the suspension medium when activated. By measurements with an energy dispersive X-ray SEM system in combination with laser microsonde mass analysis it could be demonstrated that the efflux comes from theinside of the spore: Furch and Pambor (unpublished). K^+ is known to be an activating cation: Pambor (1978b).

Further elaborate studies of the protein metabolism, gluconeogenesis, and glycolysis did not indicate further important heat mediated events in *Phycomyces blakesleeanus*: Van Assche and Carlier (1973); Van Laere and Carlier (1975); Van Assche *et al.* (1972); see fig. 4. According to Van Assche *et al.* (1977) and Van Assche and Carlier (1973), gene activation as trigger of germination can be discarded. A blockage of protein biosynthesis by inhibitors, and by damage of the genetic material by radiation could not render heat activation impossible.

CONCLUDING REMARKS AND CONSIDERATIONS ABOUT THE NATURE OF THE TRIGGER MECHANISM(S) OF HEAT MEDIATED INDUCTION OF GERMINATION

There are three fungal objects mainly, the spores of *Dictyostelium*, *Neurospora*, and *Phycomyces*, which have been used to study the mode of action of the cessation of constitutive dormancy by heat activation. Especially with Cotter's conception on the thermodynamics of reversible activation, the causal analysis of germination induction by heat and certain chemicals has progressed, such that there could several models of germination initiation be put forward by different authors.

Considering thermodynamic aspects, Cotter (1973) and Cotter and George (1975) conclude that for *Dictyostelium* heat activation, as well as treatment with activating agents such as urea and DMSO, denaturate or otherwise change a regulative or control-protein in the phosphorylation complex of terminal oxidation in dormant spores. Thus full phosphorylation capacity is achieved, which in turn makes germination possible. From the observation that a supra-optimal heating of spores destroys the inner mitochondrial membrane they conclude that a protein of the respiratory chain be the first target site of activation. Van Thevelein *et al.* (1979) and Van Cauwelaert and Verbeke (1979) extend Cotter's hypothesis on *Phycomyces* in that they suggest that activation of spores is achieved by a conformational change of a protein, which may be hydrophobic in nature. Their concept is supported by high pressure experiments and by the mode of effect of certain alcohols in connection with heat. They argue that according to Heremans (1978) the effect of high pressure on membrane phase transitions of about 20 k/1000 bar is quite different from the effect of pressure on protein conformation changes of 2 to 6 K/1000 bar. Obtaining a value of 5 K/1000 bar, they can discard the possibility of membrane phase transitions being the first target site of activation.

Moreover calculations of free energies of adsorption of various alcohols seem to indicate a protein to be the first site of activation.

This hypothesis is also supported by differential scanning calorimetric observations of the activation mechanism by Cauwelaert and Verbeke (1979). From their scanning spectra, which show that activation procedure influence the subsequent protein denaturation pattern, they conclude a protein change to be involved in the activation mechanism. Their arguments are underlined in that they can show that the appearance of the dormancy pattern and the activation pattern of their scanning lines correlates with the cyclic phenomenon of activation and deactivation of the spores.

None of the authors, however, could detect the proteins position in the spore. Studies of Hohl *et al.* (1978) on *Dictyostelium* give some clue in that they can show that at the end of the activation process the plasma membrane particles seen on the EF and PF faces display a drastically changed distribution pattern as compared to the dormant state. As reported, Furch and Pambor (unpublished) as well as Malhotra and Tewari (1973) obtain very similar pictures, with *Phycomyces* under the same conditions. Thus this object, too, gives evidence for the plasma membrane being involved in the activation process.

Sussman (1976) proposes phospholipids to be involved in germination initiation. The activating treatment may induce phase transitions of membrane lipids which leads to a liberation of membrane bound enzymes (integral or peripheric) which begin to operate and effect a higher catabolic rate. He points out that, contrary to Cotter's opinion, the temperatures applied for activation lie within the rate in which phase transitions of some common membrane lipids from gel-like to liquid take place. Catabolic enzymes could thus directly be activated by a phase transition of the phospholipids. In further support of this hypothesis are the findings of Machtinger and Fox (1973) who demonstrated that many enzymes indeed require lipids to become active. According to Sussman (1976) and Anderson (1978), the activating stimulus could furthermore effect a change in membrane permeability. Thus a separation of one or more starter enzymes from its/their substrate(s) and/or their effectors might be overcome, thus allowing for contact of low molecular compounds with katabolic enzymes. Mandels and Maguire (1972) consider such a mechanism of activation for the spores of *Myrothecium verrucaria*, which are susceptible to heat activation. Hecker and Sussman (1973) have strong evidence for a participation of phospholipids in the activation of *Neurospora* spores. To summarize their conclusions: 'heat activation could be accomplished through a change in the permeability of the membrane separating the cytoplasm from the mural compartment containing trehalase, thereby allowing access of substrate to the enzyme'. Such considerations are made by others, too. Hohl *et al.*

(1978), measuring efflux rates of ^{14}C-labelled compounds, could show that only the activated spores of *Dictyostelium* were permeable for these substances. Furch (1979) presents a great number of leakage measurements.Various substances, mainly amino acids, but also sugars, alcohols and acids are shown to leak out of activated spores only. It remains to be shown from which compartment these substances derive (subcytomembraneous vesicles? small vesicles? vacuoles like those in *Saccharomyces cerevisiae*:Wiemken and Dürr (1974)?)

Should the leakage of substances out of their compartments be of certain importance in starting the activation of *Phycomyces* spores, then glutamine, which is stored in great quantities of up to 2% (w/w) in dormant spores, might act as one starter substrate, for it can yield large amounts of reduction equivalents and energy immediately if metabolized via gamma-aminobutyric acid shunt. Moreover it occupies a central position in many different metabolic pathways.

At present it seems impossible to decide which of the hypotheses on the mode of action of spore activation is to prefer. However, the findings are similar with different objects studied. Thus the trigger mechanism could also be similar to one another. May be one should not insist on one hypothesis only. At present we tend to think along with Hohl *et al.* (1978) that 'the heat treatment could effect the properties of the lipid phase or the configuration of the membrane proteins, or a combination of both processes might be responsible for the morphological changes observed'.

ACKNOWLEDGMENTS. Thanks to Dr. Leonie Pambor who provided unpublished material and to the Deutsche Forschungsgemeinschaft supporting the investigations.

REFERENCES

Anderson, J.G. (1978). Temperature-induced fungal development. *In* "The Filamentous Fungi" (Ed. J.E. Smith and D.R. Berry). pp. 358-375. Edward Arnold, London.

Bergman, K., Burke, P.V., Cerdá-Olmedo, E., David, C.N., Delbrück, M., Foster, K.W., Goodell, E.W., Heisenberg, M., Meissner, G., Zalokar, M., Dennison, D.S. and Shropshire, W. (1969). *Phycomyces*. *Bacteriological Reviews* 33, 99-157.

Borchert, R. (1962). Über die Azetat-Aktivierung der Sporangiosporen von *Phycomyces blakesleeanus*. *Beiträge zur Biologie der Pflanzen* 38, 31-61.

Byrne, B.J. (1975). Freeze-etch and thin section studies of *Neurospora crassa* ascospores. *Protoplasma* 84, 357-377.

Cohen, R.J. (1974). Cyclic AMP levels in *Phycomyces blakesleeanus*. *Journal of Biological Chemistry* 249, 1973-1979.

Cotter, D.A. (1973). Spore germination in *Dictyostelium discoideum*. *Journal of Theoretical Biology* 41, 41-51.

Cotter, D.A. and George, R.P. (1975). Germination and mitochondrial damage in spores of *Dictyostelium discoideum* following supraoptimal heating. *Archives of Microbiology* 103, 163-168.

Farkas, V. (1979). Biosynthesis of cell walls of fungi. *Microbiological Reviews* 43, 117-144.

Furch, B. (1971). Über die Beeinflussung des Ruhestoffwechsels der Sporen von *Phycomyces blakesleeanus*. *Archiv für Mikrobiologie* 80, 293-299.

Furch, B. (1972). Zur Wärmeaktivierung der Sporen von *Phycomyces blakesleeanus*. *Protoplasma* 75, 371-379.

Furch, B. (1974a). Über die Funktion des Vorlaufes bei Untersuchung des Phänomens der Wärmeaktivierung von *Phycomyces blakesleeanus*. *Biochemie und Physiologie der Pflanzen* 166, 363-365.

Furch, B. (1974b). Changes in the free nucleotide pattern of spores from *Phycomyces* in relation to heat-induced germination. *Archives of Microbiology* 98, 77-84.

Furch, B., Poltz, J. and Rudolph, H. (1976). Lipids of spores from *Phycomyces* in relation to heat-induced germination. *Biochemie und Physiologie der Pflanzen* 169, 249-256.

Furch, B. (1978). On the water potential, osmotical pressure and water uptake of dormant sporangiospores of *Phycomyces blakesleeanus*. *Zeitschrift für Pflanzenphysiologie* 88, 269-272.

Furch, B. (1979). Untersuchungen zur Keimruhe und deren Aufhebung durch eine Wärmebehandlung bei den Sporangiosporen von *Phycomyces blakesleeanus* BURGEFF, Stamm K1+ (Sammlung HALBSGUTH). *Habilitationsschrift, Botanisches Institut der Universität Kiel*, 1-51.

Furch, B. and Gooday, G.W. (1978). Sporopollenin in *Phycomyces blakesleeanus*. *Transactions of the British Mycological Society* 70, 307-309.

Furch, B. and Pambor, L. (1978a). Cell wall constituents of *Phycomyces blakesleeanus* 2. *Microbios Letters* 4, 211-219.

Furch, B. and Pambor, L. (1978b). Cell wall constituents of *Phycomyces blakesleeanus* 3. *Microbios Letters* 8, 71-80.

Galle, H. (1964). Untersuchungen über die Entwicklung von *Phycomyces blakesleeanus* unter Anwendung des Mikrozeitrafferfilmes. *Protoplasma* 59, 423-471.

Gould, G.W. and Hurst, A. (1969). "The Bacterial Spore". Academic Press, New York and London.

Halbsguth, W. and Rudolph, H. (1958). Untersuchungen über die Wärmeaktivierung der Sporangiosporen von *Phycomyces blakesleeanus*. *Archiv für Mikrobiologie* 32, 296-308.

Hawker, L.E. and Madelin, M.F. (1976). The dormant spore. *In* " The Fungal Spore" (Ed. D. Weber and W.M. Hess). pp. 1-72. John Wiley and Sons, New York, London, Sydney, Toronto.

Hecker, L.I. and Sussman A.S. (1973). Localization of trehalase in the ascospores of *Neurospora*. *Journal of Bacteriology* 115, 592-599.

Heremans, K. (1978). Pressure effects on biochemical systems. *In* "High Pressure Chemistry" (Ed. H. Kelm). pp. 467-487. Reidel, Dordrecht.

Hess, W.M. (1978). Ultrastructure of dried spores. *In* "Dry Biological Systems" (Ed. J.H. Crowe and J.S. Clegg). pp. 75-96. Academic Press, New York and London.

Hill, E.P. and Sussman, A.S. (1964). Development of trehalase and invertase activity in *Neurospora*. *Journal of Bacteriology* 88, 1556-1566.

Hohl, H.R., Bühlmann, M. and Wehrli, E. (1978). Plasma membrane alterations as a result of heat activation in *Dictyostelium* spores. *Archives of Microbiology* 116, 239-244.

Jervis, H.H. (1978). Isopycnic centrifugation of *Neurospora crassa* conidia. *Experimental Mycology* 2, 295-298.

Khan, A.A. (1977). "The Physiology and Biochemistry of Seed Dormancy and Germination". North-Holland Publishing Company, Amsterdam.

Lingappa, B.T., Lingappa, Y. and Turian, G. (1970). Phenethyl alcohol induced germination of ascospores of *Neurospora*. *Archiv für Mikrobiologie* 72, 97-105.

Lowry, R.J. and Sussman, A.S. (1968). Ultrastructural changes during germination of ascospores of *Neurospora tetrasperma*. *Journal of General Microbiology* 51, 403-409.

Machtinger, N.A. and Fox, C.F. (1973). Biochemistry of bacterial membranes. *Annual Review of Biochemistry* 42, 575-600.

Madelin, M.F. (1966). "The Fungus Spore". Colston Papers 18. Butterworths, London.

Malhotra, S.K. (1978). Molecular structure of biological membranes: functional characterization. *Subcellular Biochemistry* 5, 221-259.

Malhotra, S.K. and Tewari, J.P. (1973). Molecular alterations in the plasma membrane of sporangiospores of *Phycomyces* related to germination. *Proceedings of the Royal Society of London, B.* 184, 207-216.

Mandels, G.R. and Maguire, A. (1972). Endogenous metabolism of fungal spores. Stimulation by physical and chemical means. *Plant Physiology* 50, 525-531.

Manners, J.G. (1966). Assessment of germination. *In* "The Fungus Spore" (Ed. M.F. Madelin). Colston Papers 18, pp. 165-174. Butterworths, London.

Pambor, L. (1978a). Development of mitochondria in dormant and germinating sporangiospores of *Phycomyces blakesleeanus*. *Microbios Letters* 8, 143-153.

Pambor, L. (1978b). Ionenabhängige Keimungsaktivierung der Sporangiosporen von *Phycomyces blakesleeanus* und *Phycomyces nitens*, demonstriert an verschiedenen Stämmen. *Archives of Microbiology* 117, 35-40.

Rudolph, H. and Ochsen, B. (1969). Trehalose-Umsatz wärmeaktivierter Sporen von *Phycomyces blakesleeanus*. *Archiv für Mikrobiologie* 65, 163-171.

Rudolph, H. and Furch, B. (1970). Untersuchungen zur Aktivierung von Sporenhomogenaten durch Wärmebehandlung. *Archiv für Mikrobiologie* 72, 175-181.

Rudolph, H. (1972). Über die Wirkung von D-Glycerat-3-phosphat und D-Glycerat-2-phosphat-Zusätzen auf Sporenhomogenate von *Phycomyces blakesleeanus*. *Zeitschrift für Pflanzenphysiologie* 67, 212-222.

Smith, J.E., Gull, K., Anderson, J.G. and Deans, S. G. (1976). Organelle changes during fungal spore germination. *In* "The fungal Spore" (Ed. D.J. Weber and W.M. Hess).pp. 301-354. John Wiley and Sons, New York, London, Sydney, Toronto.

Sommer, L. and Halbsguth, W. (1957). Grundlegende Versuche zur Keimungsphysiologie von Pilzsporen. *Forschungsberichte des Wirtschafts- und Verkehrsministeriums von Nordrhein-Westfalen* 411, 1-90.

Sussman, A.S. (1966). Dormancy and spore germination. *In* "The Fungi" (Ed. G.C. Ainsworth and A.S. Sussman) Vol. II, pp. 733-764. Academic Press, New York and London.

Sussman, A.S. and Douthit, H.A. (1973). Dormancy in microbial spores. *Annual Review of Plant Physiology* 24, 311-352.

Sussman, A.S. (1976). Activators of fungal spore germination. *In* "The fungal Spore" (Ed. D.J. Weber and W.M. Hess). pp. 101-139. John Wiley and Sons, New York, London, Sydney, Toronto.

Tu, J.C. and Malhotra, S.K. (1973). Histochemical localization of adenyl cyclase in the fungus *Phycomyces blakesleeanus*. *Journal of Histochemistry and Cytochemistry* 21, 1041-1046.

Tu, J.C. and Malhotra, S.K. (1976). The formation of sporangiospores in *Phycomyces*. *Microbios* 15, 15-25.

Tu, J.C. and Malhotra, S.K. (1977). Plasma membrane of *Phycomyces*: sequential changes in the structure during spore formation. *Cytobios* 20, 121-132.

Van Assche, J.A., Carlier, A.R. and Dekeersmaker, H.I. (1972). Trehalase activity in dormant and activated spores of *Phycomyces blakesleeanus*. *Archives of Microbiology* 93, 129-136.

Van Assche, J.A. and Carlier, A.R. (1973). The pattern of protein and nucleic acid synthesis in germinating spores of *Phycomyces blakesleeanus*. *Archives of Microbiology* 93, 129-136.

Van Assche, J.A. and Carlier, A.R. (1975). Some properties of trehalase of *Phycomyces blakesleeanus*. *Biochimica Biophysica Acta* 391, 154-161.

Van Assche, J.A., Carlier, A.R. and Van Tiegham, L. C. (1977). The effect of gamma radiation of breaking dormancy of *Phycomyces* spores. *Archives of Microbiology* 113, 95-97.

Van Assche, J.A., Van Laere, A.J. and Carlier, A.R. (1978). Trehalose metabolism in dormant and activated spores of *Phycomyces blakesleeanus* BURGEFF. *Planta* 139, 171-176.

Van Cauwelaert, F.H. and Verbeke, M.N. (1979). Differential scanning calorimetric observations concerning the activation mechanism of spores of *Phycomyces blakesleeanus*. *Biochemical and Biophysical Research Communications* 89, 414-419.

Van Laere, A.J. and Carlier, A.R. (1975). Glucose metabolism of dormant and heat-activated spores of *Phycomyces blakesleeanus* BURGEFF. *Planta* 125, 217-225.

Van Laere, A.J., Carlier, A.R. and Van Assche, J.A. (1976). Effect of 5-fluorouracil and cycloheximid on the early development of *Phycomyces blakesleeanus* spores and the activity of N-Acetylglucosamine synthesizing enzymes. *Archives of Microbiology* 108, 113-116.

Van Laere, A.J. and Carlier, A.R. (1978). Synthesis and proteolytic activation of chitin synthetase in *Phycomyces blakesleeanus* BURGEFF. *Archives of Microbiology* 116, 181-184.

Van Laere, A.J., Van Assche, J.A. and Carlier, A.R. (1980). Density changes of *Phycomyces* spores after reversible and irreversible activation. *Archives of Microbiology* 124, 289-291.

Van Thevelein, J.M., Van Assche, J.A., Carlier, A. R. and Heremans, K. (1979). Heat activation of *Phycomyces blakesleeanus* spores: thermodynamics and effect of alcohols, furfural, and high pressure. *Journal of Bacteriology* 139, 478-485.

Wiemken, A. and Dürr, M. (1974). Characterization of amino acid pools in the vacuolar compartments of *Saccharomyces cerevisiae*. *Archives of Microbiology* 101, 45-57.

LIGHT AND MYCOSPORINES

N. Arpin and M.L. Bouillant

Département de Biologie Végétale, Laboratoire de Mycochimie (LA 44), Université Claude Bernard, 43 Bd du 11 nov. 1918. 69622 Villeurbanne-F.

INTRODUCTION

By its profound effects on the growth regulation (stimulation, inhibition or orientation) and on development process (sexual, asexual reproduction and sclerotium formation) the light affects decisively the life of fungi. Notice here that often fungi need light and darkness during their several stages of development. Two main orientations of research have been developed; they first concern the nature of photoreceptor and secondly the primary and secondary effects of light absorption. The important literature dealing with fungal responses has been reviewed several times and the reader is suggested to refer to the recent reviews of Durand (1976) and Tan (1978).

From the point of view of action spectra, three kinds of photoresponses are observable in fungi: photoresponses only to UV light, to UV and blue light and to wavelengths higher than blue (Durand, 1976; Tan, 1978). Moreover, reversible photoreactions have been observed; sporulation photoinduced by near-ultraviolet (NUV) can be reversed by blue light exposure. A mycochrome system mediates photooxidation reduction reactions (see review articles of Tan, 1978; Kumagai, 1978, 1979). Recently, Valadon *et al.* (1979) assert that a phytochrome system was involved in *Verticillium agaricinum* carotenogenesis.

Biochemical events occurring after photoinduction first deal with modifications of photoreceptors and then with metabolite synthesis: nucleotides, nucleic acids, proteins, enzymes and several other products (see Tan, 1978).

Since Tan's review, new results have been reported; they concern: regulation of morphogenesis (Moore-Landecker, 1979a, b), regulation of growth (Lynch and Geoghegan, 1979; Saltarelli and Coppola, 1979), cell division and amino acid transport (Edmunds *et al.*,1979), photostimulated or inhibited synthesis (Haggblöm and Unestam, 1979; Lynch and Geoghegan 1979; Saltarelli and Coppola, 1979; Osman and Valadon, 1978) and photoregulation of asparagin metabolism (Jerebzoff-Quintin and Jerebzoff, 1980).

Among several problems, due to the light effects, the effects on sporulation have received great attention. Leach (1961, 1962a and b, Leach and Trione 1965) carefully studied the sporulation of diverse species of fungi under ultra-violet radiations; action spectra for sporulation of *Ascochyta pisi, Alternaria dauci* and *Pleospora herbarum* exhibit a maximum of effectiveness at about 290 nm. Recently, very similar action spectra for photoinduced conidiation in *Helminthosporium oryzae* (Yamamura *et al.*, 1977) and in *Botrytis cinerea* (Honda and Yunoki, 1978) were obtained showing that basically the same receptor is involved in all these fungi.

Exploring the possible mechanism of this ultraviolet photosporogenesis, Leach (1964, 1965) observed that photoinduction of species strictly photodependent for conidiation (*Alternaria chrysanthemi, Ascochyta pisi, Pleospora herbarum*) or for sexual reproduction (*Ophiobolus graminis* and *Pyronema omphalodes*) is accompanied by synthesis of compounds responsible for increased absorption at wavelengths between 280 and 350 nm, first called P310 (Leach, 1965) and now mycosporines (Favre-Bonvin *et al.*, 1976).

Structures, distribution, photoinductibility and sporogenic activity of these compounds will constitute the main paragraphs of this chapter.

STRUCTURES

Mycosporines sensu stricto.

Some physical and chemical characteristics of P310 have been described (Trione and Leach, 1966, 1969) but the first chemical structure of a compound of this family: mycosporine-1 (fig.1) from *Stereum hirsutum* (Arpin *et al.*, 1977a) was only elucidated four years ago, in our laboratory (Favre-Bonvin *et al.* 1976).

Fractionation of the aqueous extracts of the frozen carpophores of this Basidiomycete on repeated ion exchange columns provided a pure, colorless, amorphous substance, very soluble in water and showing a strong absorption max. at 310 nm. As the first described P310, this substance is relatively stable at high temperature and in acidic or alkaline solutions. Spectral studies in 1-H and 13-C NMR and MS of the natural and derived compounds have shown that myco-

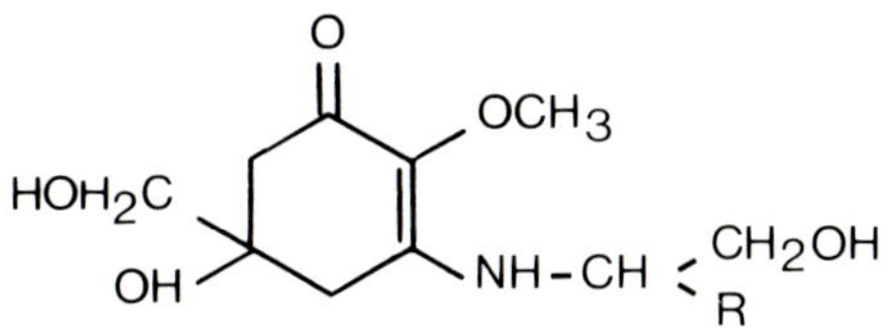

Mycosporine 1, R = CH_2OH *Stereum hirsutum*, Carp., Arpin *et al.* 1977a Favre-Bonvin *et al.* 1976.

Mycosporine 3 = mycosporine-serine, R = COOH, *Stereum hirsutum*, Carp., Arpin *et al.* 1979.

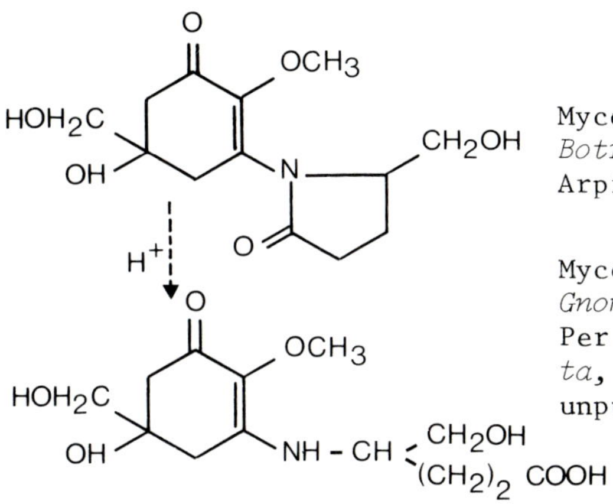

Mycosporine 2 ring form, *Botrytis cinerea*, Coni., Arpin *et al.* 1977b.

Mycosporine 2 open form, *Gnomonia leptostyla*, Coni. Per. and *Morchella esculenta*, Ap., Fayret *et al.* unpublished.

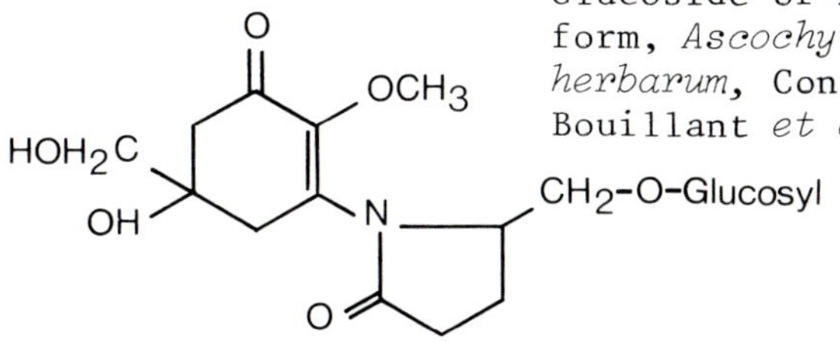

Glucoside of Mycosporine 2, ring form, *Ascochyta pisi*, *Cladosporium herbarum*, Coni., Leach 1965, Bouillant *et al.* unpublished.

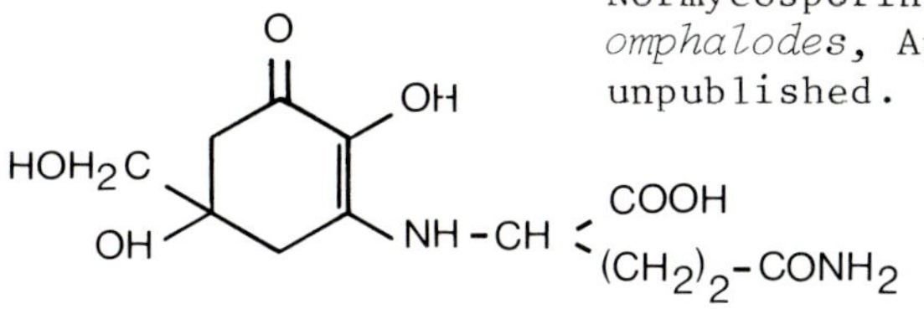

Normycosporine-glutamine, *Pyronema omphalodes*, Ap., Lunel *et al.* unpublished.

Carp.= Carpophores, Coni.= Conidiospores, Per.= Perithecia Ap.= Apothecia.

Fig. 1 - Mycosporines: Structures, occurrence, localisation and references.

sporine 1 was characterized by: a typical cyclohexenone ring bound to reduced serine moitey.

Later, this characteristic carbon skeleton has been found again in other fungi, but with different amino substituents.

In *Stereum hirsutum*, mixed with mycosporine 1 (95% of the 310 absorbance), another compound exists: mycosporine 3 (5%), in which the amino moiety is serine. This substance is analogous to mycosporine-glycine, isolated from the zoanthid *Palythoa tuberculosa*. Both these products are remarkably susceptible to hydrolysis, typical behaviour of mycosporines with a free amino acid substituent.

Arpin *et al.* (1977b) have isolated and elucidated the structure of mycosporine 2 from *Botrytis cinerea*. In this compound the amino part was a derivative of glutamic acid, in which the acid function was reduced. The mycosporine 2, that seems the most widespread in the fungal field, exists in two forms: a ring form, in which a dehydratation occurs between the δ acid group and the amino group. This structure is the

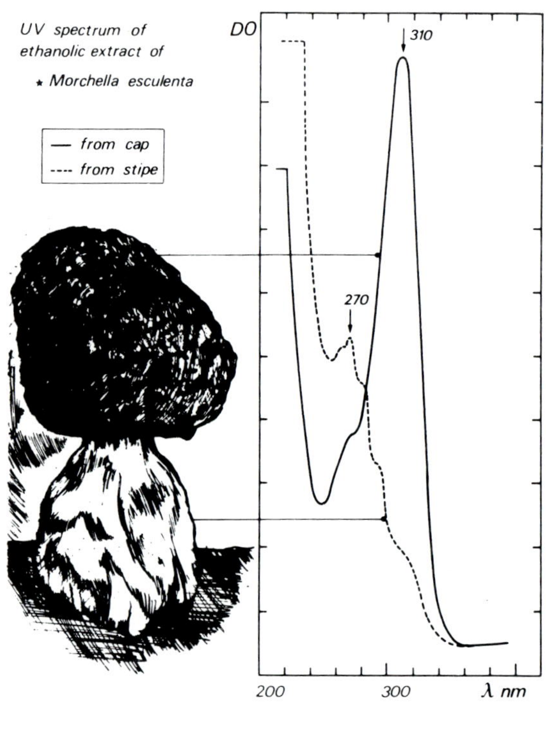

Fig. 2. UV spectra of ethanolic extracts from Morchella esculenta.

main, *in vivo*, in *Botrytis cinerea*, but, in acidic medium or with peculiar cultural conditions, the ring can be hydrated leading to an open form. The isolation of this form, after cationic ion exchange chromatography, can be artefactual; nevertheless, studying the natural mycosporines of *Gnomonia leptostyla* (Fayret *et al.*, unpublished), its co-occurrence *in vivo* with ring form was demonstrated. Besides in our laboratory, we recently observed that open form of mycosporine 2 represents the main mycosporine from *Morchella esculenta* receptacle (Fig. 2).

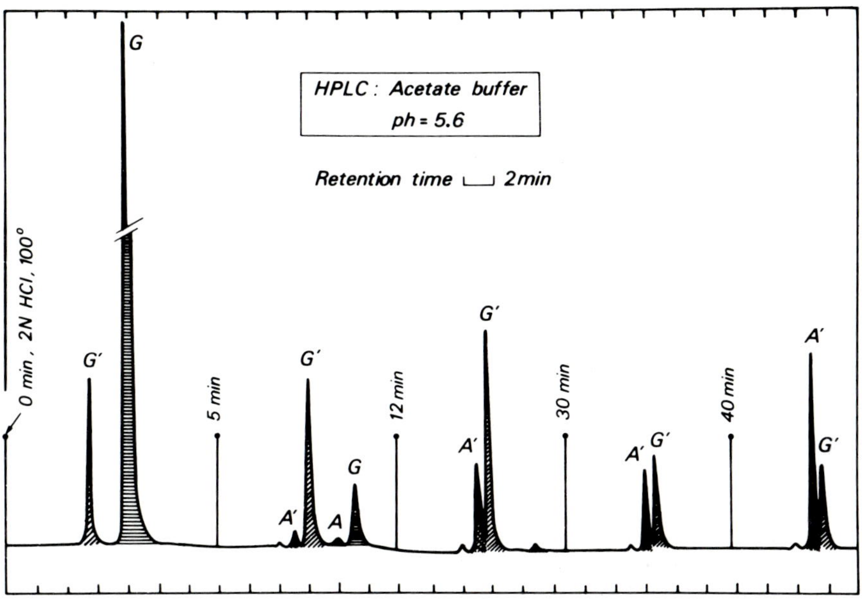

Fig. 3. Acid hydrolysis of glucosides of mycosporine 2, analysed by HPLC; G and G': glucosides of ring and open forms of mycosporine 2; A and A': aglycones, ring and open forms.

Here it is good to talk about HPLC, that is a new very important technology to study mycosporines. Using bounded phase (microbondapack C18) eluted with pH 5.6 acetate buffer, native compound can be observed: ring and open mycosporines 2 can well be distinguished (Fig. 3, A and A').

Lately, in our laboratory (unpublished), the structure of P310 substance isolated by Leach from *Ascochyta pisi* have been elucidated and, at the same time, the same compound was isolated from the conidiospores of *Cladosporium herbarum*.

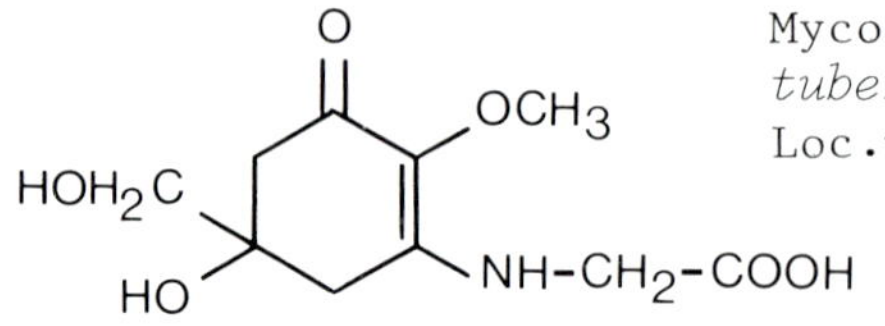

Mycosporine-glycine, *Palythoa tuberculosa* § Ito et Hirata 1977, Loc.n.s.

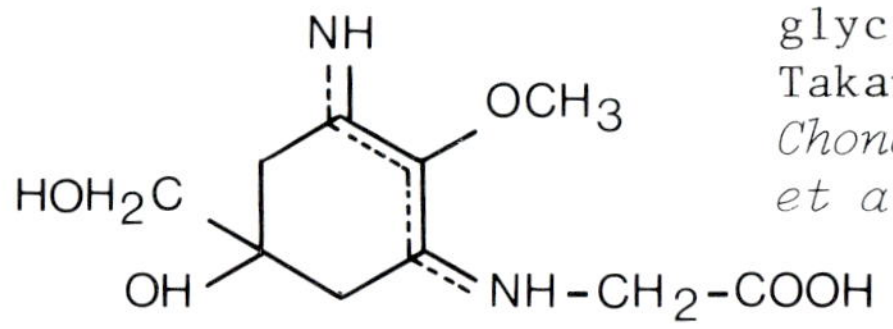

Palythine = Iminomycosporine-glycine, *Palithoa tuberculosa* § Takano *et al.* 1978a, Loc.n.s. *Chondrus yendoi* §§ Tsujino *et al* 1978. Loc.n.s.

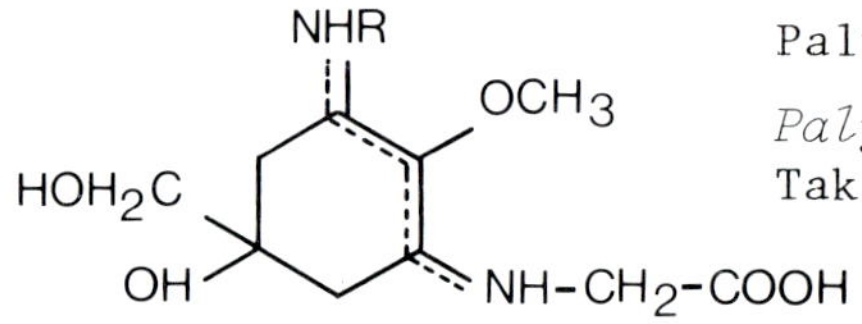

Palythinol: R = $CH(CH_3)CH_2OH$

Palythene: R = $CH = CH - CH_3$

Palythoa tuberculosa § Takano *et al.* 1978b, Loc.n.s.

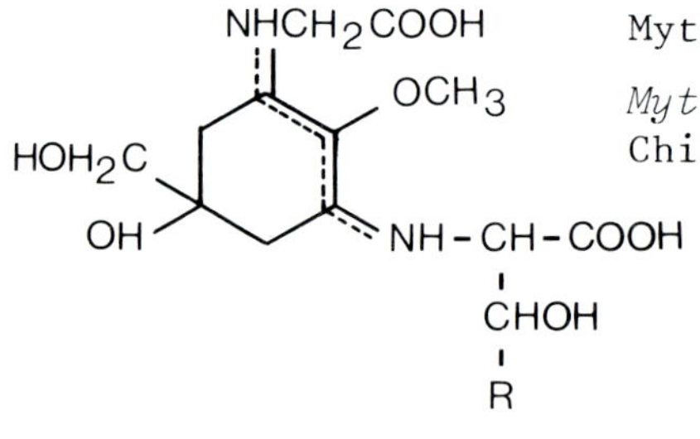

Mytilins A (R=H), B (R=CH_3)

Mytilus galloprovincialis §§§ Chioccara *et al.* 1979, Loc.n.s.

Loc.n.s. = Localisation no specified
§ Coelenterata. -§§ Algae. -§§§ Mollusca.

Fig. 1 - Continued.

Today, we can say that the main compound of *Ascochyta pisi* is the glucoside of mycosporine 2 (sugar bound to the CH_2OH corresponding to reduced α acid function).

As the aglycone, the glucoside can occur in two forms *in vivo*, but chiefly in ring form, as observed in *Cl. herbarum*. After chromatographic fractionations, Leach had obtained 3 P310 substances from *A. pisi*, but, in our work, the HPLC with the afore-mentioned conditions only shows the glucoside ring form. Therefore, it can be concluded that on ion exchange columns (acidic medium) artefacts were obtained: the glucoside could be opened and (or) partially hydrolysed. The results of an *in vitro* acid hydrolysis of the mycosporine 2 glucosides from *Cl. herbarum* observed in HPLC, are summarized in fig. 3.

In fungi, others mycosporines (P310), different from the already known structures have been detected in UV and HPLC screenings from aqueous extracts of several species: their chemical structures are under investigation.

Normycosporine.

In his previous work, Leach related the presence in the Ascomycete *Pyronema omphalodes* of P310 substances. Howewer, looking at the spectrum of the extracts of this fungus, another λ max must be read: 320 nm. In our laboratory (Lunel *et al.* unpublished), this strong absorption at 320 nm substance was found again and isolated from sporulating mycelia of *P. omphalodes*; its structure was elucidated in spite of hard difficulties in the process of its purification.

In this compound (fig. 1), both cyclohexenone and amino moiety are different in reference to mycosporines *sensu stricto*: a hydroxyl group supplies the methoxyl ("nor" prefix), and the amino part is glutamine. Both these peculiarities give to this substance, named "normycosporine-glutamine", a great lability. The free amino acid glutamine has been identified from its easy hydrolysis by a mild acid treatment.

A probable relationship between mycosporine *sensu stricto* and normycosporine could be expected, and, it is tempting to imagine a metabolic relationship between normycosporine and 5-dehydroquinic acid (a metabolite of the shikimate pathway).

Iminomycosporines.

Although not concerning the fungal field, we summarize here the work about mycosporine-like compounds, isolated from marine plants (Algae) and animals (Coelenterae, mollusca) (see structures and references in fig. 1).

Among these compounds, one has a mycosporine *sensu stricto* structure: mycosporine-glycine. All the others are substances in which the CO group is substitute by an imino group. Works on that object were mainly structural investigations and their role and biogenesis *in vivo* are still unknown. It is probable that they must be different in Algae and in animals, in which mycosporines could be ingestion substances, metabolized by a digestion system for the removal from their organism.

DISTRIBUTION OF MYCOSPORINES IN FUNGI.

Zygomycetes

Circinella (Mucor) minor	Arpin *et al.* unpublished
Cunninghamella elegans	" " "

Deuteromycetes

Alternaria alternata	Arpin *et al.*, unpublished
Alternaria chrysanthemi	Leach 1965;Trione *et al.*1966
Alternaria dauci	" " " " "
Ascochyta pisi	Leach 1965;
" "	Bouillant *et al.*, unpublish.
Aureobasidium pullulans	Arpin *et al.*, unpublished
Beauveria bassiana	" " "
Botryotrichum piluliferum(?)	" " "
Botrytis allii	" " "
Botrytis cinerea	Hite 1973;Tan and Epton 1974
" "	Arpin *et al.*, 1977b
Calcarisporium arbuscula (?)	Arpin *et al.*, unpublished
Cladosporium cladosporioides	" " "
Cladosporium herbarum	Bouillant *et al.*, unpublish.
Cladosporium macrocarpum	Arpin *et al.*, unpublished
Cladosporium sphaerospermum(?)	" " "
Colletotrichum dermatium (?)	" " "
Colletotrichum nasae(?)	" " "
Dactylaria purpurella	" " "
Epicoccum purpurescens(?)	" " "
Helminthosporium halodes	" " "
Helminthosporium dictyoides	Vargas and Wilcoxon, 1969
Monilia fructicola	Trione *et al.*, 1966
Stemphylium sarcinaeforme	" " "
Phoma herbarum(var.*medicaginis*)	" " "
Tubercularia vulgaris	Arpin *et al.*, unpublished

Ascomycetes	
Aleuria aurantia	Arpin *et al.*, 1979
Chaetomium globosum(?)	Arpin *et al.*, unpublished
Chaetomium indicum(?)	" " "
Chaetomium seminudum(?)	" " "
Helvella leucomelas	" " "
Leptosphaerulina australis	Trione *et al.*, 1966
Leptosphaerulina trifolii	" " "
Leptosphaerulina briosiana	Moyer and Leath, 1976
Morchella esculenta	Arpin *et al.*, unpublished
Mycosphaerella pinodes	Trione *et al.*, 1966
Nectria cinnabarina	Arpin *et al.*, 1979
Nectria galligena	Dehorter, 1976
Ophiobolus graminis	Leach, 1965
Plectania coccinea	Arpin *et al.*, 1979
Pleospora herbarum	Leach, 1965
Sclerotinia fructicola	Ende V.Den and Cornelis 1970
Scutellinia scutellata	Arpin *et al.*, 1979
Trichoscypha calycina	" " "

Basidiomycetes	
Auricularia mesenterica	Arpin *et al.*, 1979
Byssomerulius corium	" " "
Calocera viscosa	" " "
Cerrena unicolor	" " "
Chondrostereum purpureum	" " "
Eichleriella spinulosa	" " "
Gloeophyllum odoratum	" " "
Gloeophyllum sepiarium	" " "
Hypholoma fasciculare(C)	" " "
Lopharia spadicea	" " "
Panaeolus foenisecii	" " "
Peniophora incarnata	" " "
Peniophora lycii	" " "
Peniophora quercina	" " "
Phanerochaete roungerii	" " "
Pholiota aurivella(C)	" " "
Schizophora paradoxa	" " "
Steccherinum ochraceum	" " "
Stereum gaussapatum	" " "
Stereum hirsutum	Favre-Bonvin *et al.*, 1976
" "	Arpin *et al.*, 1977a
Stereum rugosum	Arpin *et al.*, 1979
Stereum sanguinolentum	" " "
Stereum subtomentosum	" " "
Stereum sulphuratum	" " "

Tilletia controversa	Trione *et al.*, 1966
Tremella mesenterica	Arpin *et al.*, 1979
Vuilleminia comedens	" " "

(*?*) in UV spectra, a shoulder at 310-320 nm
(*C*) from cultures

PHOTOINDUCIBILITY OF MYCOSPORINES

In his first investigation concerning P310 occurrence, Leach (1965) noticed that these water soluble compounds were present in extracts obtained from irradiated sporulating colonies of several fungi, but not detectable or in very low amount in the aqueous extracts from dark grown non sporulating cultures.

Leach (1965) also showed that amount of P310 produced by colonies of *Ascochyta pisi*, which at first were grown 6 days in darkness, depended on the length of exposure: the rate production increased on linear way between 6 and 22 hours of irradiation; in fact Leach demonstrated that light was needed solely to initiate the process: once stimulated - about 6 hrs of NUV (Leach, 1965) or even less (15 mn, Trione and Leach, 1969) - the production of P310 continued, in dark as in light at a steadily increasing rate. In this case, there was a parallel between P310 synthesis and pycnidial formation.

From one isolate of *A. pisi*, n°274, able to sporulate profusely in darkness on special medium, Leach (1965) observed a great amount of mycosporine from both exposed and unexposed colonies.

The relationships between light and P310 synthesis were later on studied by several researchers:

Van den Ende and Cornelis (1970) observed that production of P310 was much higher in light grown *Sclerotinia fructicola* cultures than in darkness, and this production increased with the light intensity.

Kuss (1970) marked that absorbance in the 290-320 nm region increased in cultures of *Stemphylium botryosum* after light exposure, and P310 accumulated in dark after photoinduction.

Hite (1973), Tan and Epton (1974) noticed a dramatic increase of P310 level with time of irradiation from *Botrytis cinerea* cultures.

Moyer and Leath (1976) were able to extract P310 compounds from two isolates of *Leptosphaerulina briosiana*; isolate 650 produced ascostromata in darkness or in light with a P310 level of 1.5 to 2 times higher in light than in dark; isolate 931 required NUV light for spore formation and produced stromata without asci or ascospores in dark: with light treatment P310 synthesis was 2 to 3 times higher. In any case P310 compounds were much more abundant from ascostromata than from mycelial mats.

Dehorter (1976) pointed out from *Nectria galligena* that both formation of ascospores and of mycosporine required similar light conditions (wavelengths shorter than 600 nm); if sexual reproduction was also dependent upon other external factors, mycosporine synthesis only needed light (this fact may be brought together with Trione and Leach's observations (1969) that photoinduction could take place at 0°C). P310 synthesis began after 12 hours of light exposure, increased with irradiation and, as already afore mentioned, continued in darkness.

Recently, Fayret and Arpin (unpublished) observed that, in *Gnomonia leptostyla*, both mycelia producing conidia and perithecia elaborate the same mycosporine (mycosporine 2, ring form *vide supra* fig 1) but in higher amount with conidia than

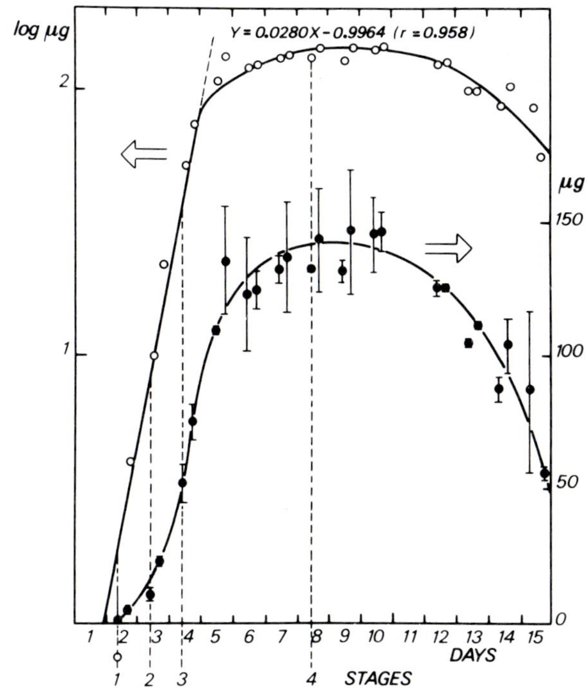

Fig. 4. Normycosporine-glutamine synthesis by sporulating mycelia of Pyronema omphalodes.

with perithecia (0.6% of dry weight *versus* 0.2%).

From grown light cultures (12 hrs light/ 12 hrs dark) of *Pyronema omphalodes*, Lunel (unpublished) has followed the normycosporine-glutamine synthesis (*alias* P320, the main mycosporine of this species). As showed in Fig. 4, synthesis begins after 36-48 hrs, increases exponentialy during development process - sexual organs then apothecia formation - reaches the highest level on the seventh day as asci become dehiscent, then decreases as spores discharge.

P320 amount synthesized, at the highest level, has been estimated at about 150 μg per petri dish (Φ: 100 mm) (*ca.* 1.2% of dry weight).

From all these observations, it appears that if often mycosporines need light for synthesis, in some cases, their synthesis can also take place in darkness, merely when sporogenesis occurs without light. Does the medium composition interfer in mycosporine production? Bouillant and Arpin have

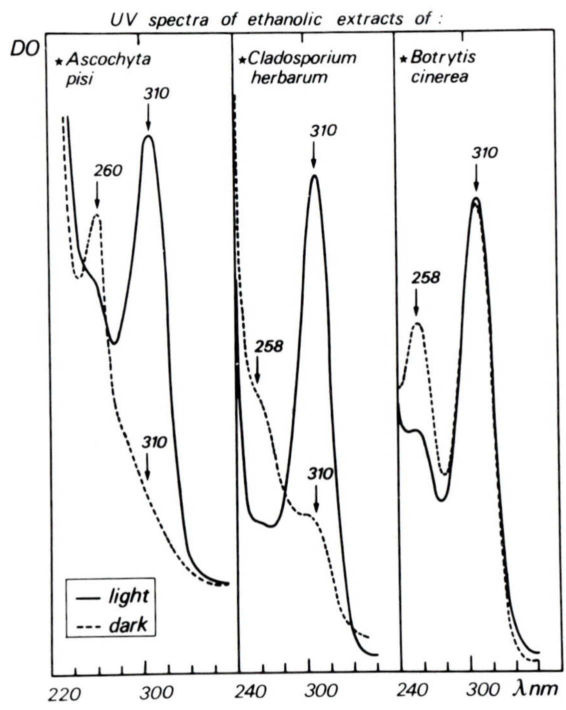

Fig. 5. Comparative spectra from dark and light cultures of A. pisi, Cl. herbarum *and* B. cinerea *grown on different C sources: glucose, saccharose and glycerol respectively.*

TABLE 1

Variations in mycosporine content of Botrytis cinerea *and* Cladosporium herbarum *conidiospores in function of culture conditions (light/dark; different carbone sources).*[1]

			B. cinerea		*Cl. herbarum*	
			% myc./ dry w.	%L/%D [2]	% myc./ dry w.	%L/%D [2]
Malt	8g/1	L	2.3	1.8	3.6	3.0
		D	1.3		1.2	
	15g/1	L	2.6	2.2	3.1	3.4
		D	1.2		0.9	
	30g/1	L	5.1	1.9	2.7	2.0
		D	2.7		1.3	
Glucose	10g/1	L	3.3	3.0	2.8	5.6
		D	1.1		0.5	
	40g/1	L	1.8	1.6	1.7	2.4
		D	1.1		0.7	
Glycerol	10g/1	L	2.5	1.6	0.7	14
		D	1.6		0.05	
	50g/1	L	2.0	1.0	0.4	13
		D	2.0		0.03	
Saccharose	5g/1	L	1.6	1.3	2.5	5
		D	1.2		0.5	
	15g/1	L	1.8	1.2	2.1	3
		D	1.5		0.7	
	30g/1	D	0.7	1	2.7	3
		L	0.7		0.9	

1 - Mineral medium: NH_4NO_3: 1.8g/1; $Ca(NO_3)_2$,4 H_2O: 0.6g/1; KNO_3: 0.2g/1; KH_2PO_4: 0.6g/1; $(NH_4)_2SO_4$: 0.12g/1; $MgSO_4$, 7 H_2O: 0.5g/1 + oligoelements; agar: 1.5%; pH:6.7; Temperature: 20 ± 1°C.; L: 12 hrs light (2500 ergs.cm^2/s) 12 hrs dark; D: continuous dark.

2 - % mycosporine:dry weight in light/ % mycosporine:dry weight in dark.

studied the variations in mycosporine content in *Botrytis cinerea* and *Cladosporium herbarum* conidiospores as a function of culture conditions: these fungi were grown on several carbon sources, in light or in darkness. The results (Tab.1) show that...a) Mycosporine content of conidiospores varies in an important way according to the C sources, for exemple from 0.75 to 5.1% of dry weight in *B. cinerea* conidiospores, in the light, with saccharose 30 g/l and malt 30 g/l...b) in any case, the mycosporine percentage in respect to the dry weight of spores is higher in light than in darkness: the percentage ratio is between 1 to 2 with *B. cinerea*, always superior to 2 with *Cladosporium herbarum* where it reaches 14 with glycerol medium (Fig. 5). In this fig. we also can see a strong absorption at 258 nm in extract of *B. cinerea* grown in darkness on saccharose medium; this feature is still more evident in dark *A. pisi* extracts. So we believe that P258 compounds, at least partly, can be related with mycosporine products and now we attempt to verify this hypothesis. Moreover, we think that it will be very instructive to look again for mycosporine content as a function of nitrogenated sources: this work is now undertaken.

So, presently, we conclude that mycosporines cannot fundamentally be considered as photoproducts on account of the fact that in some cases (isòlate 274 of *A. pisi; B. cinerea* and *Cl. herbarum*) they occur in complete darkness if given the right nutritional conditions. However, light always favours their synthesis and each time when photosporogenesis is strictly needed, mycosporines only occur after light exposure. So mycosporines could be regarded as metabolites closely associated with the reproduction process that, very often, needs light.

SPOROGENESIS AND MYCOSPORINES

Relationships between sporogenesis and mycosporines synthesis

As already mentioned above, Leach (1965) observed that P310 compounds were abundant in sporulating and absent in non sporulating mycelia of *A. pisi*; he was able, too, to characterize P310 in conidiospores of this species. With the exception of Van den Ende and Cornelis (1970) who found mycosporine in non sporulating mycelia (and not in sporulating) all other researchers substantiate Leach's results: P310 synthesis in irradiated colonies of *Stemphylium botryosum* well correlated with sporulation for various nitrogenated sources (Kuss, 1970); Tan and Epton (1974) also found a good correlation between the two processes, but with old irradiated cultures which produce few spores according to amount of P310 production. In our laboratory, Arpin *et al.* (1979) established a linear relationship between sporulation and P310 synthesis during *B. cinerea* conidiogenesis on a definite medium. We found that, in this species and in *Cl. herbarum* too, all mycosporine was localized on spores; likewise, we were able to characterize mycosporines from *Stereum hirsutum* basidiospores (Arpin *et al.*, 1977a), and from *Plectania coccinea* ascospores (unpublished data). Fig. 2 shows that mycosporine (here mycosporine 2, open form) is only produced in the receptacle of *Morchella esculenta*; the little shoulder at 310 nm observed with the extract from the stipe may be explained by delivered ascospores. Finally, in *Pyronema omphalodes* cultures, Lunel (unpublished) found P320 inside apothecia; a very few amount only is detected in mycelium. However, when ascospores are freed they only contain P310 minor compounds; so, we cannot decide now where precisely P320 is, but, in any case, this normycosporine-glutamine, very labile, is quickly hydrolysed when spores become free.

So there is no doubt that mycosporines are metabolites closely associated with sporulation *sensu lato* process and localized inside reproduction organs. This does not necessarily mean that their synthesis occurs inside spores: mycosporines could be synthesized inside mycelia and conidiophores and, given their low molecular weight and water solubility, they could be easily transportable inside spores where they accumulate. In support of that, we have noticed (Bouillant and Arpin, unpublished) that when *B. cinerea* was grown on malt-acetate mixture (8:2 g/l), conidiogenesis frequently aborted: above conidiophores apex there were big droplets containing, *inter alia*, the same mycosporine than in conidiospores. Another excretion process has been also observed with *Byssomerulius corium* (Basidiomycete) cultures by Lunel (unpublished): under certain culture conditions, she observed drops of excretion liquid above mycelium hyphae, in which mycosporine 1, the same than the one inside basidiospores,was dissolved. Is it an excretion process related with attempts of differentiation? We are now trying to clarify the significance of these phenomenons.

Sporogenic activity of mycosporines

The first tests to detect P310 activity were still undertaken by Leach (1965) with semipurified water soluble fractions of extracts from irradiated and non irradiated colonies of *A. pisi* (therefore with and without P310). These tests were inconclusive because sporulation was induced in both cases and even with distillated water. In 1966, Trione *et al.* succeeded to induce sexual (perithecia of *Pleospora herbarum*) and asexual reproduction (conidiospores of *P. herbarum* and of *A. pisi*) with P310B and C extracted from *A. pisi*: so, these compounds have been considered as sporogens. Later, several authors questioned this biological activity: neither Vargas and Wilcoxon (1969) with *Helminthosporium dictyoides*, nor Van den Ende and Cornelis (1970) with *Sclerotinia fructicola*, nor Hite (1973), Tan and Epton (1974) with *B. cinerea*, nor Moyer and Leath (1976) with *Leptosphaerulina briosiana* were able to confirm P310 as sporogens.

However Dehorter (1976) showed that *Nectria galligena* P310 (mycosporine 1) was photomimetic since its adding to the culture medium induced fertile perithecia in this species in continuous darkness: the amount of fructifications depended on P310 concentration until optimum; beyond, he observed a partial inhibition.

Recently, Dehorter *et al.* (1980) have established an interesting relationship between nature and content of free sterols and different degrees of sexual morphogenesis induced either by light or by mycosporine addition. As showed in table 2, P310 addition to the dark grown cultures induces a drastic decrease of sterols content to the same level as in light. This result is of a great significance, for it is the first biochemical event known after P310 addition : in this case, mycosporine seems to be a biochemical intermediate between light and sexual morphogenesis.

TABLE 2

Free sterols content (in mg per 100 g of lyophilised mycelium) isolated from Nectria galligena *in diverse culture conditions (from Dehorter* et al., *1980).*

Cultural conditions	7d. D	7d. L	13d. D	13d. L	7d.+ 6d. D L	13d. (D + P310)
Sterols	210	130	270	140	80	80

L = light (12 hrs light + 12 hrs dark a day)
D = dark; d. = days

To act upon sterol content and composition P310 added has necessarily to penetrate from medium into mycelium; Pittet (unpublished work in our laboratory) has just showed that it is the case with *A. pisi*; in this fungus added P310 (here glucoside of mycosporine 2, ring form, *vide supra* Fig. 1) really begins to penetrate the seventh day; this fact could explain, at least in part, the inconclusive reports afore mentioned.

SUMMARY AND CONCLUDING REMARKS

In 1965 Leach asserted that "the next important step in elucidating the possible role of P310 in the light mechanism would seem to be chemical identification of this substance". This step is now over, but still much work remains to be done about mycosporines especially about their biogenesis, their position in metabolism, their accurate localisation and their biological and physiological role.

Nevertheless, we have improved our knowledge about these compounds on the following points:

a) *their structure*: these small, hydrophilic and nitrogenated molecules characterized by intense and unique UV absorption (310, 320..nm) arise from condensation between a cyclohexenone ring and a nitrogenated residue: amino acid or reduced amino acid. Very often, in fungi, the COOH group of amino acid is reduced in primary alcohol; significance of this reduction still remains unknown.

b) *their distribution*: the mycosporine synthesis probably corresponds to a widespread metabolism because they are produced by very different fungi; in fact, they occur in Zygomycetes, Ascomycetes, Basidiomycetes and Deuteromycetes. Their detection in a fungus can often escape the researcher on account of low quantities or also because they are linked to a short period of fungus life.

c) *their localisation*: mycosporines are very often linked to reproduction process *sensu lato*, on a level of organs producing either sexual spores or conidiospores.

d) *their photoinductibility*: mycosporines cannot be fundamentally considered as photoproducts; in some cases they are synthesized in complete darkness, but always in lower amounts than in light. However, when sporogenesis absolutely needs light, mycosporines are also absent in darkness.

e) *their biological activity*: "sporogenicity" of mycosporines remains questioned; few new experiments have showed that they can penetrate into mycelium and strongly modify the sterol metabolism.

REFERENCES

Arpin, N., Thivend, S. and Favre-Bonvin, J. (1977a). Substances photoabsorbantes de *Stereum hirsutum* (Willd ex Fr.)Fr. Isolement, purification et caractéristiques de P310 (mycosporine 1), *Bull. Soc. Myc. Fr.* 93, 39-52.

Arpin, N., Favre-Bonvin, J. and Thivend, S. (1977b). Structure de la mycosporine 2, nouvelle molécule isolée de *Botrytis cinerea, Tetrahedron Letters* (10), 819-20.

Arpin, N., Curt, R. and Favre-Bonvin, J. (1979). Mycosporines Mise au point et données nouvelles concernant leurs structures, leur distribution, leur localisation et leur biogenèse, *Revue de Mycologie* 43, 247-257.

Chioccara, F., Misuraca, G., Novellino, E. and Prota, G. (1979). Occurrence of two new mycosporine-like amino-acids mytilins A and B in the edible mussel, *Mytilus galloprovincialis, Tetrahedron Letters* (34), 3181-2.

Dehorter, B. (1976). Induction des perithèces de *Nectria galligena* Bres. par un photocomposé mycélien absorbant à 310 nm, *Can. J. Bot.*, 54, 600-4.

Dehorter, B., Brocquet, M.F., Lacoste, L., Alais, J., Lablache-Combier, A., Maquestiau, A., Van Harverbeke, Y., Flammang, R. and Mispreuve, H. (1980). Influence de la lumière et de la mycosporine sur la composition stérolique du *Nectria galligena* Bres., au cours de la morphogenèse sexuée, *Phytochemistry*, in press.

Durand, R. (1976). Influence des radiations lumineuses sur les processus de reproduction des champignons; hypothèses sur l'identité des photorécepteurs. Revue bibliographique, *Mycopathologia*, 60, 3-16.

Edmunds, L.N., Apter, R.I., Rosenthal, P.J., Shen, W-K. and Woodward, J.R. (1979). Light effects in yeast: persisting oscillations in cell division activity and amino acid transport in cultures of *Saccharomyces cerevisiae* entrained by light-dark cycles, *Photochemistry and Photobiology*, 30, 595-601.

Ende, G. van den and Cornelis, J.J. (1970). The induction of sporulation in *Sclerotinia fructicola* and some other fungi and the production of P310, *Neth. J. Plant Pathol.* 76, 183-91.

Favre-Bonvin, J., Arpin, N. and Brevard, C. (1976). Structure de la mycosporine (P310), *Can. J. of Chem.* 54, 1105-13.

Häggblom, P. and Unestam, T. (1979). Blue light inhibits mycotoxin production and increases total lipids and pigmentation in *Alternaria alternata, Applied and Environnemental Microbiology*, 38, 1074-77.

Hite, R.E. (1973). Substances from *Botrytis cinerea* associated with sporulation and exposure to near-ultraviolet radiation, *Plant Dis. Rept.*, 57, 760-4.

Honda, Y. and Yunoki, T. (1978). Action spectrum for photosporogenesis in *Botrytis cinerea* Pers. ex Fr., *Plant Physiol.*, 61, 711-3.

Ito, S. and Hirata, Y. (1977). Isolation and structure of a mycosporine from the zoanthid *Palythoa tuberculosa*, *Tetrahedron Letters*, (28), 2429-30.

Jerebzoff-Quintin, S. and Jerebzoff, S. (1980). Metabolisme de l'asparagine et rythme endogène de sporulation chez le *Leptosphaeria michotii* (West) Sacc., *Physiol. Veg.*, 18, 147-56.

Kumagai, T. (1978). Mycochrome system and conidial development in certain fungi imperfecti, *Phytochemistry and Photobiology*, 27, 371-9.

Kumagai, T. (1979). Near ultraviolet light-induced absorbance changes and mycochrome system in the Fungus *Alternaria tomato*, *Agric. Biol. Chem.* 43, 1733-7.

Kuss, F.R. (1970). Light regulated morphogenesis of *Stemphylium botryosum*, Ph. D. Thesis, Cornell University, USA.

Leach, C.M. (1961). The sporulation of *Helminthosporium orizae* as infected by exposure to near UV radiation and dark periods, *Canad. J. Bot.*, 39, 705-15.

Leach, C.M. (1962a).Sporulation of divers species of fungi under near UV radiation, *Can. J. Bot.*, 40, 151-61.

Leach, C.M. (1962b). The quantitative and qualitative relationships of UV and visible radiation to the induction of reproduction in *Ascochyta pisi*, *Can. J. Bot.*, 40, 1577-1602.

Leach, C.M. (1964). The relationship of visible and ultraviolet light to sporulation of *Alternaria chrysanthemi*, *Transactions of the British Mycological Society*, 47, 153-8

Leach, C.M. (1965). UV absorbing substances associated with light induced sporulation in fungi, *Can. J. Bot.*, 43, 187-200.

Leach, C.M. and Trione, E.J. (1965). An action spectrum for light induced sporulation in the fungus *Ascochyta pisi*, *Plant. Physiol.*, 40, 808-11.

Lynch, F.J. and Geoghegan, M.J. (1979). Regulation of growth and cercosporin photoinduction in *Cercospora beticola*, *Trans. Br. Mycol. Soc.*, 73, (2), 311-27.

Moore-Landecker, E. (1979a). Effect of light regimens and intensities on morphogenesis of the discomycete *Pyronema domesticum*, *Mycologia*, 71, 699-712.

Moore-Landecker, E. (1979b). Effect of cultural age and a single photoperiod on morphogenesis of the discomycete: *Pyronema domesticum*, *Can. J. Bot.*, 57, 1541-9.

Moyer, B.G. and Leath, K.T. (1976). Sporogens sterols and sporulation of *Leptosphaerulina briosiana*, *Can. J. Bot.*

54, 1839-45.

Osman, M. and Valadon, L.R.G. (1978). Studies on the near UV effect on carotenogenesis in *Verticillium agaricinum*, *Microbios*, 23, 53-64.

Saltarelli, C.G. and Coppola, C.L. (1979). Effect of light on growth and metabolite synthesis in *Candida albicans*, *Mycologia*, 71, 773-85.

Takano, S., Uemura, D. and Hirata, Y. (1978a). Isolation and structure of a new amino acid, palythine, from the zoanthid *Palythoa tuberculosa*, *Tetrahedron Letters*, (26), 2299-300.

Takano, S., Uemura, D. and Hirata, Y. (1978b). Isolation and structure of two new amino acids, palythinol and palythene from the zoanthid *Palythoa tuberculosa*, *Tetrahedron Letters*, (49), 4902-12.

Tan, K.K. (1978). Light induced fungal development, *in* "The Filamentous Fungi" (Eds. J.E. Smith and D.R. Bercy). Vol. III Developmental Mycology, pp. 334-57. Edward Arnold, London.

Tan, K.K. and Epton, H.A.S. (1974). Ultraviolet absorbing compounds associated with sporulation in *Botrytis cinerea*, *Trans. Br. Mycol. Soc.*, 63, 157-67.

Trione, E.J., Leach, C.M. and Mutch, J.T. (1966). Sporogenic substances isolated from fungi, *Nature*, 212, 163-4.

Trione, E.J. and Leach, C.M. (1969). Light induced sporulation and sporogenic substances in fungi, *Phytopathol.*, 59, 1077-83.

Tsujino, I., Yabe, K., Sekikawa, I. and Hamanaka, N. (1978). Isolation and structure of a mycosporine from the red alga *Chondrus yendoi*, *Tetrahedron Letters*, (16), 1401-2.

Valadon, L.R.G., Osman, M. and Mummery, R.S. (1979). Phytochrome mediated carotenoid synthesis in the fungus *Verticillium agaricinum*, *Photochemistry and Photobiology*, 29, 605-7.

Vargas, J.M. and Wilcoxon, R.D. (1969). Some effects of temperature and radiation on sporulation by *Helminthosporium dictyoïdes* in agar media, *Phytopathol.*, 59, 1706-12.

Yamamura, S., Kumagai, T. and Oda, Y. (1977). An action spectrum for photoinduced conidiation in *Helminthosporium orizae*, *Plant and Cell. Physiol.*, 18, 1163-66.

VII. METABOLIC CONTROLS

BIOCHEMICAL EVENTS AND CONTROLS: DISCUSSANT'S INTRODUCTION

S. Bartnicki-Garcia

Department of Plant Pathology
University of California
Riverside, CA 92521

Some of the most interesting aspects of fungal morphogenesis revolve about spores: their formation and their germination. What are the key biochemical events and controls in these processes?

Despite considerable progress since the first spore symposium (Madelin, 1966), as evidenced by the papers presented at the second symposium (Weber and Hess, 1976) and herein, a satisfying understanding of the biochemistry of sporogenesis and spore germination is not yet in sight and will probably have to await a major breakthrough in developmental biology. Perhaps we need new concepts and techniques that will allow us to comprehend the complex interactions among the thousands of different chemical components present in a living cell. We can only marvel at the intricate network of temporal and spatial relationships that must exist in the microcosmos of a living cell. During morphogenesis, the transition from one stage in the life cycle to the next, we can expect a precise readjustment of the exceedingly intricate biochemical machinery, so that a cell with a different appearance and properties is produced.

In the last 3 decades, enormous strides have been made in unraveling the biosynthesis of macromolecules. Our current knowledge of the chain of molecular information [DNA $\longrightarrow$ RNA $\longrightarrow$ protein] is most impressive. Yet, we should not let our enthusiasm for the accomplishments of molecular genetics lure us into believing that we can explain the "causes" or "controls" of morphogenesis in terms of selective or sequential genome read-out (see Wright,

1973). Essential as gene expression is, it is only a part of the total intricacy of the living cell. Without the participation of, and interplay with, other equally important extranuclear structures and events, genome read-out is biologically meaningless. Imagine, for instance, the entire DNA of a cell being placed into an appropriate test tube and supplied with all the micro- and macromolecular ingredients needed for transcribing and translating this DNA. Can we expect a living cell to result from this concoction? Highly unlikely. As a way of redressing the present bias in biology, it might be useful to describe the nucleus not as the governing center of a cell, but rather as an obedient keeper and copier of the myriad of blueprints required to make all the catalysts needed by the cell.

Much research in fungal morphogenesis deals with experimental perturbations of the phenotype or genotype that change or disturb morphological development. The biochemical, physiological and/or cytological effects elicited by these perturbations have been analyzed to various degrees. However, so far, no good "explanation" of the basis of morphogenesis has emerged from these studies. Perhaps the most significant conclusion is the realization that identifying either an external agent that elicits a morphogenetic transition, or a genetic lesion that impairs normal morphogenesis in a mutant, will not disclose what exactly regulates morphogenesis. For example, studies on mycelial-yeast dimorphism have shown that a great variety of unrelated environmental agents can determine whether a fungal spore would germinate into a mycelium or into a budding yeast cell. Moreover, an agent that elicits yeast development in one fungus, may elicit mycelial growth in another. Despite much progress in defining dimorphism (see review by Stewart and Rogers, 1978), this seemingly simple morphogenetic transformation has not succumbed to reductionism. The same is true for genetic manipulations as shown by Brody (1973) in his identification of genetic loci responsible for the aberrant "colonial" morphology of a number of mutants of *Neurospora crassa*. Some of these mutations were traced to defects in central enzymes of carbohydrate metabolism (e.g., glucose-6-phosphate dehydrogenase, phosphoglucomutase or 6-phosphogluconic acid dehydrogenase). Disclosure of the mutational lesion, however, did not lead to a ready understanding of its morphological consequences. It did lead, however, to the realization that such mutations produce a host of pleiotropic effects on many and diverse aspects of cellular metabolism and, hence, it underscored the futility of trying to analyze the biochemistry of morphogenesis in terms of single causes and relationships.

The biochemical activities of a cell constitute an enormous complex of extensively interconnected events and pathways. Metabolites of one pathway may regulate enzymes of the same and other pathways. Metabolism is thus under the control of a complex network of feed-back loops. Seemingly minor perturbations in one pathway can produce major and unsuspected effects in other parts of the metabolic system and _somehow_ the sum total of these effects affects or directs morphogenesis.

At the risk of sounding overly pessimistic, I feel there is scant chance that the reductionistic studies that have been done in the past and are presently being conducted will unlock the fundamental secrets of fungal morphogenesis. Nevertheless, I am not advocating a mass exodus from reductionism. On the contrary, I see this simplistic approach as an obligatory and productive stage in the development of biological knowledge--just as the painstaking cataloguing of extant and extinct species was a necessary preamble to the theory of evolution. By collecting basic information about many isolated parts of the biochemical and cytological apparatus of the cell, we may be accumulating the assortment of individual bricks with which, eventually, the edifice of morphogenesis will be put together.

At the same time, holistic moves away from reductionism must be undertaken with greater vigor. A notable pioneering effort in this direction can be found in the studies of B. Wright and coworkers, who have undertaken to deal simultaneously with the multiple and changing biochemical parameters during differentiation of _Dictyostelium discoideum_ by means of computer simulations (Wright, 1973; Kelly _et al_., 1979).

And so, rather than searching for specific, isolated causes or controls of morphogenesis, it might be more pertinent to view morphological development in the life cycle of a fungus as a sequence of well defined, steady-state episodes. In each episode, the flux of biochemical and cytological activities of the fungus is in a state of equilibrium. In each episode, a characteristic set of genes is turned on and a corresponding cast of enzymes produces and consumes metabolites at rates defined by internal and external conditions. In turn, the quality and concentration of these metabolites maintains the biochemical steady-state through feed-back regulation, of enzyme activity, enzyme activation, translation and/or transcription. The biochemical equilibrium produces a steady flow of precursors for the orderly construction of subcellular structures that are potential determinants of cell wall morphogenesis (vesicles, membranes, microfilaments, microtubules, etc). Each episode is initially staged by events that take place during the collapse of the preceding episode. An episode will last as long as conditions in the external and/or internal milieu remain relatively constant. Eventually, however, a point is reached where changes in either the external environment or in the internal milieu, would cause the steady-state to collapse. The change from one episode to the next need not be a gradual process but might be a rather abrupt change in a manner described by "catastrophe" theory (Zeeman, 1976). The collapse of an episode sets the biochemical and ultrastructural stage for the next episode, and so on, until a full circle is completed. With its system precariously poised at the verge of catastrophic transitions, a fungus may be massively affected by seemingly minor and unrelated changes in the external milieu. Some of these agents might appear as "triggers" or "causes" of morphogenesis but they are probably no more significant than the proverbial "straw that broke the camel's back."

A preliminary attempt to subdivide the asexual life cycle of *Mucor rouxii* into 13 distinct episodes is shown in Fig.1.

REFERENCES

Brody, S. (1973). Metabolism, cell walls, and morphogenesis. *In* "Developmental Regulation" (Ed. S. Coward) pp. 107-154. Academic Press, New York.

Katz, D., Goldstein, D. and Rosenberger, R. F. (1972). Model for branch initiation in *Aspergillus nidulans* based on measurements of growth parameters. *J. Bacteriol.* *109*, 1097-1100.

Kelly, P. J., Kellener, J. K. and Wright, B. E. (1979). Glycogen phosphorylase from *Dictyostelium*: a kinetic analysis by computer simulation. *Biosystems* *11*, 55-63.

Madelin, M. F. (1966). "The Fungus Spore" Butterworths, London.

Stewart, P. R. and Rogers, P. J. (1978). Fungal dimorphism: a particular expression of cell wall morphogenesis. *In* "The Filamentous Fungi" Vol. 3. (Ed. J. E. Smith and D. R. Berry). pp. 164-196. Arnold, London.

Weber, D. J. and Hess, W. M. (1976). "The Fungal Spore" Wiley, New York.
Wright, B. E. (1973). "Critical variables in Differentiation" . Prentice-Hall, Englewood Cliff, New Jersey.
Zeeman, E. C. (1976). Catastrophe theory. Sci. Amer. 234, 65-83.

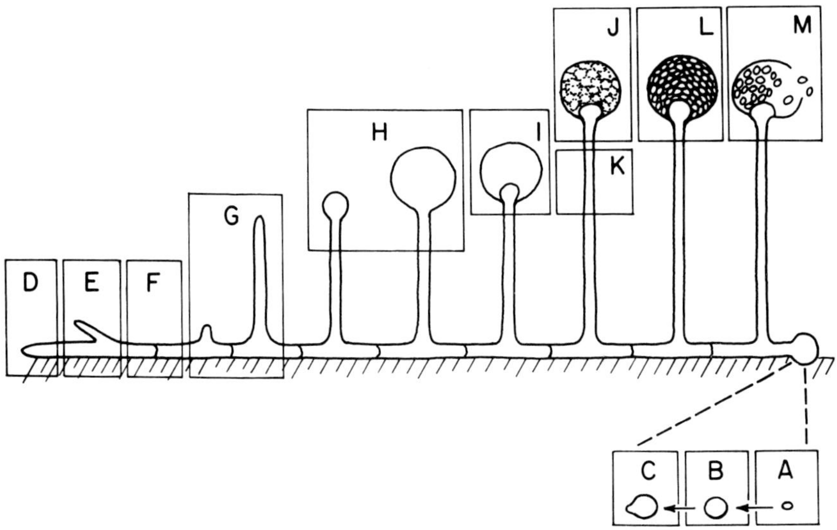

Fig. 1. Differentation episodes in the asexual life cycle of Mucor rouxii. Idealized cross-section of a colony on agar.

Spore germination	Sporulation
A. Initiation of germination	G. Sporangiophore development
B. Spherical growth	H. Sporangium development
C. Germ tube emergence	I. Sporangium delimitation
Mycelium development	J. Protoplasmic cleavage
D. Apical growth	K. Sporangiophore elongation
E. Hyphal branching	L. Spore formation
F. Hyphal septation	M. Spore liberation

Each episode represents a steady state of biochemical and cytological activity. Changes from one episode to the next are brought about by collapse of the steady state. For instance, branching occurs when a hypha is no longer capable of sustaining an initial exponential rate of elongation (re Katz et al., 1972). Presumably, the ensuing accumulation of apical-growth precursors destroys the steady state and creates conditions that lead to lateral branch formation at some critical distance from the primary apex.

LIPIDS IN FUNGAL GROWTH AND REPRODUCTION

J.D. Weete

Department of Botany, Plant Pathology, and Microbiology
Auburn University, Alabama 36849 U.S.A.

INTRODUCTION

Although some species have a particularly high capacity for lipid production, most fungi contain between 3 and 32% lipid depending on the culture conditions, stage of growth, and species. The principal components of this lipid are generally triacylglycerides and glycerophospholipids (phospholipids) which may be accompanied by sterols and their esters, fatty acids, sphingolipids, hydrocarbons, etc. Fatty acids potentially constitute the largest class of lipids in nature, but they occur mainly in the esterified form, e.g. glycerolipids and sterol esters. Fatty acid chain-lengths range from about C_{12} to C_{24} with even-numbered carbon chains predominant. Palmitic (C_{16}) is usually the principal saturated acid, and oleic ($C_{18:1}$) and linoleic ($C_{18:2}$) acids are the major unsaturated acids. $C_{18:3}$ also occurs in fungi, with the ω3 isomer (α-linolenic acid) being found in the higher fungi (Fungi Imperfecti, Ascomycetes, and Basidiomycetes.) There are reports of α-linolenic acid in lower fungi, but the ω6 isomer (γ-linolenic acid) is the predominant C_{18} triene in these fungi.

The principal phospholipids in fungi are phosphatidylcholine and phosphatidylethanolamine which are usually accompanied by lesser amounts of phosphatidylserine, phosphatidylinositol, cardiolipin, and others. Phospholipids are regarded as the major lipid structural components of biomembranes, but they may be accompanied by other lipids. Phospholipids, particularly phosphatidylcholine, are required for the activity of certain membrane-bound enzymes (Coleman, 1973). The lipid composition of fungal membranes has been reviewed by Rose (1977) and Weete (1980).

Quantitatively, the most important lipids of many fungi are triacylglycerides which are usually regarded as storage materials. Other neutral lipids may function in this capacity in some species such as sterol and wax (not widely distributed in fungi) esters. Under certain circumstances, phospholipids may be utilized. Ergosterol is the major sterol in most fungi, but is absent from lower Phycomycetes and rust fungi where it is replaced by other sterols. Sterols are not produced by pythiaceous fungi. They also occur as esters and in a bound form in many species. The principal roles of sterols have been described as membrane components and precursors of steroid hormones (Heftmann, 1971). Numerous other lipids occur in fungi, some of which have well-defined functions. For example, polyprenol phosphates serve as carrier-intermediates in the transfer of monosaccharide units from sugar nucleotides to oligosaccharides during glycan biosynthesis, a subject that has been reviewed by Hemming (1974). However, the carrier role in chitin biosynthesis by *Blastocladiella emersonii* is satisfied by a monoglucosyldiacylglycerol (Mills and Cantino, 1980). The topic of fungal lipids has been covered in more detail by Weete (1974, 1980).

Lipids were discussed at the First International Fungal Spore Symposium in Bristol, England (Madelin, 1966). Professor David Gottlieb presented the state of our knowledge of lipid metabolism during fungal spore germination, and Professor Virgil G. Lilly discussed the effects of sterols and light on the production and germination of *Phytophthora* spores. At the Second International Fungal Spore Symposium in Provo, Utah, in 1975, (Weber and Hess, 1976), Professor Hans J. Reisener considered lipid metabolism of fungal spores during their formation and germination.

This chapter will be concerned generally with the metabolism of lipids during vegetative and reproductive growth of filamentous fungi, and how lipids, particularly sterols, may be involved in reproductive differentiation in these organisms. However, the chapter is supplemented with relevant information on yeasts. Because of the limited space available, this cannot be a comprehensive review of the topics covered; therefore it represents an overview of the subject, citing relevant articles, and emphasizing particularly important deficiencies in our knowledge of the role of lipids in the physiology of fungal growth and reproduction.

SPORE GERMINATION, HYPHAL GROWTH, AND LIPID METABOLISM

There has been relatively little advancement in our knowledge of lipid metabolism during fungal spore germination since the topic was reviewed by Professor Gottlieb (Gottlieb, 1966); therefore it will be outlined only briefly here. The lipid content of spores of most fungi generally ranges between 5 and 17% of their dry weight, but spores of some species such as rusts may contain up to 35%. Lipid metabolism during germination has been studied mainly in species with spores having a high lipid content. The importance of lipid as a reserve material in fungal spores has been illustrated by respiratory quotient (RQ) measurements of 0.6 to 0.7 for dormant and germinating spores of several rust species (Shu *et al.*, 1954), *Fusarium solani* (Cochrane *et al.*, 1963), *Neurospora crassa* (Turian and Bianchi, 1972), and *N. sitophila* (Owens, 1955). However, the RQ values change during the germination of wheat stem rust uredospores suggesting that the endogenous substrate is changing with time (Maheshwari and Sussman, 1970). There is evidence that a change from mainly carbohydrate to lipid utilization occurs (Daly *et al.*, 1967). Further evidence for the utilization of lipid during fungal spore germination is provided by the numerous electron microscope observations of the disappearance of lipid globules. However, in yeast ascospore germination, some investigators have suggested that the disappearance of lipid is too slow to account for it being an important reserve material (Steele and Miller, 1974). Others, however, have reported that lipids are important in germination (Rousseau and Halvorson, 1973). In spite of the fact that *S. cerevisiae* ascospores have a relatively high neutral lipid content, carbohydrate, mainly glycogen followed by trehalose, has been suggested as the principal endogenous substrate for germination (Sols *et al.*, 1971).

Acyl lipids (mainly triglycerides and phospholipids) are the principal forms of lipid present in spores; therefore, the first step in their utilization would be cleavage of ester bonds with fatty acids being released and made available for oxidation. Lipase has been isolated from germinating uredospores of *Uromyces phaseoli* (Knoche and Horner, 1970). The activity of phospholipases was suggested by the rapid decrease in phospholipid content in the very early stages of germination by *Aspergillus niger* conidia (Nishi, 1961) and *U. phaseoli* (Langenbach and Knoche, 1971). Furthermore, phospholipid catabolism in *A. niger* was accompanied by an increase in sugar phosphates and nucleotides, with some of the phosphorus being incorporated into RNA.

The next steps in lipid degradation result in the formation of acetyl-CoA via the β-oxidation of fatty acids. Acetyl-CoA may be oxidized in the TCA cycle, or converted to malate and subsequently carbohydrate via the glyoxylate pathway. The fatty acid content falls during rust spore germination (Daly *et al.*, 1967), and evidence that these substances are being used to generate energy is that [$^{14}CO_2$] is liberated from radiolabeled, short-chain fatty acids supplied to germinating spores (Reisener *et al.*, 1963). Fatty acid degradation products were used also for the synthetic processes during germination, since radioactivity from the short-chain acids appeared in amino acids, proteins, and soluble and insoluble carbohydrates. Furthermore, the enzymes of the glyoxylate pathway, isocitrate lyase and malate synthetase, have been detected in rust spores (Frear and Johnson, 1961; Caltrider *et al.*, 1963).

Although lipid catabolism is very active, lipid synthesis in most fungi studied is generally very slow or absent in the early stages of spore germination. In species where lipid synthesis has been detected, radiolabeled-carbon from exogenous [^{14}C] labeled CO_2, acetate, glucose, and alanine appeared in lipophilic substances. In those rust species studied, fatty acid synthesis appeared to be absent initially (Daly *et al.*, 1967). The non-saponifiable lipid content increased in some *Penicillium* species, but this was not due principally to ergosterol which increased only slightly (Gottlieb and Van Etten 1964). [^{14}C] Acetate was incorporated into diacylglycerides, phospholipids, and sterols during *Ustilago maydis* teliospore germination (Bushnell, 1972). As noted above, spore phospholipid content decreased during the very early stages (0 to 2 hours) of germination of *A. niger* conidia; whereas there was a progressive increase in phospholipid during the first 2 to 6 hours of *Melampsora lini* uredospore germination (Jackson and Frear, 1967). It is not known whether the absence or reduced levels of lipid synthesis are due to the repression of enzymes (acetyl-CoA carboxylase, fatty acid synthetase) required for fatty acid synthesis, the activities of the enzymes are inhibited, or perhaps the substrates for synthesis are unavailable. It might be expected that the earliest requirement for phospholipid during germination would be for the formation of membranes. This would require *de novo* fatty acid synthesis, or transesterification with triacylglycerides providing acyl groups for phospholipid synthesis. Whether (or the extent to which) triacylglycerides serve this purpose, under conditions of net lipid breakdown and subsequent fatty acid oxidation typically present in germinating spores utilizing lipid, is not known (see below). It might be expected also that spores of low lipid content would require lipid synthesis at an earlier stage of germination than spores with high lipid content. This appears to be true at least for sporangiospores of *Rhizopus stolonifer* which requires *de novo* fatty acid synthesis as early as the swelling stage of germination as well as at germ tube formation, and probably mitochondrial biogenesis (Nickerson and Leastman, 1978).

The transition from spore germination to hyphal growth is marked by a dependence on exogenous nutrients, and is followed by a period of exponential growth if these nutrients are not limiting. This transition is characterized by a shift from predominantly lipid utilization to rapid production. It appears that the enzymes necessary for lipid synthesis are present and active in virtually all fungi at the onset of and throughout logarithmic growth and sporulation. Fungi vary in their ability to produce lipid and, within intrinsically determined limits, lipid production is subject to influence by nutritional and environmental factors. Nevertheless, there is a rapid production of glycerolipid during exponential growth of most fungi. Some of the principal factors influencing the degree of lipid production are the nature of carbon (C) and nitrogen (N) sources, and their relative proportions (C:N ratio) (Chesters and Peberdy, 1965). In fungi grown on media having a high C:N ratio, more carbon would be expected to be available for lipid synthesis and it is well-established that lipid production is favored under this condition. Otherwise, little is known about possible regulatory factors that coordinate the distribution of exogenous carbon between fatty acids and other products such as carbohydrates and proteins.

The first committed steps in lipid production are catalyzed sequentially by acetyl-CoA carboxylase and fatty acid synthetase leading to the formation of fatty acids in the form of coenzyme A derivatives. Two acyl groups are then transferred from coenzyme A to *sn*-glycero-3-phosphate to form phosphatidic acid which serves as the precursor common to other phospholipids as well as triacylglycerides. In yeast, phospholipids are produced in both the microsomal cell fraction and mitochondria (Cobon *et al.*, 1974). Mitochondria have the highest specific activity for phospholipid synthesis, and different individual lipids are produced by the two fractions. There appears to be a relatively high rate of phospholipid turnover during vegetative growth, at least in *Glomerella cingulata* in which 56% of the radioactivity from a pulse label remained after 15 minutes and only 23% after 45 minutes (Jack and Anekwe, 1972). On the basis of function, it might be expected that the net production of membranes would establish the minimum requirement for the amount and type (phospholipid>triacylglyceride) of glycerolipid produced during this growth period. In yeast, the amount of phospholipid might be expected to remain relatively constant on a dry weight basis throughout exponential growth. For example in *Cryptococcus neoformans*, the levels of major phospholipids remained constant throughout the cell growth, but the distribution pattern of radioactivity from

[^{32}P] in individual phospholipids changed considerably during the growth cycle (Itoh and Kaneko, 1977). Measurements of changes in radioactivities in the phospholipids showed that phosphatidic acid was one of the most active metabolites. Since growth is restricted to hyphal tips in filamentous fungi, the amount of phospholipid expressed on a dry weight basis may remain constant during growth, or it may decrease. Except for net synthesis, in the very early stages of growth, this generally appears to be the case (Haley and Jack, 1977; Gunasekaran *et al.*, 1972). Triacylglycerides are formed via acylation of the corresponding diacylglyceride at the *sn*-3 position after dephosphorylation of phosphatidic acid. Although phospholipids might be expected to be functionally more important during logarithmic growth, triacylglycerides are more abundant than polar lipids in most fungi. One viewpoint is that phospholipid synthesis is under rigid control and that the facility for triacylglyceride synthesis accommodates the production of more fatty acids (as CoA derivatives) than is required for synthesis of polar lipids. When grown in media having high C:N ratios, fungi would be expected to show reduced production of nitrogenous substances (serine, ethanolamine, choline) and cofactors (CTP) resulting in reduced membrane phospholipid synthesis, and a corresponding increase in triacylglyceride synthesis. The fact that the neutral lipid content is less in yeast grown in minimal medium than when grown on a very rich medium (Keith *et al.*, 1973), and that neutral lipid accumulates in choline starved mutants of *N. crassa* (Juretic, 1977), is evidence for this. Also, triacylglyceride accumulation may be favored by a low rate of breakdown, which might be expected under conditions favorable to rapid growth in the presence of abundant exogenous carbohydrate. Otherwise, it is not known precisely how the distribution of phosphatidic acid between the neutral and polar acyl lipids is regulated so that the rate of phospholipid synthesis, and hence membrane formation, is compatible with membrane synthesis and the growth rate as established by nutritional and environmental factors.

The role of triacylglycerides in the metabolic activities of fungi during the logarithmic growth is not well-established. Endogenous reserves in fungi appear to be in a dynamic state of flux rather than isolated from the metabolic machinery of the cell; and the rate of endogenous respiration in filamentous fungi is high relative to respiration in the presence of an exogenous substrate (Blumenthal, 1977). In *Aspergillus sojae* grown in a medium with a high C:N ratio, carbohydrate is the first endogenous substrate utilized, followed by nitrogenous substances (lipids have not been considered), whereas under low C:N conditions nitrogenous substances are utilized first (Mizunuma, 1966). Under high C:N conditions, the mycelium would be expected to have a relatively high lipid (mainly triacylglyceride) as well as carbohydrate content. Although it is not known for many species, the turnover rate of triacylglycerides appears to be relatively high. For example, in *G. cingulata*, 50% of the radioactivity from a 5 minute pulse label remained in triacylglyceride after 15 minutes, whereas 48% remained after 45 minutes (Jack and Anekwe, 1972). In triacylglycerides of *Lipomyces lipofer*, greater changes in the fatty acid composition occurs at the *sn*-3 position as a function of culture age compared to the *sn*-1 and *sn*-2 positions (Phornpiboonya and Jack, 1980). In yeast, about 70% of the triacylglycerol lipase was associated with membranes, about 60% of which was associated with mitochondria (Schousboe, 1976). Compartmentalization of triacylglycerides occurs in some organisms such as the protozoan *Tetrahymena pyriformis* in which newly labeled material is metabolized more rapidly than older material (Borowitz and Blum, 1976), but whether it occurs in fungi is not known.

Acyl exchange between triacylglycerides and phospholipids is well-known in many organisms, but little is known about the interaction between these two types of lipid during fungal growth. It has been suggested that acyl exchange occurs between these lipids in *G. cingulata* (Haley and Jack, 1977). In 90 minutes, 85% of the radioactivity of triacylglyceride turned over in *T. pyriformis*, most of which appeared in phospholipid (Borowitz and Blum,1976). In this case, not all the label appearing in phospholipid was due to acyl exchange; some originated from the glycerol moiety suggesting that triacylglycerides are direct precursors of phospholipids in this organism. Perhaps this is true in fungi as well since it is the *sn*-3 position of triacylglycerides that undergoes the greatest change in fatty acid composition during growth, at least in *L. lipofer*, but there is no direct evidence for this.

The cessation of exponential growth occurs with the depletion of a component of the medium essential for cell division or hyphal tip growth. Depending on the conditions, this may initiate reproductive differentiation (see below) or a stationary phase. With regard to lipid metabolism during the stationary phase, fungi vary depending on the species and nutritional state. The most studied of the nutritional parameters related to lipid metabolism during the stationary phase is nitrogen deficiency. As mentioned above, a high C:N ratio favors lipid production. Under this condition, logarithmic growth is usually terminated by depletion of nitrogen from the medium while some of the carbon source remains. Some fungi are able to continue assimilating sugar, and may accumulate 50 to 70% of their dry weight as lipid (mainly triacylglyceride). Lipid production during the stationary phase by oleaginous fungi, at least by *Candida* 107 and perhaps by *Rhodotorula glutinis* and *Mucor circinelloides*, is due to the interaction of several metabolic activities initiated by the depletion of cellular AMP under nitrogen-deficient conditions (Botham and Ratledge, 1979). NAD^+-dependent isocitrate dehydrogenase of *Candida* 107, but not the non-oleaginous *C. utilis*, requires AMP for activity. Therefore, citrate accumulates under nitrogen-deficient conditions, and passes into the cytosol from the mitochondria where it is degraded to oxalacetate and acetyl-CoA through the action of ATP:citrate lyase. Acetyl-CoA is then incorporated into fatty acids, which requires NADPH believed to be derived from the metabolism of sugar via the pentose phosphate pathway, resulting in lipid accumulation.

One of the prerequisites for lipid accumulation during the stationary phase is that the active enzymes for fatty acid synthesis be present in the hyphae. A non-oleaginous filamentous species tested incorporated less radiolabeled substrate with increasing age. For example, the greatest capacity for lipid production by mycelium of *G. cingulata*, taken at 24 hour intervals during the growth cycle and given [^{14}C] acetate or glucose for 5 minutes, coincided with logarithmic growth (Jack and Anekwe, 1972).

Although not so quantitatively important as acyl lipids, the free sterol content of fungi also increases progressively during logarithmic growth. The pattern of change in free sterol content appears to be similar in several species studied. For example, the sterol content of *Neurospora crassa*, peach-fluffy-cot mutant, (Ballou and Bianchi, 1978), a wild-type *N. crassa* strain (Elliott *et al.*, 1974), and *Rhizopus arrhizus* (Weete *et al.*, 1973) increased during exponential growth then decreased during late logarithmic growth or early in the stationary phase. In the wild-type *Neurospora* and *Rhizopus* species, the free sterol content increased rapidly late in the stationary phase. The increase in sterol content may reflect a shift from fermentative to a more oxidative mode of growth as the sugar is depleted from the medium. For example, sterol synthesis in *Saccharomyces cerevisiae* is subject to catabolite repression when grown with adequate fermentable sugar in the medium (Quain and Haslam, 1979; Taketani *et al.*, 1978). When grown under conditions favorable to catabolite derepression (oxidative mode of growth) sterol synthesis was stimulated four-fold. Higher sterol levels have been also observed in *Lipomyces starkeyi* grown under glucose-deficient conditions (Suzuki and Hasegawa, 1974). In *S. cerevisiae*, the stimulation of sterol synthesis under these conditions is due to increased activity of β-hydroxy-β-methylglutaryl-CoA reductase (Quain and Haslam, 1979). Under conditions favoring catabolite derepression the sterols formed are almost exclusively converted to esters of monoenoic acids. Increased sterol ester content occurred during the stationary phase of *S. cerevisiae* (Bailey and Parks, 1975) due to increased sterol esterase activity which became increasingly active at the end of logarithmic growth (Parks and Stromberg 1978). It is well-known that under anaerobic conditions sterol synthesis is inhibited, and sterol esters are hydrolyzed (Taketani *et al.*, 1978).

LIPID METABOLISM DURING SPORULATION

The transition from vegetative to reproductive growth involves a complex series of metabolic changes resulting in the formation of spores and other structures associated with reproductive differentiation. One of the main problems in studying the metabolic aspects of sporulation is in separating events associated with the vegetative and reproductive growth phases, i.e. distinguishing "sporulation-specific" metabolism. One of the most common approaches to studying metabolism associated with sporulation is to measure changes as a function of time in a fungus undergoing the transition from vegetative to reproductive growth. In most cases, vegetative and sporulation-specific metabolism overlap, the former often masking the latter, making it difficult to identify that aspect of metabolism specific to and perhaps required for reproduction. Ascospore formation in *S. cerevisiae* is a good system to study sporulation-related metabolic activities because 80 to 90% of the vegetative cells incubated in a medium favoring sporulation may undergo ascospore formation. Ascospores are visible after about 24 hours in the sporulation medium (deficient in nitrogen with acetate as the carbon source) and spore formation is complete after 48 hours. There are two periods of lipid synthesis during yeast sporulation, between 0 and 18 hours in the sporulation medium and between 24 and 30 hours (Illingworth *et al.*, 1973). After 48 hours in the sporulation medium, the lipid content was four-fold greater in sporulating cells than in vegetative cells. Using a diploid sporulating strain of yeast and haploid non-sporulating strain, it was shown that the initial period of lipid synthesis occurs in both whereas the second period occurs only in the one undergoing sporulation (Henry and Halvorson, 1973).

All lipids increased during the first 24 hours in the sporulation medium, but it was increased synthesis of triacylglycerides and sterol esters that was mainly responsible for the second period of lipid production (Illingworth *et al.*, 1973). These two types of neutral lipid represented 66% of the total lipid after 120 hours in the sporulation medium compared to 19% in vegetative cells. The degree of unsaturation also increased drastically during ascospore development. Extensive membrane formation occurs during ascospore development, and it appears that the initial increase in sterol content parallels this process. However, after 120 hours in the sporulation medium, the sterol content of asci containing fully developed ascospores (complete after 48 hours) was ten-fold that of vegetative cells and represented about 4% of the total lipid. The increase in phospholipid content was about 171% greater than could be accounted for by membrane formation; it was speculated that the excess lipid was in electron-transparent vesicles. Fatty acid synthesis was required for ascospore formation in *S. cerevisiae* which was shown with the fatty acid synthesis inhibitor cerulenin (Ohno *et al.*, 1976). The fatty acid requirement was met during the first 6 hours in the sporulation medium since no inhibition occurred when the inhibitor was added after that time. Since the sporulation-specific increase in neutral lipid was not inhibited by cerulenin, it appears that acyl groups are derived from endogeneous pools or perhaps from a second fatty acid synthetase insensitive to cerulenin during the second period of lipid synthesis (Ohno *et al.*, 1976).

There are relatively few studies on the relationship between lipid metabolism and sporogenesis in filamentous fungi, particularly in cases where some degree of synchrony in the sporulation process was achieved through manipulation of the culture medium or through light stimulation. However, sporulation in *Achlya*, which begins 20 hours from spore germination, was accompanied by an increase of about 28% followed by a decrease in lipid content (Law and Burton, 1976a). Although lipid increased in the form of neutral glycerides (Law and Burton, 1976b), increased activity of the fatty acid synthetase during sporulation appeared to be responsible for the overall lipid increase. The phospholipid content of mycelium declined slowly during sporulation. Triacylglycerides increased during vegetative growth of *N. crassa*, but the increase was more pronounced (24.3% over pre-microconidial cultures) during microconidial maturation (Ballou and Bianchi, 1978).

One of the most comprehensive studies of lipids relating to reproductive differentiation was conducted by Fayret

(1975) with *Gnomonia leptostyla*. Four stages of reproductive development can be selectively obtained through the manipulation of culture conditions i.e., growth at 20 C in the light results in macroconidia formation; 10 C in light, microconidia; 20 C in dark, sterile perithecia; 10 C in dark, fertile perithecia. There was a reduction in lipid content during photo-induced conidiogenesis; but lipid synthesis was very active during ascocarp formation, followed by lipid degradation in the dikaryotic phase and in sterile perithecia. Also, there were qualitative differences in lipid composition of perithecia and vegetative mycelium (see below for differences in sterols).

The pattern of change in sterol content during growth appears to be similar in several filamentous fungi studied (see above). There is an interruption of free sterol accumulation during sporangiospore formation in *R. arrhizus* (Weete *et al.*, 1973), microconidia formation in the peach-fluffy cot mutant of *N. crassa* (Ballou and Bianchi, 1978), and protoperithecium formation in a wild-type strain of *N. crassa* (Elliott *et al.*, 1974). It is recognized that there may be several explanations for the interruption of sterol metabolism during reproduction which are unrelated to the sporulation process *per se*; however, Ballou and Bianchi (1978) suggested that when a critical level of sterol is reached no new sterols are produced, and that the critical concentration of sterols may contribute to the metabolic requirements for microconidiation.

Sporulation also appears to involve increased lipid utilization. Asexual reproduction in a yeast-like fungus was accompanied by increased lipase activity, lipid utilization, and reduced RQ (Lloyd *et al.*, 1971). In the same study, similar results were obtained with *Aspergillus niger*; increased lipase activity coincided with decreased intracellular lipid during the late stages of conidiation. The increased lipase (esterase) activity detected cytochemically occurred in the conidiophore tip prior to the formation of the vesicles and phialides, and in the phialides during their formation (Lloyd *et al.*, 1972). Further evidence for the utilization of lipids during asexual reproduction was obtained with *Achlya* (Law and Burton, 1976b). Increased oxidation of exogenously added [^{14}C] labeled fatty acids occurred with the onset of sporulation. In *G. leptostyla*, lipid utilization appeared to be more important in the maturation of perithecia than in the development of conidia (Fayret, 1975).

MEMBRANE PROPERTIES AND FUNGAL REPRODUCTION

Lipids are required for vegetative and reproductive growth in fungi. The roles commonly ascribed to different lipids, as outlined in the Introduction of this chapter, probably do not change during the fungal life cycle. There may, however, be a change in degree to meet the needs of the fungus as it varies with changing nutritional conditions and differentiation. Recent advances in our understanding of the physical properties of membranes, and insight into the relationship between these properties and membrane-associated activities, provide us with an additional dimension with which to view the role of lipids as membrane components. Membranes regulate the flux of substances between the cell and its surrounding environment, and between intracellular compartments. Membranes also are sites of biochemical and electrochemical activities. Permeability and activities associated with membranes are subject to influence, and perhaps regulation, by the physical properties of membrane lipids (Cronan and Gelmann, 1975). Consideration must be given to the possibility of a connection between the physical state of membrane lipid and more complex phenomena such as respiration and even reproductive differentiation. We have a general understanding of the physical properties of membranes, particularly artificial membranes, and how certain changes in lipid composition affect these properties; but little definite is known about how changes in, or perhaps the regulation of, these properties are translated to physiological and morphogenetic events.

A biological membrane is visualized as a lipid bilayer imbedded with globular protein which may be restricted to one side or extend through the entire thickness of the bilayer (Singer and Nicolson, 1972). The matrix of a homogeneous bilayer is composed of the mobile, hydrophobic component (acyl chains) of the lipid forming the membrane. The lipid (mostly phospholipid) can undergo physical state changes in the form of reversible thermotropic and lyotropic gel to liquid-crystalline phase transitions. Both physical states can exist simultaneously in a membrane. The relative proportion of these physical states is the result of a complex interaction of many factors, mainly the qualitative and quantitative composition of lipid and non-lipid substances and the temperature. The relative motion of hydrocarbon chains constituting the bilayer matrix is determined by the physical state of the lipid, and dictates the degree of fluidity of the lipid. Fluidity is a measurable property that takes into account all of the factors determining the physical state of the membrane lipid. Two factors that have been studied extensively as they relate to membrane properties are the sterol composition and degree of unsaturation of membrane lipid. The inclusion of a sterol (cholesterol) in a phospholipid bilayer appears to have simultaneous opposing effects on membrane fluidity; it tends both to "fluidize" (liquifying effect) the gel areas and to decrease the fluidity (condensing effect) in the liquid-crystalline areas (Oldfield and Chapman, 1972; Demel and DeKruyff, 1976). Increasing unsaturation of membrane lipid lowers the transition temperature (temperature at which gel to liquid-crystal transition occurs) of lipids and allows greater freedom of molecular motion of the hydrocarbon chains comprising the bilayer matrix, and hence increases fluidity.

Since membrane-associated processes are subject to influence by the physical state of membrane lipid, there might be expected to be certain degrees of fluidity for optimum activities of these processes. Furthermore, fungi as poikilothermic organisms would be expected to be able to resist changes in membrane fluidity due to alterations in temperature or other environmental conditions, and during a particular stage of growth. This might be accomplished by regulating the degree of lipid unsaturation and/or the composition of the membrane. There is evidence that fungi maintain the degree of fluidity of their membranes in response to temperature by adjusting the degree of lipid unsaturation. It is well-documented that the degree of unsaturation increases in fungi grown at temperatures below that optimum for growth (Weete, 1980). The converse is also true. Increases in the rate of desaturation relative to fatty acid oxidation (Kates and Baxter, 1962), and higher oxygen tension which favors desaturation, have been offered as explanations for increased unsaturation at low temperatures. The substrate for desaturation in *Candida lipolytica* appears to be temperature-dependent (Pugh and Kates, 1975). Microsomal desaturases from cells grown at 10 C preferred oleyl-CoA as the substrate, whereas the phosphatidylcholine derivative (oleyl-PC) was the preferred substrate for the enzyme from cells grown at 20 C. Evidence for a mechanism whereby the degree of lipid unsaturation is self-regulating has been obtained with *Tetrahymena pyriformis* (Martin *et al.*, 1976; Kasai *et al.*, 1976). During temperature acclimation in this protozoan, membrane-bound desaturase activity was reduced with increasing fluidity at higher temperatures, and the reduced enzyme activity was due to changes in fluidity rather than temperature. Furthermore, membrane fluidity of cells grown in a medium supplemented with polyunsaturated fatty acids remained unchanged due to increased *de novo* fatty acid synthesis and incorporation of saturated fatty acids into phospholipids. The relation between fluidity and desaturase activity has not been tested directly in fungi.

The only lipids for which there is evidence for a specific role in fungal reproduction are the sterols. Sterols serve as precursors of steroid hormones involved in sexual differentiation. In *Achlya*, antheridiol, derived biosynthetically from fucosterol (Popplestone and Unrau, 1974), induces antheridia and antheridial branch formation, and oogoniol (McMorris *et al.*, 1975) induces oogonia formation. There is indirect evidence that cholesterol is the precursor of an unidentified steroid hormone(s) that controls oospore development in *Phytophthora cactorum* (Elliott and Sansome, 1977). Oogonia formation proceeded normally in the presence of cholesterol or cholestanol, but oospores developed only in the presence of the former sterol suggesting that it was a precursor to a steroid hormone controlling meiosis. Otherwise, there is little evidence for sterols serving as precursors to steroid hormones that are involved in fungal reproduction. Sexual reproduction by *Achlya*, and the involvement of non-steroid hormones in fungal reproduction have been reviewed by Bu'Lock (1977).

Further evidence for the involvement of sterols in fungal reproduction is that they are required for oospore and zoosporanium formation in pythiaceous fungi, which do not produce these substances. This discovery in the early 1960's stimulated considerable interest in sterols as the possible key to understanding the regulation of fungal reproduction. The large body of information generated has been reviewed by Hendrix (1971) and was included in a review by Elliott (1977). The metabolism of exogenous sterols by pythiaceous species has been studied and the structural requirements for reproduction have been determined, but no steroid hormones have been identified in these organisms. There is limited evidence for the requirement of sterols in the reproduction by sterol-producing fungi. For example, perithecial development was inhibited in *Cochliobolus carbonum* by the sterol synthesis inhibitor SKF-3301-A at levels which allowed growth (Nelson *et al.*, 1967), but these results could not be repeated with *Sordaria fimicola* (Elliott, 1969). However, perithecia production was prevented in *S. fimicola* by a second sterol synthesis inhibitor, AY-9944, at concentrations which allowed growth. Apart from being a precursor to steroid hormone(s) in *Achlya*, it is not known precisely how sterols are involved in fungal reproduction.

Sterols might be expected to affect any membrane-associated activity subject to influence by changes in membrane properties, some of which appear to be vital to survival of the organism. Although pythiaceous fungi have developed a means of growth without them, sterols are required for the growth of sterol-producing fungi. This has been demonstrated by means of sterol synthesis inhibitors for numerous fungi (Ragsdale, 1975; Sherald and Sisler, 1975). Also, it is well-known that *S. cerevisiae* cannot grow anaerobically without sterol supplements in the growth medium (Andreasen and Stier, 1953). While the sterol requirement for growth by yeast was not absolute for ergosterol (the principal sterol of this organism) based on growth response, a marked dependence on precise structural features has been demonstrated (Nes *et al.*, 1978). The 3-hydroxyl group was required and other structural aspects that distinguish ergosterol were of the following relative importance: 24β-methyl-Δ^{22}>24β-methyl>$\Delta^{5,7}$-diene=$\Delta^{5}\simeq$ no double bond. In this connection, it is well-established that sterols play an important role in mitochondrial function, as discussed by Parks *et al.* (1978). Sterols and mitochondria are linked both biosynthetically and functionally. Although most of the enzymes of ergosterol biosynthesis are soluble, or located in the microsomal cell fraction, in yeast S-adenosylmethionine:Δ^{24} sterol methyltransferase is located in the mitochondria (Bailey *et al.*, 1974). Sterol formation is dependent on respiratory cytochrome synthesis (Astin *et al.*, 1977), but ergosterol is not required for cytochrome oxidase activity *in vitro* (Thompson and Parks, 1977). Respiratory incompetence can be reversed in yeast with ergosterol (Parks *et al.*, 1978). Yeast mitochondrial ATPase shows a progressive decrease in the transition temperature with increasing temperature (Cobon and Haslam, 1973), and sterol depletion from mitochondria leads to a loss of coupled phosphorylation due to increased passive permeability to protons (Astin and Haslam, 1977). The relation between mitochondrial function and sterol content has not been studied as thoroughly in filamentous fungi as in yeast. However, respiratory deficiency in *A. niger* has been associated with the depletion of ergosterol (and cardiolipin) from mitochondria (Mandal *et al.*, 1978).

It is well-established that reproductive development is preceded by an intense peak of respiratory activity (Hawker and Hepden, 1962), and has a high requirement for energy relative to vegetative growth. The efficiency of energy production would be expected to be subject to influence by the physical state of mitochondrial membrane lipid, and there may be a critical degree of fluidity for energy production at levels required for reproduction. Sterols may be regulatory agents used by fungi to control the degree of mitochondrial membrane fluidity, and allow maximum efficiency of energy production. The central point of this hypothesis is that, other factors being equal, the physical properties, not sterols *per se*, may be the most important factors in determining the level of energy production. This may explain how apparently unrelated factors such as light, temperature change, physical contact with a fixed object, etc. stimulate sporulation.

Certainly, there are a variety of ways in which sterols may be involved in fungal sporulation. However, although based on limited evidence, the above hypothesis offers a plausible alternative explanation of how sterols may function in satisfying one of the prerequisites for sporulation.

SUMMARY AND FUTURE PROSPECTS

In recent years, there have been considerable advances in our knowledge of the chemistry and biosynthesis of lipids in fungi. Although relatively few of the vast number of species have been analyzed, we have a good general knowledge of the lipid composition of these organisms. While more complete information on the species studied is needed and data on additional species will be useful, it would not be expected to change our current views on the general distribution of lipids in fungi. The biosynthetic pathways of the major lipids are reasonably well-established in fungi; most of this work being conducted with *S. cerevisiae* and, with the exception of *N. crassa*, the filamentous fungi have received relatively little attention in this area. However, there is no evidence that the basic pathways of lipid metabolism in yeasts and filamentous fungi differ greatly. The roles of lipids in fungi are generally similar to those in other organisms, but how these roles are integrated into the complex metabolic and physiological processes as they occur in the fungal life-cycle are not well-established.

Like other intermediary metabolites, glycerolipids are active metabolically during fungal growth and reproduction. Their metabolism is subject to regulation to meet the needs of the fungus, and they in turn may be involved in the regulation of physiological and morphogenetic processes. There are certain key points in pathways of lipid metabolism for which little is known about the regulatory mechanisms in relation to that of other major intermediary metabolites during each of the various stages of growth, and the transition between them. For example, the factors governing the distribution of exogenous carbon between carbohydrates, protein, and lipids during the fungal growth cycle are not well established. One important aspect of this is the regulation of fatty acid synthesis in relation to the fungal growth cycle.

Another area of lipid metabolism that has particular

significance to the physiology of fungi is the control of phospholipid biosynthesis, which would be expected to relate directly to the regulation of membrane biogenesis. One aspect of this is the control of the relative proportions of triacylglycerides and phospholipids formed during different stages of the life cycle. Furthermore, although the role of triacylglycerides as storage substances is well-established, little is known about their metabolic interaction with phospholipids in relation to the regulation of membrane properties and membrane-associated activities as a function of fungal growth and development.

Of course, there are other important aspects of lipid metabolism that a better understanding of which might provide more insight into their roles in fungi. Probably the most important clues to unraveling the complex control systems of fungal growth and differentiation can be obtained through a better understanding of the relationship between the physical properties of membranes and membrane-associated activities. Undoubtedly, the regulation of lipid metabolism is closely linked to the control of membrane biogenesis and physical properties which in turn serves to modulate membrane-associated activities to suit the needs of the fungus in various stages of growth as determined by environmental and genetic restraints. Perhaps some important clues of how biochemical activities are translated into morphogenetic events can be obtained from better understanding these relationships.

ACKNOWLEDGMENTS

This work was supported by Alabama State Research Project Ala-921.

REFERENCES

Andreasen, A.A. and Stier, T.J.B. (1953). Anaerobic nutrition of *Saccharomyces cerevisiae*. *J. Cell. Comp. Physiol.* 41, 23-26.

Astin, A.M. and Haslam, J.M. (1977). The effects of altered membrane sterol composition on oxidative phosphorylation in a haem mutant of *Saccharomyces cerevisiae*. *Biochem. J.* 166, 287-298.

Astin, A.M., Haslam, J.M., and Woods, R.A. (1977). The manipulation of cellular cytochrome and lipid composition in a haem mutant of *Saccharomyces cerevisiae*. *Biochem. J.* 166, 275-286.

Bailey, R.B., Thompson, E.D., and Parks, L.W. (1974). Kinetic properties of S-adenosylmethionine: Δ^{24}sterol methyltransferase enzyme(s) in mitochondrial structures of *Saccharomyces cerevisiae*. *Biochim. Biophys. Acta* 334, 127-136.

Bailey, R.B. and Parks, L.W. (1975). Yeast sterol esters and their relationship to the growth of yeast. *J. Bacteriol.* 124, 606-612.

Ballou, L.R. and Bianchi, P.E. (1978). Lipids associated with microconidial differentiation in a mutant of *Neurospora crassa*. *Current Microbiol.* 1, 111-115.

Blumenthal, H.J. (1977). Reserve carbohydrates in fungi. *In* "The Filamentous Fungi" (Ed. J.E. Smith and D.R. Berry). pp. 292-307. John Wiley and Sons, New York.

Borowitz, M.J. and Blum, J.J. (1976). Triacylglycerol turnover in *Tetrahymena pyriformis*: Relation to phospholipid synthesis. *Biochim. Biophys. Acta* 424, 114-124.

Botham, P.A. and Ratledge, C. (1979). A biochemical explanation for lipid accumulation in *Candida* 107 and other oleaginous microorganisms. *J. Gen. Microbiol.* 114, 361-375.

Bu'Lock, J.D.(1977). Hormones in fungi. *In* "The Filamentous Fungi" (Ed. J.E. Smith and D.R. Berry). Vol. II, pp. 345-368.

Bushnell, J.L. (1972). Lipids of fungal spores. Identification and metabolism. M.S. Thesis. Brigham Young University, Provo, Utah.

Caltrider, P.G., Ramachandran, S., and Gottlieb, D. (1963). Metabolism during germination and function of glyoxylate enzymes in uredospores of rust fungi. *Phytopathology* 53, 86-92.

Chesters, C.G.C. and Peberdy, J.F. (1965). Nutritional factor in relation to growth and fat synthesis in *Mortierella vinacea*. *J. Gen. Microbiol.* 41, 127-134.

Cobon, G.S. and Haslam, J.M. (1973). The effect of altered membrane sterol composition on the temperature dependence of yeast mitochondrial ATPase. *Biochem. Biophys. Res. Commun.* 52, 320-326.

Cobon, G.S., Crowfoot, P.D., and Linnane, A.W. (1974). Biogenesis of mitochondria. *Biochem. J.* 144, 265-275.

Cochrane, V.W., Cochrane, J.C., Collins, C.B., and Serafin, F.G. (1963). Spore germination and metabolism in *Fusarium solani*. *Amer. J. Bot.* 50, 806-814.

Coleman, R. (1973). Membrane-bound enzymes and membrane ultrastructure. *Biochim. Biophys. Acta* 300, 1-30.

Cronan, J.E. and Gelmann, E.P. (1975). Physical properties of membrane lipids: Biological relevance and regulation. *Bacteriol. Rev.* 39, 232-256.

Daly, J.M., Knoche, H.W., and Wiese, M.W. (1967). Carbohydrate and lipid metabolism during germination of uredospores of *Puccinia graminis tritica*. *Plant Physiol.* 42, 1633-1642.

Demel, R.A. and DeKruyff, B. (1976). The functions of sterols in membranes. *Biochim. Biophys. Acta* 457, 109-132.

Elliott, C.G. (1969). Effects of inhibitors of sterol synthesis on growth of *Sordaria* and *Phytophthora*. *J. Gen. Microbiol.* 56, 331-343.

Elliott, C.G. (1977). Sterols in fungi: Their functions in growth and reproduction. *Adv. Microbiol. Physiol.* 15, 121-173.

Elliott, C.G., Knights, B.A., and Freeland, J.A. (1974). Sterols of *Neurospora crassa* and the pattern of their binding during the growth cycle. *Biochim. Biophys. Acta* 360, 339-347.

Elliott, C.G. and Sansome, E. (1977). The influence of sterols on meiosis in *Phytopthora cactorum*. *J. Gen. Microbiol.* 98, 141-145.

Fayret, J. (1975). Etude du cycle de reproduction de *Gnomonia leptostyla*. Thesis. Univ. Paul Sabatier de Toulouse, France.

Fayret, J., Lacoste, L., Alais, J., La Blache-Combrer, A., Maquestiau, A., Van Harverbeke, G., Flammang, R., and Mispreuve, H. (1979). Composition stérolique et morphogenèses reproductrices chez *Gnomonia leptostyla*. *Phytochemistry* 18, 431-435.

Frear, D.S. and Johnson, M.A. (1961). Enzymes of the glyoxylate cycle in germinating uredospores of *Melampsora lini* (Pres.) Lev. *Biochim. Biophys. Acta.* 47, 419-421.

Gottlieb, D. (1966). *In* "The Fungus Spore" (Ed. M.F. Madelin). p. 217. Butterworths, London.

Gottlieb, D. and Van Etten, J.L. (1964). Biochemical changes during the growth of fungi II. Ergosterol and fatty acids in *Penicillium atrovenetum*. *J. Bacteriol.* 88, 114-121.

Gunasekaran, M., Weber, D.J., and Hess, S.L. (1972). Total, polar, and neutral lipids of *Rhizopus arrhizus* Fischer. *Lipids* 7, 430-434.

Haley, N.J. and Jack, R.C. (1977). Influence of α-DL-tocopherol on fungal lipid composition. *Biochim. Biophys. Acta.* 489, 207-213.

Hawker, L.E. and Hepden, P.M. (1962). Sporulation in *Rhizopus sexualis* and some other fungi following a period of intense respiration. *Ann. Bot. N.S.* 26, 619-632.

Heftmann, E. (1971). Functions of sterols in plants. *Lipids* 6, 128-133.

Hemming, F.W. (1974). Lipids in glucan synthesis. *In* "Biochemistry of Lipids" (Ed. T.W. Goodwin). Vol. 4, pp. 39-97.University Park Press, Baltimore, MD.

Hendrix, J.W. (1971). Sterols in growth and reproduction of fungi. *Annu. Rev. Phytopath.* 8, 111-130.

Henry, S.A. and Halvorson, H.O. (1973). Lipid synthesis during sporulation of *Saccharomyces cerevisiae*. *J. Bacteriol.* 114, 1158-1163.

Illingworth, R.F., Rose, A.H., and Beckett, A. (1973). Changes in the lipid composition and fine structure of *Saccharomyces cerevisiae* during ascus formation. *J. Bacteriol.* 113, 373-386.

Itoh, T. and Kaneko, H. (1977). The *in vivo* incorporation of [^{32}P]-labeled orthophosphate into pyrophosphatidic acid and other phospholipids of *Cryptococcus neoformans* through cell growth. *Lipids* 12, 809-813.

Jack, R.C. and Anekwe, G.E. (1972). Incorporation of labelled precursors into, and accumulation of, triglycerides and phospholipids at selected growth stages. *Physiol. Plant.* 26, 166-169.

Jackson, L.L. and Frear, D.S. (1967). Lipids of rust fungi. I. Lipid metabolism of germinating flax rust uredospores. *Can. J. Biochem.* 45, 1309-1315.

Juretic, D. (1977). The effect of phosphatidylcholine depletion on biochemical and physical properties of *Neurospora crassa* membrane mutant. *Biochim. Biophys. Acta* 469, 137-150.

Kasai, R., Kitajima, Y., Martin, C.E., Nozawa, Y., Skriver, L., and Thompson, G.A. (1976). Molecular control of membrane properties during temperature acclimation. Membrane fluidity regulation of fatty acid desaturase action. *Biochemistry* 15, 5228-5233.

Kates, M. and Baxter, R.M. (1962). Lipid composition of mesophilic and psychrophylic yeasts (*Candida* species) as influenced by environmental temperature. *Can. J. Biochem. Physiol.* 40, 1213-1227.

Keith, A.D., Wisnieski, B.J., Henry, S., and Williams, J.C. (1973). Membranes of yeast and *Neurospora*: Lipid mutants and physical studies. *In* "Lipids and Biomembranes of Eukaryotic Microorganisms" (Ed. J.A. Erwin). pp. 259-321. Academic Press, New York.

Knoche, H.W. and Horner, T.L. (1970). Some characteristics of a lipase preparation from the uredospores of *Puccinia graminis tritici*. *Plant Physiol.* 46, 401-405.

Langenbach, R.J. and Knoche, H.W. (1971). Phospholipids in the uredospores of *Uromyces phaseoli*. *Plant Physiol.* 48, 735-739.

Law, S.W.T. and Burton, D.N. (1976a). Lipid metabolism in

Achlya: Studies of lipid turnover during development. *Can. J. Microbiol.* 22, 1710-1715.

Law, S.W.T. and Burton, D.N. (1976b). Lipid metabolism in *Achlya* sp. Changes in lipid composition during development. *Can. J. Microbiol.* 22, 1716-1719.

Lloyd, G.I., Morris, E.O., and Smith, J.E. (1971). A study of the esterases and their function in *Candida lipolytica, Aspergillus niger,* and a yeast-like fungus. *J. Gen. Microbiol.* 63, 141-150.

Lloyd, G.I., Anderson, J.G., Smith, J.E., and Morris, E.O. (1972). Conidiation and esterase synthesis in *Aspergillus niger*. *Trans. Br. Mycol. Soc.* 59, 63-70.

Madelin, M.F. (Ed.). (1966). "The Fungus Spore". Colston Papers 18. Butterworths, London.

Maheshwari, R. and Sussman, A.S. (1970). Respiratory changes during germination of urediospores of *Puccinia graminis* f. sp. *tritici*. *Phytopathology* 60, 1357-1364.

Mandal, S.B., Sen, P.S., and Charkrabarti, P. (1978). Effect of respiratory deficiency and temperature on the mitochondrial lipid metabolism of *Aspergillus niger*. *Can. J. Microbiol.* 24, 586-592.

Martin, C.F., Hiramatsu, K., Kitajima, Y., Nozawa, Y., Skriver, L., and Thompson, G.A. (1976). Molecular control of membrane properties during temperature acclimation. Fatty acid desaturase regulation of membrane fluidity in acclimating *Tetrahymena* cells. *Biochemistry* 15, 5218-5227.

McMorris, T.C., Seshadri, R., Weihe, G.R., Arsenault, G.R., and Barksdale, A.W. (1975). Structure of oogonial-1,-2 and -3, steroidal sex hormones of the water-mold, *Achlya*. *J. Amer. Chem. Soc.* 97, 2544-2545.

Mills, G.L. and Cantino, E.C. (1980). The glycolipid involved in chitin synthesis by zoospores of *Blastocladiella emersonii*. *Experimental Mycol.* 4 (in press).

Mizunuma, T. (1966). Studies on the metabolism of *Aspergilli*. *Agr. Biol. Chem.* 30, 742-749.

Nelson, R.R., Huisingh, D., and Webster, R.K. (1967). Sexual differentiation in *Cochliobolus carbonum* as influenced by inhibition and repair of steroid biosynthesis. *Phytopathology* 57, 1081-1085.

Nes, W.R., Sekula, B.C., Nes, W.D., Adler, J.H. (1978). The functional importance of structural features of ergosterol in yeast. *J. Biol. Chem.* 253, 6218-6225.

Nickerson, K.W. and Leastman, E. (1978). Cerulenin inhibition of spore germination in *Rhizopus stolonifer*.

Experimental Mycol. 2, 26-31.

Nishi, A. (1961). Role of polyphosphate and phospholipid in germinating spores of *Aspergillus niger*. *J. Bacteriol.* 81, 10-19.

Ohno, T., Awaya, J., and Omura, S. (1976). Inhibition of sporulation by cerulenin and its reversion by exogenous fatty acids in *Saccharomyces cerevisiae*. *Antimicrobial Agents Chemother.* 9, 42-48.

Oldfield, E. and Chapman, D. (1972). Dynamics of lipids in membranes: Heterogeneity and the role of cholesterol. *FEBS Letters* 23, 285-297.

Owens, R.G. (1955). Metabolism of fungus spores. I. Oxidation and accumulation of organic acids by conidia of *Neurospora sitophila*. *Contri. Boyce Thompson Inst.* 18, 125-144.

Parks, L.W., MacLean-Bowen, C., Taylor, F.R., and Hough, S. (1978). Sterols in yeast subcellular fractions. *Lipids* 13, 730-735.

Parks, L.W. and Stromberg, V.K. (1978). Measurement *in vitro* of the esterification of yeast sterols. *Lipids* 13, 29-33.

Phornpiboonya, Y. and Jack, R.C. (1980). Changes in stereospecific distribution of yeast fatty acids with age. *Chem. Phys. Lipids* 26, 57-66.

Popplestone, C.R. and Unrau, A.M. (1974). Studies on the biosynthesis of antheridiol. *Can. J. Chem.* 52, 462-468.

Pugh, E.L. and Kates, M. (1975). Desaturation of phosphatidylcholine and phosphatidylethanolamine by a microsomal enzyme system from *Candida lipolytica*. *Biochim. Biophys. Acta.* 316, 305-316

Quain, D.E. and Haslam, J.M. (1979). The effects of catabolite derepression on the accumulation of sterol esters and the activity of β-hydroxymethylglutaryl-CoA reductase in *Saccharomyces cerevisiae*. *J. Gen. Microbiol.* 111, 343-351.

Ragsdale, N.N. (1975). Specific effects of triarimol on sterol biosynthesis in *Ustilago maydis*. *Biochim. Biophys. Acta.* 380, 81-96.

Reisener, H.J. (1976). Lipid metabolism of fungal spores during sporogenesis and germination. *In* "The Fungal Spore". pp. 165-187. John Wiley and Sons, New York.

Reisener, J.H., Finlayson, A.J., McConnell, W.B., and Ledingham, G.A. (1963). The metabolism of propionate by wheat stem rust uredospores. *Can. J. Biochem. Physiol.* 41, 737-743.

Rose, A.H. (1977). Chemical nature of membrane components. *In* "The Filamentous Fungi" (Ed. J.E. Smith and D.R. Berry). pp. 308-327. John Wiley and Sons, New York.

Rousseau, P. and Halvorson, H.O. (1973). Physiological changes following the breaking of dormancy of *Saccharomyces cerevisiae* ascospores. *Can. J. Microbiol.* 19, 547-555.

Schousboe, I. (1976). Triacylglycerol lipase activity in baker's yeast (*Saccharomyces cerevisiae*). *Biochim. Biophys. Acta.* 424, 366-375.

Sherald, J.L. and Sisler, H.D. (1975). Antifungal mode of action of triforine. *Pest. Biochem. Physiol.* 5, 477-488.

Shu, P., Tanner, K.G., and Ledingham, G.A. (1954). Studies on the respiration of resting and germinating uredospores of wheat stem rust. *Can. J. Bot.* 32, 16-23.

Singer, S.J. and Nicolson, G.L. (1972). The fluid mosaic model of the structure of cell membranes. *Science* 175, 720-731

Sols, A., Gancedo, C., and Delafuente, G. (1971). Energy-yielding metabolism in yeasts. *In* The yeasts. (Ed. Rose and Harrison). Vol. II. pp. 271-307. Academic Press, New York.

Steele, S.D. and Miller, J.J. (1974). Ultrastructural changes in germinating ascospores of *Saccharomyces cerevisiae*. *Can. J. Microbiol.* 20, 929-933.

Suzuki, T. and Hasegawa, K. (1974). Lipid molecular species of *Lipomyces starkeyi*. *Agr. Biol. Chem.* 38, 1485-1492.

Taketani, S., Nagai, J., and Katsuki, H. (1978). Quantitative aspects of free and esterified sterol in *Saccharomyces* under various conditions. *Biochim. Biophys. Acta.* 528, 416-423.

Thompson, E.D. and Parks, L.W. (1977). Lipids associated with cytochrome oxidase derived from yeast mitochondria. *Biochim. Biophys. Acta* 260, 601-607.

Turian, G. and Bianchi, D.E. (1972). Conidiation in *Neurospora*. *Bot. Rev.* 38, 119-154.

Weber, D.J. and Hess, W.M. (1976). The Fungal Spore: Form and Function. John Wiley & Sons (Wiley-Interscience), New York.

Weete, J.D. (1974). Fungal Lipid Biochemistry. Plenum Publishing Corporation, New York.

Weete, J.D. (1980). Lipid Biochemistry of Fungi and Other Organisms. Plenum Publishing Corporation, New York.

Weete, J.D., Lawler, G.C., and Laseter, J.L. (1973). Total lipid and sterol components of *Rhizopus arrhizus*: Identification and metabolism. *Arch. Biochem. Biophys.* 155, 411-419.

BIOGENESIS OF SPOROPOLLENIN IN FUNGAL SPORE WALLS

G.W. Gooday

Department of Microbiology, University, Aberdeen AB9 1AS, UK

INTRODUCTION

Sporopollenin is the remarkably resistant material characterising the exine of pollen grains of spermatophytes and spores of pteridophytes. Until recently it was thought to be confined to these plants, but now it has been found in fungi, algae and bacteria. Its chemical and physical resilience implies that fungal spores that contain it in their walls should have added resistance to adverse chemical and physical conditions.

CHEMISTRY OF SPOROPOLLENIN

The term sporopollenin was coined by Fritz Zetsche, who first established the identity of this group of biopolymers at the University of Berne between 1919 and 1936. Zetsche and Vicari (1931) demonstrated that the 'sporonin' of lycopodium spore exines was very similar to the 'pollenin' of pollen grain exines - hence 'sporopollenin'.

Sporopollenin, like chitin and cellulose, has no effective non-destructive solvent, and so it has to be investigated by dissolving, hydrolysing or degrading the material that accompanies it in the cell wall. Str organic bases, such as 2-aminoethanol dissolve some samples of sporopollenin, at least on a small scale on a microscope slide (Southworth, 1974), but this has not been investigated chemically. It is a polymer of indeterminate molecular weight. Following the work of Zetsche, empirical formulae are arbitrarily expressed as C90 units. It consists of carbon, hydrogen and oxygen, with a characteristically low hydrogen content (Table 1). Empirical formulae vary for sporopollenins of different sources.

TABLE 1

*Compositions of Sporopollenins**

Species	Source	Formula
Mucor mucedo	Zygospore	$C_{90}H_{130}O_{33}$
Neurospora crassa	Ascospore	$C_{90}H_{128}O_{63}$
Lycopodium clavatum	Clubmoss spore	$C_{90}H_{144}O_{27}$
Lilium henryi	Pollen grain	$C_{90}H_{142}O_{36}$
Chara corallina	Algal oospore	$C_{90}H_{117}O_{28}$
Tasmanites punctatus	Fossil alga	$C_{90}H_{132}O_{16}$
Polymer of β-carotene, with O_2		$C_{90}H_{130}O_{30}$

*From Shaw (1971); Brooks (1971); Gooday *et al.*(1974).

Sporopollenin contains hydroxyl groups, typically about thirteen for each C90 unit, and C-methyl groups, typically about three for each C90 unit (Shaw, 1971). It is highly unsaturated, and so absorbs ultraviolet light (Southworth, 1969). It is resistant to degradation by alkali and by strong acids such as sulphuric, hydrochloric, hydrofluoric and phosphoric acids. It can be degraded readily only by strong oxidising agents, such as ozone followed by hydrogen peroxide, sodium hypochlorite, chromic acid, and strong nitric acid. The ozone/peroxide treatment gives simple dicarboxylic acids (C3, C4, C5, C6), accounting for about 40% of the total structure, together with simple long straight and branched chain monocarboxylic acids, and smaller amounts of longer chain dicarboxylic acids. Potash fusion at 250°C gives the same pattern of dicarboxylic acids, and phenolic acids, mainly *p*- and *m*-hydroxybenzoic acids (Shaw and Yeadon, 1966). (The phenolic acids suggested that sporopollenin might have, in part, a lignin-like structure, but Shaw and Yeadon discounted this idea by showing that prior treatment with diazomethane did not give rise to the methoxy aromatic acids that would be expected from lignin structures).

Brooks and Shaw (1968a) have rationalised these structural details as being totally consistent with sporopollenin being an oxidative polymer of carotenoids. Such a polymer would be expected to retain substantial amounts of unsaturation, to contain hydroxyl and C-methyl groups, and to give the carboxylic acids after ozonolysis and the phenolic acids after potash fusion. Polymerisation of carotenoids with oxygen, in the presence of the ionic catalyst boron trifluoride, gives materials with all of the characteristics of natural sporopollenin (e.g. Table 1).

Brooks and Shaw (1978) illustrate possible molecular structures for sporopollenins derived from carotenoids and from carotenoid esters (cf. Fig.11).

CHARACTERISATION OF SPOROPOLLENIN

In order to characterise sporopollenin in biological material it is necessary to extract the sample exhaustively with organic solvents (to remove lipids), water (to remove carbohydrates and amino acids), alkali (to remove proteins), acid (to remove polysaccharides by hydrolysis), and finally by acetolysis (with acetic anhydride: sulphuric acid at 100°C)(Gooday *et al.*,1974). In particular, survival of material after acetolysis is indicative of sporopollenin. This residue can then be investigated by chemical and physical techniques and compared with other known sporopollenin preparations. As well as the chemical characteristics described in the previous section, the sporopollenins can be 'fingerprinted' by their infrared spectra and their pyrolysis gas chromatograms (Figs. 1,2).

The presence of sporopollenin can also be inferred by its dissolution by strong oxidising agents such as hypochlorite (e.g. 'Chloros') or hydrogen peroxide. Such treatments of course will affect other cell components, for example melanin.

There is no specific cytochemical stain which will identify sporopollenin by microscopy. Pollen grain exines are autofluorescent in ultraviolet light (Willemse, 1971) and stain with the fluorochrome primuline (Waterkeyn and Bienfait, 1971)(as do many other cell components), and with Fast blue B (Ahokas, 1978). The basophilia,

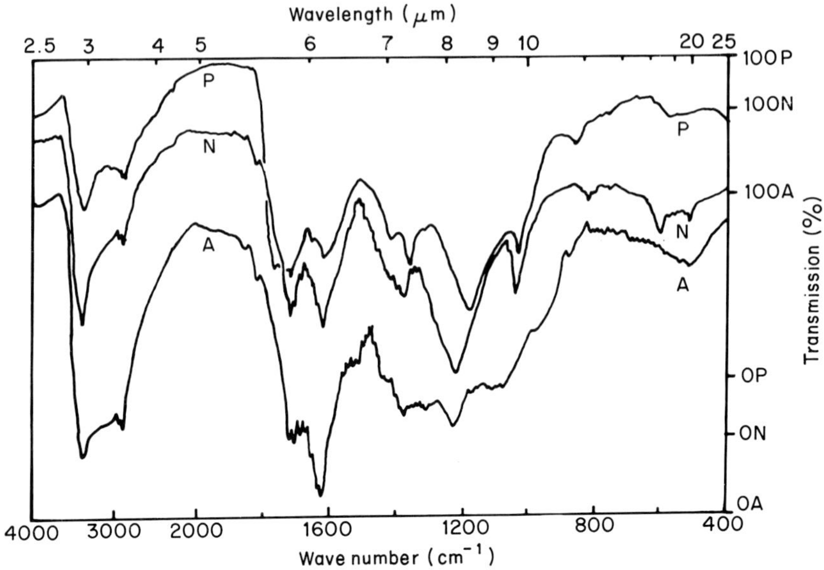

Fig.1. Infrared spectra (KBr discs) of sporopollenin preparations from: P Phycomyces blakesleeanus *sporangiospores; N* Neurospora crassa *ascospores; A* Ascobolus crenulatus *ascospores. All prepared by extraction with solvents, alkali and phosphoric acid; N also by acetolysis. For probable assignments of bands, see Furch & Gooday (1978), Wickramasinghe* et al. *(1977).*

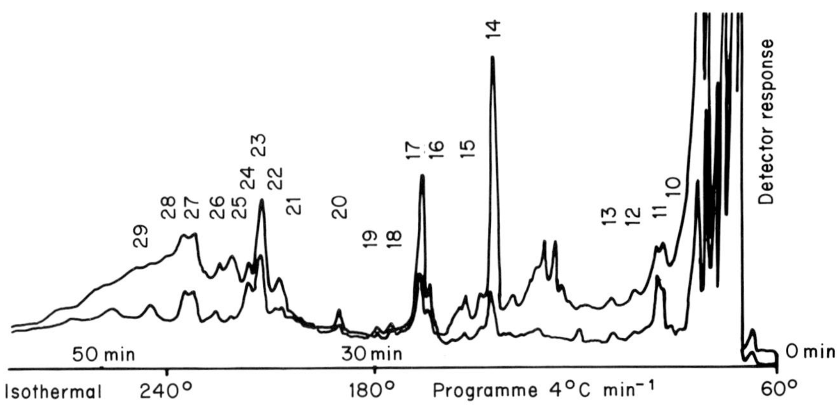

Fig.2. Capillary column pyrolysis gas chromatographs of (upper trace) sporopollenin from Mucor mucedo, *and (lower trace) a synthetic oxidative polymer of* β-*carotene. Characteristic peaks are numbered from origin at right hand side. (30m x 0.76mm stainless steel WCOT column, flow 6ml/min at 20°C; doubling in sensitivity after peak 13).*

osmophilia, and red colour with sulphuric acid, that are shown by pollen exines, are considered by Rowley and Prijanto (1977) to be caused by anionic material, probably acidic polysaccharides, being entrapped in the sporopollenin during its deposition, and they conclude that its reactivity to stains is less than has been thought.

OCCURRENCE OF SPOROPOLLENIN

Where its cellular site has been identified, sporopollenin has been found only as a component of the walls of spores or vegetative cells. It is not known to form intracellular structures. At present, its distribution appears sporadic through nature, but this is through want of a thorough systematic search for it, and certainly it is more widespread than this account will suggest. It has not been found in any organism that totally lacks C40 carotenoids.

Prokaryotes

Carotenoids are sporadically distributed among bacteria (Goodwin, 1972), where they play roles in photosynthesis and photoprotection (Krinsky, 1978). Strohl *et al.* (1977) have initiated a seach for sporopollenin in bacteria. In a survey including, among others, species of *Myxococcus*, *Cytophaga*, *Micrococcus* and *Bacillus*, only *Myxococcus* strains contained sporopollenin. Both vegetative cells and myxospores of these carotenogenic myxobacteria contained material resistant to hydrolysis, that stained with primuline, that was digestible in chromic acid, and which had infrared spectral characteristics of sporopollenin.

Zygomycotina

Gooday *et al.* (1973) demonstrated the presence of sporopollenin in the zygospore wall of *Mucor mucedo*. This material was prepared by the exhaustive extraction and acetolysis procedure. It has the following properties in common with other sporopollenin preparations:
a) a low hydrogen content, with only C,H and O (Table 1);
b) the infrared spectrum;
c) the pyrolysis gas chromatogram pattern (Fig.2);
d) dicarboxylic acids, 3-6 carbon atoms, as major products after ozonolysis;
e) *m*- and *p*-hydroxybenzoic acids as major phenolic acids after potash fusion.

Of the two major layers of the zygospore wall (cf. Hawker and Beckett, 1971) the following cytochemical investigation places the sporopollenin in the outer layer. The inner layer is translucent, elastic, and strongly birefringent in polarized light. It dissolves readily in 80% phosphoric acid. It does not stain for chitosan (I_2/KI in 1% H_2SO_4), except after treatment with alkali (23M-KOH, 120°C, 3h; which deacetylates chitin to chitosan), when it stains strongly purple. It thus appears to be substantially chitin.

The outer layer is black, warty and brittle. Its constituents include chitin, chitosan, melanin and sporopollenin. The warty spines contain chitin and chitosan (staining procedure as for inner layer). The black pigment is melanin, as shown by its bleaching with H_2O_2 (30 vol., 48h, 20°C) or 0.1% $KMnO_4$ (16h, 20°C, followed by washing with water and 1% oxalic acid), its staining green with Fe^{2+} (Pearse, 1972), and its dissolution in alkali (10% aqueous KOH or 6% methanolic KOH; both 16h, 20°C). The peroxide and permanganate treatments lead to disintegration of this outer layer (Fig.3) and as these oxidations would degrade sporopollenin this suggests that the sporopollenin may play a structural role in holding this layer together. Treatment of the zygospore wall with 80% phosphoric acid and acetolysis, however, causes a progressive disintegration of the outer layer (Fig.4), first into individual warty spines, and finally to give sporopollenin as an amorphous material about 3% dry weight of the original intact zygospore. The sporopollenin thus does not form a complete coherent layer.

Vegetative hyphae, sporangiophores, sporangiospores and zygophores of *Mucor mucedo*, and zygospores of *Rhizopus sexualis* are all completely dissolved by the extraction procedure, and so contain no detectable sporopollenin. The absence of sporopollenin from *Rhizopus sexualis* correlates with the lack of detectable carotenoids in this fungus.

The investigation of *Phycomyces blakesleeanus* has shown that, as expected(by analogy with *Mucor mucedo)* for this strongly carotenogenic organism, the zygospore wall contains sporopollenin (at about 1.4% dry weight of the intact spores) Also, however, sporopollenin has been detected in the sporangiospores at about 0.6%, the sporangiophores at about 3.6%, and the thorns on the suspensor cells subtending the zygospores, but not in the vegetative hyphae (Furch and Gooday, 1978). Furch and Pambor (1978) have localized

Figs 3-10. Nomarski interference micrographs. Scale bars 100μm (Figs 3,4) and 10μm. (Figs 5-10).

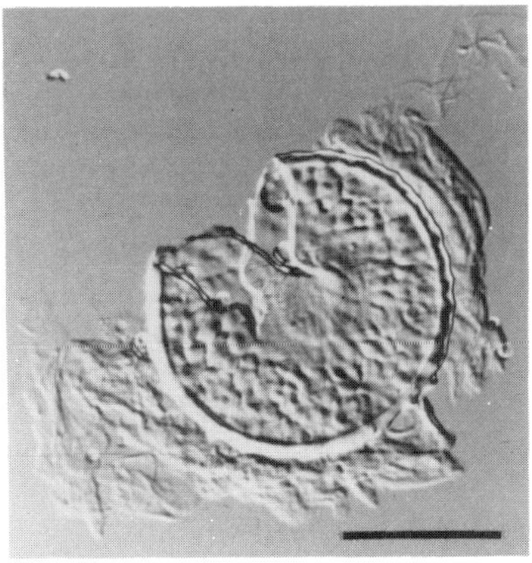

Fig.3. Zygospore of Mucor mucedo *treated with hydrogen peroxide (30vol., 20°C, 24h). Outer warty layer has disintegrated.*

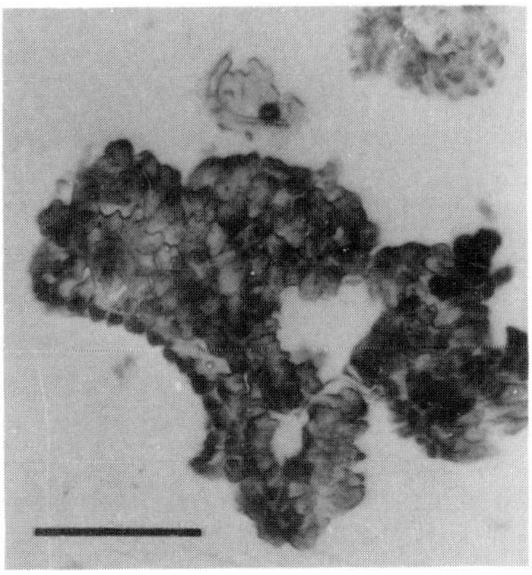

Fig.4. Zygospore of M.mucedo *treated with phosphoric acid (80%, 50°C, 24h). Inner smooth layer of wall has dissolved, outer warty layer is disintegrating.*

Fig.5. Sporopollenin preparation of ascospores of Neurospora tetrasperma.

Fig.6. Sporopollenin preparation of ascospores of Neurospora crassa.

the sporopollenin as being in the thick warty secondary layer of the zygospore wall (cf. description of zygosporogenesis by O'Donnell *et al.*, 1976); and in the perispore and adjacent part of the epispore of the sporangiospore wall, appearing as a wrinkled shell after acetolysis (cf. description of sporogenesis by Tu and Malhotra, 1976). During the first four hours of germination, Furch and Pambor estimate the sporopollenin as remaining at about 10% of the spore wall, in contrast to changes in other wall components, and their micrograph shows the perispore layer disrupted, as has been described for germinating spores by Grove (1976). Preparations from the sporangiophores appear as thin translucent sheets of broken irregular outline, but it is unclear whether or not this represents an intact layer around the living sporangiophore.

Ascomycotina

Ascospores of the pyrenomycetes *Neurospora crassa* and *Neurospora tetrasperma* have high contents of sporopollenin, at about 12% of their dry weight (Gooday *et al.*, 1974). The exhaustive extractions of other cell materials leave coherent, brown, sculptured shells (Figs.5 and 6), which are resistant to further treatment by acetolysis, and have the chemical properties of sporopollenin, such as empirical formula (Table 1) and infrared spectrum (Fig.1). The wall structure of *Neurospora* species has been investigated in detail by Sussman (1966), Lowry and Sussman (1968) and Austin *et al.* (1974) and from their descriptions it is clear that the sporopollenin is localized in the outer ribbed layer, the perispore layer, which is removed readily from spores by treatment with oxidizing agents such as 1% sodium hypochlorite solution (e.g. Fig.9 of Austin *et al.*). Beckett (1976) has investigated the wall structure of the pyrenomycete *Daldinia concentrica*. He finds that two of the outer wall layers, designated W2 and W3, have properties that suggest they contain sporopollenin, as they withstand the exhaustive extractions and acetolysis to give 'spore-shaped shells with a regular split along the line of the germ fissure', but are disintegrated by treatment with 1% sodium hypochlorite.

Ascospores of the discomycete *Ascobolus crenulatus* (Fig. 7) have sporopollenin in their walls, as evidenced by the resistance of the ribbed outer layer to the exhaustive extraction and to acetolysis (Figs.8 and 9), and the infra-red spectrum of the resultant preparation (Fig.1). This ribbed layer, the perispore layer (cf. description of

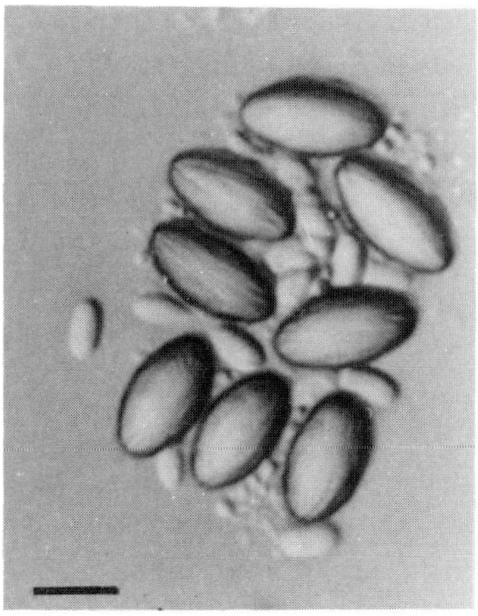

Fig.7. Ascospores and oidia of Ascobolus crenulatus.

Fig.8. Sporopollenin prep. (organic solvents, alkali, phosphoric acid) of ascospores of A.crenulatus.

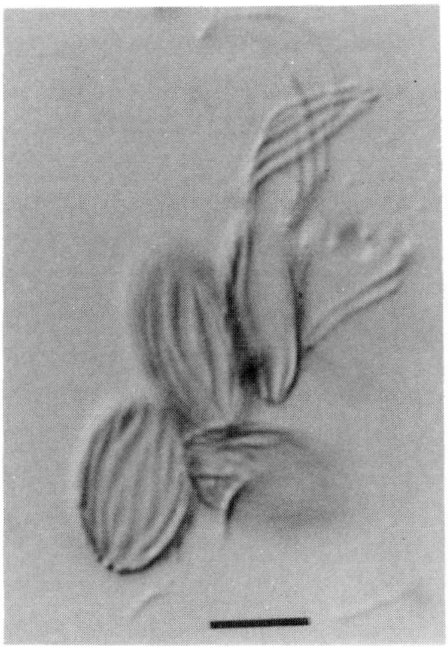

Fig.9 Sporopollenin prep. (as Fig.8, also acetolysis) of ascospores of Ascobolus crenulatus.

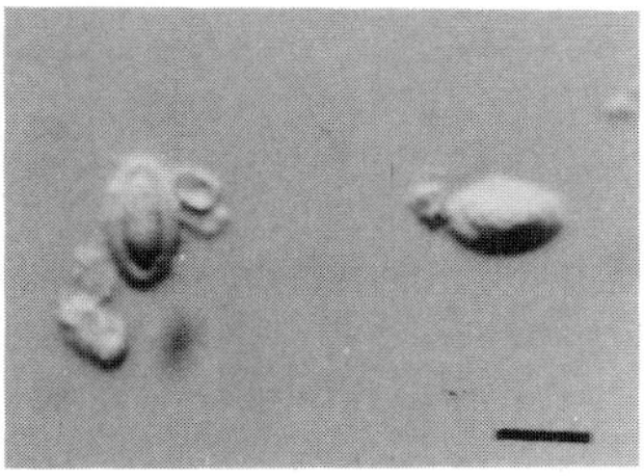

Fig.10. Ascospores of A. crenulatus *treated with* H_2O_2 *(30vol., 20°C,24h). Outer ribbed wall layer has dissolved.*

ascospore formation in *Ascobolus stercorarius* by Wells, 1972), is rapidly degraded by treatment with oxidizing agents (Fig.10). Ascospores of the carotene-rich discomycete *Aleuria aurantia*, however, are solubilised completely by the extraction procedure, as are oidia of *Ascobolus crenulatus* (Fig.7), and so do not contain sporopollenin. Brooks and Shaw (1977) note its presence in spores (conidia ?) of *Aspergillus niger* and *Penicillium brevi*.

Other fungi

There is currently no evidence for sporopollenin in other fungi. No systematic search has been instituted, however. Samples of RS plants of *Blastocladiella emersonii* and *Blastocladiella britannica*, and plants of *B. emersonii BEM* (a mutant unable to produce RS plants but still producing γ-carotene)(all kindly supplied by Professor E.C. Cantino) were totally soluble in the phosphoric acid treatment, and so did not contain sporopollenin.

Algae

Sporopollenin has been identified in the cell walls of some strains of species of the Chlorophycean genera *Chlorella*, *Coelastrum*, *Scenedesmus*, *Prototheca*, *Pediastrum*, and *Phycopeltis* (Atkinson *et al.*, 1972; Marchant, 1977; Pickett-Heaps and Staehelin, 1975; Brooks and Shaw, 1977; Good and Chapman, 1978). It is characterised as a trilaminar structure in electron micrographs of cell walls that have been subjected to acetolysis.

Sporopollenin is also present in the oospore walls of *Chara vulgaris* and *Chara corallina*. The exhaustive extraction procedure and acetolysis gave a residue of brittle sculptured fragments of these shells, at about 4% of the original dry weight of the oospores of *Chara vulgaris*, with the typical chemical properties of sporopollenin, such as empirical formula (Table 1). Unfertilized oogonia were digested totally, consistent with the sporopollenin being deposited after fertilization as a component of the black decorated layer of the oospore wall.

Plants

Sporopollenin is present in the exine of pollen grains (Heslop-Harrison, 1976), and of the spores of pteridophytes, such as ferns and clubmosses (Brooks and Shaw, 1977). Its structure gives them their characteristic sculpturing, and so enables identifications of plant species to be made in the pollen analysis of sediments and peat deposits. As might be expected from such a polymer, comparative studies (e.g. Shaw, 1971) have suggested that the exine sporopollenins from different species are not identical, but show a range of chemical properties. The sporopollenins of a single exine also show different properties (Heslop-Harrison, 1971), for example that of the outer sculpturing being degraded more readily than that of the inner smoother layer (Southworth, 1974).

Fossils

The remarkable resistance of sporopollenin to non-oxidative chemical degradation means that structures containing it can be visualised in sedimentary rocks (Chaloner, 1976), thus contributing in large degree to the science of palynology. Of the structures discussed previously, oospores of *Chara* species, colonial shells of *Pediastrum* species, spores of pteridophytes and pollen grains of spermatophytes are identified commonly in fossil deposits. The microfossils, *Tasmanites* species, find their closest living counterparts in the cyst walls of the prasinophycean planktonic algae, *Pachysphaera* species. Sporopollenin can be identified chemically from such fossils, and from some 'kerogens' from sedimentary rocks (Brooks and Shaw, 1968b, 1970).

Following the finding of sporopollenin in fungal spores, it is to be expected that the structures of such spores should have survived to be recognized in fossiliferous rocks. Unfortunately, however, the fossil fungus spore is the Cinderella of palynology, and usually is ignored, or at best assessed in a footnote to the main descriptions of microfossils of a particular sediment.

Brooks and Shaw (1969, 1973) suggest, from chemical analyses of the organic matter, that sporopollenin may be a constituent of carbonaceous meteorites, and thus spores may be carried by such chondrites as 'panspermia',an extraterrestrial origin of life on earth. Wickramasinghe et al. (1977) draw attention to the similarity of the infrared spectra of interstellar dusts to those of sporopollenins, but Sagan and Khare (1979) conclude that sporopollenin itself is very unlikely to be a constituent of the galactic clouds.

BIOSYNTHESIS OF SPOROPOLLENIN

Brooks and Shaw (1968a) provide the only plausible suggestion for the origin of sporopollenin - that it is formed by the oxidative polymerisation of carotenoids and carotenoid esters. Fig.11 illustrates possible sites of oxygen insertion that are required, and is based on a suggested structure for sporopollenin presented by Brooks and Shaw (1977). Some of the evidence supporting this origin is as follows.

(a) In anthers, sporopollenin deposition in the pollen exines follow closely on carotenoid accumulation. In *Mucor mucedo*, sporopollenin deposition in zygospore walls follows on the accumulation of carotene in the mating mycelium consequent on the action of the sex hormone, trisporic acid (Gooday *et al.*,1973; Jones *et al.*, 1980).

(b) Synthetic polymers can be made by reacting solutions of carotenoids with oxygen in the presence of the inorganic catalyst, boron trifluoride. These polymers show a close resemblence to natural sporopollenins (Brooks, 1971; Shaw, 1971). For example, the polymerised β-carotene is very similar to the sporopollenin from zygospores of *Mucor mucedo*, and β-carotene is the major carotenoid of this fungus.

(c) Sporopollenin becomes labelled after feeding tissues with radioactive carotenoids, or their precurosors acetate and mevalonate: in pollen grains (Brooks and Shaw, 1978); in zygospores of *Mucor mucedo* (Gooday *et al.*, 1973); and in ascospores of *Neurospora crassa* (Gooday *et al.*, 1974). The efficiences of incorporations in these experiments have been low, but the carotenoids are exceedingly insoluble in water and their uptake into cells from the emulsions that have been used is probably very low.

As stated earlier, sporopollenin has not been reported from any organism totally devoid of C40 carotenoids. Albino mutants of *Neurospora crassa* produced sporopollenin in their ascospores (Gooday et al., 1974), but such mutants still produce C40 carotenoids, with altered ratios of pool sizes of colourless precursors (Subden and Turian, 1970). No totally caroteneless mutants are known of the carotene-producing fungi that have been investigated, such as *Neurospora crassa* and *Phycomyces blakesleeanus* (and this itself suggests a more fundamental role of carotenoids in these organisms than has hitherto been supposed).

Nothing is known of the biochemistry of sporopollenin formation. Techniques of hydrophobic biochemistry will have to be developed to investigate the oxidation of carotenoids. In *Mucor mucedo*, β-carotene is present in oil droplets and as crystals, especially in zygophores and developing zygospores. Clearly, any enzymic activity will have to act on carotenoids in a lipophilic phase, presumably within membrane matrices. One proposed oxidation of the C40 β-carotene in *Mucor mucedo* is its cleavage to C20 retinal, via a specific dioxygenase, as the first step in the biosynthesis of the sex hormone, trisporic acid (Jones *et al.*,1980), but this reaction has no obvious relationship to the formation of sporopollenin. Further oxidations of the apocarotenoids during the biosynthesis of trisporic acid are at position 4 (shown oxidized in Fig.11) and at one of the gem-dimethyl groups of the β-ionone ring (not shown oxidized in Fig.11) and it is possible that oxygenases involved in trisporate biosynthesis could act during sporopollenin formation in this fungus. The major carotenoid of *Neurospora crassa* is usually the C35 acid, neurosporaxanthin, the formation of which must also involve oxygenase attack on a C40 precursor at position 4. Another oxidation of carotenoids, occurring in chloroplasts, is epoxide formation, of no known function, for example of zeaxanthin to violaxanthin (Krinsky, 1978). Carotenoids are also powerful quenchers of singlet oxygen, but only a small proportion of such molecules react chemically with the oxygen to yield oxidation products (Krinsky, 1978). Thus at present there is no clue as to whether the cross-linking of the hydrocarbon chains shown in Fig.11 could be via epoxides (formed as a result of the action of oxygenases and molecular oxygen), or via peroxides (analogous to the process of autoxidation of unsaturated fatty acids by lipoxygenases), or perhaps via a direct cross-linking mediated by oxygenases.

The only evidence of the cytological ontogeny on sporopollenin comes from detailed accounts of exine formation in pollen grains, where its deposition occurs on membranous lamellae and vesicles (Dickinson, 1976).

Sporopollenin appears to be associated with acidic polysaccharides: in pollen exines (Rowley and Prijanto, 1977); in the perispore layer of ascospores of *Neurospora crassa* (Sussman, 1966); and in the warty layer of zygospores of *Mucor mucedo*, which stains for acidic polysaccharides, probably mucoran, the polyglucuronide described by Bartnicki-Garcia and Reyes (1968), by the Alcian blue technique.

FUNCTIONS OF SPOROPOLLENIN

The remarkable resilience of sporopollenin can be expected to confer increased resistance to the following adverse environmental factors:
(a) chemical attack, for example by organic solvents or heavy metals;
(b) physical disruption;
(c) desiccation;
(d) ultraviolet light;
(e) enzymic lysis;
(f) microbial degradation.

Thus among the fungi it is an ideal component for the wall of a dormant spore. The ascospores of *Neurospora* species are very resistant to adverse agents, and both they (Sussman, 1966) and the sporangiospores of *Phycomyces blakesleeanus* (B.Furch, this volume) need a heat shock for germination. The sporopollenin in the wall of the sporangiophore of *Phycomyces blakesleeanus* must serve to reduce desiccation of this remarkable aerial structure, while the zygospores of *Mucor* species are dormant spores *par excellence*, and will not germinate in response to any known stimulus (cf. Gauger, 1965).

Among the algae the following properties may be ascribed to the presence of sporopollenin: strains of *Chlorella* species containing it in their walls were extracted poorly by ethanol and failed to stain with aniline blue, in contrast to those lacking it (Syrett and Thomas, 1973), and were in general more acceptable as symbionts of the green Hydra (Jolley and Smith, 1980); and strains of *Pediastrum* species containing it were resistant to microbia decomposition in soil (Gunnison and Alexander, 1975), and were exceptionally tolerant to mercury pollution (Richardso *et al.*, 1975). Such ecologically relevant examples can be expected in the future from fungal studies.

ACKNOWLEDGEMENTS

I thank Dr G. Shaw and his colleagues for chemical collaborations, and the many friends who have sent me cultures, samples and reprints.

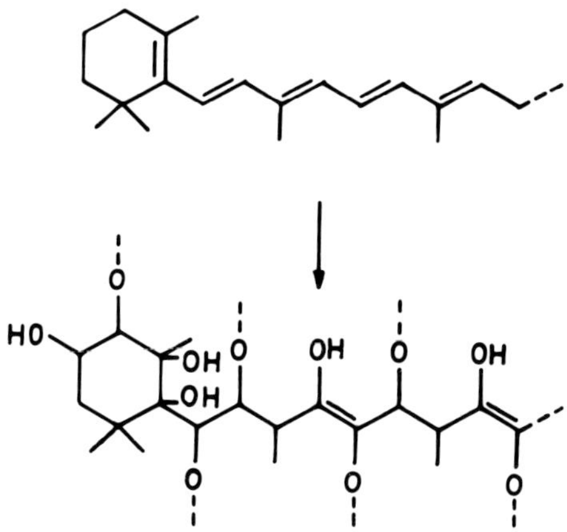

Fig. 11 . Possible sites of oxidation of β-carotene (half of molecule shown only) for polymerisation to sporopollenin (one small fragment shown only). Based on structure of sporopollenin suggested by Brooks and Shaw (1977).

REFERENCES

Ahokas, A.W. (1978). Cytoplasmic male sterility in barley II. Physiology and anther cytology of *msm* . *Hereditas* 89, 7-21.

Atkinson, A.W., Gunning, B.E.S. and John, P.C.L. (1972). Sporopollenin in the cell wall of *Chlorella* and other algae: ultrastructure, chemistry, and incorporation of 14C-acetate, studied in synchronous cultures. *Planta* 107, 1-32.

Austin, W.L., Frederick, L. and Roth, I.L. (1974). Scanning electron microscope studies on ascospores of homothallic species of *Neurospora*. *Mycologia* 66, 130-138.

Bartnicki-Garcia, S. and Reyes, E. (1968). Polyuronides in the cell walls of *Mucor rouxii*. *Biochimica et Biophysica Acta* 170, 54-62.

Beckett, A. (1976). Ultrastructural studies on exogenously dormant ascospores of *Daldinia concentrica*. *Canadian Journal of Botany* 54, 689-697.

Brooks, J. (1971). Some chemical and geochemical studies on sporopollenin. *In* "Sporopollenin" (Ed. J. Brooks, P.R. Grant, M. Muir, P. van Gijzel and G. Shaw), pp. 351-407. Academic Press, London.

Brooks, J. and Shaw, G. (1968a). Chemical structure of the exine of pollen walls and a new function for carotenoids in nature. *Nature, London* 219, 532-533.

Brooks, J. and Shaw, G. (1968b). Identity of sporopollenin with older kerogen and new evidence for the possible biological source of chemicals in sedimentary rocks. *Nature, London* 220, 678-679.

Brooks, J. and Shaw, G. (1969). Evidence for extraterrestrial life: identity of sporopollenin with the insoluble organic matter present in the Orgueil and Murray meteorites and also in some terrestrial microfossils. *Nature, London* 223, 756.

Brooks, J. and Shaw, G (1970). Relationship of kerogen and sporopollenin - a reply. *Nature, London* 227, 1975-196.

Brooks, J. and Shaw, G. (1973). "Origin and Development of Living Systems". Academic Press, London.

Brooks, J. and Shaw, G. (1977). Recent advances in the chemistry and geochemistry of pollen and spore walls. *Transactions of the Bose Research Institute* 40, 19-38.

Brooks, J. and Shaw, G. (1978). Sporopollenin: a review of its chemistry, palaeochemistry and geochemistry. *Grana* 17, 91-97.

Chaloner, W.G. (1976). The evolution of adaptive features in fossil exines. *In* "The Evolutionary Significance of the Exine" (Ed. I.K. Ferguson and J. Muller). pp. 1-14. Academic Press, London.

Dickinson, H.G. (1976). Common factors in exine deposition. *In* "The Evolutionary Significance of the Exine" (Ed. I.K. Ferguson and J. Muller). pp. 67-89.

Furch, B. and Gooday, G.W. (1978). Sporopollenin in *Phycomyces blakesleeanus*. *Transactions of the British Mycological Society* 70, 207-309.

Furch, B. and Pambor, L. (1978). Cell wall constituents of *Phycomyces blakesleeanus* 2. Localization of sporopollenin in the zygospores. *Microbios Letters* 4, 211-219.

Gauger, W.L. (1965). The germination of zygospores of *Mucor hiemalis*. *Mycologia* 57, 634-641.

Good, B.H. and Chapman, R.L. (1978). The ultrastructure of *Phycopeltis* (Chroolepidaceae:Chlorophyta) 1. Sporopollenin in the cell walls. *American Journal of Botany* 65, 27-33.

Gooday, G.W., Fawcett, P., Green, D. and Shaw, G. (1973). The formation of fungal sporopollenin in the zygospore wall of *Mucor mucedo* : a role for the sexual carotenogenesis in the Mucorales. *Journal of General*

Microbiology 74, 233-239.

Gooday, G.W., Green, D., Fawcett, P. and Shaw, G. (1974). Sporopollenin formation in the ascospore wall of *Neurospora crassa*. *Archives of Microbiology* 101, 145-151.

Goodwin, T.W. (1972). Carotenoids in fungi and non-photosynthetic bacteria. *Progress in Industrial Microbiology* 11, 31-88.

Grove, S.N. (1976). Form and function in zygomycete spores. *In* "The Fungal Spore : Form and Function" (Ed. D.J. Weber and W.M. Hess). pp.559-392. John Wiley, New York.

Gunnison, D. and Alexander, M. (1975). Basis for the resistance of several algae to microbial decomposition. *Applied Microbiology* 29, 729-738.

Hawker, L.E. and Beckett, A. (1971). Fine structure and development of the zygospore of *Rhizopus sexualis* (Smith) Callen. *Philosophical Transactions of the Royal Society of London B* 263 71-100.

Heslop-Harrison, J. (1971). The pollen wall: structure and development. *In* "Pollen: Development and Physiology" (Ed. J. Heslop-Harrison), pp. 75-98. Butterworths, London.

Heslop-Harrison, J. (1976). The adaptive significance of the exine. *In* "The Evolutionary Significance of the Exine". (Ed. I.K. Ferguson and J. Muller). pp. 27-37. Academic Press, London.

Jolley, E. and Smith, D.C. (1980). The green hydra symbiosis II. The biology of the establishment of the association. *Proceedings of the Royal Society of London B* 207, 311-333.

Jones, B.E., Williamson, I.P. and Gooday, G.W. (1980). Sex pheromones in Mucor. *In* "Sexual Interactions in Eukaryotic Microbes" (Ed. D.H. O'Day and P.A. Horgen), Academic Press, New York.

Krinsky, N.I. (1978). Non-photosynthetic functions of carotenoids. *Philosophical Transactions of the Royal Society of London B* 284, 581-590.

Lowry, R.J. and Sussman, A.S. (1968). Ultrastructural changes during germination of ascospores of *Neurospora tetrasperma*. *Journal of General Microbiology* 51, 403-409.

Marchant, H.J. (1977). Cell division and colony formation in the green alga *Coelastrum* (Chlorococcales). *Journal of Phycology* 13, 102-110.

O'Donnell, K.L., Hooper, G.R. and Fields, W.G. (1976). Zygosporogenesis in *Phycomyces blakesleeanus*. *Canadian Journal of Botany* 54, 2573-2586.

Pearse, A.G.E. (1972). "Histochemistry, Theoretical and Applied". Vol. 2. 3rd.Ed. Churchill, London.

Pickett-Heaps, J.D. and Staehelin, L.A. (1975). The ultrastructure of *Scenedesmus* (Chlorophyceae) II. Cell division and colony formation. *Journal of Phycology* 11, 186-202.

Richardson, T.R., Millington, W.F. and Miles, H.M. (1975). Mercury accumulation in *Pediastrum boryanum* (Chloropyceae) *Journal of Phycology* 11, 320-323.

Rowley, J.R. and Prijanto, B. (1977). Selective destruction of the exine of pollen grains. *Geophytology* 7, 1-23.

Sagan, C. and Khare, B.N. (1979). Tholins: organic chemistry of interstellar grains and gas. *Nature, London* 277, 102-107.

Shaw, G. (1971). The chemistry of sporopollenin. *In* "Sporopollenin" (Ed. J. Brooks, P.R. Grant, M. Muir, P. van Gijzel and G. Shaw), pp. 305-350. Academic Press, London.

Shaw, G. and Yeadon, A. (1966). Chemical studies on the constitution of some pollen and spore membranes. *Journal of the Chemical Society C*, 16-22.

Southworth, D. (1969). Ultraviolet absorption spectra of pollen and spore walls. *Grana Palynologica* 9, 5-15.

Southworth, D. (1974). Solubility of pollen exines. *American Journal of Botany* 61, 36-44.

Strohl, W.R., Larkin, J.M., Good, B.H. and Chapman, R.L. (1977). Isolation of sporopollenin from four myxobacteria. *Canadian Journal of Microbiology* 23, 1080-1083.

Subden, R.E. and Turian, G. (1970). A mechanism for carotenoid synthesis in fungi involving a multifunctional enzyme complex. *Molecular and General Genetics* 108, 358-364.

Sussman, A.S. (1966). Types of dormancy as represented by conidia and ascospores of *Neurospora In* "The Fungus Spore" (Ed. M.F. Madelin). pp. 235-257. Butterworths, London.

Syrett, P.J. and Thomas, E.M. (1973). The assay of nitrate reductase in whole cells of *Chlorella:* strain differences and the effect of cell walls. *New Phytologist* 72, 1307-1310.

Tu, J.C. and Malhotra, S.K. (1976). The formation of sporangiospores in *Phycomyces*. *Microbios* 15, 15-25.

Waterkeyn, L. and Bienfait, A. (1971). Primuline induced fluorescence of the first exine elements and Ubisch bodies in *Ipomoea* and *Lilium*, *In* "Sporopollenin" (Ed. J. Brooks, P.R. Grant, M. Muir, P. van Gijzel and G. Shaw), pp. 108-129. Academic Press, London.

Wells, K. (1972). Light and electron microscopic studies of *Ascobolus stercorarius* II. Ascus and ascospore ontogeny. *University of California Publications in Botany* 62, 1-93.

Wickramasinghe, N.C., Hoyle, F., Brooks, J. and Shaw, G. (1977). Prebiotic polymers and infrared spectra of galactic sources. *Nature, London* 269, 674-676.

Willemse, M.T.M. (1971). Morphological and fluorescence microscopical investigation on sporopollenin formation in *Pinus sylvestris* and *Gasteria verrucosa*. *In* "Sporopollenin" (Ed. J. Brooks, P.R. Grant, M. Muir, P. van Gijzel and G. Shaw), pp. 68-107. Academic Press, London.

Zetsche, F. and Vicari, H. (1931). Untersuchungen über die Membran der Sporen und Pollen. *Helvetica Chimica Acta* 14, 58-78.

MELANINS

D.M. Rast, H. Stüssi, H. Hegnauer, and L.E. Nyhlén

Institute of Plant Biology, University of Zürich
Zollikerstrasse 107, CH-8008 Zürich, Switzerland

INTRODUCTION

Originally, the term melanin was restricted to dark-colored substances resulting from tyrosine oxidation (Thomson, 1962) whilst, later, pigments formed by the oxidative polymerization of non-nitrogenous phenolics were also included in this class of compounds (Nicolaus, 1968). They are insoluble in water, mineral acids and the usual organic solvents — but mostly soluble in aqueous alkali — are bleached by oxidizing agents and reduce ammoniacal silver nitrate (Nicolaus *et al.*, 1964). Melanins have remained somewhat ill-defined since even in a recent review article they are circumscribed as "products of high molecular weight formed by enzymic oxidation of phenols" (Thomson, 1976). Degradation, by alkali fusion or permanganate oxidation, followed by identification of the resulting products is still the only method to characterize a melanin chemically — a difficult task if this turns out to be of an unknown type. The solution of the problem may, however, be made easier if at least circumstantial evidence on the nature of the substrate of melanogenesis in the organism concerned can be revealed by physiological and biochemical studies, thus repeating the historical approach to the elucidation of the structural features of dopa melanin which has been unusual in that — in contrast to most other natural products — the starting material and some of the intermediates were discovered before the chemical structure of the product itself (Swan, 1974).

As is generally known, animals synthesize melanin from tyrosine, starting with the transformation of the monomer to dopa and dopaquinone. Peroxidase and/or phenolase (also called tyrosinase) catalyze these reactions (cf. Swan, 1974). The unstable quinone is presumed to be non-enzymatically converted into more complex quinones, such as dopachrome and indole-5,6-quinone, which condense in a proportionate fashion to form the melanin polymer (for the pathway of reactions, see e.g. Farishian and Whittaker, 1979). In Vertebrates, the entire sequence proceeds, for the most part, in discrete, membrane-limited particles called melanosomes. In the course of the melanization process these organelles are transformed into pigment granules (Seiji *et al.*, 1963 ; Jimbow and Tacheuchi, 1979).

In fungi, melanin appears to be confined to the cell wall region. Reports are numerous of its occurrence as electron-dense grains (Durrel, 1964 ; Tsao and Tsao, 1970 ; Griffiths and Swart, 1974 ; Ellis and Griffiths, 1974, 1975 ; Berthe *et al.*, 1976 ; Wheeler *et al.*, 1976 ; Reisinger *et al.* 1977 ; a.o.) which display a size range of about 30-150 nm, whereas melanin formed by autoxidation or by enzymatic polymerization of phenols other than the corresponding natural substrates of melanogenesis (see below) is amorphous (Wheeler *et al.*, 1978; Wheeler and Stipanovic, 1979). The few studies that have been carried out on the chemical composition of fungal melanins indicate that they might differ from the animal melanins inasmuch as the variety of phenolic monomers is concerned. Thus, dopa is but one of the substrates of melanogenesis in these organisms, e.g. in *Aspergillus nidulans* (Bull, 1970a). The assumption that dopa is also the precursor of the dark pigments of *Epicoccum nigrum, Humicola grisea, Amorphotheca resinae, Colletotrichum coccodes, Verticillium dahliae* and *Phomopsis* sp. (Ellis and Griffiths, 1974,1975) has been questioned (Bell *et al.*, 1976a ; Wheeler *et al.*, 1978) on the basis of the non-specificity of the method used to analyze the degradation products from the alkali fusion of these melanins and on account of the results of ultrastructural studies. The *V. dahliae* pigment, at least, is very likely derived from 1,8-dihydroxynaphthalene (Bell *et al.*, 1976a,b ; Stipanovic and Bell, 1976), as are the *Daldinia concentrica* (Allport and Bu'Lock, 1958,1960), *Thielaviopsis basicola* (Wheeler and Stipanovic,1979) and probably also the *Aspergillus niger* (cf. Nicolaus, 1968 ; Swan, 1974) melanins. The dark pigment of *Ustilago maydis* represents a catechol metabolite (Piattelli *et al.* 1965). In *Cochliobolus miyabeanus*, finally, catechol and dopa might both contribute to melanin formation (Ito *et al.*, 1979). *Capnodium nerii* (Nicolaus *et al.*, 1964) and *Alternaria* sp.

(Pridham and Woodhead, 1977) contain substances belonging to as yet unknown types of melanin.

Unlike in animals, the elucidation of the pathway of melanogenesis in fungi encompasses the question of the origin of the aromatic nucleus. The shikimate-chorismate as well as the polyketide pathways have been implied in this respect (in *A. nidulans* (Bull, 1970a), on the one hand, in *D. concentrica* (Allport and Bu'Lock, 1960) and *V. dahliae* (Bell *et al.*, 1976b; cf. also Aldridge *et al.*, 1974), on the other). In *Alternaria* sp., acetate and shikimate can both serve as melanin precursors (Pridham and Woodhead, 1977). Melanogenesis *sensu stricto* may then proceed in a way analogous to the formation of dopa melanin in animals.

Little is known about the physiological role of melanins in fungi : storage of water and ions (White, 1958), conferring resistance to lysis by other microorganisms (Potgieter and Alexander, 1966 ; Kuo and Alexander, 1967 ; Bull, 1970b) and protection against solar radiation (cf. Sussman, 1968) have been suggested as their function. Any further advance in the understanding of the physiological and ecological role of melanins at the biochemical level will probably require more information than presently available on the chemical nature, relationship to other wall polymers and biogenesis of melanins. Spores would appear to be the experimental material of choice to yield such data as they display intrinsically the highest capacity of fungal cells to produce melanin and represent a site where the biological function of these pigments can be thought to have the highest degree of multiplicity. However, the melanin of spores has been studied in only four species, i.e. *E. nigrum*, *H. grisea* (Ellis and Griffiths, 1974), *A. niger* (cf. Nicolaus, 1968 ; Swan, 1974) and *U. maydis* (Piattelli *et al.*, 1965), and none of these seems to be particularly well suited for experimental studies necessitating the large-scale production and handling of spores.

The spores of *Agaricus bisporus* display certain features and advantages that would appear to make them appropriate to serve as a model system for studying the chemistry and biochemistry of melanization in fungi. The most obvious of these are : (1) the non-pathogenicity of the organism (not even one case of opportunistic infection is known — in contrast to some other easily cultivated species with dark-pigmented spores), (2) the absence of deleterious immunoreactions such as occur with the spores of the mushroom *Pleurotus florida* (Schulz *et al.*, 1974 ; Zadražil, 1974) , (3) easy collection of spores

(Rast, 1965) , (4) high spore yields (ca. 0.5 g/kg of mushrooms : Stäuble, 1972) , (5) the heavy melanization of the spore wall which contributes about one third to the cell dry weight (Rast and Hollenstein, 1977) , and (6) the likelihood that at least part of the pigment exists in a physical state that is characteristic for the melanin of other spores, namely in the form of granules (see above) — though this cannot normally be discerned in the intact spore (Greuter and Rast, 1975). Grainy material is, however, present in wall fragments that have been treated such as to remove the inner, non-pigmented spore wall layer as well as part of the outer one (Hollenstein, 1975 ; Rast and Hollenstein, 1977). Still, the evidence is not complete, and further investigations are needed to prove this point.

The aim of the experimental work presented in this paper was to gather information on the identity and biosynthesis of the phenolic precursor of the dark-brown pigment of the *A. bisporus* spore wall and to study the chemical and ultrastructural properties of native and synthetic *Agaricus* melanin.

STRUCTURE AND SYNTHESIS OF THE *AGARICUS BISPORUS* SPORE WALL MELANIN

Biochemical aspects of melanogenesis

Nature and biosynthesis of the phenolic melanin precursor Besides α-1,3- and β-1,3-glucan the mushroom spore wall contains mainly lipid, protein, chitin and melanin (Hollenstein, 1975 ; Rast and Hollenstein, 1977). This contributes at least 12% to the dry weight of the wall (Stüssi, 1979). The elemental analysis of the dark amorphous powder obtained after removal of all other wall components by exhaustive extraction with solvents and drastic acid hydrolysis (a procedure that, though causing partial decarboxylation and deamination, does not basically alter the structure of the pigment : Nicolaus *et al.*, 1964) showed it to be a nitrogenous polymer. Its N-content (see Table 3) considerably exceeds that of other fungal melanins (Piattelli *et al.*, 1965 ; Pridham and Woodhead, 1977 ; Bull, 1970a ; Rowley and Pirt, 1972) and is even higher than that of most of the animal melanins (Nicolaus *et al.*, 1964 ; Nicolaus, 1968). For this reason, and because the patterns of the products obtained after permanganate oxidation and alkali fusion of the *A. bisporus* melanin were different from those observed after degrading dopa melanin (ibid.) the mushroom pigment was thought to be of an unknown type.

γ-Glutaminyl-4-hydroxybenzene (GHB ; Jadot *et al.*, 1960 ; Weaver *et al.*, 1971a) was considered to be the substrate of melanogenesis in *A. bisporus*, for the following reasons : (1) it is the only phenol of the mushroom fruit body present in quantity (Weaver *et al.*, 1971a) ; (2) its concentration increases steadily during the growth of the sporocarp (Rast *et al.*, 1979) ; (3) the developmentally-linked accumulation of the phenol relates mainly to the hymenium which, at the time of spore shedding, has a GHB content of more than 2% whereas the spores are almost devoid of GHB (ibid.); the corresponding diphenol, γ-glutaminyl-3,4-dihydroxybenzene (GDHB), is also present (Szent-Gyorgyi *et al.*, 1976 ; Stüssi, 1979); in contrast to GHB, GDHB occurs only after induction of fruiting (Stüssi and Rast, 1981), a fact fitting nicely with the observation that the substrate mycelium displays neither phenolase activity (Turner, 1974) nor melanization ; (6) the concentration of the diphenol — although expectedly always considerably smaller than that of the monophenol — increases with the age of the fruit body (Stüssi and Rast, 1981), the gradual building-up of the GDHB pool thus likely being responsible for the acceleration of the melanization process occurring during

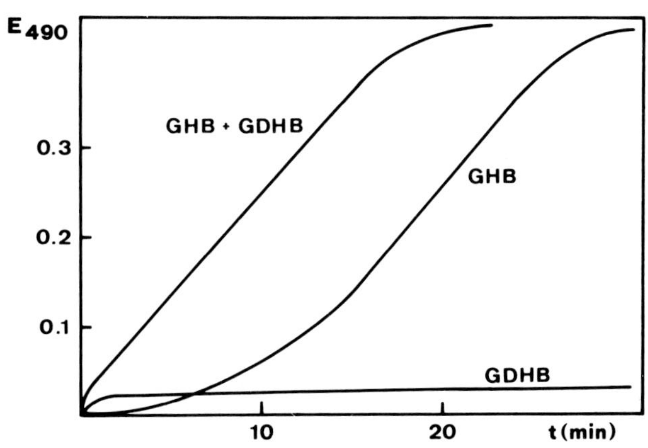

Fig. 1. Time-course of the oxidation of γ-glutaminyl-4-hydroxybenzene (GHB) by phenolase in the presence or absence of γ-glutaminyl-3,4-dihydroxybenzene (GDHB). The reaction medium consisted of GHB (2.1 mM) and/or GDHB (7.8 μM), mushroom tyrosinase (4000 U/mg ; 0.2 mg/ml), and phosphate buffer (0.05 M, pH 6.8). The reaction temperature was 22°.

sporogenesis, as the diphenol acts not only as a substrate of the phenolase but also as a co-substrate in the hydroxylation reaction (Boekelheide *et al.*, 1979) ; (7) in aqueous plectenchyme extracts the two phenols are quantitatively transformed into γ-glutaminyl-3,4-benzoquinone (GBQ : Weaver *et al.*, 1971b) which is then quickly converted into red, yellow and, finally, brownish products (Rast *et al.*, 1979) ; (8) when solutions of GHB and/or GDHB are supplemented with mushroom tyrosinase they behave in the same way as the tyrosine-dopa system : the diphenol 'activates' the enzyme (Swan, 1974 ; see also Boekelheide *et al.*, 1979) and, thus, eliminates the lag phase that is associated with the oxidation of the monophenol (Fig. 1).

To elucidate the biosynthetic pathway of the putative phenolic substrate of melanogenesis the compounds listed in Table 1 were tested for their ability to serve as GHB precursors. Shikimic acid proved to be outstandingly effective in

TABLE 1

Incorporation of labelled precursors into γ-glutaminyl-4-hydroxybenzene in the A. bisporus fruit body

Precursor $[^{14}C]$ [a]	Dilution factor [a] x
$NaH^{14}CO_3$	51282
L-Glutamine-U-^{14}C	499
L-Glutamic acid-U-^{14}C	493
L-Pyrrolidone carboxylic acid-U-^{14}C	463
Acetic acid-U-^{14}C	858
Benzoic acid-(ring)U-^{14}C	618
Glucose-U-^{14}C	781
Shikimic acid-U-^{14}C	16

[a] The administration of radioactive compounds, isolation of GHB and quantitation procedures were as described elsewhere (Stüssi and Rast, 1981).

this respect and was incorporated exclusively into the aromatic part of the molecule (Table 2). The dilution factor of glucose was relatively high, but a good proportion of the radioactivity found in GHB after application of the sugar was present in the *p*-aminophenol part. The other compounds tested are metabolically quite remote from the aromatic ring though they all contributed to its synthesis in yielding the side chain. Interestingly enough, most of the label of benzoic acid — fed as a model compound for soluble lignin breakdown products (cf. Higuchi, 1971) that might be transported from the substrate hyphae into the fruit body — was not found in the aromatic moiety of GHB, but in the side chain.

TABLE 2

Intramolecular distribution of radioactivity in γ-glutaminyl-4-hydroxybenzene after administration of various labelled precursors

Precursor $[^{14}C]^a$	Label found in Glutamyl moiety (%)[b]	Label found in Aromatic moiety (%)[b]
$NaHCO_3$	99	1
L-Glutamine	99	1
L-Glutamic acid	98	2
L-Pyrrolidone carboxylic acid	98	2
Acetic acid	95	5
Benzoic acid	94	6
Glucose	31	69
Shikimic acid	1	99

[a] For details of the tracer substances used, see Table 1.
[b] The determination of the labelling pattern was according to Stüssi and Rast (1981).

The dilution factors observed in GHB synthesis from $[^{14}C]$-shikimate, -glucose, and -acetate (Table 1) and the distribution of radioactivity in the two moieties of the phenol after administering these tracers (Table 2) thus demonstrate that GHB is derived from an early intermediate of the shikimate-chorismate pathway. It can, hence, be regarded as a metabolite of *p*-aminobenzoic acid (cf. Haslam, 1974). This is likely to be decarboxylated and hydroxylated to yield *p*-aminophenol. There is no reason to consider synthesis of GHB from acetate via a polyketide as an additional, though minor, pathway of

GHB synthesis since the small proportion of radioactivity found in the phenolic residue of GHB after feeding labelled acetate (Table 2) is probably the result of the gluconeogenetic activity of the fruit body hyphae (Wehrli and Rast, 1967). As benzoic acid undergoes ring fission (Table 2) and, hence, is very likely to be converted to succinate and acetyl-CoA (via β-ketoadipate : cf. Haslam, 1974), the slight labelling of the aromatic residue of GHB after feeding benzoic acid might — as in the case of acetate — be due to the tracer being in part metabolized to the GHB precursor glucose. Obviously, the fruit body possesses an autonomous system for the

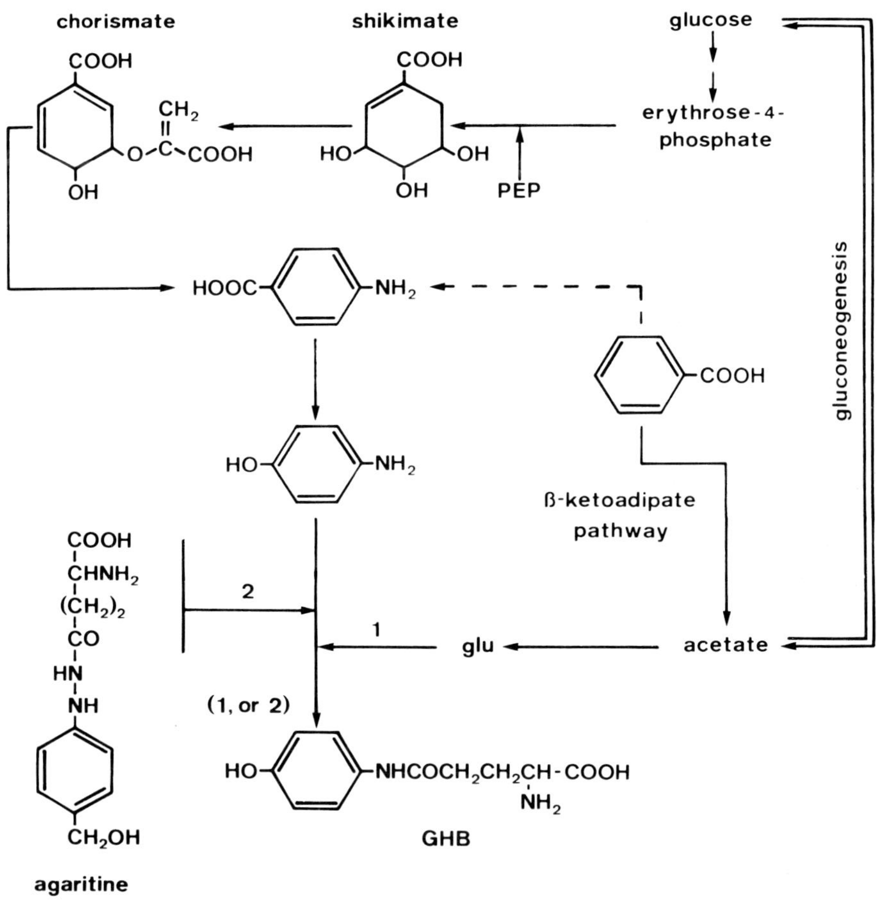

Fig. 2. Biosynthetic pathway of γ-glutaminyl-4-hydroxybenzene (GHB) in Agaricus bisporus.

sythesis of GHB, and melanogenesis does, therefore, not require a supply of preformed aromatics from the substrate hyphae. Glutamine, glutamic acid and pyrrolidone carboxylate displaying about the same dilution factor with respect to GHB synthesis (Table 1) can best be explained by assuming that they are freely interconvertible and constitute a common GHB precursor pool. Two possibilities may be considered for the formation of the peptide bond : (1) a reaction involving glutamic acid or pyrrolidone carboxylate, or (2) group transfer. The occurrence in the *A. bisporus* fruit body of agaritine (Levenberg, 1961,1964 ; Kelly *et al.*, 1962 ; for structural formula, see Fig. 2) and of a γ-glutamyltransferase with unusual substrate specificity (Gigliotti and Levenberg, 1964) that catalyzes the formation of GHB from *p*-aminophenol in the presence of agaritine would suggest (2) as the more likely possibility. A scheme summarizing the metabolic relationships between the various precursors tested and GHB is given in Fig. 2.

Synthesis and properties of mushroom melanins Melanin was prepared using synthetic GHB and commercial mushroom tyrosinase in a phosphate-buffered medium (pH 6.8) which was aerated and stirred for 30 h at room temperature (for more details, see Stüssi and Rast, 1981). As the enzyme preparation used

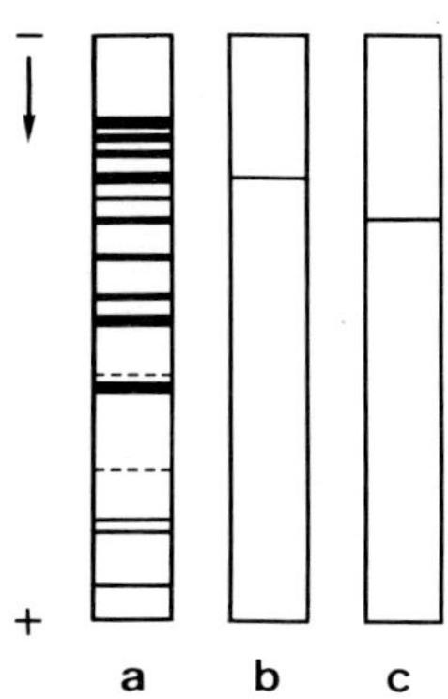

Fig. 3. Polyacrylamide gel electrophoresis of GHB oxidizing enzymes in the A. bisporus spore. a, protein pattern of extract ; b and c, gel stained for phenolase and peroxidase activities, respectively. For the preparation of the enzyme extract and conditions of analysis, see Stüssi and Rast (1981).

displayed not only phenolase but also peroxidase activity the question remains open which of the two enzymes was actually involved in the synthesis of the product thus obtained. This is a situation analogous to that existing in the field of melanogenesis in animals (see Introduction). Interestingly, the spore also contains a phenolase as well as a peroxidase (see Fig. 3).

TABLE 3

Properties of A. bisporus melanins

Property tested		Spore wall melanin	GHB melanin
Solubility in	KOH	+	+
	HCl	-	-
	org. solvents	-	-
Elemental composition* (C,H,N,S)		C 57.9;H 3.8 N 8.8;S 3.1	C 53.4;H 5.2 N 10.3

*Calc. on ash-free basis.

A comparison (Table 3) of some properties of GHB melanin and the melanin of mushroom spores (obtained according to the procedure used by Piattelli *et al.* (1965) to isolate pure *U. maydis* melanin) shows that the two substances have identical solubility characteristics with respect to KOH, HCl and a variety of organic solvents tested. They also have a similar elemental composition. The high sulfur content of the native *A. bisporus* melanin is noteworthy. As the drastic purification method used must have eliminated protein, the sulfur probably comes from an amino acid — likely enough cystein — that is linked to the melanin and originally served to bind it to protein, as in the melanin of cuttlefish (Piattelli *et al.*,1963). The IR-spectra of the two melanins are practically identical (Fig. 4). Though little information on the structural details of melanins can be gained by IR measurements, one point may be worth mentioning : the strong band occurring at about 1100 cm^{-1} is not present in the spectrum of any other fungal melanin (whether of animal, higher plant or fungal origin : see

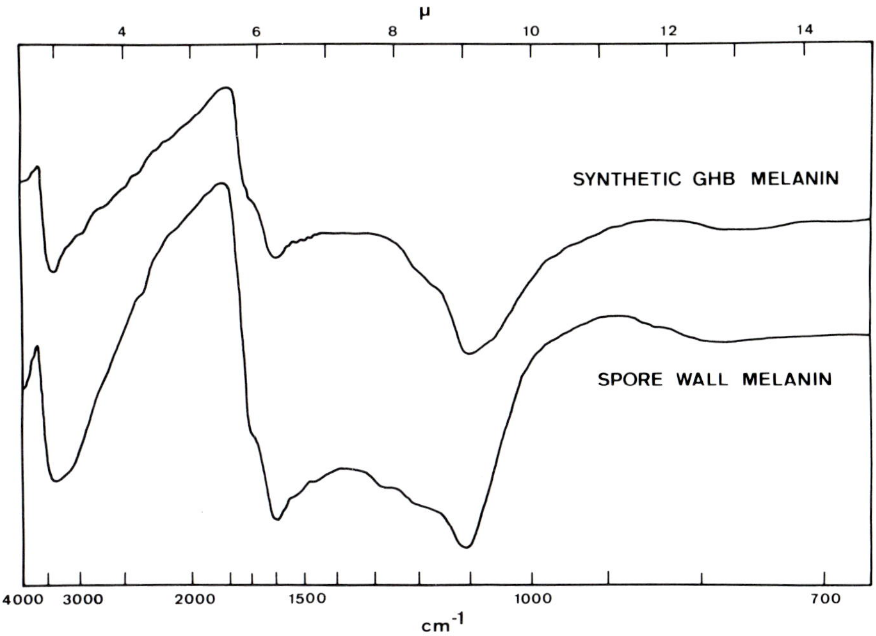

Fig. 4. Infra-red spectra of native and synthetic A. bisporus melanins. GHB, γ-glutaminyl-4-hydroxybenzene.

Nicolaus, 1968 ; Andrews and Pridham, 1967), which is an additional observation favoring the hypothesis (see above) that the substrate of melanogenesis in *A. bisporus* is not identical with any of the precursors of the melanins described till now.

According to the evidence available to-date, melanogenesis in the common mushroom follows the pathway(s) outlined in Fig. 5 : *o*-hydroxylation of GHB , followed by dehydrogenation of the diphenol and polymerization of GBQ and of quinoid products of the latter. However, it is also possible that the γ-glutamyl moiety of GHB does not enter the melanin but is removed prior to polymerization, since the mushroom fruit body contains a γ-glutamyltransferase that has the ability to transfer the γ-glutamyl residue of GHB, and undoubtedly also that of GDHB, to a variety of acceptors (Gigliotti and Levenberg, 1964), thus liberating 4-aminophenol and 4-aminocatechol, respectively, which can undergo enzymatic and/or nonenzymatic conversion into very reactive oxidized intermediates, such as 2-hydroxy-4-iminoquinone (Boekelheide *et al.*, 1980), that polymerize to yield melanin. GHB would, thus, function as a relatively stable storage product for the highly unstable monomer 4-aminocatechol, and γ-glutamyltransferase could, hence, play the role of a starter enzyme for melanogenesis. Finally, the possibility cannot be ruled out that the *A. bisporus* melanin is a mixed polymer derived from GDHB as well as 4-aminocatechol. — Thioether linkages can be thought to be formed between the pigment and protein (see also above), resulting from interaction of sulfhydryl groups of the latter with the melanin itself and/or with quinones generated during melanogenesis.

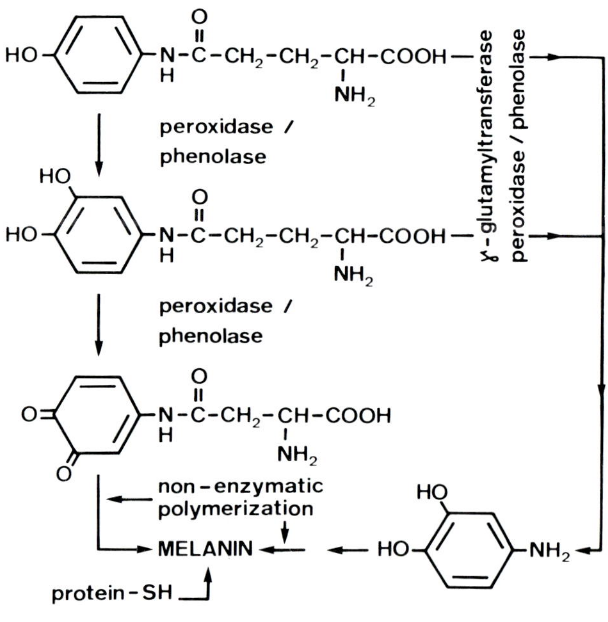

Fig. 5. The melanin biosynthetic pathway in the A. bisporus spore (according to Stüssi, 1979). For explanations, see text.

Ultrastructural aspects of melanogenesis

Any hypothesis on the mechanism of melanogenesis in fungi will, eventually, have to provide an explanation also for the fact that, at least, part of the pigment seems to be present as granules (with diameter of 30-150 nm : see Introduction). The claim made above that *A. bisporus* could represent a model system for studying melanogenesis in fungi appears, therefore, fully justified only in the case that, here too, melanin exists in granular form.

Presence of melanin in the mushroom spore wall Neither thin-sectioning nor freeze-etching (Figs. 6 and 7) yielded more than an indication for the existence of spherical bodies in the wall. However, the dark-brown residue of spores treated with diethyl ether, acetone, methanol, methanolic KOH, water, HCl and H_3PO_4 according to the procedure used by Gooday *et al.* (1973) to isolate sporopollenin (except that phosphorolysis was not allowed to proceed to completion) did clearly display the presence of granules having the proper dimensions (Figs. 8a and 8b). The surface thus exposed is very similar to that of the melanized microsclerotial cell of *V. dahliae* (Wheeler *et al.*, 1976). The release of spherical particles of the same size when digesting untreated wall fragments with lipase (Fig. 8c) was considered proof for the genuine presence of granules in the wall, i.e. representing evidence against the objection of their being an artefact of extraction. The inability to unequivocally demonstrate the existence of melanin granules in the untreated spore may be due to a high degree of their interspersion and intermingling with the fibrillar as well as with the amorphous material of the wall (see also Rast and Hollenstein, 1977).

Ultrastructure of chemically isolated mushroom spore wall melanin The purified pigment consists of plate-like particles that vary greatly in form and size, and fairly uniform granules with diameter of 50-100 nm (Figs. 9a, 9b). The spherical bodies are very electron-dense and display no substructure (Fig. 9c). The plate-like particles are of medium electron density (Fig. 9d) and appear to contain melanin in form of granules (Fig. 9d, arrows) which have about the same size as the spherical bodies. The lower electron density of the granular material present in the plate-like particles could be due to the pigment being less concentrated in them.

Ultrastructure of mushroom melanin synthesized in vitro Unexpectedly enough, synthetic GHB melanin (purified in the same way as the spore wall melanin) was found not to be amorphous, but to consist of round bodies (with a size range of 30-200 nm) accompanied by irregularly shaped material (Figs. 10a,10b). There is a remarkable similarity between the ultrastructural appearance of the thin-sectioned synthetic GHB melanin (Fig. 10d) and the melanin present in thin-sectioned HCl residues of *Dendryphiella vinosa* walls (Reisinger and Bonaly, 1972). Furthermore, it appears that the seemingly amorphous matrix into

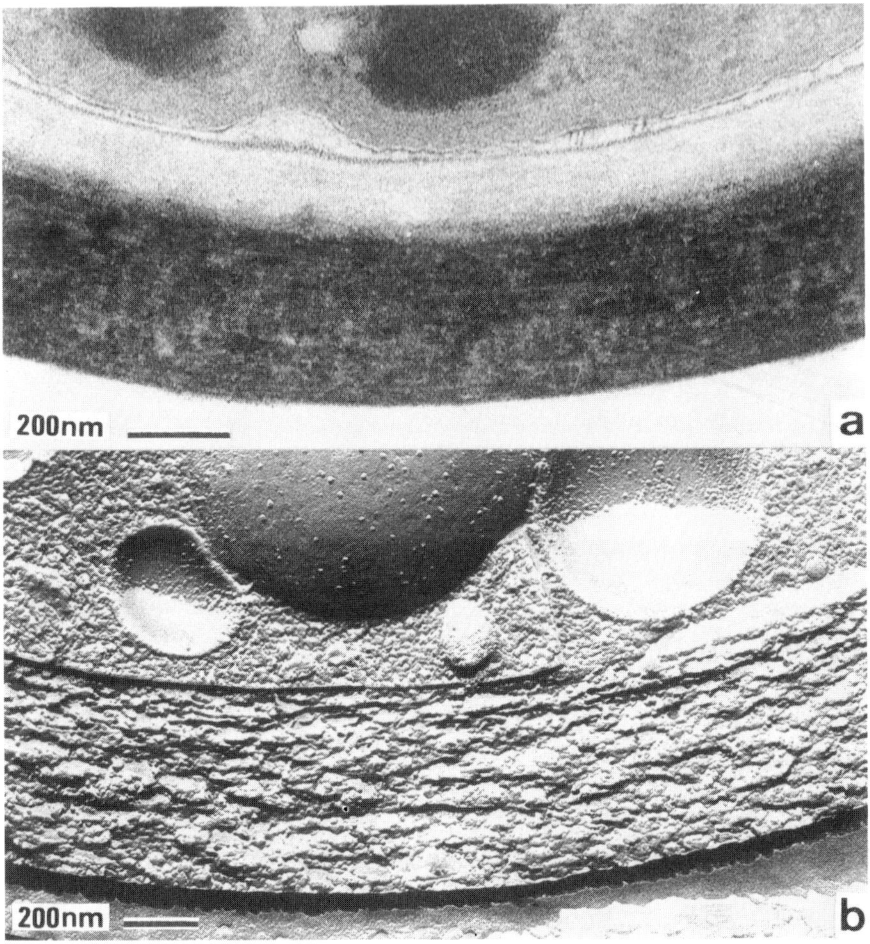

Fig.6. TEM micrographs of the A. bisporus spore wall. a, section, fixed in glutaraldehyde and osmium tetroxide, embedded in Spurr's low viscosity resin, and poststained with uranyl acetate and lead citrate ; b, radial freeze-fracture.

which the electron-dense round bodies are embedded (Figs.10c, 10d) also consists of granules, though less melanized ones. A plausible interpretation of the observation that the native as well as the synthetic melanin are present at least partly in granular form, would be that the phenolase is incorporated into spherical bodies becoming transformed into pigment granules during the melanization process. Indeed, the enzyme

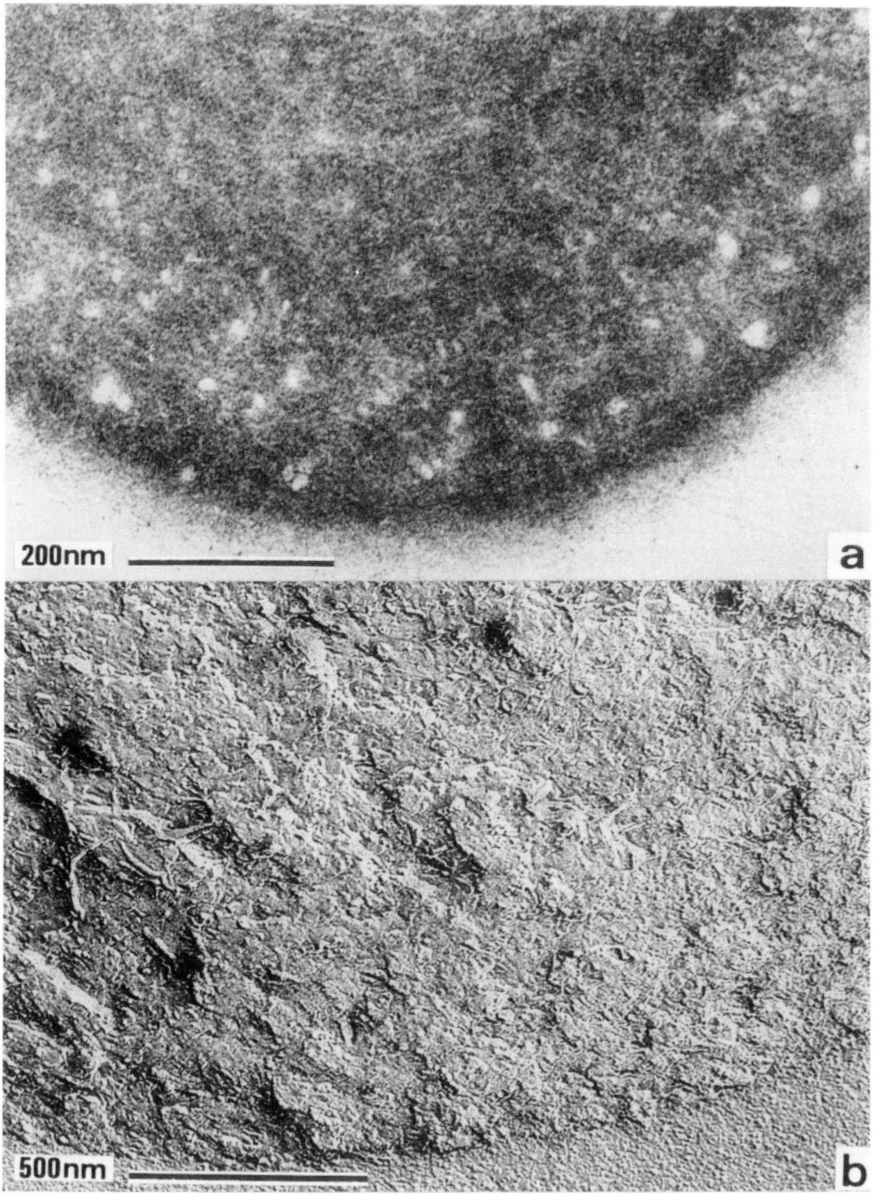

Fig.7. TEM micrographs of the A. bisporus spore wall. a, tangential section, displaying patches of electron-dense material within a fibrillar network ; b, tangential freeze-fracture, showing depressions and elevations which may correspond to particles with a diameter of 50-100 nm.

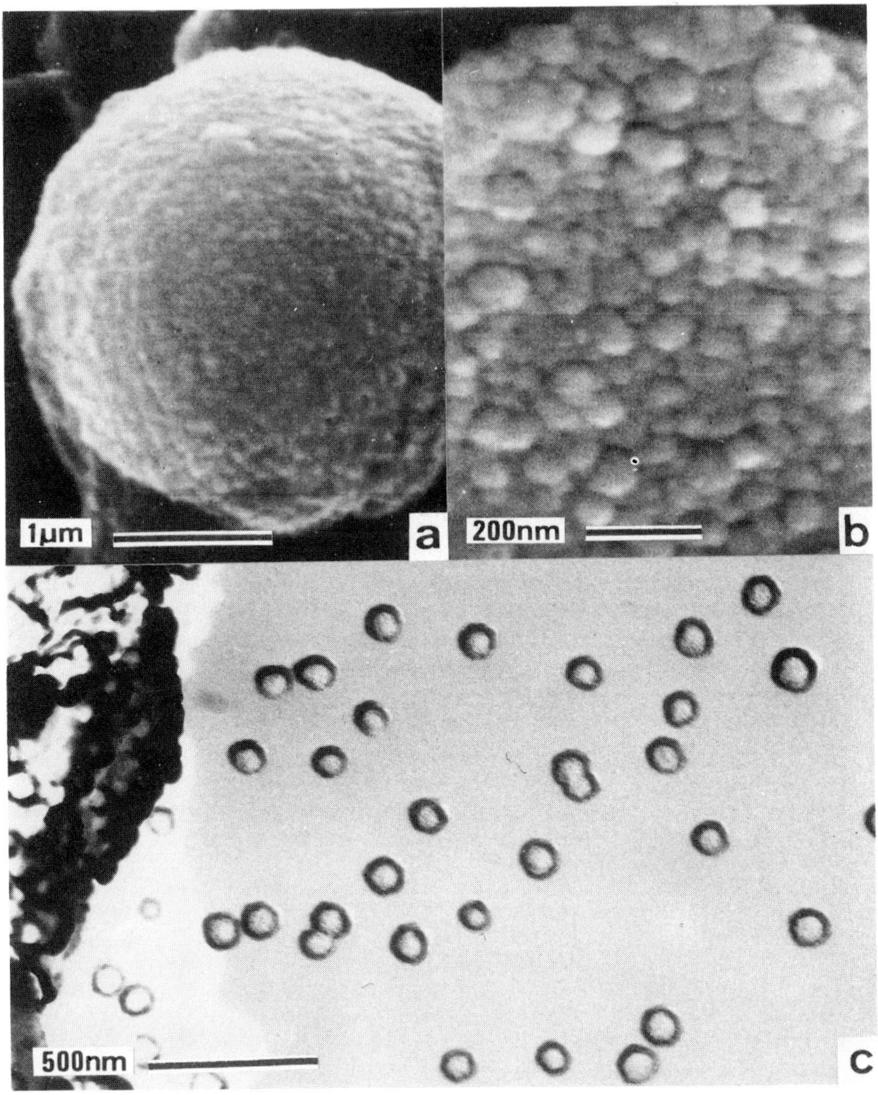

Fig.8. Presence of spherical particles (50-100 nm) in the A. bisporus spore wall after chemical or enzymic digestion. a,b: SEM micrographs ; a, surface of a spore extracted with organnic solvents and acids (cf. text) ; b, detail of a ; c, TEM micrograph (from Hollenstein, 1975) : replica of a wall fragment after treatment with lipase, demonstrating the release of globular bodies under the influence of the enzyme.

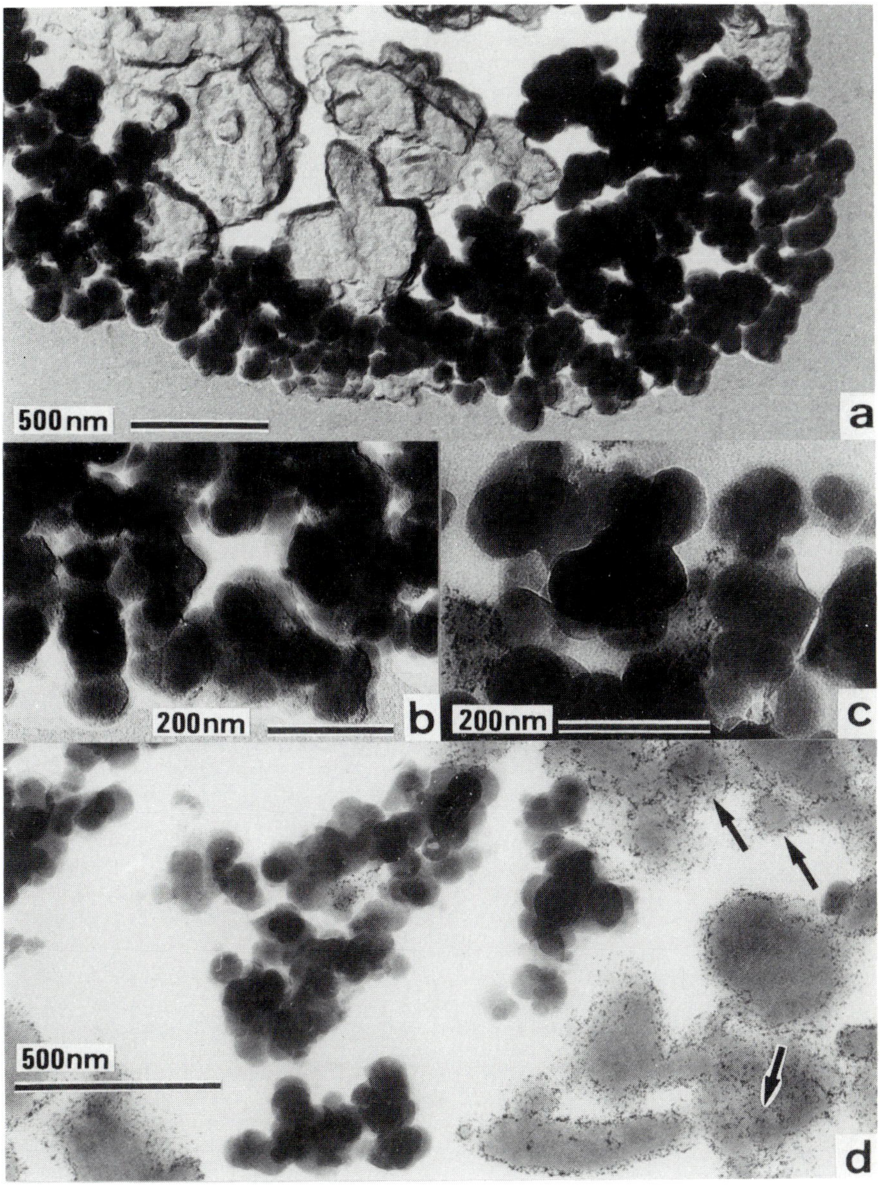

Fig.9. TEM micrographs demonstrating morphological heterogeneity of the A. bisporus spore melanin (isolated and purified according to Piattelli et al., 1965). a, replica (the spherical bodies have not been removed by the chromic acid washing procedure) ; b, detail of a ; c, detail of d ; d, thin section.

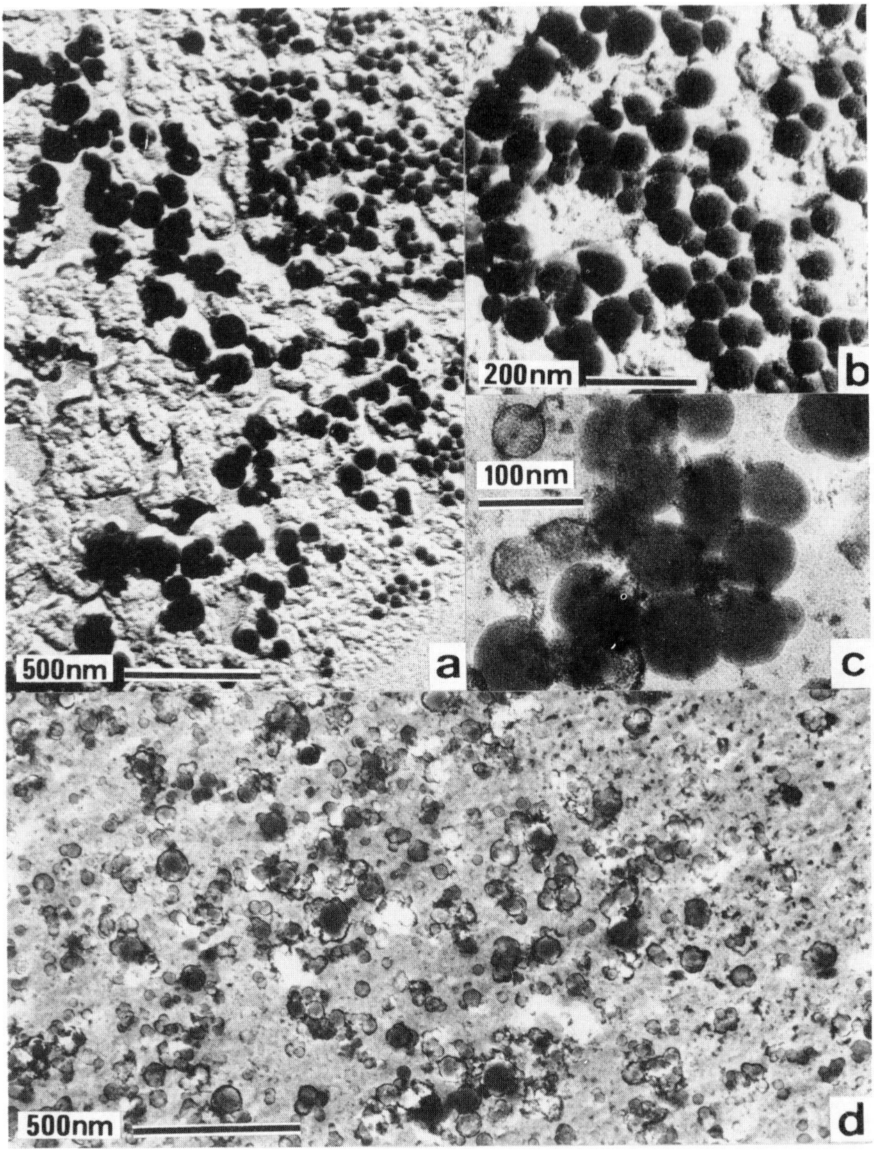

Fig.10. TEM micrographs revealing a highly structured organization of synthetic A. bisporus melanin (prepared according to Stüssi and Rast, 1981). a,b: surface structure of pigment preparation (here, too, the spherical bodies have not been removed from the replica by the acid washing procedure: cf. Fig.9); c, detail of d; d, section.

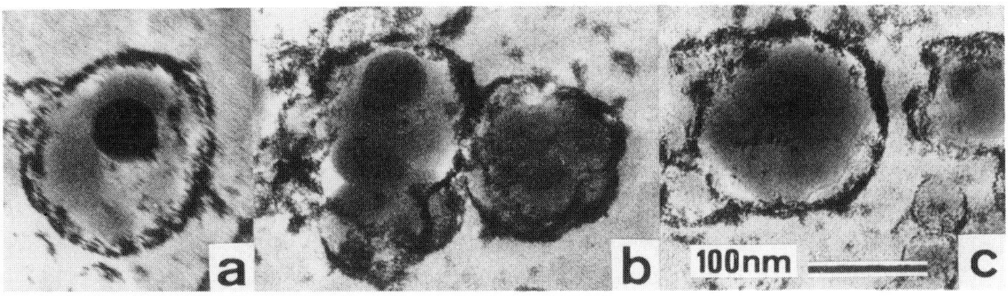

Fig. 11. Thin sections of synthetic A. bisporus melanin (see also caption to Fig. 10) showing the presence in the pigment preparation of round particles (ca. 200 nm) delimited by a zone of very high electron density and containing spheres (about 50 nm) that display various degrees of electron density (a,b). Sometimes, these granular materials seem to fuse together and, thus, give the impression the contents of the large round bodies to be amorphous (c).

preparation used to synthesize the melanin consists, to a considerable proportion, of spherical particles with diameter of 50-100 nm (Nyhlén and Rast, unpublished). In addition to granules (Fig. 10c), the synthetic melanin also contains large round bodies (about 200 nm : see Figs. 10d and 11) that are bound by very electron-dense material and contain smaller, less electron-dense spheres (approximately 50 nm). The sequence of pictures shown in Fig. 11 (a-c) is very much reminiscent of the course of melanization of pheomelanosomes in animals (cf. Jimbow and Tacheuchi, 1979).

GENERAL CONCLUSIONS

1,8-Dihydroxynaphthalene (DHN), γ-glutaminyl-3,4-dihydroxybenzene, catechol, dopa and possibly other phenols (see Introduction) can all be regarded as natural melanin precursors in fungi. This fact is evidence against a recent suggestion that DHN could be a general intermediate of melanogenesis in these organisms and indicates that the structural diversity of fungal melanins is by far greater than that of animal melanins (cf. Swan, 1974). The occurrence in a fungus of a nitrogenous melanin not derived from dopa and containing sulfur (this paper) renders impracticable the further use of Nicolaus's (1968) classification of the melanins into eumelanins, allomelanins and pheomelanins. Depending on the species, the shikimate-chorismate as well as the polyketide pathway can be used for melanin synthesis. Considering the presence of granular melanin in a large variety of fungi (see Introduction) and the results of the present study on the ultrastructure of chemically isolated native and synthetic mushroom melanins, the presence in fungi of intracellular compartments for the processing of granular melanin is postulated here. They might be analogous to the melanosomes of animals. Finally, it is thought that further investigations following the strategy used in this paper will not only reveal more details on the mechanism of melanogenesis in fungi but also provide an experimental basis to study the synthesis of melanin-protein-chitin complexes as present in *Mucor rouxii* (Bartnicki-Garcia and Reyes, 1964), *A. bisporus* (Rast and Hollenstein, 1977) and presumably in spores of many other fungi.

ABBREVIATIONS

GHB, γ-L-glutaminyl-4-hydroxybenzene ; GDHB, γ-L-glutaminyl-3,4-dihydroxybenzene ; GBQ, γ-L-glutaminyl-3,4-benzoquinone.

ACKNOWLEDGEMENTS

Financial support of this work by the Swiss National Science Foundation (grants nos. 3.001-0.76 and 3.576-0.79) is gratefully acknowledged. Work with the scanning electron microscope was performed by Mrs. D. Metzger of the Cytological Laboratory of our Institute. It is also a pleasure to thank Mr. P.O. Hollenstein, O.S.B., for permission to reproduce one of the electron micrographs of his M.Sc. thesis.

REFERENCES

Aldridge, D.C., Davies, A.B., Jackson, M.R. and Turner, W.B. (1974). Pentaketide metabolites of the fungus *Phialophora lagerbergii*. *J. Chem. Soc. Perkin I*, 1540-1541.

Allport, D.C. and Bu'Lock, J.D. (1958). The pigmentation and cell-wall material of *Daldinia* Sp. *J. Chem. Soc.*, 4090-4094.

Allport, D.C. and Bu'Lock, J.D. (1960). Biosynthetic pathways in *Daldinia concentrica*. *J. Chem. Soc.*, 654-662.

Andrews, R.S. and Pridham, J.B. (1967). Melanins from dopa-containing plants. *Phytochemistry* 6, 13-18.

Bartnicki-Garcia, S. and Reyes, E. (1964). Chemistry of spore wall differentiation in *Mucor rouxii*. *Arch. Biochem. Biophys.* 108, 125-133.

Bell, A.A., Puhalla, J.E., Tolmsoff, W.J. and Stipanovic, R. D. (1976a). Use of mutants to establish (+)-scytalone as an intermediate in melanin biosynthesis by *Verticillium dahliae*. *Can. J. Microbiol.* 22, 787-799.

Bell, A.A., Stipanovic, R.D. and Puhalla, J.E. (1976b). Pentaketide metabolites of *Verticillium dahliae*. Identification of (+)-scytalone as a natural precursor to melanin. *Tetrahedron* 32, 1353-1356.

Berthe, M.C., Bonaly, R. et Reisinger, O. (1976). Dégradation chimique et enzymatique des parois de *Helminthosporium spiciferum*. Rôle protecteur des pigments. *Can. J. Microbiol.* 22, 929-936.

Boekelheide, K., Graham, D.G., Mize, P.D., Anderson, C.W. and Jeffs, P.W. (1979). Synthesis of γ-L-glutaminyl-[3,5-^{3}H]4-hydroxybenzene and the study of the reactions catalyzed by the tyrosinase of *Agaricus bisporus*. *J. Biol. Chem.* 254, 12185-12191.

Boekelheide, K., Graham, D.G., Mize, P.D. and Jeffs, P.W. (1980). The metabolic pathway catalyzed by the tyrosinase of *Agaricus bisporus*. *J. Biol. Chem.* 255, 4766-4771.

Bull, A.T. (1970a). Chemical composition of wild-type and mutant *Aspergillus nidulans* cell walls. The nature of polysaccharide and melanin constituents. *J. Gen. Microbiol.* 63, 75-94.

Bull, A.T. (1970b). Inhibition of polysaccharases by melanin: enzyme inhibition in relation to mycolysis. *Arch. Biochem. Biophys.* 137, 345-356.

Durrell, L.W. (1964). The composition and structure of walls of dark fungus spores. *Mycopath. Mycol. Appl.* 23, 339-345.

Ellis, D.H. and Griffiths, D.A. (1974). The location and analysis of melanins in the cell walls of some soil fungi. *Can. J. Microbiol.* 20, 1379-1386.

Ellis, D.H. and Griffiths, D.A. (1975). Melanin deposition in the hyphae of a species of *Phomopsis*. *Can. J. Microbiol.* 21, 442-452.

Farishian, R.A. and Whittaker, J.R. (1979). Tyrosine utilization by cultured melanoma cells : analysis of melanin biosynthesis using [^{14}C]tyrosine and [^{14}C]thiouracil. *Arch. Biochem. Biophys.* 198, 449-461.

Gigliotti, H.J. and Levenberg, B. (1964). Studies on the γ-glutamyltransferase of *Agaricus bisporus*. *J. Biol. Chem.*

239, 2274-2284 (1964).

Gooday, G.W., Fawcett, P., Green, D. and Shaw, G. (1973). The formation of fungal sporopollenin in the zygospore wall of *Mucor mucedo* : a role for the sexual carotenogenesis in the Mucorales. *J. Gen. Microbiol.* 74, 233-239.

Greuter, B. and Rast, D. (1975). Ultrastructure of the dormant *Agaricus bisporus* spore. *Can. J. Bot.* 53, 2096-2101.

Griffiths, D.A. and Swart, H.J. (1974). Conidial structure in two species of *Pestalotiopsis*. *Trans. Br. mycol. Soc.* 62, 295-304.

Haslam, E. (1974). "The Shikimate Pathway". Butterworths, London.

Higuchi, T. (1971). Formation and biological degradation of lignins. *Adv. Enzymology* 34, 207-283.

Hollenstein, G. (1975). Ultrastruktur und Chemie der Sporenwand von *Agaricus bisporus*. M.Sc. thesis, University of Zürich, Switzerland.

Ito, Y., Nanba, H. and Kuroda, H. (1979). Melanin produced by *Cochliobolus miyabeanus*. II. Gas chromatographic-mass spectrometric analysis of the degradation products by alkaline fusion. *Yakugaku Zasshi* 99, 971-975.

Jadot, J., Casimir, J. et Renard, M. (1960). Séparation et caractérisation du L(+)-γ-(p-hydroxy)anilide de l'acide glutamique à partir de *Agaricus hortensis*. *Biochim. Biophys. Acta* 43, 322-328.

Jimbow, K. and Tacheuchi, T. (1979). Ultrastructural comparison of pheo- and eumelanogenesis in animals. In "Biological Basis of Pigmentation" (Ed. S.N. Klaus). Pigment Cell, Vol. 4, pp. 308-317. Karger, Basel.

Kelly, R.B., Daniels, E.G. and Hinman, J.W. (1962). Agaritine: isolation, degradation, and synthesis. *J. Org. Chem.* 27, 3229-3231.

Kuo, M.-J. and Alexander, M. (1967). Inhibition of the lysis of fungi by melanins. *J. Bact.* 94, 624-629.

Levenberg, B. (1961). Structure and enzymatic cleavage of agaritine, a phenylhydrazide of L-glutamic acid isolated from Agaricaceae. *J. Am. Chem. Soc.* 83, 503-504.

Levenberg, B. (1964). Isolation and structure of agaritine, a γ-glutamyl-substituted arylhydrazine derivative from Agaricaeae. *J. Biol. Chem.* 239, 2267-2273.

Nicolaus, R.A. (1968). "Melanins". Hermann, Paris.

Nicolaus, R.A., Piattelli, M. and Fattorusso, E. (1964). The structure of melanins and melanogenesis — IV. On some natural melanins. *Tetrahedron* 20, 1163-1172.

Piattelli, M., Fattorusso, E., Magno, S. and Nicolaus, R.A. (1963). The structure of melanins and melanogenesis. – III. The structure of sepiomelanin. *Tetrahedron* 19, 2061-2072.
Piattelli, M., Fattorusso, E., Nicolaus, R.A. and Magno, S. (1965). The structure of melanins and melanogenesis. – V. Ustilagomelanin. *Tetrahedron* 21, 3229-3236.
Potgieter, H.J. and Alexander, M. (1966). Susceptibility and resistance of several fungi to microbial lysis. *J. Bact.* 91, 1526-1532.
Pridham, J.B. and Woodhead, S. (1977). The biosynthesis of melanin in *Alternaria*. *Phytochemistry* 16, 903-906.
Rast, D. (1965). Zur stoffwechselphysiologischen Bedeutung von Mannit und Trehalose in *Agaricus bisporus* (eine gaschromatographische Studie). *Planta* 64, 81-93.
Rast, D. and Hollenstein, G.O. (1977). Architecture of the *Agaricus bisporus* spore wall. *Can. J. Bot.* 55, 2251-2262.
Rast, D., Stüssi, H. and Zobrist, P. (1979). Self-inhibition of the *Agaricus bisporus* spore by CO_2 and/or γ-glutaminyl-4-hydroxybenzene and γ-glutaminyl-3,4-benzoquinone : a biochemical analysis. *Phys. Plant.* 46, 227-234.
Reisinger, O. et Bonaly, R. (1972). Etudes des modifications ultrastructurales des parois de *Dendryphiella vinosa* (Berk. et Curt.) Reisinger après traitements chimiques et enzymatique. *C. R. Acad. Sc. Paris* 274, 50-53.
Reisinger, O., Kiffer, E., Mangenot, F. et Olah, G.H. (1977). Ultrastructure, cytochimie et microdissection de la paroi des hyphes et des propagules exogènes des Ascomycètes et Basidiomycètes. *Revue de Mycologie* 41, 91-117.
Rowley, B.I. and Pirt, S.J. (1972). Melanin production by *Aspergillus nidulans* in batch and chemostat cultures. *J. Gen. Microbiol.* 72, 553-563.
Schulz, K.H., Felten, G. and Hausen, B.M. (1974). Allergy to the spores of *Pleurotus florida*. *The Lancet* 5, 29.
Seiji, M., Fitzpatrick, T.B., Simpson, R.T. and Birbeck, M.S. C. (1963). Chemical composition and terminology of specialized organelles (melanosomes and melanin granules) in mammalian melanocytes. *Nature* (London) 197, 1082-1084.
Stipanovic, R.D. and Bell, A.A. (1976). Pentaketide metabolites of *Verticillium dahliae*. 3. Identification of (–)-3,4-dihydro-3,8-dihydroxy-1(2*H*)naphthalenone[(–)-vermelone] as a precursor to melanin. *J. Org. Chem.* 41, 2468-2469.
Stäuble, E.J. (1972). Vorkommen und Stoffwechsel leichtflüchtiger Substanzen von bekannter oder vermuteter physiologischer Wirksamkeit in *Agaricus bisporus*. Ph.D. thesis, Univ.

of Zürich, Switzerland. Juris Druck + Verlag, Zürich. ISBN 3 260 03242 8.

Stüssi, H. (1979). Vorkommen, Biogenese und Bedeutung von γ-Glutamylderivaten in *Agaricus bisporus*. Ph.D. thesis, Univ. of Zürich, Switzerland. Juris Druck + Verlag, Zürich. ISBN 3 260 04608 9.

Stüssi, H. and Rast, D.M. (1981). The biosynthesis and possible function of γ-glutaminyl-4-hydroxybenzene in *Agaricus bisporus*. *Phytochemistry* (in press).

Sussman, A.S. (1968). Longevity and survivability of fungi. *In* "The Fungi" (Ed. G.C. Ainsworth and A.S. Sussman). Vol. III, pp. 447-486. Academic Press, New York and London.

Swan, G.A. (1974). Structure, chemistry, and biochemistry of the melanins. *In* "Progress in the Chemistry of Organic Natural Products" (Ed. W. Herz, H. Grisebach and G.W. Kirby). Vol. 31, pp. 521-582. Springer, New York.

Szent-Gyorgyi, A., Chung, R.H., Boyajian, M.J., Tishler, M., Arison, B.H., Schoenewalt, E.F. and Wittick, J.J. (1976). Agaridoxin, a mushroom metabolite. Isolation, structure and synthesis. *J. Org. Chem.* 41, 1603-1606.

Thomson, R.H. (1962). Melanins. *In* "Comparative Biochemistry" (Ed. M. Florkin and H.S. Mason). Vol. IIIA, pp. 727-753. Academic Press, New York.

Thomson, R.H. (1976). Miscellaneous pigments. *In* "Chemistry and Biochemistry of Plant Pigments" (Ed. T.W. Goodwin), pp. 597-632. Academic Press, London.

Tsao, P.W. and Tsao, P.H. (1970). Electron microscopic observations on the spore wall and operculum formation in chlamydospores of *Thielaviopsis basicola*. *Phytopathology* 60, 613-616.

Turner, E.M. (1974). Phenoloxidase activity in relation to substrate and development stage in the mushroom, *Agaricus bisporus*. *Trans. Br. mycol. Soc.* 63, 541-547.

Weaver, R.F., Rajagopalan, K.V., Handler, P., Rosenthal, D. and Jeffs, P.W. (1971a). Isolation from the mushroom *Agaricus bisporus* and chemical synthesis of γ-L-glutaminyl-4-hydroxybenzene. *J. Biol. Chem.* 246, 2010-2014.

Weaver, R.F., Rajagopalan, K.V., Handler, P. and Byrne, W.L. (1971b). γ-L-glutaminyl-3,4-benzoquinone. Structural studies and enzymatic synthesis. *J. Biol. Chem.* 246, 2015-2020.

Wehrli, M. und Rast, D. (1967). Stoffwechsel von $^{14}CO_2$ in Plectenchymschnitten von *Agaricus bisporus*. *Z. Pflanzenphys.* 56, 325-339.

Wheeler, M.H., Tolmsoff, W.J. and Meola, S. (1976). Ultrastructure of melanin formation in *Verticillium dahliae* with (+)-scytalone as a biosynthetic intermediate. *Can. J. Microbiol.* 22, 702-711.

Wheeler, M.H., Tolmsoff, W.J., Bell, A.A. and Mollenhauer, H. H. (1978). Ultrastructural and chemical distinction of melanins formed by *Verticillium dahliae* from (+)-scytalone, 1,8-dihydroxynaphthalene, catechol, and L-3,4-dihydroxyphenylalanine. *Can. J. Microbiol.* 24, 289-297.

Wheeler, M.H. and Stipanovic, R.D. (1979). Melanin biosynthesis in *Thielaviopsis basicola*. *Exp. Mycol.* 3, 340-350.

White, L.P. (1958). Melanin : a naturally occurring cation exchange material. *Nature* (London) 182, 1427-1428.

Zadražil, F. (1974). Pleurotus-Sporen als Allergene. *Naturwissenschaften* 61, 456-457.

CELL WALL CONSTRUCTION DURING SPORE GERMINATION IN PHYCOMYCETES

S. Bartnicki-Garcia

Department of Plant Pathology
University of California
Riverside, CA 92521, U.S.A.

1. INTRODUCTION

The purpose of this article is to discuss the process of cell wall formation by germinating spores of fungi using as specific examples zoospores and sporangiospores of phycomycetes. Spore germination entails the formation of new cell walls; necessarily, the development and/or operation of the cytological and biochemical machinery for wall formation are crucial aspects of the germination process. Despite obvious major differences between zoospores and sporangiospores, their germination process involves the synthesis of totally new cell walls and takes place in two well defined stages: I) development of a spherical wall around the spore protoplast and II) emergence of a germ tube (Fig. 1) (Bartnicki-Garcia *et al.*, 1968; 1979). Described below are some of the similarities and differences in the enzymology and cytology of wall formation among germinating sporangiospores, uniflagellate and biflagellate zoospores.

2. SPORANGIOSPORES OF MUCORALES

The germination of a sporangiospore of *Mucor rouxii* is a clear example of the two-stage germination process common in fungal spores (Bartnicki-Garcia *et al.*, 1968). Prior to germ tube emergence, the ellipsoidal spore undergoes a mandatory period of isodiametric growth and becomes a large sphere--the germ sphere (Fig. 2). This initial growth phase is commonly but misleadingly called "swelling". Although the initial increase in size of a sporangiospore has been viewed primarily as imbibition of water (e.g.

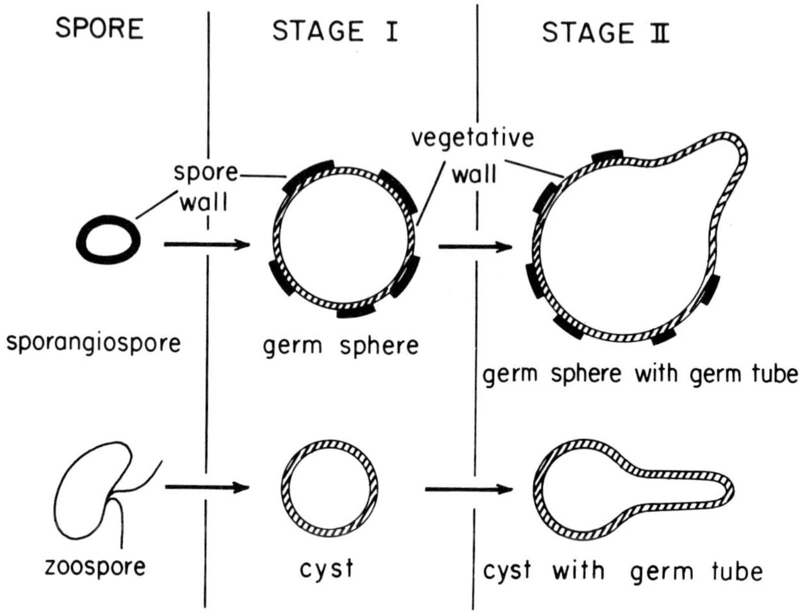

Fig. 1. The two stages of germination of sporangiospores and zoospores of Phycomycetes.

Ekundayo and Carlile, 1964), it is actually a bona fide growth phase termed spherical growth (Bartnicki-Garcia et al., 1968). In the second stage of germination, the germ spheres produce either germ tubes or spherical buds depending on environmental conditions.

In mucoraceous fungi, unlike other fungi, the spore wall composition is totally different from that of vegetative cell walls (Bartnicki-Garcia and Reyes, 1964; Van Laere et al., 1977). Germination requires a return to the ability to make vegetative wall polymers.

The sporangiospore wall does not seem to play an active role during germination (Bartnicki-Garcia et al., 1968; Van Etten et al., 1974). A vegetative wall is deposited anew under the existing sporangiospore wall (Fig. 3). This is true true in *Rhizopus* (Hawker and Abbott, 1963), *Mucor* (Bartnicki-Garcia et al., 1968) and *Gilbertella* (Bracker, 1971). In *M. rouxii*, the spore wall becomes distended and eventually fractured under the pressure of the expanding vegetative cell. The spore wall fragments adhere to the newly formed wall of the germ-sphere. Aside from this

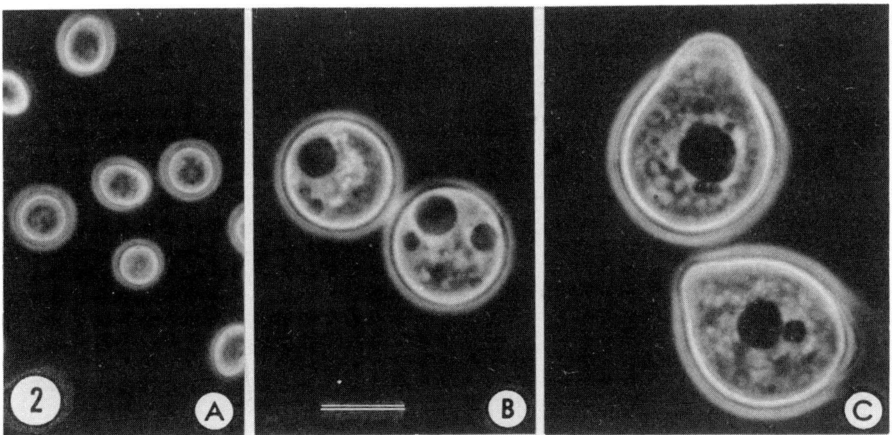

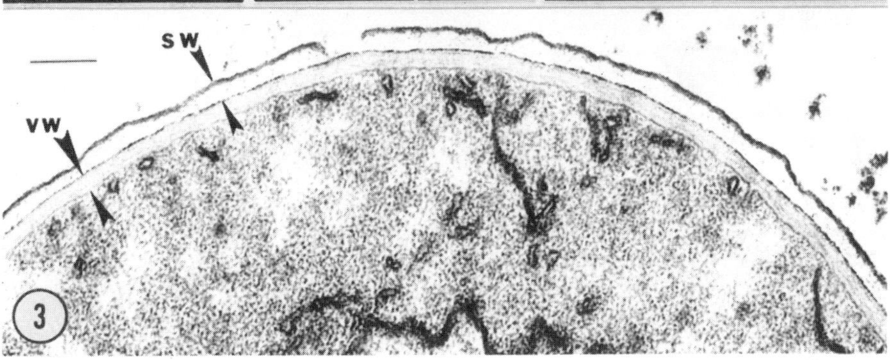

Fig. 2. Germination of Mucor rouxii sporangiospores: A. Ungerminated spores. B. Germ spheres (stage I). C. Germ spheres with incipient germ tubes (stage II). Bar = 10 μm. Fig. 3. The new vegetative wall (vw) formed under the spore wall (sw) in a germinating sporangiospore of M. rouxii. From Bartnicki-Garcia et al. (1968). Bar = 200 nm.

attachment, there is no structural continuity between spore wall and the newly formed wall of the germ sphere. [Since the spore wall is not converted into the vegetative wall, it seems inappropriate to describe wall formation during sporangiospore germination as a rebuilding or reconstruction of the spore wall (Furch and Pambor, 1978). The germ-tube wall, however, is formed as a direct extension of the germ-sphere wall. Likewise, when spherical buds rather than germ tubes are produced during germination of M. rouxii sporangiospores, the bud wall is a direct continuation of the inner layers of the germ-sphere wall (Lara and Bartnicki-Garcia, 1974).

(a) Kinetics of germination

Kinetic measurements indicate that sporangiospores of M. rouxii begin to grow exponentially upon or shortly after being inoculated into nutrient media. The same exponential growth rate is maintained during the two stages of germination. There is no apparent change in growth rate coinciding with germ tube emergence. This was established by the following parameters: dry weight increase (Fig. 4), CO_2 evolution (Bartnicki-Garcia, 1968), and accumulation of glucosamine polymers (Fig. 5) which are major components of vegetative cell walls (Bartnicki-Garcia and Lippman, 1977). Accordingly, we must conclude that there are no major changes in metabolic activity during the two stages of germination, i.e. the emergence of a germ-tube does not coincide with, or result from, a change in growth rate or a major shift in metabolism. The emergence of a germ tube is probably due to a more subtle change (see below).

(b) Enzymology of cell wall formation

Our knowledge of cell wall metabolism during spore germination of Mucorales is very scant. The fact that an entirely new wall is formed during germination poses the question of whether the enzymes needed to make this wall are preformed in the sporangiospore or not. In Phycomyces blakesleeanus, chitin synthetase was found to be nearly absent in dormant sporangiospores and to be synthesized de novo from 4 h onwards (Van Laere and Carlier, 1978). Both cycloheximide and 5-fluorouracil prevented synthesis of both zymogen and active enzyme indicating that the protein was synthesized de novo on newly formed m-RNA.

Although we have no critical kinetic information on the production of wall-synthesizing enzymes during spore germination of M. rouxii, the finding that germinating spores grow and synthesize glucosamine polymers exponentially indicates that cell-wall synthesizing enzymes must be produced in parallel exponential fashion during the entire germination process. An initial supply of preformed wall-synthesizing enzymes, if present, might initiate the development of the germ sphere but could not sustain the exponential rate of wall formation. It is not surprising, therefore, that protein synthesis inhibitors can block sporangiospore germination completely (Orlowski and Sypherd, 1978).

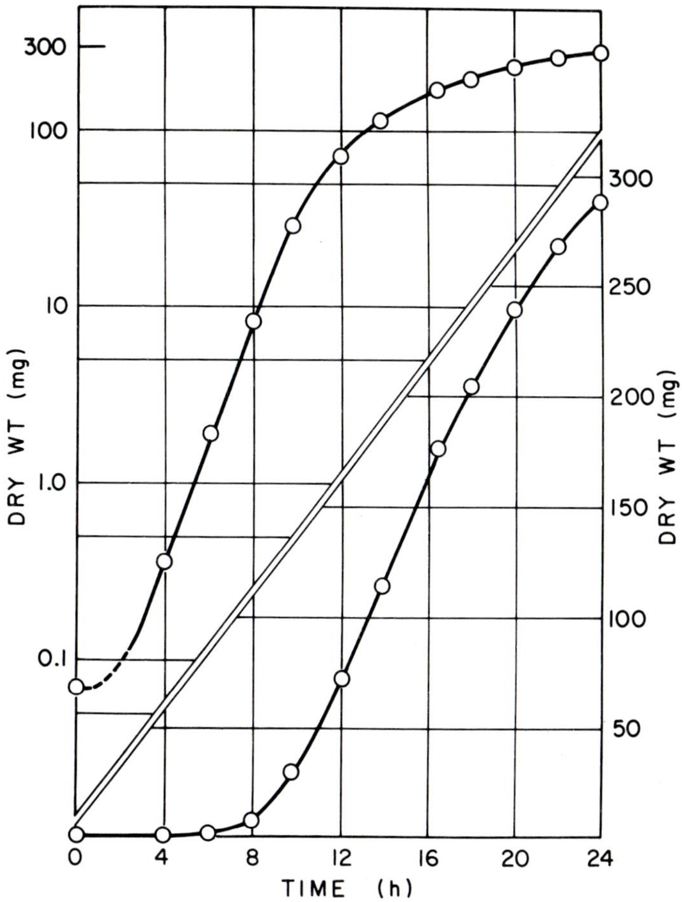

Fig. 4. Kinetics of germination of Mucor rouxii sporangiospores. Spores ($2x10^6$) were germinated aerobically in 50 ml of standard yeast extract-peptone-glucose medium (Bartnicki-Garcia, 1968) and the dry weight determined after filtration and drying at 80° C. For comparison, the same data are plotted on logarithmic and linear coordinates.

Note. The apparent "rapid" or "sudden" increase in cell mass after 6 h, which coincides with the time of germ tube emergence, is an illusion created by plotting an exponential function on a linear coordinate. Likewise, the literature on fungal spore germination contains many examples of biochemical activities or events that supposedly begin or accelerate rapidly at about the time of germ tube emergence. Such purported correlations should be reexamined carefully for they might merely be false impressions produced by plotting exponential functions on linear coordinates.

Fig. 5. Synthesis of glucosamine polymers by germinating spores of M. rouxii. Values in parenthesis indicate the % of cells with germ tubes. Adapted from Bartnicki-Garcia and Lippman (1977).

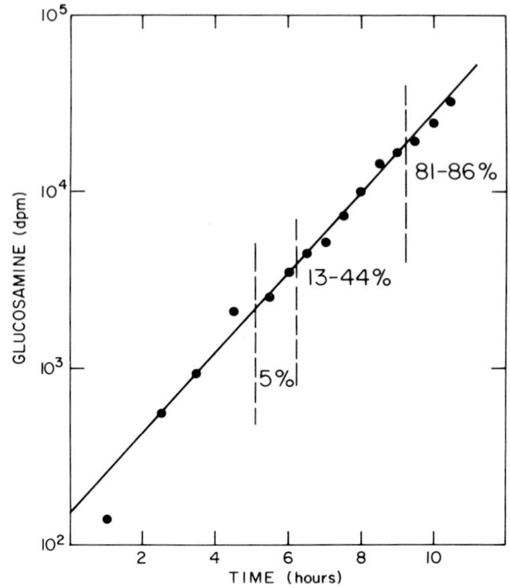

(c) Cytology of wall formation

Cytoplasmic vesicles, presumably involved in wall growth, can be seen though in small numbers, in incipient germ tubes of sporangiospores (Bracker, 1971; Hess and Weber, 1973). The absence of cytoplasmic vesicles from published electron micrographs of germ tubes of some fungi probably results from the delicate and sensitive nature of these vesicles (Bracker, 1971). That few vesicles are seen in the tips of incipient or young germ tubes, compared to those in established hyphae, need not be surprising. If, as we believe, vesicles are directly involved in apical growth, it follows that adult hyphae with a linear rate of elongation many times higher than that of the emerging tube (Trinci, 1971) should have a proportionally greater number of vesicles in their apices.

The population of cytoplasmic vesicles in a fungal cell is heterogenous. On size alone, at least two distinct kinds can be readily recognized. Vesicle heterogeneity may reflect the need for different specialized biochemical functions during cell wall construction (Bartnicki-Garcia, 1973). We now have biochemical and cytological evidence indicating that microvesicles participate in the elaboration of chitinous microfibrils. These microvesicles, which we have named "chitosomes", appear to be the main cytoplasmic containers and conveyors of chitin synthetase in fungi

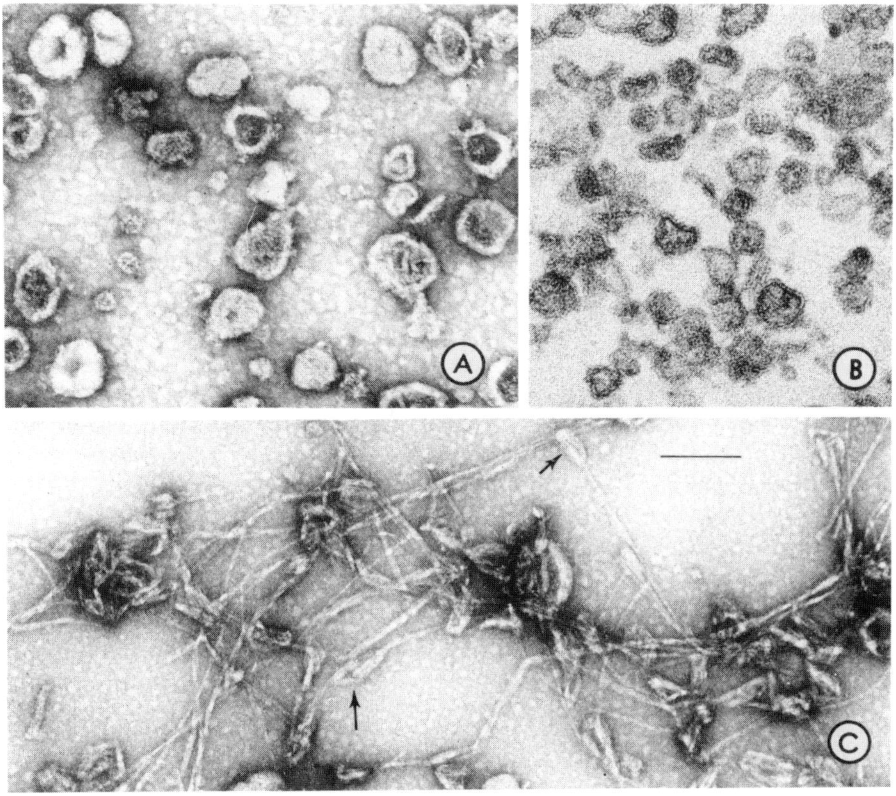

Fig. 6. Purified chitosomes from Mucor rouxii in negative staining (A) and thin section (B). Microfibrils produced after incubating chitosomes with substrate and activators (C). Arrows point to fibroid configurations at the ends of some microfibrils. (See Bracker et al., 1976). Bar = 100 nm.

(Fig. 6) (Bracker et al., 1976; Bartnicki-Garcia et al., 1978). Chitosomes are the cellular units for the synthesis of chitin microfibrils. It has been proposed that one chitosome contains and delivers to the cell surface an organized assemblage of zymogenic chitin synthetase molecules needed to synthesize one microfibril (Bartnicki-Garcia et al., 1979).

(d) Initiation of germination

Sporangiospores of M. rouxii do not seem to have an intrinsic impediment to their germination and are ready to resume vegetative growth upon falling into a suitable nutrient environment. Our kinetic measurements (Fig. 4) leave the possibility of a short lag phase before the spore embarks on its exponential growth. The facility of germination of M. rouxii sporangiospores is not a universal feature of all mucorales. Thus, sporangiospores of P. blakesleeanus manifest constitutive dormancy which can be broken by chemicals, heat shock or gamma radiation (Van Assche et al., 1977).

The sporangiospores of P. blakesleeanus (Furch & Gooday, 1978) but not Mucor mucedo (Gooday et al., 1973) have a small amount of sporopollenin in the walls. Furch & Gooday (1978) conjectured that it may have implications in spore dormancy by being an extra barrier between cytoplasm and environment. But once the impediments to germination are overcome, the spores of Phycomyces seem as ready to germinate as those of Mucor or Rhizopus. Sporangiospores of P. blakesleeanus (heat activated), Rhizopus stolonifer and Mucor racemosus begin RNA and protein synthesis immediately upon introduction to nutrient medium (Van Assche and Carlier, 1973; Van Etten et al., 1974; Orlowski and Sypherd, 1978). In sporangiospores, there is no need, for new mRNA synthesis in order to start protein synthesis.

(e) Germ-tube emergence: polarization of wall synthesis

An autoradiographic study of patterns of wall growth in M. rouxii showed what might be the key determinant of germ tube emergence, namely polarization of wall synthesis (Bartnicki-Garcia and Lippman, 1977). During the first stage of germination, the expanding germ sphere shows a uniformly dispersed pattern of wall synthesis (Fig. 7A). The pattern becomes polarized even before there is any morphological evidence of a germ tube (Fig. 7B,C). The autoradiographs of germ spheres suggest that polarization is a gradual process. The wall synthesizing activity of a germ sphere congregates on a progressively smaller area. A sharp radial gradient of wall synthesis is thus established (Bartnicki-Garcia and Lippman, 1969); subsequently, a protuberance appears at the target area. The continuous polarized growth of the cell wall produces a long hyphal tube (Fig. 7D,E). In essence, the second stage of germination consists of the transfer of the wall synthesizing activity from germ sphere to emerging germ tube. The germ sphere is left with little wall-synthesizing capacity (Bartnicki-Garcia and Lippman, 1977).

Although the mechanism underlying polarization is not understood, it seems reasonable to assume that polarization of wall synthesis is brought about by a directional change in the pattern of migration of chitosomes and other vesicles involved in wall growth. Accordingly, during the first stage of germination, chitosomes and other wall-destined vesicles would migrate with equal frequency to any point on the cell surface. After a period of spherical growth, vesicle migration would become polarized. The polarity signal would guide vesicles toward a target area on the cell surface where the tube would later emerge. This redirection of cell-wall synthesizing vesicles would cause the germ sphere to cease, or vastly reduce, its own expansion and produce instead an outgrowth. If polarization is maintained after emergence, a typical hyphal tube would be generated; if polarization disappears, the outgrowth would expand into spherical bud.

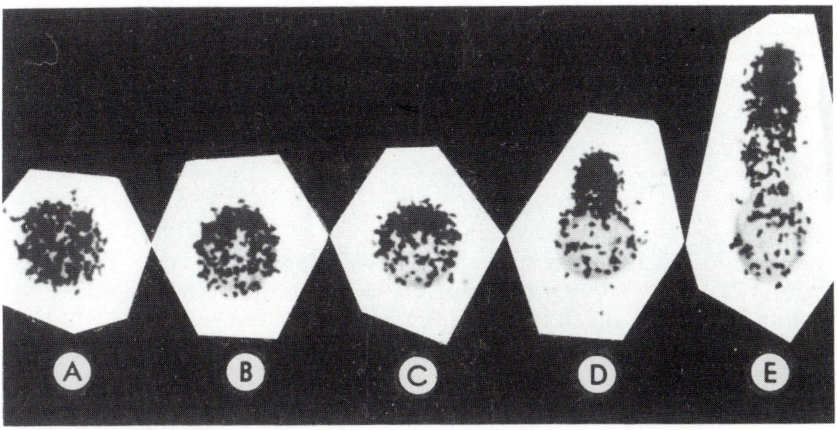

Fig. 7. Selected autoradiographs of germinating spores of M. rouxii arranged to depict progressive stages in the polarization of cell wall synthesis (Bartnicki-Garcia and Lippman, 1977).

What controls the onset of polarity? There is no evidence that germ-sphere size or incubation time alone decides the time of appearance of a germ tube (germ tubes are produced from large as well as small germ spheres over a period of several hours) (Bartnicki-Garcia and Lippman, 1977). The key factor(s) responsible for polarity is not known. The possible participation of structural elements, e.g. microfilaments and microtubules, in vesicle migration may provide a structural basis for directional control. There is evidence, largely from studies on the zygote of the alga *Fucus*, that cell polarity may result from the establishment of an electric potential and ensuing current (Jaffe, 1968).

During spherical growth of *M. racemosus* sporangiospores, the levels of cyclic AMP increase 10-fold (Paznokas and Sypherd, 1975) but, before germ-tube emergence, the concentrations decrease to the lower levels characteristic of hyphal cells. Cyclic AMP, added to germinating sporangiospores of *M. racemosus* during the spherical growth stage, prevented the emergence of germ tubes, and the germ spheres "swelled" indefinitely into giant cells. This correlation between cyclic AMP and germ tube emergence suggests the possible involvement of this nucleotide in the onset of polarity. The observation that fungal membrane depolarization affects the intracellular level of cyclic AMP (Trevillyan and Pall, 1979) suggests a plausible link between cyclic AMP and germ tube emergence via an alteration of plasma membrane potential.

3. ZOOSPORES OF BLASTOCLADIALES

(a) *Enzymology of encystment*

Available evidence indicates that the enzymatic apparatus required for encystment is already preformed in the zoospore, i.e. the zoospore has no need to synthesize any of the enzymes required for encystment. Thus Lovett (1968) found that cycloheximide inhibits protein synthesis but does not block zoospore encystment of *Blastocladiella emersonii*. The same is true for zoospores of *Allomyces macrogynus* (Burke *et al.*, 1972). All enzymes in the pathway of chitin synthesis from glucosamine synthetase to chitin synthetase have been found in cell-free extracts of *B. emersonii* zoospores (Camargo *et al.*, 1967; Myers and Cantino, 1974: Norrman *et al.*, 1973; Giddings and Cantino, 1974; Selitrennikoff *et al.*, 1976; Selitrennikoff and Sonneborn, 1976).

This enzymic sufficiency brings to fore the question of what triggers zoospore encystment. Zoospores can swim a long time without encysting. If the zoospore is fully equipped enzymatically to make its cyst wall, what then prevents encystment? Three explanations may be advanced: i) substrate deficiency, ii) enzyme latency, and iii) enzyme/ substrate compartmentation. The experimental evidence at hand suggests that all three may apply, at least, in part.

i) Substrate deficiency. Zoospore extracts of B. emersonii contain UDP-GlcNAc, the immediate precursor for chitin biosynthesis. Although substrate availability could thus not be the reason for the absence of chitin synthesis in the swimming zoospore, Selitrennikoff et al. (1976) have shown that the amount of UDP-GlcNAc in a zoospore would be insufficient by an order of magnitude to account for the chitin made during encystment. Hence, the need for active production of GlcNAc from other cell reserves (e.g. glycogen) to complete the encystment process.

According to Selitrennikoff and Sonneborn (1976), the concentration of UDP-GlcNAc in the zoospore seems sufficient to cause maximal feedback inhibition of the first enzyme in the pathway of UDP-GlcNAc biosynthesis. As soon as chitin synthesis commences, the ensuing substrate consumption would be expected to release the inhibition.

ii) Enzyme latency. Although new enzyme synthesis is not required for encystment, this does not necessarily mean that all enzymes involved in the formation of cyst wall polymers are in full readiness. Thus one or more enzymes may not be operational, and some post-translational modification of the polypeptide may be required to activate the enzyme.

Studies on fungal chitin synthetase have shown, in all cases where the issue has been criticaly studied, that this enzyme is present in a zymogen form that requires proteolytic digestion to become active (Cabib et al., 1979; Bartnicki-Garcia et al., 1978). There is no published information on this issue for the zoospores of chytridiales. However, we have recently found that the chitin synthetase from zoospores of A. macrogynus can be stimulated several fold by trypsin (W. E. Barstow and S. Bartnicki-Garcia, unpublished observations). Zymogenicity, therefore, must be an important means of keeping chitin synthetase inactive in the swimming zoospore.

iii) Enzyme/substrate compartmentation. Zoospores of Blastocladiales undergo an extensive process of intracellular differentiation during encystment; gamma particles disintegrate into numerous vesicles (Fig. 8) which migrate to and fuse with the plasma membrane (Truesdell and Cantino,

1970; Olson, 1973; Kirby *et al.*, 1974; Barstow and Pommerville, 1980). This appears to be the main mechanism by which enzymes and other materials, initially stored in the gamma particles, are delivered to the cell surface and thus initiate the formation of the cyst wall (see below). Presumably, while inside the gamma particles, wall-synthesizing enzymes are kept separated from substrate and/or activating proteases.

(b) Cytology of encystment

The involvement of gamma particles in cyst wall formation, initially postulated from cytological evidence (Truesdell and Cantino, 1970), received strong support by the finding that the chitin synthetase activity of zoospores of *B. emersonii* was located in the gamma particle (Myers and Cantino, 1974). These unique organelles, not found in the growing thallus, are formed during zoosporogenesis and disappear during zoospore encystment. Initially discovered in *B. emersonii* (Cantino and Horenstein, 1956), gamma particles have also been found in other genera: e.g. *Allomyces*, *Catenaria* (Barstow, 1979).

Zoospores of *B. emersonii* contain 5-20 gamma particles, each particle being approximately 500 nm in diameter. In thin section, gamma-particles show a cup-shaped electron-dense core inside a spheroidal vesicle (Fig. 8). The ontogeny of gamma particles was elucidated by Barstow and Lovett (1975): electron-dense granules with electron-transparent spots are produced in cisternae of the rough endoplasmic reticulum. These granules coalesce and differentiate into the characteristic horseshoe structure. Barstow (1979) found a similar ontogenetic sequence for the gamma particles of *A. macrogynous*.

According to Truesdell and Cantino (1970), gamma particles disintegrate during encystment and produce two types of vesicles. Vesicles (80 nm) appear to be formed from the inner electron-transparent side of the electron dense cup and fuse with the gamma-particle membrane. Vesicles (50-100 nm) bud off from this membrane and seemingly migrate and fuse with the zoospore plasma membrane. In this process, the gamma particle matrix is consumed. Near the end, the decayed gamma membranes fuse with one another to form multivesicular bodies. As in *B. emersonii* zoospores, the gamma particles of *A. macrogynus* mitospores produce vesicles (35-70 nm) during encystment which appear

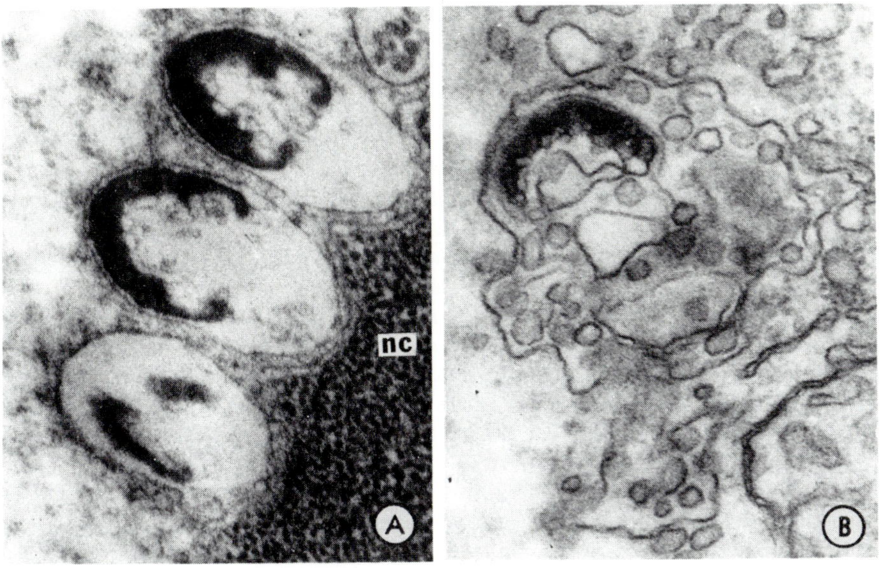

Fig. 8. Gamma particles from zoospores of Blastocladiella emersonii. A) Three gamma particles aligned next to the nuclear cap (nc) from a zoospore in the second stage of encystment. B) Vesiculation of the gamma-particle matrix during the third stage of encystment. (From Truesdell and Cantino (1970)).

to fuse with the plasma membrane (Barstow and Pommerville, 1980).

The subcellular differentiation that accompanies encystment of Blastocladiales zoospores suggests that microvesicles participate in the process of wall formation. Barstow and Pommerville (1980) and Mills and Cantino (private communication) concluded that the microvesicles derived from gamma particles of B. emersonii and A. macrogynus, respectively, are similar to the chitosomes isolated from the mycelium or yeast cells of diverse fungi (Bracker et al., 1976; Bartnicki-Garcia et al., 1978).

This proposed derivation of chitosomes from gamma-particles is supported by the recent finding that isolated gamma particles from B. emersonii can be induced to vesiculate in vitro (E. C. Cantino and G. L. Mills, private communication). The resulting populations of chitosome-like vesicles synthesize microfibrils in vitro.

3. ZOOSPORES OF OOMYCETES

(a) Cell wall architecture

The structure and composition of cyst walls of Oomycetes has been studied in Phytophtora palmivora (Tokunaga and Bartnicki-Garcia, 1971; Hegnauer and Hohl, 1973) and Aphanomyces astaci (Nyhlen and Unestam, 1978). β-glucans are the major components of cyst walls. Chemically, two types of polymers can be recognized, cellulosic glucan, made principally of β-1,4-linked chains and noncellulosic cuprammonium-insoluble polymer. Structurally, two wall components can be recognized: amorphous and microfibrillar. The amorphous component of the cyst wall of P. palmivora is easily lost and isolated cyst walls prepared either by alkaline extraction or by mechanical breakage of whole cysts show a distinctly fibrillar nature. The microfibrils form a tightly intermeshed fabric.

(b) Enzymology of cyst wall formation

Like the uniflagellate zoospore of the Chytridiomycetes, the biflagellate zoospore of the Oomycetes is also fully equipped enzymatically to encyst. Protein synthesis inhibitors, like cycloheximide, do not block the encystment process of P. palmivora (Hemmes and Hohl, 1971). Encystment is very rapid and, within 1-2 min, most zoospores of P. palmivora induced to encyst by agitation produce a recognizable microfibrillar wall (Tokunaga and Bartnicki-Garcia, 1971). Wall formation continues for about 10 min at a linear pace.

Cell-free extracts from either unencysted zoospores or cysts have the enzymatic capacity to produce alkali-insoluble β-1,3-glucans from UDP-Glc (Wang and Bartnicki-Garcia, 1971; unpublished). The glucans contain some β-1,6 bonds as is characteristic for the non-cellulosic glucan of the cyst wall. As yet, it has not been possible to reproduce the synthesis of cellulosic glucan in cell-free extracts from either zoospores or cysts, or for that matter, with any type of cell-free extract from Phytophthora spp. Either the β-1,4 glucan synthetase does not survive cell breakage or the conditions for activity have not been met experimentally.

(c) Cytology of cyst wall formation

In addition to ostensible morphological changes that are probably not involved in the development of a cyst wall, namely loss of flagella and disappearance of the water expulsion apparatus, the most conspicuous change during zoospore encystment of *P. palmivora* and *Pythium aphanidermatum* is the disappearance of the so-called peripheral vesicles and cisternae (Hemmes and Hohl, 1971; Hegnauer and Hohl, 1973; Sing and Bartnicki-Garcia, 1975a; Grove and Bracker, 1978). These structures are positioned next to the plasma membrane and most of them are consumed during encystment; they fuse with the plasma membrane and their contents are extruded. We have evidence that the material extruded from the peripheral vesicles is, at least in part, a Con A-reacting glycoprotein (Sing and Bartnicki-Garcia, 1975b). This amorphous material (Desjardins *et al.*, 1973), or cyst coat (Grove and Bracker, 1978), adheres the encysting zoospore to solid surfaces (Sing and Bartnicki-Garcia, 1975a). It may also be the material that cements, or rigidifies the wall fabric, or perhaps it might function as a matrix in which the fibrils are synthesized. The discharge of the peripheral vesicles precedes the appearance of microfibrils. Whether the amorphous material serves as a "matrix" for fibril formation or as a source of enzyme and precursors for microfibril synthesis remains to be seen. The role of peripheral cisternae in wall formation is unclear.

Oomycetous zoospores lack the characteristic gamma particles found in the zoospores of Chytridiomycetes; also, there is no evidence of any extensive vesiculation process during the encystment of oomycetous zoospores. The question of whether or not microvesicles comparable to chitosomes play a role in the development of the cellulosic microfibrils of the Oomycetes remains to be established. Microvesicles ($< 100 = $ nm) do occur in active cell wall making regions of Oomycetes e.g. hyphal tips of *Pythium* (Grove *et al.*, 1970) and *Saprolegnia* (Heath *et al.* 1971), developing oogonia and fertilization tubes of *Phytophthora* (Hemmes and Bartnicki-Garcia, 1975). Microvesicles can also be seen in thin sections of zoospores of *P. palmivora* but their sparsity and minuteness has prevented us from deciding whether or not they are consumed during encystment. An answer as to what subcellular structure, microvesicular or otherwise, is involved in the synthesis of the cellulosic microfibrils is not expected until it becomes possible to detect cellulose synthetase *in vitro* and, with this biochemical handle, isolate and identify the subcellular structure associated with this enzyme.

A detailed summary of cytological events during encystment and germination of a zoospore of P. aphanidermatum is shown in Fig. 9.

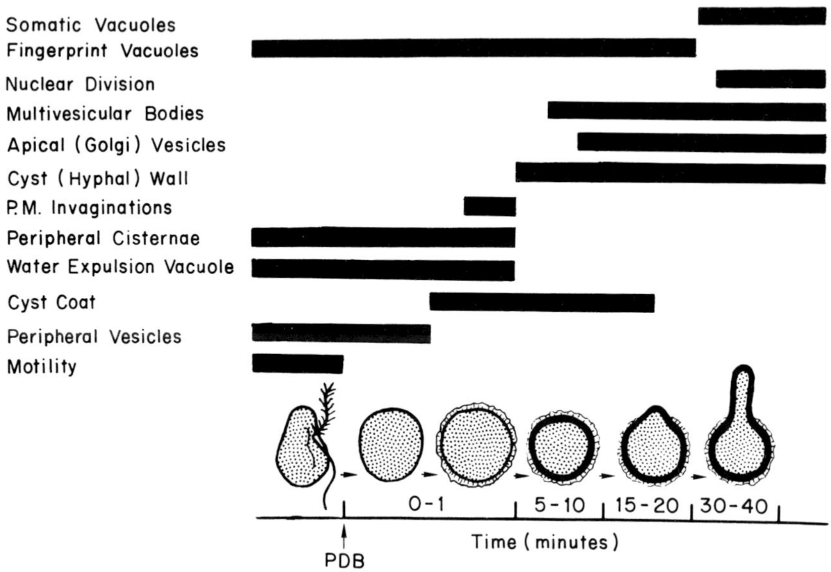

Fig. 9. Diagram of cytological features during the encystment and germination of zoospores of Pythium aphanidermatum. From Grove and Bracker (1978).

A freeze-fracture study of encysting zoospores of P. palmivora (Nogueira et al., 1977) failed to show any structure that could be connected with wall formation. The random pattern of distribution of plasma membrane particles of the zoospore remained unchanged during encystment (Fig. 10). There were no obvious structural connections between the wall microfibrils and cytoplasmic structures. This study did reveal that the plasma membrane of P. palmivora is completely covered by a uniform fuzzy coat of unknown chemical nature. The presence of this 20-25 nm thick coat (Fig. 10A), questions the current belief that a "naked" plasma membrane is the outermost boundary of these (and possibly other) zoospores. Perhaps, the coat is an integral extension of plasma membrane components, in which case, we may regard this envelope as a "reinforced" plasma membrane.

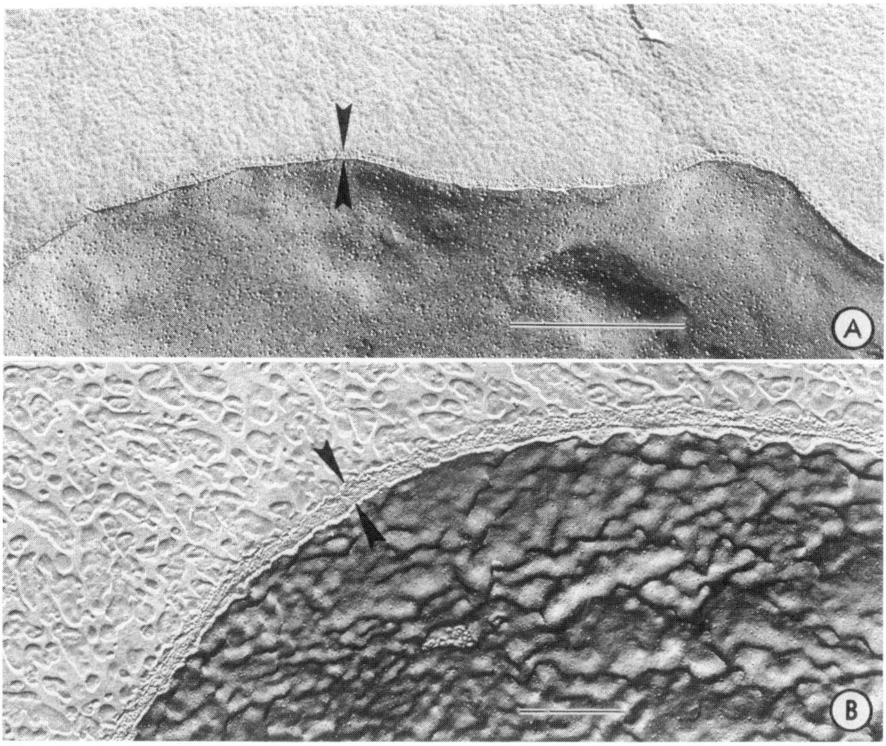

Fig. 10. Freeze fracture views of zoospores of Phytophthora palmivora before (A) and after (B) encystment. Arrows delimit the fuzzy coat covering the plasma membrane (A) and the cyst wall (B). From Nogueira et al. (1977). Bar=500nm

(d) Germ tube emergence

The participation of vesicles in germ-tube emergence is supported by the observation of vesicle clusters in cysts of P. aphanidermatum at the site where the tube is about to appear (Grove and Bracker, 1978).

Studies on the germination of cysts of P. palmivora (Tokunaga and Bartnicki-Garcia, 1971; Hegnauer and Hohl, 1973) revealed one major architectural difference between cysts and germ tube walls. The germ tube walls (washed with alkali to remove soluble components) show two distinct textures: an outer non-fibrillar texture and an inner, distinctly fibrillar texture. By contrast, cyst walls equally treated showed only a microfibrillar texture. The added complexity of the germ tube wall suggests that germ tube emergence necessitates the development of new enzymatic capacities and, perhaps, a cytological mechanism to account for the dual texture of the germ tube wall.

4. CONCLUSION

Despite our very fragmentary knowledge on the biochemistry and cytology of cell wall formation during germination of phycomycetous spores, a few general conclusions can be made. These pertain primarily to the fungi used as main examples namely *Mucor*, *Blastocladiella* and *Phytophthora*. Probably, these generalizations extend to other genera within the same orders (or classes), but exceptions cannot be ruled out.

Totally new vegetative cell walls are formed during the germination of zoospores and sporangiospores of Phycomycetes. Spore germination may be regarded as a succession of two patterns of wall synthesis. After an initial period of spherical wall deposition (stage I), an outgrowth, usually a germ tube with cylindrical walls is formed (stage II).

There is no evidence that the enzymes responsible for wall formation are deployed at the spore surface and are ready to operate at the onset of germination. Germination entails the development and/or operation of a vesicular mechanism for the orderly delivery of enzymes and other materials needed for cell wall construction. The cytological evidence at hand indicates that an active process of vesicular transport of enzymes and other materials to the cell surface (exocytosis) takes place during wall formation.

The nutritional requirements and the kinetics of germination are quite different for zoospores and sporangiospores. Zoospores can germinate in the absence of nutrients; zoospore germination is not a growth phase. Sporangiospores, on the other hand, require a full supply of nutrients for germination. Germination is a *bona fide* phase of exponential growth.

In zoospores, germination stage I (encystment), is a relatively short phase of wall synthesis by preformed enzyme which is mobilized to the cell surface (and activated) at the time of encystment. The rate of cyst wall formation obeys linear kinetics and there is no evidence of extension growth of this wall.

In sporangiospores, by contrast, stage I is a relatively long growth phase. The expanding germ sphere requires considerable extension growth of the germ sphere wall. The rate of wall formation obeys exponential kinetics. New enzyme synthesis is required to sustain these kinetics.

Germination stage II, the formation of a germ tube, clearly involves extension growth of the wall for both zoospores and sporangiospores. Protein synthesis is required at this time. In sporangiospores, the kinetics and metabolism of wall formation of stage I are simply continued, uninterruptedly, into stage II. However, for encysted zoospores, the transition from stage I to stage II, represents a change from non-extension to extension wall growth. The germ-tube wall has a greater architectural complexity than the cyst wall and is generated by a process of wall extension that probably requires the participation of lytic enzymes (Bartnicki-Garcia, 1973). The zoospore is apparently not fully equipped to elaborate the more complex germ tube wall, hence the need for new enzyme synthesis.

In germinating spores of these fungi, the switch from stage I to stage II probably begins with the introduction of a polarization signal which directs wall-synthesizing vesicles to move preferentially to a small target area on the germ sphere or cyst, where the germ tube would emerge.

5. ACKNOWLEDGEMENT

The work performed in the author's laboratory was supported in part by grants from the National Institutes of Health (AI-05540) and the National Science Foundation (BMS 72-01862; GB-32165 and OIP-7508378).

6. REFERENCES

Barstow, W. E. (1979). Ultrastructure of the formation of γ particles during zoosporogenesis in Allomyces macrogynus and Catenaria anguilullae. Exp. Mycol. 3, 28-41.

Barstow, W. E. and Lovett, J. S. (1975). Formation of gamma particles during zoosporogenesis in Blastocladiella emersonii. Mycologia 67, 518-529.

Barstow, W. E. and Pommerville, J. (1980). The ultrastructure of cell wall formation and of gamma particles during encystment of Allomyces macrogynus zoospores. Arch. Microbiol. In press.

Bartnicki-Garcia, S. (1968). Control of dimorphism in Mucor by hexoses: inhibition of hyphal morphogenesis. J. Bacteriol. 96, 1586-1594.

Bartnicki-Garcia, S. (1969). Cell wall differentiation in the Phycomycetes. Phytopathology 59, 1065-1071.

Bartnicki-Garcia, S. (1973). Fundamental aspects of hyphal morphogenesis. In "Microbial Differentiation" (Ed. J. M. Ashworth and J. E. Smith). pp. 245-267. Cambridge Univ. Press, Cambridge.

Bartnicki-Garcia, S. and Lippman, E. (1969). Fungal morphogenesis: cell wall construction in Mucor rouxii. Science 165, 302-304.

Bartnicki-Garcia, S. and Lippman, E. (1977). Polarization of cell wall synthesis during spore germination of Mucor rouxii. Exper. Mycology 1, 230-240.

Bartnicki-Garcia, S. and Reyes, E. (1964). Chemistry of spore wall differentiation in Mucor rouxii. Arch. Biochem. Biophys. 108, 125-133.

Bartnicki-Garcia, S., Nelson, N. and Cota-Robles, E. (1968). Electron microscopy of spore germination and cell wall formation in Mucor rouxii. Arch. Mikrobiol. 63, 242-255.

Bartnicki-Garcia, S., Bracker, C. E., Reyes E. and Ruiz-Herrera, J. (1978). Isolation of chitosomes from taxonomically diverse fungi and synthesis of chitin microfibrils in vitro. Exper. Mycology 2, 173-192.

Bartnicki-Garcia, S., Ruiz-Herrera, J. and Bracker, C. E. (1979). Chitosomes and chitin synthesis. In "Fungal Walls and Hyphal Growth" (Ed. J. H. Burnett and A. P. J. Trinci). pp. 149-168. Cambridge Univ. Press, Cambridge.

Bracker, C. E. (1971). Cytoplasmic vesicles in germinating spores of Gilbertella persicaria. Protoplasma 72, 381-397.

Bracker, C. E., Ruiz-Herrera J. and Bartnicki-Garcia, S. (1976). Structure and transformation of chitin synthetase particles (chitosomes) during microfibril synthesis in vitro. Proc. Nat. Acad. Sci. USA, 73, 4570-4574.

Burke, D. J., Seale, T. W. and McCarthy, B. J. (1972). Protein and ribonucleic acid synthesis during the diploid life cycle of Allomyces arbuscula. J. Bacteriol. 110, 1065-1072.

Cabib, E., Duran, A. and Bowers, B. (1979). Localized activation of chitin synthetase in the initiation of yeast septum formation. In "Fungal Walls and Hyphal Growth" (Ed. J. H. Burnett and A. P. J. Trinci). pp. 189-201. Cambridge University Press, Cambridge.

Camargo, E. P., Dietrich, C. P., Sonnerborn, D. and Strominger, J. L. (1967). Biosynthesis of chitin in spores and growing cells of Blastocladiella emersonii. J. Biol. Chem. 242, 3121-3128.

Cantino, E. C. and Horenstein, E. A. (1956). Gamma and the cytoplasmic control of differentiation in Blastocladiella emersonii. Mycologia 48, 443-446.

Desjardins, P. R., Wang, M. C., and Bartnicki-Garcia, S. (1973). Electron microscopy of zoospores and cysts of Phytophthora palmivora: morphology and surface texture. Arch. Mikrobiol. 88, 61-70.

Ekundayo, J. A. and Carlile, M. J. (1964). The germination of sporangiospores of Rhizopus arrhizus; spore swelling and germ tube emergence. J. Gen. Microbiol. 35, 261-269.

Furch, B. and Gooday, G. W. (1978). Sporopollenin in Phycomyces blakesleeanus. Trans. Br. Mycol. Soc. 70, 307-309.

Furch, B. and Pambor, L. (1978). Cell wall constituents of Phycomyces blakesleeanus. 3. Carbohydrate and protein composition of sporangiospore cell walls in relation to heat-induced germination. Microbios Letters 8, 71-80.

Giddings, T. H. and Cantino, E. C. (1974). Partial purification and properties of D-glucosamine 6-phosphate N-acetyltransferase from zoospores of Blastocladiella emersonii. J. Bacteriol. 120, 976-979.

Grove, S. N. and Bracker, C. E.. (1978). Protoplasmic changes during zoospore encystment and cyst germination in Pythium aphanidermatum. Exper. Mycology 2, 51-98.

Grove, S. N., Bracker, C. E. and Morré, D. J. (1970). An ultrastructural basis for hyphal tip growth in Pythium ultimum. Amer. J. Bot. 57, 245-266.

Hawker, L. E. and Abbott, P. M. (1963). An electron microscope study of maturation and germination of sporangiospores of two species of Rhizopus. J. Gen. Microbiol. 32, 295-298.

Heath, I. B., Gay, J. L. and Greenwood, A. D. (1971). Cell wall formation. in the Saprolegniales: cytoplasmic vesicles underlying developing walls. J. Gen. Microbiol. 65, 255-232.

Hegnauer, H. and Hohl, H. R. (1973). A Structural comparison of cyst and germ tube wall in Phytophthora palmivora. Protoplasma 77, 151-163.

Hemmes, D. E. and Bartnicki-Garcia, S. (1975). Electron microscopy of gametangial interaction and oospore development in Phytophthora capsici. Arch. Microbiol. 103, 91-112.

Hemmes, D. E. and Hohl, H. R. (1971). Ultrastructural aspects of encystation and cyst germination in Phytophthora parasitica. J. Cell Sci. 9, 175-191.

Hess, W. M. and Weber, D. J. (1973). Ultrastructure of dormant and germinated sporangiospores of Rhizopus arrhizus. Protoplasma 77, 15-33.

Jaffe, L. F. (1968). Localization in the developing Fucus egg and general role of localizing currents. Adv. Morphogenesis 7, 295-328.

Kirby, E. G., Kroh, M. and Sassen, M. M. A. (1974). Cell wall formation in zoospores of Allomyces arbuscula: I. Fine structural observations on encysting haploid zoospores. Arch. Microbiol. 98, 147-158.

Lara, S. L. and Bartnicki-Garcia, S. (1974). Cytology of budding in Mucor rouxii: Wall ontogeny. Arch. Microbiol. 97, 1-16.

Lovett, J. S. (1968). Reactivation of ribonucleic acid and protein synthesis during germination of Blastocladiella zoospores and the role of the ribosomal nuclear cap. J. Bacteriol. 96, 962-969.

Myers, R. B. and Cantino, E. C. (1974). "The gamma particle" Monographs in Developmental Biology, Vol. 8. Karger, Basel.

Nogueira, M. L. B., Pinto da Silva, P. and Bartnicki-Garcia, S. (1977). Freeze-fracture and freeze-etching study of encystment of Phytophthora palmivora zoospores. J. Gen. Microbiol. 102, 149-155.

Norrman, J., Myers, R. B., Giddings, T. H. and Cantino, E. C. (1973). Partial purification of L-glutamine: D-fructose 6-phosphate aminotransferase from zoospores of Blastocladiella emersonii. Biochim. Biophys. Acta. 302, 173-177.

Nyhlén, L. and Unestam, T. (1978). Cyst and germ tube wall strucutre in Aphanomyces astaci, Oomycetes. Can. J. Microbiol. 24, 1296-1299.

Olson, L. W. (1973). The meiospore of Allomyces. Protoplasma 78, 113-127.

Orlowski, M. and Sypherd, P. S. (1978). Regulation of macromolecular synthesis during hyphal germ tube emergence from Mucor racemosus sporangiospores. J. Bacteriol. 134, 76-83.

Paznokas, J. L. and Sypherd, P. S. (1975). Respiratory capacity, cyclic adenosine 3',5'-monophosphate, and morphogenesis in Mucor racemosus. J. Bacteriol. 124, 134-139.

Ruiz-Herrera, J., Lopez-Romero, E. and Bartnicki-Garcia, S. (1977). Poperties of chitin synthetase in isolated chitosomes from yeast cells of Mucor rouxii. J. Biol. Chem. 252, 3338-3343.

Selitrennikoff, C. P. and Sonneborn, D. R. (1976). Post-translational control of de novo cell wall formation during Blastocladiella emersonii zoospore germination. Developmental Biol. 54, 37-51.

Selitrennikoff, C. P., Allin, D. and Sonneborn, D. R. (1976). Chitin biosynthesis during Blastocladiella zoospore germination: Evidence that the hexosamine biosynthetic pathway is post-translationally activated during cell differentiation. Proc. Nat. Acad. Sci. USA 73, 534-538.

Sing, V. O. and Bartnicki-Garcia, S. (1975a). Adhesion of Phytophthora palmivora zoospores: electron microscopy of cell attachment and cyst wall fibril formation. J. Cell Sci. 18, 123-132.

Sing, V. O. and Bartnicki-Garcia, S. (1975b). Adhesion of Phytophthora palmivora zoospores: detection and ultrastructural visualization of concanavalin A-receptor sites appearing during encystment. J. Cell Sci. 19, 11-20.

Tokunaga, J. and Bartnicki-Garcia, S. (1971). Structure and differentiation of the cell wall of Phytophthora palmivora: cysts, hyphae and sporangia. Arch. Mikrobiol. 79, 293-310.

Trevillyan, J. M. and Pall, M. L. (1979). Control of cyclic adenosine 3',5'-monophosphate levels by depolarizing agents in fungi. J. Bacteriol. 138, 397-403.

Trinci, A. P. J. (1971). Exponential growth of the germ tubes of fungal spores. J. Gen. Microbiol. 67, 345-348.

Truesdell, L. C. and Cantino, E. C. (1970). Decay of γ particles in germinating zoospores of Blastocladiella emersonii. Arch. Mikrobiol. 70, 378-392.

Van Assche, J. A. and Carlier, A. R. (1973). The pattern of protein and nucleic acid synthesis in germinating spores of Phycomyces blakesleeanus. Arch. Mikrobiol. 93, 129-136.

Van Assche, J. A., Carlier, A. R. and Van Tieghem, L. L. C. (1977). The effect of gamma radiation on breaking of dormancy in Phycomyces spores. Arch. Microbiol. 113, 95-97.

Van Etten, J. L., Bulla, L. A. and St. Julian, G. (1974). Physiological and morphological correlation of Rhizopus stolonifer spore germination. J. Bacteriol. 117, 882-887.

Van Laere, A. J. and Carlier, A. R. (1978). Synthesis and proteolytic activation of chitin synthetase in Phycomyces blakesleeanus Burgeff. Arch. Microbiol. 116, 181-184.

Van Laere, A. J., Carlier, A. R. and Van Assche, J. A. (1977). Cell wall carbohydrates in Phycomyces blakesleeanus Burgeff. Arch. Microbiol. 112, 303-306.

Wang, M. C., and Bartnicki-Garcia, S. (1971). Cell wall glucan biosynthesis during zoospore encystment in Phytophthora palmivora. Bacteriol. Proc. p. 61.

ADDENDUM :

Gooday, G.W., Fawcett, P., Green, D. and Shaw, G. (1973). The formation of fungal sporopollenin in the zygospore wall of Mucor mucedo : a role for the sexual carotenogenesis in the Mucorales. J. Gen. Microbiol. 74, 233-239.

VII. METABOLIC CONTROLS

CHANGES IN THE BIOCHEMICAL REACTIONS OF FUNGAL CELLS WITH AGE : DISCUSSANT'S INTRODUCTION

David Gottlieb

Department of Plant Pathology, University of Illinois Urbana, Illinois, U.S.A.

INTRODUCTION

A discussion of metabolic controls in fungi is difficult for many reasons, but primarily for the lack of published data on that subject. Even the term itself has widely varied usage; not the least of which is its application to a vast range of compounds and processes. For example, it is used to denote compounds that enter into or are secreted by organisms. It also refers to the chemicals which are single and free or are bound as part of the cell structure. Others are elements of complex pathways in cells. The compounds can be either active in the living cell or in cell free extracts. Metabolites of these cells might be useful, innocuous, or harmful to their producers. The essential feature seems to be that in someway the term is related to the activity of the living cell.

PRESENTATION

An introduction to metabolic controls might best accomplish its aim if it establishes at the start the normal behavior of many aspects of fungal metabolism prior to the onset of morphological developments. This information could establish a base line for the effect that time has on any of the physiological or biochemical changes that occur later in the development of a microbe and are responsible for the development of the organism.

The data presented here are from the research my colleagues and I conducted on *Rhizoctonia solani* and *Sclerotium bataticola*. They show that the growth can be uniform over a long time span thus indicating no shortage of nutrients in this period of growth and that toxic metabolites are not produced in sufficient concentrations to limit the growth of the organisms.

The knowledge of metabolic changes in fungi over an early time span, during which no morphological changes are apparent, can serve as a base line for understanding future developments in the life of the microbe. Studies to this end were the aim of our program. The two fungi were used during their linear growth periods. These rates differed, *R. solani* was faster than *S. bataticola* but straight line growth occurred with each. No gross morphologic, cytologic or changes in number or percentage occurred. We could conclude that during the duration of the experiment, there was no deleterius decrease in nutrients. Furthermore, there was no production of toxic metabolites which might cause decreased growth.

Even though no microscopic or macroscopic changes were apparent in the mycelium of *R. solani* and *S. bataticola* during the period of study, there were changes on the biochemical level. Composition of the cells at various durations of time is often different, DNA, RNA, protein and ergosterol, for example, decreased with time even though the mycelium appeared to be unchanged. Total carbohydrate increased in *R. solani* but decreased in *S. bataticola*, total lipid remained relatively the same in *R. solani* but increased in *S. bataticola* (Gottlieb *et al.*, 1968; Molitoris *et al.*, 1968).

There was a decreased respiration (QO_2) in both endogeous and exogenous conditions (Van Etten *et al.*, 1966). One could reasonably attribute such changes to the cell's respiratory enzymes and their decrease as an explanation for the lowered oxygen uptake. Intensive studies of the QO_2 of 31 respiratory enzymes of the major respiration pathways did not yield the results that we had expected. We had expected to find one or more enzymes in both fungi that consistently decreased with age. Instead, sometimes there were decreases of neither, of one and not the other or its converse. But never was the situation found in which the same enzyme in both fungi decreased with time (Van Etten *et al.*, 1966). The cause of the puzzle was found after much frustration. To make conditions optimum for respiration we had been adding coenzymes to our media in the Warburg studies. Thus there would always be an excess of coenzymes. If the coenzyme necessary for the reaction was in short supply we would never detect it. We had added exogenous coenzyme to the system and thereby allowing respiration to go on even by older cells which, *in vivo*, had not had sufficient of the coenzyme for maximum respiration. This concept was put to test and the cells of different ages were analyzed for the content of coenzyme. The decreases were marked in the two fungi and were from both the pyridine nucleotide cofactors and the flavin groups. Some reductions were as much as 90% depending on the fungus and cofactor - as well as the baseline used i.e. protein, DNA, or dry weight. Adding an exogenous mixture of cofactors, allowed the cell free extracts of old cell to respire at the rate of those made from young cells (Van Etten *et al.*, 1966; Nicolas *et al.*, 1968).

The decrease in protein that had been found in mycelium of *R. solani* and *S. bataticola* was apparently due to defects in the synthesizing capacity of the cells (Gottlieb *et al.*, (1968). The reduction in protein was also found in cell free preparations in which the incorporation by young, mature and old cells of ^{14}C-phenylalanine and ^{14}C-leucine decreased in the same direction as occurred in the *in vivo* incorporation. The faults responsible for this decrease were in both the soluble and particulate fraction of *R. solani* and were additive. Each fraction was responsible for approximately 50% of the total decrease in peptide synthesis (Obrig *et al.*, 1970). The decreased *in vivo* protein synthesis was associated with phenylalanine transfer tRNA with a net decrease in phenylalanyl-tRNA. The amino acid activation step with ATP did not change; however, the charging properties of the tRNA synthetase were reduced with time under conditions where protein synthesis was blocked. The charging properties were further studied with individual amino acids. Charging decreased only with five amino acids, another five increased, two remained the same, and in two the charging was at too low a level to measure (Obrig *et al.*, 1970).

The decreases that occur in protein synthesis even when the fungi are growing in a uniform and vigorous state might not be caused by any lesion in that system but might well be caused by a fault or absence in parts of systems other than from amino acids to polypeptides. One such lesion could be the synthesis of nucleic acids. *R. solani* contains such and needs them for the formation of peptides (i.e. proteins). It has ribosomal, heavy 28S, and light 25S; there is also mitochondrial heavy, 18S; and light, 16S. Transfer RNA, 4S→5S, is also present as is messenger RNA but the turn over is very slow. ^{14}C-uracil adequately labels the nucleic acids that formed *in vivo* by the mycelium. Mycelia of the fungus were from areas of the colony that were of different ages and cut into mycelial rings of 2.0, 4.5, 8.0 cm along the radius were called young, mature, and old, respectively. The ability of these cells to synthesize various species of RNA from ^{14}C-uracil differed greatly, and generally the reduction in ability to make RNA varied conversely with age; the young had the greatest synthetic ability, the mature less and old portions of a colony the least, so that only traces of RNA synthetic ability remained in the old mycelium. The decrease occurred for all above mentioned RNA species in *R. solani* (Davies *et al.*, 1974).

The decreased protein and nucleic acids synthesis need not be ascribed only to defects in the synthesizing systems but might, for example, be caused by the breakdown of these compounds after they have been formed. Thus the rate of activity of protein synthesis might be regulated in this manner. *R. solani* has one or more ribonucleases in the various extracts using the torula RNA as the substrate for the degradation by *R. solani* RNase. Other data indicate that *R. solani* RNA is also susceptible to degradation by the homologous enzyme from *R. solani* (Gottlieb *et al.*, 1976b).

Methylated RNA is vital for the formation of some protein. The amount of and type of methylation appear to be subject to developmental changes in an organism. The degree of methylation could also be affected by the age of cells even if no changes are apparent on the microscopic or macroscopic level. Nevertheless, as can be seen in our previous studies, changes do go on during this early period. Using *Escherichia coli* tRNA as substrate, the amount methylation by *R. solani* methylases was greater by young than by old cells of the same mycelium.

Methylation of tRNA by preparations from young *R. solani* cells was also effective on homologous tRNA from the fungus (Gottlieb *et al.*, 1976a). No methylase inhibitors were present in *R. solani*.

Since protein synthesis is also dependent, in part, on the presence of ribosomes, a change in this component could alter the amount of protein produced. Though no decrease occurred in the total number of ribosomes with the age of *R. solani*, the percentage of polysomes in the cell diminished steadily with age from 80% to 15% over a 8 day period (Ricciardi *et al.*, 1974). It is obvious then that such a change would be reflected in protein formation which is carried on more efficiently on polysomes than on monosomes.

CONCLUSION

The feature that is most obvious from these studies is that all kinds of changes are taking place in fungal cells long before any results of these activities cause visual deviations in the mycelia. Care must be taken that any comparisons between or among different cultures should be based not only on macroscopic or microscopic features but also on the biochemical state in which the cells are at a given time. Different isolates of fungi, for example, grow at different rates and if they are compared merely on the age of the total mycelium one might be led astray. The biochemical reactions depend in part on the stage of development and this can be different depending on the rate of development of the cells more than on chronological age. Finally, we have not the vaguest notion which of the various biochemical reactions set off the aging process. All the gross vital reactions are going on, usually with diminished activities as chronological time proceeds. From our data one cannot support the concept of a vital key reaction and trigger mechanism which set off the aging phenomenon. The reduction in various rates of the different reactions all go at about the same time and might be caused by the same inciter.

REFERENCES

Davies, F. L., Lin, F. K. and Gottlieb, D. (1974), Synthesis of RNA with relation to aging in *Rhizoctonia solani*. *Arch. Microbiol.* 95, 145-152.

Gottlieb, D., Molitoris, H. P. and Van Etten, J. L. (1968), Changes of fungi with age. III. Incorporation of amino acids into cells of *Rhizoctonia solani* and *Sclerotium bataticola*. *Arch. Mikrobiolog.* 61, 394-398.

Gottlieb, D. and Raghu, K. (1976a), Methylation and aging in *Rhizoctonia solani*. *Mycopathologia.* 60, 57-62.

Gottlieb, D. and Sharma, V. D. (1976b), Changes in ribonuclease with age in the fungus *Rhizoctonia solani*. *Mechanisms Ageing and Development.* 5, 25-32.

Gottlieb, D. and Van Etten, J. L. (1966), Changes in fungi with age. I. Chemical composition of *Rhizoctonia solani* and *Sclerotium bataticola*. *J. Bacteriol.* 91, 161-168.

Molitoris, H. P., Van Etten, J. L. and Gottlieb, D. (1968), Altersbedingte Änderungen der Zusammensetzung und des Stoffwechsels bei Pilzen. Mushroom Science VII. Proc. Second Scientific Symp. and the Seventh Intern. Congr. on Mushroom Science, Hamburg 1968. 59-67.

Nicolas, G. and Gottlieb, D. (1968), Changes in fungi with age. IV. Role of coenzymes in the respiratory decreases in *Rhizoctonia solani* and *Sclerotium bataticola*. *J. of Gerontol.* 23, 544-550.

Obrig, T. G. and Gottlieb, D. (1970), *In vitro* protein synthesis and aging in *Rhizoctonia solani*. *Jour. Bacteriol.* 101, 755-762.

Ricciardi, R. P., Hollomon, D. W. and Gottlieb, D. (1974), Age dependent changes in fungi: Ribosomes and protein synthesis in *Rhizoctonia solani* mycelium. *Arch.*

Microbiol. 95, 325-336.

Van Etten, J. L., Molitoris, H. P. and Gottlieb, D. (1966), Changes in fungi with age. II. Respiration and respiratory enzymes of *Rhizoctonia solani* and *Sclerotium bataticola*. *J. Bacteriol.* 91, 169-175.

INHIBITORS AND STIMULANTS OF SPORE GERMINATION AND INFECTION STRUCTURE FORMATION IN FUNGI

V. Macko

Boyce Thompson Institute, Tower Road, Ithaca, NY 14853

INTRODUCTION

Spore germination in fungi is a crucial event in insuring the survival of the species. Thus, while the spore is forming on the mycelium, various reserve energy and information sources, including regulatory chemicals of low molecular weight, are laid down to be used during germination. I will emphasize those chemical agents which fit loosely into the category of "secondary metabolites". A spore carries a variety of immediate instructions besides DNA, RNA, etc. -- instructions which are essential for development of a new generation. Of particular significance for development are the special chemical substances, often small molecules such as germination self-inhibitors and stimulants, which are most likely produced during spore formation and deployed to function in control of germination. I will also discuss similar substances from other sources, e.g. germination stimulants from plants, and will also review some recent accounts on chemical stimuli of infection structure formation.

Control of spore germination by self-inhibitors and stimulants has received increasing attention during the last decade. This resurgence of interest stems, in part, from advances in analytical techniques allowing for greater sensitivity, precision and accuracy in the chemical characterization of such regulators.

Studies of regulatory factors affecting spore germination are contributing to the understanding of mechanisms operating in morphogenetic control of the germination process. Studies of hormones, growth factors, pheromones and other regulatory factors affecting higher plant, insect and mammalian physiology have contributed greatly to understanding the biology of these organisms. Natural regulators of fungal development have been reviewed by Trione (1980). The discovery of fungal growth regulators and information on their mode of action could significantly advance the understanding of developmental biology of fungi in general and morphogenetic controls of spore germination in particular.

The purpose of this present review of chemical regulators of spore germination is more to summarize some recent studies than to provide a comprehensive review of the whole field. Particularly relevant reviews on spore germination, dormancy, and chemical regulators of germination are those by Allen (1976), Sussman and Halvorson (1966) and Sussman and Douthit (1973). Other reviews of specific aspects will be cited where relevant.

SELF-INHIBITORS OF GERMINATION

Physiological aspects

Population effect. Self-inhibitors have been detected in spores of a wide range of fungi. Spores of such fungi may swell during the hydration phase of water uptake from the medium, but germination does not occur unless the inhibitor is leached away. Inhibition in these species is directly proportional to the size of the spore population. This phenomenon of self-inhibition resembles, at least in a remote way, self-regulatory signals in animal cells or animal populations. An example of self-inhibition at the cellular level is contact inhibition in fibroblasts (Abercrombie, 1979). At the level of animal populations examples include the ability of snails, mosquitoes, or tadpoles to limit their population size by self-regulation at high densities (Hwang and Mulla, 1976; Levy *et al.*, 1973). Phytophagous insects avoid crowding by regulating spacing patterns using epideictic pheromones (Prokopy, 1980). Although the social and regulatory objectives of fungal cells, metazoan cells, and animal populations differ greatly, it is not unreasonable to expect that some fundamental mechanisms of cell regulation may be universal. Recently reviewed mechanisms of intracellular localization by electrical microcurrents (Jaffe *et al.*, 1974; Jaffe and Nuccitelli, 1977; Weisenseel *et al.*, 1975) and perisemic decoding of regulatory signals in microorganisms (Reissig, 1974) seem to substantiate this notion. "Perisemic" refers to processes by which the cell envelope transduces a regulatory signal (Reissig, 1974).

Jaffe and coworkers showed that intracellular localization in differentiating cells is controlled by microcurrents. Germinating pollen grains and fucoid eggs drive rather large electrical currents through themselves. The current enters an ungerminated grain's prospective growth site and leaves its opposite end, continuing as long as the tube grows. Are germinating fungal spores also driving current through themselves? Specialized techniques such as the vibrating probe system for measuring extracellular currents (Jaffe and Nuccitelli, 1974) should be of considerable aid in helping to show how removing or adding self-inhibitors or stimulants to the germinating spore would affect the current.

Synchrony of germination. In a population of spores, appearance of the germ tubes usually is asynchronous, although hydration and many of the internal changes that are associated with the early stages of spore germination may be relatively synchronous. Synchrony of the event of germ tube formation can be achieved using germination inhibitors and stimulants. For example, rust uredospores germinate synchronously if incubated on water solutions of self-inhibitors and then either adding a germination stimulator or transferring uredospores from self-inhibitor to water (unpublished observation).

Selective advantage for fungi. Spores of many fungi germinate poorly or not at all when in dense suspensions or crowded upon a surface. The numerous reports of self-inhibition have been summarized in reviews by Allen (1976) and by Macko *et al.* (1976). The selective advantage of self-inhibition is probably to prevent germination in the sorus and to prevent rapid germination of all spores at the same time. More detailed studies of the "crowding effect" have been carried out, particularly with the rust fungi after Allen (1955) and Yarwood (1954, 1956) showed that a chemical inhibitory principle is involved. The definitive evidence supporting the existence of self-inhibitors was developed with wheat stem rust and bean rust fungi (Allen and Dunkle, 1971; Bell and Daly, 1962).

The self-inhibition phenomenon in fungal spores has been reported for a wide range of taxonomic groups comprising more than 50 fungal species (Allen, 1976). Some of the active substances have been isolated and characterized and the chemical identity of a few inhibitors is now established (Table 1 and Fig. 1).

Sources of self-inhibitors.

Rust fungi. The self-inhibitor of bean, sunflower, corn, snapdragon, peanut and stripe rust uredospores was found to

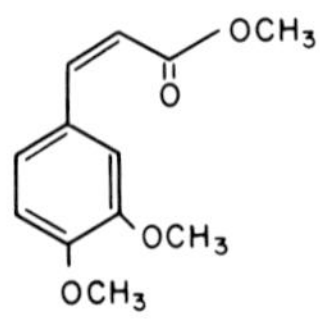

I Methyl *cis*-3,4-dimethoxycinnamate (MDC)

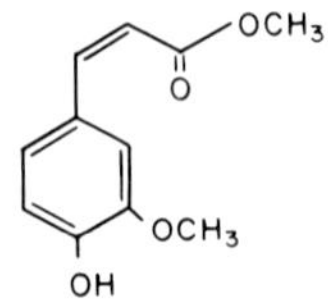

II Methyl *cis*-ferulate (MF)

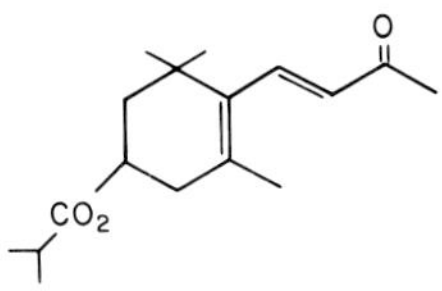

III 5-Isobutyroxy-β-ionone (quiesone)

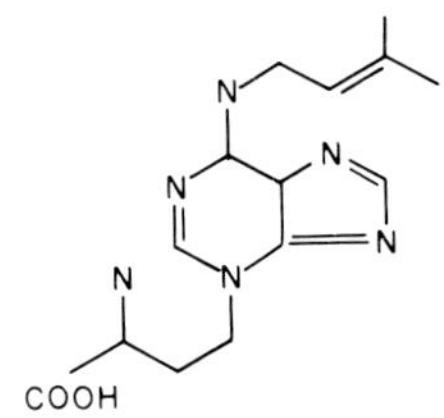

IV Discadenine

$CH_3(CH_2)_7CH_2OH$

V 1-Nonanol

$CH_3(CH_2)_7CHO$

VI Nonanal

$CH_3CH{=}CHCH{=}CHCHO$

VII 2,4-Hexadienal

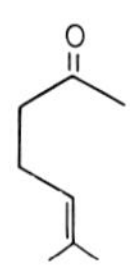

VIII 6-Methyl-5-hepten-2-on

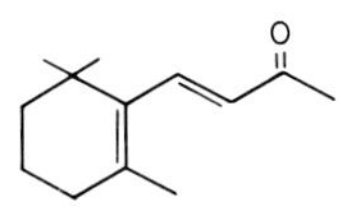

IX β-Ionone

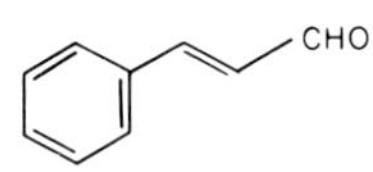

X *trans*-Cinnamaldehyde

$CH_3(C{\equiv}C)_5CH{=}CH_2$

XI 1-Tridecene-3,5,7,9,11-pentayne

$CH_3CH{=}CH(C{\equiv}C)_4CH{=}CH_2$

XII *cis* and *trans*-Trideca-1,11-dien-3,5,7,9-tetrayne

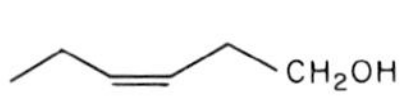

XIII *trans*-2-Hexen-1-ol

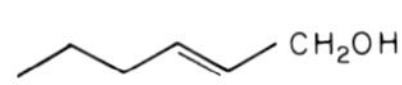

XIV *cis*-3-Hexen-1-ol

$CH_2{=}CH{-}CHO$

XV 2-Propenal

$CH_2{=}CHCOCH_3$

XVI 3-Buten-2-one

$CH_2{=}CHCOCH_2CH_3$

XVII 1-Penten-3-one

$CH_2{=}C(CH_3)CHO$

XVIII 2-Methylpropenal

Fig. 1. Formulas of germination inhibitors (I-IV), stimulants (V-XII) and compounds inducing infection structures (VIII, XIII-XVIII).

TABLE 1

Germination self-inhibitors and their relative potency

Organism	Self-inhibitor	ED_{50}(ng/ml)
Uromyces phaseoli	Methyl cis-3,4-dimethoxy-cinnamate (I)	2.7[a]
Puccinia graminis tritici	Methyl cis-ferulate (II)	0.18[a]
Peronospora tabacina	5-isobutyroxy-β-ionone, (quiesone) (III)	0.1[b]
Dictyostelium discoideum	Discadenine (IV)	10[c]

[a] Macko *et al.*, 1972
[b] Leppik *et al.*, 1972
[c] Tanaka *et al.*, 1975
Abe *et al.*, 1976

be methyl *cis*-3,4-dimethoxycinnamate (I) while that of wheat stem rust uredospores is methyl *cis*-4-hydroxy-3-methoxy-cinnamate (II) (Foudin and Macko, 1974; Macko *et al.*, 1970, 1971a, 1971b, 1972, 1977). The inhibitory concentration of the dimethoxycinnamate compound ranges from a few picograms per ml (10^{-11} M) for peanut rust uredospores to a few nanograms per ml (10^{-8} M) for bean rust. The inhibitor compounds have been synthesized, and confirmation of the chemical structure of the endogenous compound is based on identical properties of natural and synthetic compounds, including chromatography in several GLC and TLC systems, infrared, mass and U.V. spectra, ED_{50} and slopes of dosage response curves. In addition, the activity of the naturally occurring and synthetic inhibitors is identical, being confined to the same early stage of germination.

Elnaghy and Heitefuss (1976) studied the time course of the formation of the *U. phaseoli* self-inhibitor (methyl-3,4-dimethoxycinnamate) in rust-infected bean plants. Very little inhibitor was detected in resistant plant tissue where no spore formation took place but a pronounced increase in production of self-inhibitor in susceptible tissue began on the fourth day following inoculation (Fig. 2). This increase coincided with the beginning of spore formation, but it is not known if there is a dependent relationship between the formation of the self-inhibitor and sporulation.

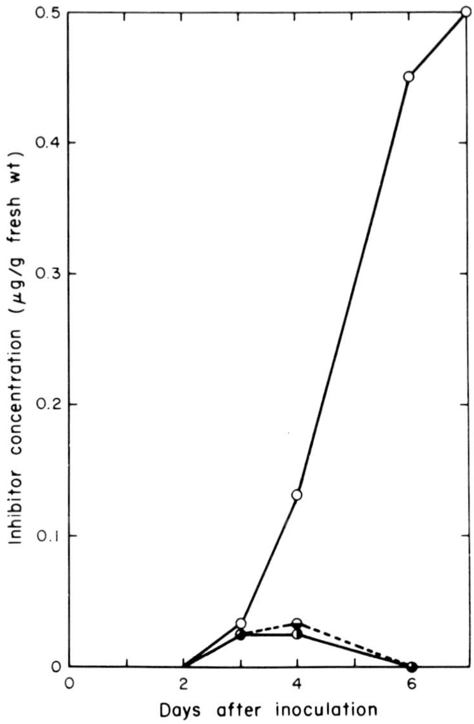

Fig. 2. Time-course of formation of the methyl-3,4-dimethoxycinnamate in bean leaves following inoculation with Uromyces phaseoli. Resistant cultivars 017 (◑——◑) and 596 (◒——◒), susceptible cultivar Favorit (○——○). Reprinted with permission from Elnaghy and Heitefuss (1976).

A similar time course was noted by Macko *et al.* (1976) on the basis of gas chromatographic analyses of rust infected bean leaves. Enzymatic evidence of self-inhibitor formation was provided by Wat and Towers (1977). Rust-infected bean leaves yielded an O-methyltransferase, which catalyzed the methylation of methyl p-coumarate, but not coumaric acid. Preparations from uninfected leaves did not methylate either methyl p-coumarate or p-coumaric acid.

As the conditions for *in vitro* culturing of rust fungi become available, a more suitable system for studies on biosynthesis, metabolism, and a possible role of germination inhibitors and stimulants in sporulation should be provided.

Peronospora tabacina. Chemical identification of the active inhibitory material was carried out starting with preparations from tobacco plants infected with downy mildew. The inhibitor from infected plants has properties similar to those of the inhibitor from conidia of *Peronospora tabacina* (Shepherd and Mandryk, 1963). The inhibitor, quiesone, was identified by Leppik *et al.* (1972) as 5-isobutyroxy-β-ionone (III). It is active at very low concentrations similar to those for inhibitors of rust uredospores (Table 1). Several compounds bearing the β-ionone moiety are natural growth regulators of higher plants, the most prominent being abscissic acid (Mandava, 1979). The quiesone molecule was synthesized by Mori (1973), and its biological and chemical properties were somewhat different from those reported for the natural product. Surprisingly, the compound prepared by Mori was a potent stimulant of bean rust uredospores (unpublished observation). β-Ionone (IX), the parent compound of quiesone, inhibits germination of *P. tabacina* conidia but stimulates the germination of bean rust uredospores (French *et al.*, 1977; Macko *et al.*, 1976).

Dictyostelium discoideum. Spores of this cellular slime mold show an inverse relationship between spore population density and germination (Cotter and Raper, 1966; Russel and Bonner, 1960). Germination occurs with the emergence of an amoeba. Several attempts were made to isolate and characterize the inhibitor from these spores (cf. Allen, 1976; Macko *et al.*, 1976) before Abe *et al.* (1976) identified 3-(3-amino-3-carboxypropyl)-6-(3-methyl-2-butenylamino) purine as the active compound and named it discadenine (IV). The level of inhibitory activity of discadenine is comparable to that of the inhibitors from other spores (Table 1).

Phomopsis viticola. A population effect has been demonstrated for germination of alpha spores of *Phomopsis viticola*, the causal agent of dead-arm disease of grapevine. Chromatographic and mass spectral evidence showed that elemental sulfur is accumulated in the cirrhus matrix of *P. viticola* and suggested that it acts as the self-inhibitor of germination of alpha spores (Pezet and Pont, 1975). The desulfuration of β-mercaptopyruvate by the enzyme β-mercaptopyruvate desulfurase could be responsible for the accumulation of free sulfur with the concomitant production of pyruvate (Pezet and Pont, 1977a). Analysis of dormant spores and sclerotia of 34 species of fungi showed that all but four species contain elemental sulfur. This suggested to Pezet and Pont (1977b) that there is a relationship between dormancy and the presence of elemental sulfur.

Mode of action of germination self-inhibitors

Problems of interpretation. Allen (1976) has clearly emphasized that the observed inhibition of a particular process in the presence of an active agent does not mean that the agent acts directly on that process. The physiological changes within the spore are interdependent. Inhibition of protein synthesis, for example, may be the ultimate result of inhibiting spore germination, but it does not necessarily mean that the self-inhibitor acts directly on protein synthesis. The same argument applies to studies with activators of germination or stimulants of infection structure formation. Often it is difficult to distinguish cause and effect and establish the proper sequence of events. In addition, each spore type has unique features, and different processes may be involved in the primary mechanism of activity. For example, the self-inhibitor in *Dictyostelium discoideum* should act on a process which is not protein synthesis but is required for swelling of spores to occur (Allen, 1976). On the other hand, in conidia of *Peronospora tabacina*, the self-inhibitor of germination appears to be a selective inhibitor of *in vivo* protein synthesis (Hollomon, 1973). One well studied example is the effect of self-inhibitor on the germ pore plug dissolution process in the uredospores of fungi (Hess *et al.*, 1975).

Mode of action of self-inhibitors in rust fungi. In a study of the mode of action of methyl *cis*-ferulate in self-inhibition of wheat stem rust uredospores, Hess *et al.* (1975) emphasized that wall material in the region of the germ pore plug was digested as germination progressed. The authors proposed that the self-inhibitor blocks dissolution of the germ pore plug and that the inhibitor can no longer be active when most of the material has been removed. Removal of the inhibitor at any time prior to disappearance of the plug material leads to reversal of the inhibition of the postulated hydrolytic enzyme and to an ensuing digestion of the plug. However, since aerobic conditions are required for digestion of the plug, the action of the inhibitor may not be directly on the digestive enzymes or their substrates (Allen, 1976).

A thorough study by Hess *et al.* (1980) demonstrated that the germ pore plug material is composed of mannoprotein. This mannoprotein represented 0.314% of the spore wall which was in the appropriate range expected if the released material was exclusively from the pore region. Hydrolytic enzymes, including pronase, chitosanase, β-glucanase and zymolase were ineffective in removing electron dense material in the region of the pore and no dissolution of the pore plug occurred when isolated walls were incubated in the presence of a spore homogenate prepared during the initial stage of dissolution (Hess *et al.*, 1980).

Despite the appeal of germ pore plug digestion as a mechanism, at least one other important process should be considered. Germ tube growth also coincides with the initiation of germ pore plug digestion, and signs of new wall formation occur at the apex of the developing germ tube shortly after germination is initiated. Cytoplasmic vesicles arranged themselves in the tip zone of the developing hypha, in a manner similar to that described by Grove and Bracker (1970) for other fungi. These vesicles are evident within a few minutes of the start of germination (Fig. 3). Thus the self-inhibitor reversibly blocks not only the germ pore plug dissolution but also the early growth of the germ tube while it is still within the confines of the spore wall.

This early growth of the germ tube was not sensitive to the presence of inhibitors of protein or RNA synthesis. For example, cycloheximide at 50 ppm did not affect the first 30 min of germ tube growth, but it effectively blocked protein synthesis (Hess *et al.*, 1975). The processes of early growth of the germ tube and dissolution of the germ pore plug material were not experimentally separated. Consequently it cannot be said with certainty which of these is the primary process affected by self-inhibitor action.

GERMINATION STIMULANTS

Self-stimulation of germination

In addition to germination inhibitors, germination promoting substances may be present in fungal spores. Nonanal (VI) was identified as an endogenous germination stimulant from uredospores of *Puccinia graminis tritici* (French and Weintraub, 1957). Subsequently the same compound was isolated from *P. coronata*, *P. sorghi*, *P. recondita*, *P. striiformis*, *P. helianthi* and *Uromyces phaseoli* (Rines *et al.*, 1974). Another germination stimulant, 6-methyl-5-hepten-2-one (VIII) was found in *P. graminis tritici* and *P. striiformis*. Over the years, French and collaborators reported that nonanal and more than 60 related compounds stimulate *P. graminis tritici*. In further screening experiments some 23 species of rust, smut and Penicillium fungi, have been stimulated by nonanal and other chemicals known

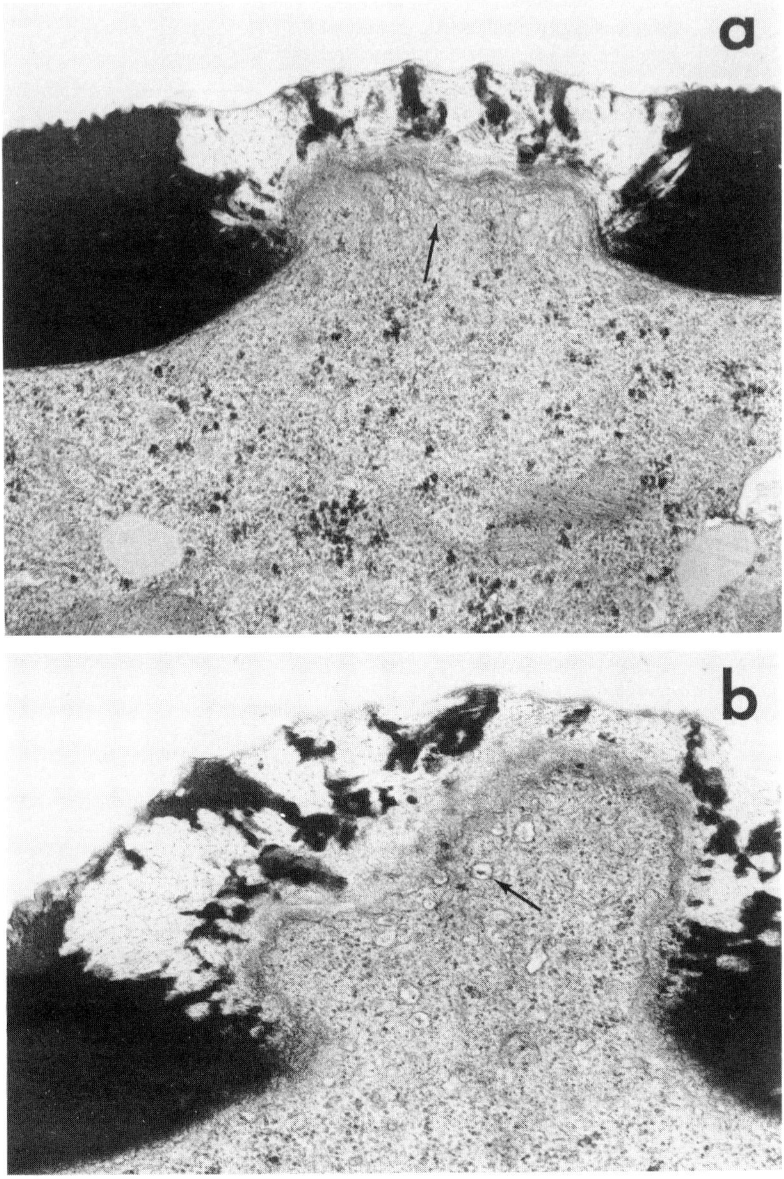

Fig. 3. Electron micrographs of thin section of a bean rust uredospore showing germ pore region a) 15 min (X28000) b) 30 min (X31000), after germination was initiated. Note vesicles in the tip zone of the developing germ tube (arrow).

previously as components of natural flavors and fragrances (French *et al.* 1978).

Some of the stimulants such as β-ionone (IX) and cinnamaldehyde (X) are structural analogs of the endogenous germination inhibitors. Very little information is available about the mode of action of these stimulants. Stimulants overcome the self-inhibition which prevails in dense populations of spores without reacting with the self-inhibitor molecule (Macko *et al.*, 1976). Nevertheless, future knowledge on site and mode of action of the self-inhibitors may provide clues about the mode of action of the stimulants.

Stimulants from plants

External stimulants are probably more important for a spore of a plant pathogen than the endogenous stimulants, because they may signal a favorable environment or a potential host plant. For example, volatile polyacetylenes from safflower, such as 1-tridecene-3,5,7,9,11-pentayne (XI) and *cis* and *trans*-trideca-1,11-dien-3,5,7,9-tetrayne (XII), stimulated teliospores of *Puccinia carthami* to germinate at concentrations of 0.15-15 ng/ml (Binder *et al.*, 1977, 1978). Volatiles from aged pea seeds stimulated germination of spores in soil-induced stasis (Harman *et al.*, 1978). It was suggested that carbonyl compounds produced in aged seeds, as a consequence of lipid peroxidation, could be responsible for this stimulation (Harman *et al.*, 1980). When tested, a range of five- to ten-carbon aldehydes, ketones and alcohols stimulated spores of *Fusarium solani* and *Alternaria alternata*, with 2,4-hexadienal (VII) being the most potent.

FORMATION OF INFECTION STRUCTURES

Fungi penetrate plants either directly through the cuticle, through natural openings such as stomata, or through wounds. Many fungal pathogens must develop specialized infection structures to penetrate a host plant. In germinating uredospores of rust fungi, these infection structures include appressorium, penetration peg, vesicle, and infection hyphae. Nuclear and cell divisions occur during the differentiation process (Maheshwari *et al.*, 1967; Staples *et al.*, 1975). The sequence of appearance of specialized structures is apparently initiated by contact of the germ tube with a guard cell of a stoma (Fig. 4; Wynn, 1976). While the regulatory mechanisms are unknown, homologous structures can be induced in the absence of the host by substitute triggers (Wynn, 1976). Several studies,

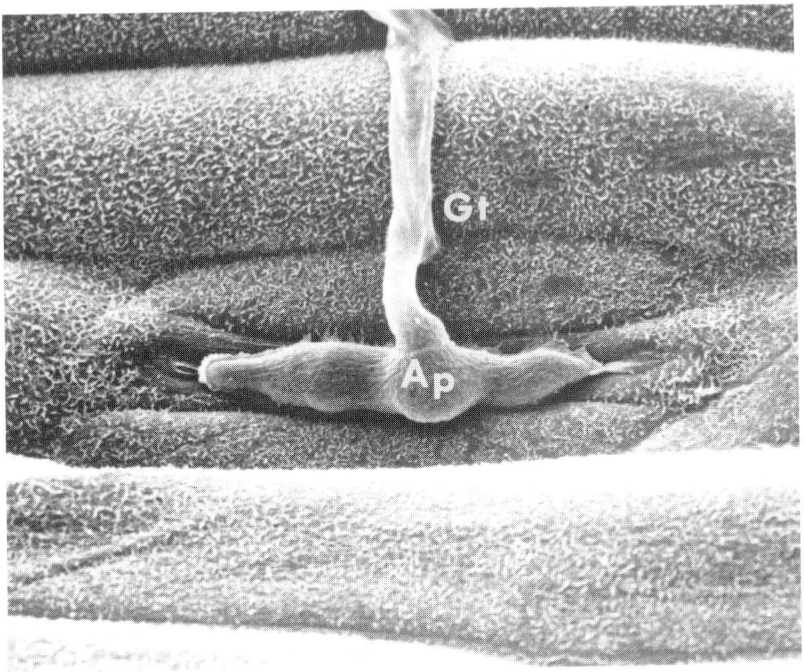

Fig. 4. Germ tube (Gt) with appressorium (Ap) of wheat stem rust fungus (Puccinia graminis f. sp. tritici) uredospore germinated on leaf of wheat. X1400. Photograph by W.K. Wynn.

recently reviewed by Allen (1976), Staples and Macko (1980), and Wynn and Staples (1980), reported experimental conditions necessary for differentiation *in vitro*. Most of the fungi studied respond to contact stimuli; however, germinating uredospores of *Puccinia graminis* f. sp. *tritici*, floating on aqueous solutions or in suspension, produce infection structures in response to heat shock or to distillates from uredospore extracts (Allen, 1957; Dunkle and Allen, 1971; French *et al.*, 1957). The latter finding suggested that a volatile chemical agent was present which initiated infection structure formation and that this agent perhaps could be an intermediary in the contact-triggered differentiation (Allen, 1976). This chemical inducer of appressoria has been identified as 2-propenal (XV) (Macko *et al.*, 1978). This compound induced formation of infection structures while the spores were germinating on a water surface (Fig. 5). It is not known how 2-propenal guides appressorium formation in the absence of physical stimulis provided by the stomatal guard cell. In nature, *Puccinia graminis* f. sp. *tritici*

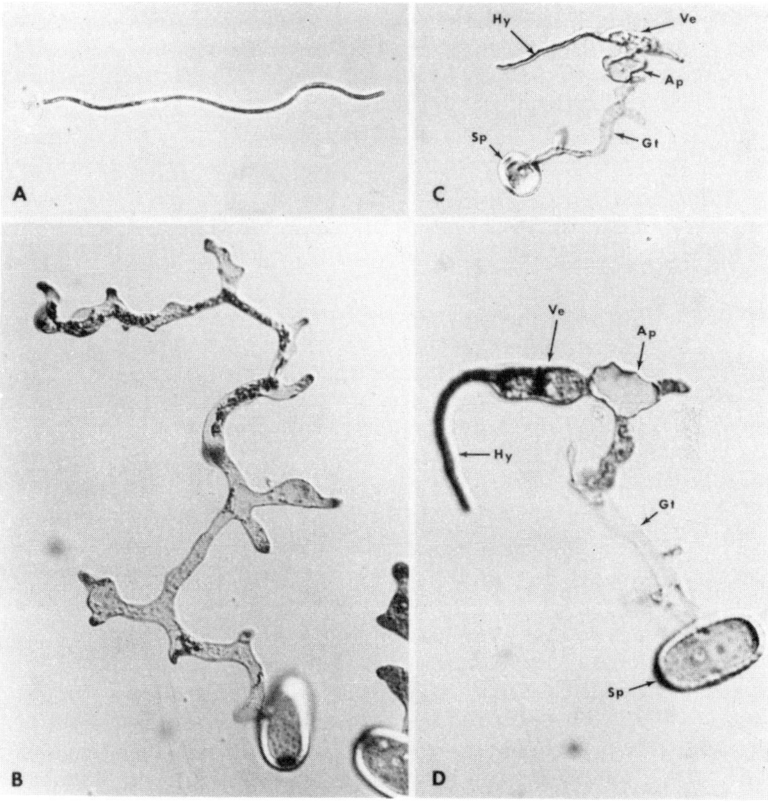

Fig. 5. Germination of uredospores of wheat stem rust (Puccinia graminis f. sp. tritici). A. Nondifferentiated germ tube, germinated on water. B. Branching in the presence of suboptimal concentration of 2-propenal. C. and D. Differentiated germ tube in the presence of optimal concentration of 2-propenal. Sp = spore, Gt = germ tube, Ap = appressorium, Ve = vesicle, Hy = hypha. The spore length is approximately 35μ in each figure.

forms appressoria only over stomatal pores (Fig. 4).

Structure-activity tests of over 60 compounds with varying structural similarities to 2-propenal indicated that differentiation was generally induced by short-chain aliphatic carbonyl compounds with a conjugated double bond (Table 2). Saturated aliphatic aldehydes and ketones, saturated and unsaturated aliphatic alcohols, and phenylacrolein (cinnamaldehyde, X), were not active.

Grambow and his associates (Grambow, 1977; Grambow and Riedel, 1977; Grambow and Grambow, 1978) found that phenols

TABLE 2

Unsaturated aliphatic aldehydes and ketones which induce infection structure formation in Puccinia graminis f. sp. tritici

Compound	nmoles[a]
2-propenal (XV)	0.4 ± 0.1
3-buten-2-one (XVI)	3.5 ± 1.5
1-penten-3-one (XVII)	4.5 ± 1.5
2-methyl propenal (XVIII)	25 ± 15

[a]Amount of compound which induces infection structures in 50-80% of germinating uredospores in a 6 ml Pyrex micro-Conway diffusion dish. Control spores germinated with long, undifferentiated, cylindrical germ tubes.

extracted from wheat leaves induce infection structure formation in uredospores of *P. graminis tritici* germinating on agar, when applied with volatile fractions from the leaves. In screening experiments, a variety of volatile plant compounds induced germinating uredospores of *P. graminis tritici* to form infection structures at a high rate (Grambow, 1977). Especially effective were *trans*-2-hexen-1-ol (XIII), *cis*-3-hexen-1-ol (XIV), and 6-methyl-5-hepten-2-one (VIII). The compounds, when tested with nonvolatile fractions from the leaves, closely reproduced the effect of the complex natural volatile fraction (Grambow, 1977).

Many volatile and nonvolatile compounds have morphogenetic effects on eukaryotic microbes *in vitro* and may or may not have a role in *in vivo* developmental processes (Fries, 1973; Hutchinson, 1973; Schindler and Sussman, 1977). While the *in vivo* role, of compounds at conditions suggested by Macko *et al.* and Grambow *et al.* is not proven, all these compounds have a definite morphogenetic effect. Nonetheless, experimental use of chemicals that induce infection structure formation should provide a means for studying the physiological response of germinating uredospores to environmental signals in determining whether cylindrical undifferentiated germ tube growth continues or the induction of infection structures occurs.

CONCLUDING REMARKS

Much of the current understanding of the various kinds of substances that play a part in control of germination in fungi has been derived from studies on rust uredospores. Germination of these and many other spores is reversibly inhibited by endogenous inhibitors within the spores that are released when the spores are floated on water. Water extracts of the spores often contain the inhibitory principle responsible for preventing premature germination. However, there may be other types of dormant spores which contain endogenous germination inhibitors that are not readily soluble in water or those which possess permeability barriers for the inhibitor (Trione, 1980). Such spores would not exhibit a population effect. Nevertheless, they may have sufficient inhibitor within each spore to cause dormancy. Teliospores of the dwarf bunt fungus, *Tilletia controversa*, is an example of the latter type (Trione, 1977).

Modern methods of chemical analysis utilizing small amounts of material have progressed to the point that they are no longer the limiting factor in structural studies of germination regulators. These methods now make increasingly possible the use of inhibitors and stimulants as experimental probes of morphogenetic controls in germinating spores. While some data are now available on the identity of chemical regulators of germination, we have little information on the primary processes controlled by them. It is difficult to distinguish primary from secondary responses, and it is usually difficult to separate growth from developmental effects. We need to know more about the biosynthesis of these regulators, how stimulants counteract self-inhibitors, and where are the inhibitors and stimulants located in the spore. Answers to these questions will help to explain the effect of inhibitors and stimulants and how they act at the molecular level.

ACKNOWLEDGEMENTS

I thank Drs. Samuel S. Hess and Gary E. Harman for the provision of manuscripts prior to publication and Willard K. Wynn for the scanning electron micrograph (Fig. 3). The author's research has been supported in part by grants from US National Science Foundation and US Department of Agriculture.

REFERENCES

Abe, H., Uchiyama, M., Tanaka, Y. and Saito, H. (1976). Structure of discadenine, a spore germination inhibitor from the cellular slime mold, *Dictyostelium discoideum*. *Tetrahedron Letters* 42, 3807-3810

Abercrombie, M. (1979). Contact inhibition and malignancy. *Nature* 281, 259-262.
Allen, P.J. (1955). The role of a self-inhibitor in the germination of rust uredospores. *Phytopathology* 45, 259-266.
Allen, P.J. (1957). Properties of a volatile fraction from uredospores of *Puccinia graminis* var. *tritici* affecting their germination and development. I. Biological activity. *Plant Physiology* 32, 385-389.
Allen, P.J. (1976). Control of spore germination and infection structure formation in the fungi. *In* "Encyclopedia of Plant Physiology" (Ed. R. Heitefuss and P.H. Williams). Vol. 4, pp.51-85. Springer-Verlag, New York.
Allen, P.J. and Dunkle, L.D. (1971). Natural activators and inhibitors of spore germination. *In* "Morphological and Biochemical Events in Plant-Parasite Interaction" (Ed. S. Akai and S. Ouchi). Pp. 23-58. The Phytopathological Society of Japan, Tokyo.
Bell, A.A. and Daly, J.M. (1962). Assay and partial purification of self-inhibitors of germination from uredospores of the bean rust fungus. *Phytopathology* 52, 261-266.
Binder, R.G., Klisiewicz, J.M. and Waiss, Jr., A.C. (1977). Stimulation of germination of *Puccinia carthami* teliospores by polyacetylenes from safflower. *Phytopathology* 67, 472-474.
Binder, R.G., Lundin, R.E., Kint, S., Klisiewicz, J.M. and Waiss, Jr., A.C. (1978). Polyacetylenes from *Carthamus tinctorius*. *Phytochemistry* 17, 315-317.
Cotter, D.A. and Raper, K.B. (1966). Spore germination in *Dictyostelium discoideum*. *Proceedings of the National Academy of Sciences, U.S.A.* 56, 880-887.
Dunkle, L.D. and Allen, P.J. (1971). Infection structure differentiation by wheat stem rust uredospores in suspension. *Phytopathology* 61, 649-652.
Elnaghy, M.A. and Heitefuss, R. (1976). Permeability changes and production of antifungal compounds in *Phaseolus vulgaris* infected with *Uromyces phaseoli*. I. Role of the spore germination self-inhibitor. *Physiological Plant Pathology* 8, 253-267.
Foudin, A.S. and Macko, V. (1974). Identification of the self-inhibitor and some germination characteristics of peanut rust uredospore germination. *Phytopathology* 64, 990-993.
French, R.C., Massey, L.M. and Weintraub, R.L. (1957). Properties of a volatile fraction from uredospores *Puccinia graminis* var. *tritici* affecting their germination

and development. II. Some physical and chemical properties. *Plant Physiology* 32, 389-393.

French, R.C. and Weintraub, R.L. (1957). Pelargonaldehyde as an endogenous germination stimulator of wheat rust spores. *Archives of Biochemistry and Biophysics* 72, 235-237.

French, R.C., Graham, C.L., Gale, A.W. and Long, R.K. (1977). Structural and exposure time requirements for chemical stimulation of germination of uredospores of *Uromyces phaseoli*. *Journal of Agricultural and Food Chemistry* 25, 84-88.

French, R.C., Long, R.K., Latterell, F.M., Graham, C.L., Smoot, J.J. and Shaw, P.E. (1978). Effect of nonanal, citral, and citrus oils on germination of conidia of *Penicillium digitatum* and *Penicillium italicum*. *Phytopathology* 68, 877-882.

Fries, N. (1973). Effects of volatile organic compounds on the growth and development of fungi. *Transactions of the British Mycological Society* 60, 1-21.

Grambow, H.J. (1977). The influence of volatile leaf constituents on the *in vitro* differentiation and growth of *Puccinia graminis* f. sp. *tritici*. *Zeitschrift für Pflanzenphysiologie* 85, 361-372.

Grambow, H.J. and Riedel, S. (1977). The effect of morphogenically active factors from host and nonhost plants on the *in vitro* differentiation of infection structures of *Puccinia graminis* f. sp. *tritici*. *Physiological Plant Pathology* 11, 213-224.

Grambow, H.J. and Grambow, G.E. (1978). The involvement of epicuticular and cell wall phenols of the host plant in the *in vitro* development of *Puccinia graminis* f. sp. *tritici*. *Zeitschrift für Pflanzenphysiologie* 90, 1-9.

Grove, S.N. and Bracker, C.E. (1970). Protoplasmic organization of hyphal tips among fungi: Vesicles and Spitzenkörper. *Journal of Bacteriology* 104, 989-1009.

Harman, G.E., Nedrow, B. and Nash, G. (1978). Stimulation of fungal spore germination by volatiles from aged seeds. *Canadian Journal of Botany* 56, 2124-2127.

Harman, G.E., Mattick, L.R., Nash, G. and Nedrow, B.L. (1980). Stimulation of fungal spore germination and inhibition of sporulation in fungal vegetative thalli by fatty acids and their volatile peroxidation products. *Canadian Journal of Botany* 58, (in press).

Hess, S.L., Allen, P.J., Nelson, D. and Lester, H. (1975). Mode of action of methyl *cis*-ferulate, the self-inhibitor of stem rust uredospore germination. *Physiological Plant Pathology* 5, 107-112.

Hess, S.L., Allen, P.J. and Lester, H.H. (1980). Release of a mannoprotein during pore dissolution in *Puccinia graminis* f. sp. *tritici* uredospores. (Submitted for publication).

Hollomon, D.W. (1973). Protein synthesis during germination of *Peronospora tabacina* conidia: an examination of the events involved in the initiation of germination. *Journal of General Microbiology* 78, 1-13.

Hutchinson, S.A. (1973). Biological activities of volatile fungal metabolites. *Annual Review of Phytopathology* 11, 223-246.

Hwang, Y.S. and Mulla, M.S. (1976). Overcrowding and its chemistry. *CHEMTECH* 6, 358-363.

Jaffe, L.F. and Nuccitelli, R. (1974). An ultrasensitive vibrating probe for measuring steady extracellular currents. *Journal of Cell Biology* 63, 614-628.

Jaffe, L.F., Robinson, K.R. and Nuccitelli, R. (1974). Local cation entry and self-electrophoresis as an intracellular localization mechanism. *Annals of the New York Academy of Sciences* 238, 372-389.

Jaffe, L.F. and Nuccitelli, R. (1977). Electrical controls of development. *Annual Review of Biophysics and Bioengineering* 6, 445-476.

Leppik, R.A., Hollomon, D.W. and Bottomley, W. (1972). Quiesone: An inhibitor of the germination of *Peronospora tabacina* conidia. *Phytochemistry* 11, 2055-2063.

Levy, M., Tunis, M. and Isseroff, H. (1973). Population control in snails by natural inhibitors. *Nature* 241, 65-66.

Macko, V., Staples, R.C., Gershon, H. and Renwick, J.A.A. (1970). Self-inhibitor of bean rust uredospores: Methyl 3,4-dimethoxycinnamate. *Science* 170, 539-540.

Macko, V., Staples, R.C., Allen, P.J. and Renwick, J.A.A. (1971a). Identification of the germination self-inhibitor from wheat stem rust uredospores. *Science* 173, 835-836.

Macko, V., Staples, R.C. and Renwick, J.A.A. (1971b). Germination self-inhibitor of sunflower and snapdragon rust uredospores. *Phytopathology* 61, 902.

Macko, V., Staples, R.C., Renwick, J.A.A. and Pirone, J. (1972). Germination self-inhibitors of rust uredospores. *Physiological Plant Pathology* 2, 347-355.

Macko, V., Staples, R.C., Yaniv, Z. and Granados, R.R. (1976). Self-inhibitors of fungal spore germination. *In* "The Fungal Spore: Form and Function" (Ed. D.J. Weber and W.M. Hess). Pp. 73-98. Wiley-Interscience, New York.

Macko, V., Trione, E.J. and Young, S.A. (1977). Identification of the germination self-inhibitor from uredospores of *Puccinia striiformis*. *Phytopathology* 67, 1473-1474.

Macko, V., Renwick, J.A.A. and Rissler, J.F. (1978). Acrolein induces differentiation of infection structures in the wheat stem rust fungus. *Science* 199, 442-443.

Maheshwari, R., Allen, P.J. and Hildebrandt, A.C. (1967). Physical and chemical factors controlling the development of infection structures from urediospore germ tubes of rust fungi. *Phytopathology* 57, 855-862.

Mandava, N.B. (1979). Natural products in plant growth regulation. *In* "Plant Growth Substances" (Ed. N.B. Mandava). Pp.135-213. American Chemical Society, Washington, D.C.

Mori, K. (1973). Synthesis of *dl*-3-isobutyroxy-β-ionone and *dl*-dehydrovomifoliol. *Agricultural and Biological Chemistry* 37, 2899-2905.

Pezet, R. and Pont, V. (1975). Elemental sulfur responsible for self-inhibition of spores of *Phomopsis viticola*. *Experientia* 31, 439.

Pezet, R. and Pont, V. (1977a). Metabolic pathway responsible for the production of elemental sulfur by *Phomopsis viticola*. *Berichte der Schweizerischen Botanischen Gesellschaft* 87, 154-162.

Pezet, R. and Pont, V. (1977b). Elemental sulfur: Accumulation in different species of fungi. *Science* 196, 428-429.

Prokopy, R.J. (1980). Epideictic pheromones regulating spacing patterns of phytophagous insects. *In* "Semiochemicals: Their Role in Pest Control". (Ed. D.A. Nordlund and R.J. Jones). In press. Wiley Press, New York.

Reissig, J.L. (1974). Decoding of regulatory signals at the microbial surface. *Current Topics in Microbiology and Immunology* 67, 43-96.

Rines, H.W., French, R.C. and Daasch, L.W. (1974). Nonanal and 6-methyl-5-hepten-2-one: Endogenous germination stimulators of uredospores of *Puccinia graminis* var. *tritici* and other rusts. *Journal of Agricultural and Food Chemistry* 22, 96-100.

Russell, G.K. and Bonner, J.T. (1960). A note on spore germination in the cellular slime mold *Dictyostelium mucoroides*. *Bulletin of the Torrey Botanical Club* 87, 187-191.

Schindler, J. and Sussman, M. (1977). Ammonia determines the choice of morphogenetic pathways in *Dictyostelium discoideum*. *Journal of Molecular Biology* 116, 161-169.

Shepherd, C.J. and Mandryk, M. (1963). Germination of conidia of *Peronospora tabacina* Adam. II. Germination *in vivo*.

Australian Journal of Biological Sciences 16, 77-87.
Staples, R.C., App, A.A. and Ricci, P. (1975). DNA synthesis and nuclear division during formation of infection structures by bean rust uredospore germlings. *Archives of Microbiology* 104, 123-127.
Staples, R.C. and Macko, V. (1980). Formation of infection structures as a recognition response in fungi. *Experimental Mycology* 4, 2-16.
Sussman, A.S. and Douthit, H.A. (1973). Dormancy in microbial spores. *Annual Review of Plant Physiology* 24, 311-352.
Sussman, A.S. and Halvorson, H.O. (1966). "Spores: Their Dormancy and Germination". Harper & Row, New York and London.
Tanaka, Y.-M., Hashimoto, Y., Yanagisawa, K., Abe, H. and Uchiyama, M. (1975). Partial structure of a spore germination inhibitor from a cellular slime mold, *Dictyostelium discoideum*. *Agricultural and Biological Chemistry* 39, 1929-1932.
Trione, E.J. (1977). Endogenous germination inhibitors in teliospores of the wheat bunt fungi. *Phytopathology* 67, 1245-1249.
Trione, E.J. (1980). Natural regulators of fungal development. *In* "Plant Disease Control: Resistance and Susceptibility" (Ed. R.C. Staples and G.H. Toenniessen). In press. Wiley Interscience, New York.
Wat, C.T. and Towers, G.H.N. (1977). Production of methylated phenolic acids by species of *Lentinus* (*Basidiomycetes*). *Phytochemistry* 16, 290-291.
Weisenseel, M.H., Nuccitelli, R. and Jaffe, L.F. (1975). Large electrical currents traverse growing pollen tubes. *Journal of Cell Biology* 66, 556-567.
Wynn, W.K. (1976). Appressorium formation over stomates by the bean rust fungus: Response to a surface contact stimulus. *Phytopathology* 66: 136-146.
Wynn, W.K. and Staples, R.C. (1980). Tropisms of fungi in host recognition. *In* "Plant Disease Control: Resistance and Susceptibility" (Ed. R.C. Staples and G.H. Toenniessen). In press. Wiley Interscience, New York.
Yarwood, C.E. (1954). Mechanism of acquired immunity to a plant rust. *Proceedings of the National Academy of Sciences, U.S.A.* 40, 374-377.
Yarwood, C.E. (1956). Simultaneous self-stimulation and self-inhibition of uredospore germination. *Mycologia* 48, 20-24.

RESPIRATION AND MITOCHONDRIAL BIOGENESIS DURING FUNGAL SPORE GERMINATION

Robert Brambl

Department of Plant Pathology
The University of Minnesota
Saint Paul, Minnesota 55108, U.S.A.

1. INTRODUCTION

The germination of dormant fungal spores into rapidly growing cells characteristically includes the elaboration of a strong aerobic respiratory capacity. This respiration, along with certain other metabolic activities, seems to be essential for spore germination. The major question I will deal with in this chapter is how development of this respiratory competence is regulated early in spore germination. In particular, I will review our research into whether the mitochondrial respiratory membrane of dormant spores is structurally and functionally complete or whether this membrane, with its ordered array of respiratory enzyme complexes, must be modified somehow or reassembled to provide for the respiratory demand of the spores as they convert from metabolic quiescence to rapid growth.

The study of the respiratory physiology of dormant[1] and germinating spores is hardly a new topic; until about fifteen years ago, it probably was the most intensively examined aspect of spore biochemistry. Many of these earlier studies [which have been reviewed by Allen (1965), Cochrane (1966), Sussman (1966), Sussman and Douthit (1973), and Gottlieb (1976)] concentrated on pathways of carbohydrate metabolism, such as the tricarboxylic acid cycle, glycolysis, and related pathways; and they included analyses of several activities of the mitochondrial cytochrome electron transport system. However, since these earlier studies, new information has accumulated about the general organization and biogenesis of the cytochrome electron transport system, and it has become possible to test new questions about the function and assembly of this respiratory membrane in fungal spores.

The mitochondrial cytochrome electron transport system functions in respiration by transport of electron equivalents (donated by NADH or flavoprotein) through a series of carriers of increasing redox potential consisting of a quinone and cytochromes *a*, *b*, and *c* to reduce oxygen to water. The energy of this electron flux through the redox carriers is conserved at three sites of the electron transport chain (where relatively large decreases in free energy occur) by the establishment of a proton gradient which in turn drives synthesis of ATP via ATP synthase (an enzyme which is better known by its hydrolytic activity as ATPase). Cytochrome electron transport is blocked by several well-characterized inhibitors such as cyanide and antimycin A; when the ATPase is bound to the respiratory membrane, it is sensitive to oligomycin and dicyclohexylcarbodiimide (DCCD).

Within recent years considerable interest has developed in another pathway of mitochondrial electron transport mediated by a poorly characterized alternate oxidase. The electron flux through this alternate oxidase is insensitive to cyanide or antimycin A, but is sensitive to certain aromatic hydroxamates such as salicylhydroxamic acid (SHAM). The alternate oxidase pathway is believed to branch from the cytochrome-mediated pathway in the region of the quinone and eventually reacts with molecular oxygen, by-passing sites II and III of oxidative phosphorylation. This pathway is relatively unproductive energetically, and the physiological function of the alternate oxidase is not at all clear. This pathway usually functions only weakly in fungi unless the biosynthesis or function of the cytochrome-mediated electron transport system is disrupted by mutation or chemical inhibition. [The volume edited by Degn *et al*. (1978) contains citations to recent research on this topic.]

The roles of the mitochondrial and nuclear genetic systems in the biosynthesis and assembly of the mitochondrial respiratory membrane have been studied intensively within the past ten years [representative reviews include Schatz and Mason (1974), Weiss *et al*. (1975), Mahler *et al*. (1975), Borst and Grivell (1978), and Tzagoloff *et al*.

(1979)]. Much of this research has concentrated on the organization of the mitochondrial genetic system and its function in the synthesis of fewer than a dozen polypeptides, most of which are subunit components of three enzyme complexes of the respiratory membrane, coenzyme Q-cytochrome *c* reductase, cytochrome *c* oxidase, and the oligomycin-sensitive ATPase. Of the several subunit polypeptides that comprise each enzyme complex, some are known to be products of nuclear genes and cytoplasmic protein synthesis, whereas other subunits are products of mitochondrial protein synthesis. The two cellular genetic systems, therefore, each contribute to the synthesis and assembly of the mitochondrial respiratory membrane, but it is not yet understood how the two systems are coordinated biochemically in this function, how the cytoplasmically synthesized proteins are transported into the mitochondria, or how the mitochondrially synthesized polypeptides serve to integrate the other subunits and prosthetic groups into active complexes. It is not even clear why the mitochondrial genetic system has been maintained through evolution and why its functions could not be served as well by the nuclear genetic system. Nevertheless, coordinate expression of the mitochondrial and nuclear genetic systems is essential for synthesis and assembly of the mitochondrial respiratory membrane, and a particularly attractive situation in which to examine this membrane biogenesis is in germinating fungal spores, where one might expect a relatively synchronized and experimentally distinguishable series of steps in the functional activation or the synthesis and assembly of components of the mitochondrial electron transport system.

2. ULTRASTRUCTURAL STUDIES OF SPORE MITOCHONDRIA

Ultrastructural analyses of dormant and germinating spores in some instances have shown distinct differences in mitochondrial size and in inner membrane complexity, but in other instances no striking differences have been observed. These observations could help determine a possible resemblance of some dormant spore mitochondria to the rudimentary "promitochondria" of anaerobically grown yeast cells (Criddle and Schatz, 1969). Such a resemblance would contain implications about the degree of biochemical and physical organization of the respiratory membrane of the dormant spores.

Mitochondria isolated from protoplasts of ungerminated *Neurospora crassa* conidia structurally resemble those of physiologically mature cells (Weiss, 1963); however, other workers reported that mitochondria observed *in situ* in such conidia are somewhat smaller and had more regular shapes than those of rapidly growing cells but otherwise are similar in their structural complexity (Beck and Greenawalt, 1976). Ortega-Perez *et al.* (1977) showed that mitochondria of the dormant conidia of *Neurospora* have a higher density than those of the vegetative mycelium. The zoospores of *Blastocladiella emersonii* contain a single giant mitochondrion which changes shape and apparently fragments in the course of spore germination (Cantino *et al.*, 1963; Soll *et al.*, 1969). The dormant conidia of *Botryodiplodia theobromae* contain mitochondria which resemble closely their counterparts in germinated spores (Wergin *et al.*, 1973; Brambl *et al.*, 1978).

A careful examination of the spores of *Trichoderma viride* showed that the mitochondria change from a regular, ovoid shape in the ungerminated spores to enlarged, highly convoluted, cup-shaped structures during germination (Rosen *et al.*, 1974). The mitochondria of chlamydospores of *Fusarium oxysporum* enlarge and become more irregular in shape upon germination (Stevenson and Becker, 1979). In most cases, the mitochondria may increase in size and number during germination and may change shape, but there usually is no great difference in the structural complexity of the inner membrane between dormant and germinated spores.

3. EFFECTS OF SPECIFIC CHEMICAL INHIBITORS UPON GERMINATION

Simple experiments that determine the effects of specific chemical inhibitors upon spore germination can provide useful predictive information, although the results of such an approach cannot be accepted uncritically. Several laboratories have employed such inhibitors of protein synthesis and respiration in attempts to understand which metabolic processes are essential to spore germination, and some of these results have become the basis of other, more definitive experimental approaches.

The conidiospores of *Botryodiplodia* germinate to about 95% during the last 90 min of a 240 min germination sequence in an aerated nutrient medium. This germination is blocked if the cytoplasmic protein synthesis inhibitor cycloheximide is in the medium during the first 135 min of incubation, but several inhibitors of the mitochondrial genetic system (chloramphenicol, erythromycin, ethidium bromide, and acriflavin) have little or no effect upon the rate or extent of germination (Brambl, 1975b). Direct evidence is available that these latter inhibitors, while not blocking germination, do inhibit mitochondrial DNA and protein synthesis in these spores. *Neurospora* conidia show a similar response to these same inhibitors of cytoplasmic and mitochondrial protein synthesis (Stade and Brambl, 1981). The major conclusion drawn from such results is that in these spores cytoplasmic protein synthesis is essential to germination, while function of the mitochondrial genetic system is not required for germination and (by implication) for development of processes such as aerobic respiration during germination. However, cycloheximide and chloramphenicol each block germination of spores of *Dictyostelium discoideum* (Cotter and Raper, 1970; Kobilinsky and Beattie, 1977).

Other experiments with inhibitors of the mitochondrial electron transport system, such as cyanide and antimycin A, have shown that these drugs completely inhibit germination of spores of *Botryodiplodia* and *Neurospora*, whereas SHAM (an inhibitor of the alternate oxidase) has no effect upon germination kinetics of these spores (Brambl, 1975b; Stade and Brambl, 1981). The respiratory poison azide also blocks germination of *Botryodiplodia* (Cotter and Raper, 1968) and *Phycomyces blakesleeanus* (Van Laere *et al*., 1980). The germination of spores of *Aspergillus oryzae*, on the other hand, is retarded (but not blocked) by antimycin A, and subsequent mycelial growth is unaffected by the continued presence of the inhibitor (Kawakita, 1970). This observation deserves further experimentation, particularly in view of a possible function of an alternate oxidase in these germinating *Aspergillus* spores (see below). The major conclusion from most of these (and many other) studies is that spore germination is a process which is strictly aerobic and which is dependent upon function of the mitochondrial cytochrome electron transport system.

4. MEASUREMENTS OF RESPIRATION DURING SPORE GERMINATION

The oxygen electrode allows convenient, rapid measurement of the kinetics of oxygen consumption of germinating spores under different regimens of growth or chemical inhibition. These measurements provide useful, more direct information about the function of the electron transport system and the involvement of cytoplasmic and mitochondrial protein synthesis in the development of respiratory activity.

In some initial experiments we established that the oxygen uptake by germinating spores of *Botryodiplodia* is nearly abolished by cyanide and antimycin A and that it is unaffected by SHAM (Brambl, 1975b). *Neurospora* conidia show identical responses to these respiratory inhibitors (Stade and Brambl, 1981), and we believe that respiration via the cytochrome electron transport system is essential to the germination of spores of these two fungi. Function of the alternate oxidase during germination seems to us to be negligible, in contrast to the conclusions of others (Schmit and Brody, 1976).

To test this last point further, we obtained three mutants of *Neurospora* (*mi*-3, *cya*-4-23, and *cya*-5-34) which have been characterized by others as being cytochrome *c* oxidase-deficient. The log-phase cells lack cytochrome *a* and cytochrome *c* oxidase activity, and the enzyme has an altered subunit polypeptide composition; it is believed that respiration in these cells depends almost entirely upon the alternate oxidase (Bertrand and Werner, 1977; Bertrand and Collins, 1978; Nargang *et al*., 1978; Bertrand and Werner, 1979; Werner and Bertrand, 1979). We found, however, that germination and oxygen uptake during germination of the conidia of these mutant strains are dependent completely upon a functional cytochrome electron transport system and (as described below) a functional cytochrome oxidase which is preserved in the dormant spores; the rates of cyanide-insensitive, SHAM-sensitive oxygen uptake are elevated only slightly above the low rates found in wild-type *Neurospora* conidia (Stade and Brambl, 1981). This result does not mean, of course, that all types of conidiospores require function of the cytochrome electron transport system for respiration during germination, but that a clear exception has not yet been found.

The spores of *Phycomyces* exhibit a rapid oxygen uptake within minutes after activation; this respiration is not completely inhibited by azide but is nearly abolished by azide and SHAM together, although SHAM alone had no effect (Van Laere *et al*., 1980). The oxygen uptake rates of *Aspergillus* spores incubated from the beginning of germination in antimycin A were very similar to those of the untreated spores (Kawakita, 1970), but it was not established whether this oxygen consumption was due to an alternate oxidase. Germination and respiration of ascospores of wild-type *Saccharomyces cerevisiae* are inhibited by azide and antimycin A (Rousseau and Halvorson, 1973). But it has also been reported that ascospores of *petite* strains of *Saccharomyces*, in which cells of the entire culture were mutagenized briefly just before sporulation and which presumably have no functional mitochondrial DNA, germinate with near normal kinetics (Tingle *et al*., 1974). Supposedly, these spores are devoid of the products of the mitochondrial genetic system which are required for function of the mitochondrial respiratory system. Although there is no accelerating elaboration of cytochrome *c* oxidase activity during germination, these *petite* spores contain about one-third of the enzyme activity of the wild-type spores. It seems possible that the cytochrome oxidase activity present in the *petite* spores could have been synthesized and assembled in the spores before chemical mutagenesis of the mitochondrial DNA and preserved during dormancy as a residual activity which is later expressed early in spore germination.

A question which has interested us in our laboratory is whether the development of respiration in germinating spores depends upon function of the nuclear and mitochondrial genetic systems (including protein synthesis) or whether the respiratory apparatus which functions in the germinating spores is preserved intact during dormancy. One of our first approaches to this question involved measurement of the effects of specific chemical inhibitors upon rates of oxygen uptake during spore germination. Cycloheximide completely blocked oxygen consumption in spores of *Botryodiplodia*; ethidium bromide, chloramphenicol, and acriflavin had no effect upon the initial acceleration of respiratory rates, but they did cause large reductions in oxygen consumption near the time at which all the spores had germinated (Brambl, 1975b). These results suggested that cytoplasmic protein synthesis is essential both for germination and respiration and that function of the mitochondrial genetic system is not necessary, possibly because the latter system's products which are required for function of the respiratory membrane are preserved in the spores during dormancy for function early in germination. The inhibitors of the mitochondrial genetic system similarly did not inhibit respiration of *Neurospora* conidia; however, treatment of these spores with cycloheximide did not block completely the development of a cyanide-sensitive respiration (Stade and Brambl, 1981). In this instance, it seemed likely (as shown directly later) that products of cytoplasmic protein synthesis required for activity of the respiratory membrane also were preserved in the dormant spores for function early in germination. Our studies were conducted with *Neurospora* conidia which had been harvested with a paraffinic hydrocarbon (Brambl, 1975a) under conditions which precluded water activation of protein synthesis (*unpublished results*).

Results of oxygen uptake experiments with *Botryodiplodia* spores incubated under conditions of reversible inhibition with cycloheximide and chloramphenicol suggested that the two protein synthesis systems either respond independently to a single stimulus for germination or that there are two separate stimuli which act on the two systems independently (Brambl, 1975b). Each of these inhibitors causes an accumulation of unused products of the opposite, uninhibited protein synthesis system. These experiments also provide firmer support for the conclusion that products of mitochondrial protein synthesis are preserved during dormancy, whereas newly synthesized cytoplasmic proteins are essential for the development of respiration in *Botryodiplodia*.

5. RESPIRATORY CYTOCHROMES OF DORMANT AND GERMINATING SPORES

The most simple, direct means of determining whether dormant spores contain a complete cytochrome electron transport system involves analysis of the mitochondria of these cells by absorption difference spectrophotometry. Yet this approach has been difficult to employ because of interfering pigments which are present in high concentration in mitochondrial fractions of the dormant spores. More recently, however, it has become possible to perform these analyses of the mitochondrial cytochromes through fractionation of the mitochondrial membranes with nonionic detergents.

One of the first attempts at this kind of experiment circumvented the masking pigment problem by use of an albino strain of *Phycomyces*, and it was reported that these dormant spores contained a full complement of cytochromes *b*, *c*, and *a* (Keyhani *et al.*, 1972). However, the published spectra leave some doubt as to whether all these cytochromes were present, and this question deserves reexamination. It has been reported that the dormant spores of *Dictyostelium* contain spectrally identifiable cytochromes *b*, *c*, and *a* (Kobilinsky and Beattie, 1977).

In initial studies we found that the spectra of respiratory cytochromes of whole mitochondria of dormant *Botryodiplodia* spores were too indistinct to give a reliable qualitative estimate of their cytochrome content, although the respiratory cytochromes were seen readily as they were elaborated in later stages of germination and vegetative growth (Brambl, 1975b). Subsequently, we fractionated these mitochondria of dormant spores with sodium cholate and $(NH_4)_2SO_4$ to obtain a preparation enriched in the cytochromes, and (by use of 77° K spectrophotometry) we saw that spectra of the mitochondrial fractions contained α, β and Soret bands of cytochrome *c* and one of two expected cytochromes *b*, but these mitochondria did not contain cytochrome *a*, and they did not contain heme *b*, as determined by analysis of pyridine hemochromogen preparations (Brambl and Josephson, 1977). This finding of a missing cytochrome *a* prompted other studies (described below) into the means by which this respiratory system component is assembled into the mitochondria during spore germination. Incidental observations made in these studies also showed that the mitochondria of *Botryodiplodia* spores germinated in the presence of chloramphenicol or ethidium bromide contain an excess of cytochrome *c* in the absence or near absence of cytochromes *b* and *a* (Brambl, 1975b; Brambl and Josephson, 1977). In addition, incubation of these spores with cerulenin, a reversible inhibitor of fatty acid and sterol synthesis, causes a three-fold accumulation specifically of cytochrome *a*, while the cellular concentrations of cytochromes *b* and *c* are unaffected for reasons not yet understood (Brambl *et al.*, 1978).

In contrast to *Botryodiplodia*, the mitochondria of dormant conidiospores of *Neurospora* contain a full complement of respiratory cytochromes, as measured by room temperature difference spectrophotometry (Stade and Brambl, 1981). This finding, supported by other evidence (see below), has led us to the conclusion that *Neurospora* spores contain a complete, assembled respiratory membrane which becomes functional when the spores are suspended in an aqueous incubation medium.

6. CATALYTIC ACTIVITY OF CYTOCHROME *C* OXIDASE AND OLIGOMYCIN-SENSITIVE ATPase IN DORMANT AND GERMINATING SPORES

The finding that the dormant spores of *Botryodiplodia* do not contain cytochrome *a* or heme *a* led us to a series of experiments to determine whether the spores also were deficient in cytochrome *c* oxidase activity and how the development of this enzyme activity was regulated during spore germination.

Analyses showed that the mitochondria of these dormant spores do not contain an active cytochrome *c* oxidase, but that the enzyme activity is elaborated rapidly during germination to a near-maximum by 240 min (Brambl, 1977). Subsequent work demonstrated directly that all components of the functional enzyme are not present in these mitochondria.

Elaboration of the cytochrome *c* oxidase activity is blocked if cycloheximide is present in the medium at or before 95 min of germination, but elaboration of the enzyme activity initially is insensitive to inhibitors of the mitochondrial genetic system. These observations suggest that function of cytoplasmic ribosomes, but not that of the mitochondrial ribosomes, is required for development of this respiratory enzyme, probably because the contributions of the latter system are preserved in the spores during dormancy for function upon initiation of germination. Incubation of the spores with antagonistic amino acid analogs severely disrupts the elaboration of cytochrome oxidase in the spores, thereby suggesting that most of the enzyme activity appeared because of *de novo* synthesis (Brambl, 1977). In the earlier study we could not find an active enzyme in the spore mitochondria between 30 and 120 min of germination. But because of the cyanide-sensitive oxygen uptake occurring during this interval, one would expect at least low amounts of an active enzyme in these spores early in germination, and in a subsequent study we did detect this enzyme activity during this early germination interval and before the rapid elaboration after 120 min (Brambl, 1980).

The clear absence of cytochrome *c* oxidase in the dormant spore mitochondria of *Botryodiplodia* may be relatively unusual, because we found that large amounts of an active enzyme were present in asexual spores of *Rhizopus stolonifer*, *Puccinia graminis* f. sp. *tritici* and *Neurospora* (Brambl, 1977; Stade and Brambl, 1981). The ungerminated spores of *Dictyostelium* also contain cytochrome c oxidase (Kobilinsky and Beattie, 1977).

We have asked whether the absence of cytochrome *c* oxidase in *Botryodiplodia* spores is a single, specific defect of this respiratory membrane or whether certain other enzymes also may be absent from the dormant spore mitochondria. We found that no activity of the oligomycin-sensitive ATPase is present in the dormant spore mitochondria and, like that of cytochrome *c* oxidase, it is elaborated rapidly after 120 min of the 240 min germination sequence through a series of steps sensitive to cycloheximide but initially insensitive to inhibitors of the mitochondrial genetic system (Wenzler and Brambl, 1981). We believe that this enzyme initially develops in the germinating spores because of nucleocytoplasmic protein synthesis, but function of the mitochondrial genetic system is not required because the components of the enzyme derived from the mitochondrial genetic system may be present in the dormant spore mitochondrial respiratory membrane. In contrast, we also found in this study that the dormant conidia of *Neurospora* contain an oligomycin-sensitive ATPase of high specific activity (Wenzler and Brambl, 1981).

Given the information about the elaboration of activities of cytochrome *c* oxidase and ATPase in mitochondria of germinating spores of *Botryodiplodia*, we have sought to establish whether the requirement for cytoplasmic ribosome function (for elaboration of these enzyme activities) is based upon translation of the nucleocytoplasmic mRNA which is preserved in these spores (Brambl and Van Etten, 1970; Wenzler and Brambl, 1978), or whether it depends upon early synthesis of a new mRNA. Equally important is the question of whether this early mRNA translation contributes directly to the structural assembly of these enzymes or whether it instead is required indirectly for a non-biosynthetic step of enzyme activation or rearrangement. Answers to these questions have come from subsequent studies of the biogenesis of these two enzymes in *Botryodiplodia*.

7. BIOGENESIS AND ASSEMBLY OF CYTOCHROME *C* OXIDASE

The purified, catalytically active cytochrome *c* oxidase of *Botryodiplodia* consists of seven distinct subunit polypeptides (of 41,000 to 9,300 molecular weight); the three largest subunits are products of mitochondrial ribosomes and the four remaining subunits are products of cytoplasmic ribosomes (Josephson and Brambl, 1980).

In a study of the biosynthesis of these enzyme subunits during *Botryodiplodia* spore germination (using antisera directed against the purified enzyme), we found that only the mitochondrial subunits are synthesized during the first 60 min of spore incubation, and during the second 60-min interval (of the 240 min germination sequence) the cytoplasmic subunits are first synthesized (Brambl, 1980). Chemical inhibition of one ribosome system does not affect synthesis of subunits by the opposite ribosome system; in the presence of cycloheximide, the mitochondrial subunits are synthesized, and in the presence of chloramphenicol, the cytoplasmic subunits are synthesized. This implies that the two protein synthesis systems independently initiate their functions early in spore germination. The temporal separation in appearance of the two groups of subunits in the germinating spore mitochondria reflects a temporal separation of synthesis (rather than assembly into the membrane) because there is no accumulation of the cytoplasmic subunits in the extra-mitochondrial cytoplasm before their appearance in the mitochondria.

We sought to determine the point at which the heme *a* moiety was incorporated into the developing cytochrome *c* oxidase complex of the germinating spores, and for this purpose we employed ^{55}Fe to label protoporphyrin IX which subsequently is incorporated into the enzyme as heme *a*. The ^{55}Fe is incorporated into cytochrome *c* oxidase at very rapid rates between 30 and 60 min of germination, and the rates drop almost to zero and then increase rapidly again after 180 min when [^{3}H]leucine also is incorporated rapidly (Brambl, 1980). This initial burst of incorporation was unexpected, and it was difficult to explain how the heme *a* could have been incorporated into a complex which presumably consists only of the three mitochondrial subunits during the first 60 min of germination. Subsequent experiments showed, however, that this presumption was incorrect.

Our earlier studies of cytochrome *c* oxidase activity in the germinating spores of *Botryodiplodia* suggested that components of the enzyme synthesized through the mitochondrial genetic system are preserved during dormancy and are available for function after the requisite synthesis of other enzyme components on cytoplasmic ribosomes (Brambl, 1977). We tested this hypothesis directly by labeling the mycelium of sporulating cultures with ^{3}H-amino acids to obtain generally labeled dormant spores from which the mitochondria were isolated. Immunoprecipitation assays of these labeled mitochondria showed that, as predicted, they contain only the mitochondrial subunits of the enzyme. However, an immunoprecipitation assay of the post-ribosomal supernatant fluid of these same spores showed that this fraction contains large amounts of four labeled peptides which are recognized by the antiserum to the cytochrome *c* oxidase and which closely resemble the cytoplasmic subunits of the enzyme in molecular weight (Brambl, 1980). This finding suggests that the two groups of subunit polypeptides of cytochrome *c* oxidase could be compartmentalized in two separate cellular fractions during spore dormancy and that the preserved components of the enzyme are mobilized from these compartments for assembly into a functional enzyme early in the course of spore germination. Subsequent tests of this hypothesis showed directly that these preserved cytoplasmic subunits of cytochrome *c* oxidase do, in fact, become functionally associated with the preserved mitochondrial subunits (and with heme *a*) during the first 30 to 60 min of germination (Brambl, 1980). The conclusions drawn from these experiments are that the *de novo* syntheses of enzyme subunits early in spore germination probably are irrelevant to the reactivation of the germinating spore respiratory system and that the newly synthesized subunits contribute to a respiratory membrane which has already been reassembled from the components preserved during spore dormancy. The role of the requisite cytoplasmic protein synthesis in the assembly or function of cytochrome *c* oxidase early in germination remains unexplained, but it seems likely that it contributes in an indirect (but essential) way to the reassembly of this complex. We believe, however, that the preserved mRNA in these spores (which is translated early in germination) does contribute eventually to synthesis of the respiratory membrane of the germinated spores, because our *in vitro* translation of the mRNA yields polypeptides which resemble the cytoplasmic subunits of cytochrome *c* oxidase or possible precursors to these subunits (Wenzler and Brambl, *unpublished*).

We have extended this type of analysis to the conidiospores of *Neurospora* (Stade and Brambl, 1981). These spores (harvested under conditions which prevent their hydration and simultaneous activation of protein systhesis) contain a functional, catalytically active cytochrome *c* oxidase and (not surprisingly) the mitochondria of these dormant spores contain all seven subunits of the enzyme assembled into a complex with heme *a*. Although the spores contain an active cytochrome *c* oxidase, they nevertheless initiate synthesis of the subunits of the enzyme within minutes after they are suspended in a germination medium, and syntheses of all subunits are begun simultaneously. Presumably, these newly synthesized subunits, as in *Botryodiplodia*, are used not for the earliest respiration of spore germination, but instead they contribute to assembly of mitochondrial respiratory membranes that are organized in later steps of germination.

8. BIOGENESIS AND ASSEMBLY OF OLIGOMYCIN-SENSITIVE ATPase

The mitochondrial ATPase is a complex of a number of identical and non-identical subunit polypeptides which are assembled into two sectors: the F_0 membrane sector, a hydrophobic, intrinsic membrane component, and the F_1 sector, a hydrophilic component which protrudes above the plane of the membrane and which contains the active site of the enzyme. The activity of the F_1, including its sensitivity to certain inhibitors, is modulated in part by its assembly with the F_0 sector and another subunit, the oligomycin sensitivity-conferring protein. We have used antisera prepared against the F_1-ATPase and against one of the F_0 sector subunits to study the structure, biosynthesis, and assembly of the entire complex in the mitochondria of the germinating spores of *Botryodiplodia* (Wenzler and Brambl, 1981).

The enzyme complex in *Botryodiplodia* consists of twelve distinct subunit polypeptides with molecular weights of 58,000 to 8,000; eleven of these are products of the cytoplasmic ribosomes, whereas only one is derived from the mitochondrial ribosomes. The hydrophobic subunit 12 proteolipid (which binds the enzyme inhibitor DCCD) of *Botryodiplodia* is a product of cytoplasmic ribosomes as it is in *Neurospora* but unlike the proteolipid of yeast, which is a product of mitochondrial ribosomes (Tzagoloff *et al*., 1979).

We found that syntheses of all subunits of the enzyme begin together during the first 45 min of spore germination and that the rates of synthesis accelerate rapidly later in germination. The fact that these spores synthesize the enzyme subunits during this first 45 min interval is significant, because during this period new mRNA synthesis has not yet begun, and the protein synthesis in these spores at this time depends upon translation of the mRNA preserved during dormancy [see Brambl, Dunkle and Van Etten (1978) for review]. To test the possibility that the dormant spore mRNA does code for subunits of the F_1-ATPase, we isolated and translated the polyadenylated mRNA from the dormant spores (Wenzler and Brambl, 1978), and by immunochemical techniques we found that the products of this *in vitro* translation do include peptides related immunogenically and in molecular weight to authentic subunits of the ATPase (Wenzler and Brambl, 1981). These observations indicate that one possible function of the mRNA preserved in dormant spores is to contribute to the reassembly of the mitochondrial respiratory system through its translation into subunits of the mitochondrial ATPase.

Although the ATPase is synthesized very early in germination, the question remains whether this *de novo* synthesis actually is essential to development of mitochondrial respiration or whether elaboration of an active ATPase depends upon a reassembly or modification process. We tested this question directly by immunochemical assay of mitochondria of radiolabeled dormant spores using antisera against the F_1-ATPase and the DCCD-binding proteolipid of the membrane sector. We found that only two or three subunits of the enzyme complex are preserved in the dormant spores, that these subunits are not associated into an enzyme complex, and that they are contained only within the mitochondria (none of the subunits were in the cytoplasmic fraction). These findings (Wenzler and Brambl, 1981) make it likely, therefore, that the development of mitochondrial ATPase activity in the germinating spores depends not upon rearrangement or modification of preserved components of the enzyme, but instead upon *de novo* synthesis of the enzyme via translation of the preserved mRNA.

9. CONCLUSIONS

It seems certain that function of the mitochondrial cytochrome-mediated electron transport system is one of several metabolic activities which are essential for spore germination. The respiratory membrane required for germination may be assembled into a functional form before spore dormancy (as it is in *Neurospora*), or it may be structurally and functionally incomplete in the dormant spores (as in *Botryodiplodia*) with cellular mechanisms for rapid assembly of components of the membrane both from preserved, compartmentalized enzyme subunit polypeptides (cytochrome *c* oxidase) and from translation of the preserved mRNA into required enzyme subunits (ATPase). The non-cytochrome pathway of electron transport, the mitochondrial alternate oxidase, seems not to function significantly in spore germination.

Whereas function of the respiratory membrane is necessary for spore germination, it almost certainly is not sufficient for this step. For example, assembly and function of cytochrome *c* oxidase are required if spores of *Botryodiplodia* are to germinate, but these steps in themselves will not ensure germination. When incubated in water, these spores conduct a cyanide-sensitive respiration and they synthesize and assemble subunits of cytochrome c oxidase, but they do not germinate (*unpublished*). The results of these studies also illustrate again the point that there likely is no single event initiating spore germination (a master switch) from which all other steps of germination follow in an ordered sequence. Instead, it is more likely that germination results from several essential and independent metabolic activities which must proceed in parallel such as cytoplasmic protein synthesis, lipid synthesis, and aerobic respiration. Some (but not all) of these activities may be independent in their initiation but interdependent in their effects. The aerobic respiration of germinating spores may be an example of one such activity among several which influence each other and which together are indispensable for spore germination.

In our work we have sought to study development of respiratory competence in spores not only for its own sake, but also as a means of measurement of activities of certain specific genes and gene products which are involved in a requisite process of cell development. We hope to understand better the cellular regulation of such individual, paradigmatic metabolic processes in order to improve eventually our comprehension of more general aspects of developmental biology.

FOOTNOTE

I use the term dormancy as a symbol for a phenomenon which may have several valid definitions. "Dormancy" is used here in Sussman's (1966) inclusive sense as any rest period or reversible interruption of the phenotypic development of an organism.

ACKNOWLEDGEMENTS

Mark Josephson, Herman Wenzler, Susan Stade, and Nora Vig are colleagues whose work and keen abilities have been invaluable in this research and preparation of this review. I thank Prof. William Bushnell for his continuing interest in these studies. I acknowledge with gratitude the financial support of this research from the National Institute of General Medical Sciences through NIH Research Grant GM-19398.

REFERENCES

Allen, P.J. (1965). Metabolic aspects of spore germination in fungi, *Ann. Rev. Phytopathol.* 3, 313-342.

Beck, D.P. and Greenawalt, J.W. (1976). Biogenesis of mitochondrial membranes in *Neurospora crassa* during cellular differentiation: Ultrastructural changes accompanying differentiation, *J. Gen. Microbiol.* 92, 97-110.

Bertrand, H. and Collins, R.A. (1978). A regulatory system controlling the production of cytochrome aa_3 in *Neurospora crassa*, *Molec. Gen. Genet.* 166, 1-13.

Bertrand, H. and Werner, S. (1977). Deficiency of subunit two of cytochrome oxidase in the *mi-3* cytoplasmic mutant of *Neurospora crassa*, *Eur. J. Biochem.* 79, 599-606.

Bertrand, H. and Werner, S. (1979). Cytochrome *c* oxidase subunits in nuclear and extranuclear cytochrome-aa_3-deficient mutants of *Neurospora crassa*, *Eur. J. Biochem.* 98, 9-18.

Borst, P. and Grivell, L.A. (1978). The mitochondrial genome of yeast, *Cell* 15, 705-723.

Brambl, R. (1975a). Presence of polyribosomes in conidiospores of *Botryodiplodia theobromae* harvested with non-aqueous solvents, *J. Bacteriol.* 122, 1394-1396.

Brambl, R. (1975b). Characteristics of developing mitochondrial genetic and respiratory activities in germinating fungal spores, *Biochem. Biophys. Acta* 396, 175-186.

Brambl, R. (1977). Mitochondrial biogenesis during fungal spore germination: Development of cytochrome *c* oxidase activity, *Arch. Biochem. Biophys.* 182, 273-281.

Brambl, R. (1980). Mitochondrial biogenesis during fungal spore germination: Biosynthesis and assembly of cytochrome *c* oxidase in *Botryodiplodia theobromae*, *J. Biol. Chem.* (*In press*).

Brambl, R., Dunkle, L.D. and Van Etten, J.L. (1978). Nucleic acid and protein synthesis during fungal spore germination. *In* "The Filamentous Fungi: Developmental Mycology" (Ed. J.E. Smith and D.R. Berry). Vol. 3, pp. 94-118. Edward Arnold, London.

Brambl, R. and Josephson, M. (1977). Mitochondrial biogenesis during fungal spore germination: Respiratory cytochromes of dormant and germinating spores of *Botryodiplodia*, *J. Bacteriol.* 129, 291-297.

Brambl, R. and Van Etten, J.L. (1970). Protein synthesis during fungal spore germination. Evidence that the ungerminated conidiospores of *Botryodiplodia theobromae* contain a messenger ribonucleic acid, *Arch. Biochem. Biophys.* 137, 442-452.

Brambl, R., Wenzler, H. and Josephson, M. (1978). Mitochondrial biogenesis during fungal spores germination: Effects of the antilipogenic antibiotic cerulenin upon *Botryodiplodia* spores, *J. Bacteriol.* 135, 211-317.

Cantino, E.C., Lovett, J.S., Leak, L.V. and Lythgoe, J. (1963). The single mitochondrion, fine structure and germination of the spore of *Blastocladiella emersonii*, *J. Gen. Microbiol.* 31, 393-404.

Cochrane, V.W. (1966). Respiration and spore germination. *In* "The Fungus Spore" (Ed. M.F. Madelin). pp. 201-213. Butterworths, London.

Cotter, D.A. and Raper, R.B. 1968. Properties of germinating spores of *Dictyostelium discoideum*, *J. Bacteriol.* 96, 45-126.

Cotter, D.A. and Raper, R.B. (1970). Spore germination in *Dictyostelium discoideum* : Trehalase and the requirement for protein synthesis, *Dev. Biol.* 22, 112-118.

Criddle, R.S. and Schatz, G. (1969). Promitochondria of anaerobically grown yeast. I. Isolation and biochemical properties, *Biochemistry* 8, 322-334.

Degn, H., Lloyd, D. and Hill, G.C. (1978). "Functions of Alternative Terminal Oxidases" Pergamon Press, Oxford.

Gottlieb, D. (1976). Carbohydrate metabolism and spore germination. *In* "The Fungal Spore. Form and Function" (Ed. D.J. Weber and W.M. Hess). pp. 141-163. John Wiley and Sons, New York.

Josephson, M. and Brambl, R. (1980). Mitochondrial biogenesis during fungal spore germination: Purification, properties, and biosynthesis of cytochrome oxidase from *Botryodiplodia theobromae*, *Biochem. Biophys. Acta* 606, 125-137.

Kawakita, M. (1970). Studies on the respiratory system of *Aspergillus oryzae* I. Development of respiratory activity during germination in the presence and absence of antimycin A. *Plant and Cell Physiol.* 11, 377-384.

Keyhani, J., Keyhani, E. and Goodgal, S.H. (1972). Studies on the cytochrome content of *Phycomyces* spores during germination, *Eur. J. Biochem.* 27, 527-534.

Kobilinsky, L. and Beattie, D.S. (1977). Respiratory competence of *Dictyostelium discoideum* spores, *J. Bacteriol.* 132, 113-117.

Mahler, H.R., Bastos, R.N., Flury, U., Lin, C.C. and Phan, S.H. (1975). Mitochondrial biogenesis in fungi. *In* "Genetics and Biogenesis of Mitochondria and Chloroplasts" (Ed. C.W. Birky, Jr., P.S. Perlman and T.J. Byers). pp. 66-116. Ohio State University Press, Columbus.

Nargang, F.E., Bertrand, H. and Werner, S. (1978). A nuclear mutant of *Neurospora crassa* lacking subunit 1 of cytochrome oxidase, *J. Biol. Chem.* 253, 6364-6369.

Ortega-Perez, R.D., Khandjian, E.W., Rougemont, A., Ojha, M. and Turian, G. (1977). Changes of mitochondrial density during differentiation of *Neurospora crassa*, *FEMS Microbiol. Lett.* 2, 207-209.

Rosen, D., Edelman, M., Galun, E. and Danon, D. (1974). Biogenesis of mitochondria in *Trichoderma viride*: Structural changes in mitochondria and other spore constituents during conidium maturation and germination, *J. Gen. Microbiol.* 83, 31-49.

Rousseau, P. and Halvorson, H.O. (1973). Effect of metabolic inhibitors on germination and outgrowth of *Saccharomyces cerevisiae* ascospores. *Can. J. Microbiol.* 19, 1311-1318.

Schatz, G. and Mason, T.L. (1974). The biosynthesis of mitochondrial proteins. *Ann. Rev. Biochem.* 43, 51-87.

Schmit, J.C. and Brody, S. (1976). Biochemical genetics of *Neurospora crassa* conidial germination, *Bacteriol. Rev.* 40, 1-41.

Soll, D.R., Bromberg, R. and Sonneborn, D.R. (1969). Zoospore germination in the water mold *Blastocladiella emersonii*. I. Measurement of germination and sequence of subcellular morphological changes, *Dev. Biol.* 20, 183-217.

Stade, S. and Brambl, R. (1981). Mitochondrial biogenesis during fungal spore germination: Respiration and cytochrome oxidase in spores of *Neurospora*. (*Submitted*).

Stevenson, I.L. and Becker, S.A.W.E. (1979). The fine structure of mature and germinating chlamydospores of *Fusarium oxysporum*. *Can. J. Microbiol.* 25, 808-817.

Sussman, A.S. (1966). Dormancy and spore germination. *In* "The Fungi" (Ed. G.C. Ainsworth and A.S. Sussman). Vol. 2, pp. 733-764. Academic Press, New York.

Sussman, A.S. and Douthit, H.A. (1973). Dormancy in microbial spores, *Ann. Rev. Plant Physiol.* 24, 311-352.

Tingle, M.A., Küenzi, M.T. and Halvorson, H.O. (1974). Germination of yeast spores lacking mitochondrial deoxyribonucleic acid, *J. Bacteriol.* 117, 89-93.

Tzagoloff, A., Macino, G. and Sebald, W. (1979). Mitochondrial genes and translation products, *Ann. Rev. Biochem.* 48, 419-441.

Van Laere, A.J., Van Assche, J.A. and Carlier, A.R. (1980). Reversible and irreversible activation of *Phycomyces blakesleeanus* spores. *Exptl. Mycol.* 4, 96-104.

Weiss, B. (1963). An electron microscope and biochemical study of *Neurospora crassa* during development, *J. Gen. Microbiol.* 39, 85-94.

Weiss, H., Schwab, A.J. and Werner, S. (1975). Biogenesis of cytochrome oxidase and cytochrome *b* in *Neurospora crassa. In* "Membrane Biogenesis" (Ed. A. Tzagoloff). pp. 125-153. Plenum, New York.

Wenzler, H. and Brambl, R. (1978). *In vitro* translation of polyadenylate-containing RNAs from dormant and germinating spores of the fungus *Botryodiplodia theobromae*, *J. Bacteriol.* 135, 1-9.

Wenzler, H. and Brambl, R. (1981). Mitochondrial biogenesis during fungal spore germination : Catalytic activity, composition, and subunit biosynthesis of oligomycin-sensitive ATPase in *Botryodiplodia. (Submitted)*.

Wergin, W.P., Dunkle, L.D. Van Etten, J.L. St. Julian, G. and Bulla, Jr., L.A. (1973). Microscopic observations of germination and septum formation in pycnidiospores of *Botryodiplodia theobromae*, *Dev. Biol.* 32, 1-14.

Werner, S. and Bertrand, H. (1979). Conversion of a mitochondrial precursor polypeptide into subunit 1 of cytochrome oxidase in the *mi-3* mutant of *Neurospora crassa*, *Eur. J. Biochem.* 99, 463-470.

GENETIC AND BIOCHEMICAL STUDIES ON *NEUROSPORA* CONIDIA GERMINATION AND FORMATION

S. Brody

Biology Department
University of California, La Jolla, CA 92037, U.S.A.

A. INTRODUCTION

The investigation of spore formation and germination has required the use of a variety of techniques. In particular, the biochemical and genetic approaches have been useful in this area, and it is precisely these two approaches that will be stressed here. The objectives of this chapter are to summarize some of the new observations about *Neurospora* conidial germination since a review appeared in 1976 (Schmit and Brody), and to present our initial genetic and biochemical studies on *Neurospora* conidiation. It seems appropriate for *Neurospora* to be discussed in this context, due to its close association with the start of biochemical genetics and its subsequent intensive analysis.
This article will confine itself to the macroconidia of *Neurospora crassa*, unless specifically mentioned otherwise.

B. STUDIES ON GERMINATION AND CONIDIATION

1. *An update on our hypothesis for the early events in conidial germination.*

In 1975, a proposal was put forth (Schmit *et al.*) stating that three events which were observed during *Neurospora* conidial germination occur in a related sequence and that these events might comprise a "trigger pathway" for germination. These events were: 1) the rapid disappearance of a large glutamic acid pool in conidia, with glutamic acid decarboxylase probably being the first step since γ-aminobutyric acid (GABA) appeared; 2) a transient surge in the pyridine nucleotide redox ratio, i.e., a temporary increase in the amounts of NADH and NADPH, and 3) the conversion of oxidized glutathione to its reduced form. The proposal stated that these events occurred in the above order and that the glutamic acid pool served as a source of "reducing power" for the conidia. The observed increase in the "reducing power" was employed for many purposes, one of which was to enzymatically reduce the oxidized glutathione which had been slowly accumulating in conidia over a period of time. The proposal also suggested the possibility that the change in the ratio of oxidized:reduced glutathione might influence the status of some protein sulfhydryl groups. Although this sequence of events all took place in the first 10-15 minutes after the hydration of conidia, one could not unambiguously state that these events were the initial events in germination, just that they occurred quite early in the sequence. Even though no reports have appeared documenting earlier events, this question still remains an open one. The proposal also pointed out that these events were unique to the start of germination and did not appear to be continuing events.

It was also reported that these early events required only water (Schmit and Brody, 1976). In other words, no carbon or nitrogen source was needed, and a protein synthesis inhibitor did not interfere, etc. Subsequent work has shown that these events will occur when conidia are maintained at high humidity (Fahey *et al.*, 1978). This suggested that extensive washing, of the kind needed to remove germination inhibitors, was not required. More recently, we have observed that even hydration in cold (4°C) water triggers these events, although the processes were slower. Another experimental approach we have recently employed was to try to deplete the conidia of their large glutamate pool in the hope that this would make conidial germination dependent upon exogenous glutamate. These experiments involved incubation of conidia in water for many days with the predicted results of a drop in viability and a large drop in the glutamate pool. However, when media and a carbon source were added to the low viability conidial suspension, they were still able to do the early events (GABA formation and redox shifts) although at a decreased rate. The addition of glutamate did not increase these responses. Therefore, our conclusion is that conidia retain their capacity to do these early steps independently of the apparent size of the glutamate pool.

2. *Generality of this hypothesis.*

The finding of high levels of glutamic acid in *Neurospora* conidia prompted a search of the literature on germination events in other organisms. It was found that glutamate accumulated in high levels in a variety of different organisms, i.e., the spores of *Bacillus subtilis* and other bacilli (Nelson and Kornberg, 1970), that large amounts of proline were found in yeast ascospores (Rousseau and Halvorson, 1973), and that polyols accumulated in the spores of many fungi (Niederpruem and Hunt, 1967). These seemingly unrelated compounds could all serve as a source of reducing power during germination. The requirements of certain organisms for exogenous compounds in order to germinate fully, such as proline in *Aspergillus* (Yanagita, 1957), *Rhizopus* (Weber and Ogawa, 1965) and *Candida* (Dabrowa *et al.*, 1976), or glutamate in *Candida* (Dabrowa *et al.*, 1976) would also come under this category.

Subsequent experimental work which supported this hypothesis has come from several sources. In *B. megaterium*, glutamic acid was shown to be degraded during germination, the redox ratio changed very early in germination, and oxidized thiols were converted to their reduced form (Setlow and Setlow 1977a,b) also very early in germination. This sequence of events is basically identical to the *Neurospora* sequence, with the exception that the principal thiol was coenzyme A (CoA) and not glutathione. Another supportive study on bacterial spore germination was the work of Ando (1979) where he showed a rapid change in the redox state of *Clostridia* spores during anaerobic germination. More recently, we have investigated the germination of wheat embryos and have found the same sequence of events, i.e., glutamate conversion to GABA and redox changes (Mikolajczyk and Brody, manuscript in preparation). Fahey *et al.*, 1980 had previously shown that oxidized glutathione was converted to its reduced form in wheat embryos. His laboratory has also shown (Fig. 1) the occurrence of oxidized glutathione in many dormant, dry structures (Fahey 1977). Previous studies on the seed germination of many different types of plants by Collins and Wilson (1972) and Spedding and Wilson (1968), employing the novel technique of hydration in tritiated water had shown that GABA was formed very early in the germination sequence of these seeds. Our recent data confirm and extend these findings in a quantitative way.

Conidia have been found to contain high levels (10 fold > mycelia) of the enzyme necessary for the conversion of glutamate to GABA, i.e., glutamic acid decarboxylase (Christensen and Schmit, in press). This activity is rapidly lost during the germination process, in agreement with its postulated role just for the germination process. Analogous findings, i.e., elevated levels of this enzyme, have previously been reported for the spores of bacilli (Aronson *et al.*, 1975; Singh *et al.*, 1977) and for wheat

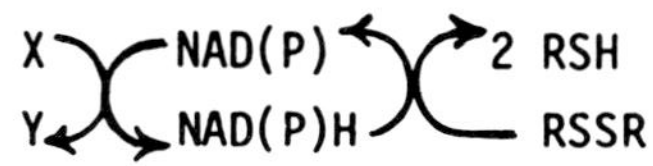

	ORGANISM	GENERATOR COMPOUND (X)	PYRIDINE NUCLEOTIDE CHANGES	THIOL COMPOUND (R)
1)	*Neurospora crassa* macroconidia	Glutamic acid (endogenous)	NAD(P)H ↑	Glutathione
2)	*Bacillus megaterium* spores	3-phosphoglyceric acid and/or Glutamic acid	NAD(P)H ↑	CoA
3)	Wheat embryos	Endogenous Glutamic acid	NAD(P)H ↑	Glutathione
4)	Spores from 10 species of fungi	Endogenous polyols	?	?
5)	Bacterial spores	Glutamic acid	?	?
6)	*Saccharomyces* ascospores	Proline	?	?
7)	Artemia cysts	?	?	Glutathione
8)	Barley embryos	?	?	Glutathione
9)	Dried yeast cells	?	?	Glutathione
10)	*Aspergillus niger* conidia	exogenous proline	?	?
11)	*Rhizopus stolonifer* conidia	exogenous proline	?	?

Fig. 1. Schematic representation of early metabolic events in germination. For references, see text.

embryos (Linko and Milner 1959), thus indicating another level of similarity between these diverse systems. Because of the supportive evidence obtained in the bacterial and plant systems for this hypothesis, we feel that it may not be just confined to *Neurospora* conidial germination and may have a more general significance.

3. *Modifications of the hypothesis.*

It also seems useful to make some slight changes in the scheme and three are proposed here (see Fig. 1). One, we have changed the specific thiol, glutathione, to a general thiol compound, in order to accommodate the finding of CoA in the bacilli. Two, we are proposing that those compounds, endogenous or exogenous, which are catabolized during germination to generate NAD(P)H, are to be designated as "generator compounds". Obviously, compounds which are required for other germination events, permeability changes, etc., and are not catabolized are not included in this grouping. Thirdly, we have eliminated from our scheme the thiol-induced changes in the overall protein sulfhydryl status as a mechanism which affects all or most of the protein disulfide bonds. The reasons for this change are elaborated in the next section.

4. *Studies on protein sulfhydryl status.*

We have found that when conidial proteins are extracted *en masse*, they have 5-6 fold more disulfide bonds than mycelial proteins, and also a slightly higher cysteine content (Mikolajcyzk and Brody, manuscript in preparation). The crude conidial extracts can be fractionated by several different methods and it was found that a significant portion of the disulfide residues in the crude conidial extract were due to a low molecular-weight, disulfide-rich protein. No comparable protein was found in the mycelial extracts down to the limits of our resolution (<2% of conidial level), so it would appear that this protein is either spore-specific or its level is amplified considerably in conidia. Beever *et al.*, (1979) have suggested that this protein may be a precursor to the rodlet layer protein found in conidia, based on the high cysteine content of both. The mole percents for the other amino acids are also similar for both of these proteins, and considerably different from the amino acid composition of a crude conidial protein extract. We are now testing this possibility of a precursor relationship in the *eas* mutant, which lacks the rodlet layer. We do not know the function of this protein, although we have found that the purified high-disulfide protein has protease activity associated with it. This is another possibility for its function, since bacterial sporulation is associated with an increase in protease activities (Dancer and Mandelstam, 1975).

As initially reported at the Tampa conference (Brody 1977), the level of the total disulfide content of protein does not decrease during germination. This is in contrast to the soluble thiols which are converted from disulfides to sulfhydryls early in germination. The apparent inability of the soluble thiols to alter the state of the bulk of the protein thiols has lead us to revise that part of the scheme which proposed large changes in the thiol status of all of the conidial proteins upon germination. The possibility still exists however that changes in the thiol status of a few critical proteins could take place during germination and could affect the activity of these proteins. We particularly have in mind the effects of thiols on the initiation complexes for protein synthesis, since many labs have shown effects of thiols on these complexes (Ernst *et al.*, 1979; Kosower *et al.*, 1972). If this were the case then the sequence of events would be the generation of reducing power, the change in soluble thiol status, and then the activation of protein synthesis. Some evidence in this direction has recently appeared (Fahey *et al.*, 1980), but more definitive studies are still needed.

5. *Selected aspects of conidial composition and germination.*

A fairly complete comparison of conidia vs. mycelia was previously reported (Schmit and Brody, 1976), as were processes occurring during germination. Since then several additional interesting reports have appeared on both of these subjects. Since many of the pertinent reports on *Neurospora* conidial germination are covered by the other contributors to these proceedings, only certain selected aspects will be presented here.

Conidia were found to have a system for transporting iron which consists of a certain type of cyclic hexapeptides, called siderochromes (Horowitz *et al.*, 1976). This system appears to be necessary during germination, since conidia missing these molecules germinate poorly, if at all.

TABLE 1

Comparison Between Mycelia and Conidia*

	Shaker Culture	Conidia
Glutamic Acid	60-100	150-300
Arginine	30-40	1-10
Ornithine	15-25	1-5
Histidine	6-12	6-10
Siderochromes	0.1-0.2	3-4
Ergothioneine	?	3-5
Lysine	7-16	4-5

*For references, see text

Two additional points on this subject: the level of siderochrome is considerably higher in conidia than mycelia (see Table I). A mutant strain (Schmit and Brody, manuscript in preparation) which was temperature-sensitive for germination lost some of its siderochrome content at the nonpermissive temperature (Charlang and Williams, 1977) and was unable to continue the germination process. The finding of a conidial-specific uptake system fits in nicely with other reports of different permease systems between conidia and mycelia (Pendyala and Wellman, 1977; Schmit and Brody, 1976).

Additional studies worth briefly mentioning have been the studies of Mirkes and collaborators (Mirkes, 1977; Farach *et al.*, 1979) on the role of a carbon and nitrogen source on polysome formation, germ tube formation, and germ tube elongation. An interesting paper by Serna and Stadler (1978) describing DNA synthesis during germination in a temperature-sensitive strain led to the conclusion that conidia are arrested at various points in the division cycle. Previously, Loo (1976) had shown germ tube formation in the absence of DNA synthesis. The independence of these processes has been illustrated by the effects of certain compounds which inhibit germ tube formation but still allow conidial swelling and mass increase (Eherhart, 1977). *Neurospora* conidia have also been investigated with respect to their cyclic AMP and GMP content during germination of the wild type strain (Rosenberg and Pall, 1978) and mutant strains (Rosenberg and Pall, 1979); their sensitivity to photodynamic inactivation (Shimizu *et al.*, 1979); their content of "extrinsic" RNA (Stuart and Woodward, 1977); the changes that occur during germination in their mitochondria (Beck and Greenawalt, 1976a,b) and their lipids (Beck and Greenawalt, 1977).

6. Genetic and biochemical studies on conidiation.

Our studies on the early events in germination have naturally led us to an interest in conidiation since those events that occur in the first few minutes of germination were "pre-programmed" into the conidia during its formation. The study of conidiation can be approached by asking three sets of questions, i.e., general, operational and specific, about the process. Some of the general questions are: What biochemical event(s) initiate this process? What biochemical events are necessary subsequently? Of the biochemical events that have been observed, which ones are conidiation specific, and which ones are just a response to any type of starvation? Are there certain biochemical events, i.e., an increase in the level of an enzyme, which occur during conidiation which are not really necessary for conidiation but are part of the pre-programming for germination? Doubtlessly, one could pose other general questions, and one could make the questions listed above somewhat more explicit as well.

The operational questions are also straightforward, but again difficult to answer definitively. They involve asking what organism to choose for the study, what particular mode of conidiation shall be employed, i.e., microcycle or whatever, and what type of approach should be selected as the principal tool, i.e., biochemical, genetic, etc. Naturally, one cannot find the "perfect" system and some compromises must be made. In addition, familiarity plays a large role in one's choice. *Neurospora* is the one we have chosen, due to familiarity and the unusual amount of information already known about it. We have previously employed (Schmit and Brody, 1975b) the *bd* strain to study conidiation since it conidiates abundantly in closed petri dishes, apparently because it is resistant to carbon dioxide. It also has the advantage that it will conidiate at the edge of a growing culture and one does not have to employ starvation conditions. Finally, we have employed both the biochemical and genetic approaches to this problem. The advantages to the genetic approach have already been pointed out by Matsuyama *et al.*, (1974) for *Neurospora* and Clutterbuck (1978) for *Aspergillus*.

The type of questions addressed to specific events will naturally encompass a broad array of events. We have decided to focus on a particular area of nitrogen metabolism, specifically amino acid metabolism. Our interests here have to do with the mechanisms by which the large pool of glutamate accumulates in the conidia. The questions that can be posed as specific ones are :

Which pathway or pathways are involved in the synthesis of glutamate in the conidia ?

What prevents the high levels of glutamate in the conidia from being degraded before germination ?

Will interference with glutamate synthesis or degradation affect the conidiation process ?

As an introduction to this area, there are listed on the following pages schematic diagrams of some of the principal reactions involved in glutamate synthesis (Fig. 2) and degradation (Fig. 3) in *Neurospora*.

Below each figure, for the sake of brevity, a designation is provided for each of these reactions in terms of a common or trivial name, rather than the official E.C. name and number. In addition, known mutant loci are included thus illustrating that one advantage of *Neurospora* is the availability of a large number of different mutants.

The reference to the mutant loci are given either in Davis (1979), Fincham *et al.* (1979), or in the *Neurospora* Newsletter No. 27 (1980).

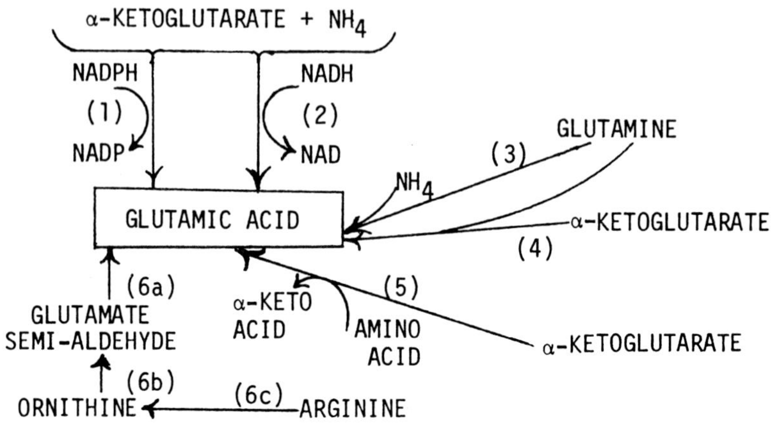

Fig. 2. Schematic diagram of the possible reactions involved in glutamic acid synthesis.

TABLE 2

Enzymes Involved in Glutamate Synthesis

Reaction	Common Name	Mutant Loci*
(1)	NADP-Dependent Glutamate Dehydrogenase	AM
(2)	NAD-Dependent Glutamate Dehydrogenase	
(3)	Glutaminase	
(4)	Glutamate Synthase or GOGAT	EN-AM
(5)	Transaminase (GOT)	
(6a)	Pyrroline-Carboxylate Dehydrogenase	
(6b)	Ornithine Transaminase	OTA
(6c)	Arginase	AGA

*For references, see text

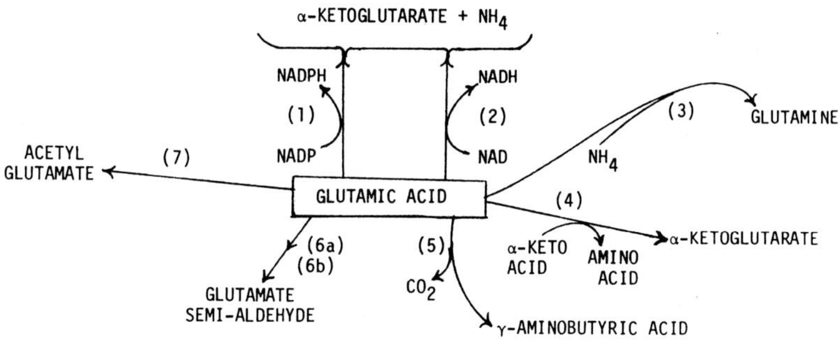

Fig. 3. Schematic diagram of the possible reactions involved in the conversion of glutamate.

TABLE 3

Enzymes Involved in Glutamate Conversion

Reaction	Common Name	Mutant Loci*
(1)	NADP-Dependent Glutamic Dehydrogenase	AM
(2)	NAD-Dependent Glutamic Dehydrogenase	
(3)	Glutamine Synthetase	GLM
(4)	Transaminase (GOT)	
(5)	Glutamate Decarboxylase	
(6a)	Glutamate Kinase	PRO-3 (ARG-8)
(6b)	Glutamate Phosphate Reductase	PRO-4 (ARG-9)
(7)	Acetyl Glutamate Synthase	ARG-14?

*For references, see text

Recently, Dr. Roland Guignard, in my lab,* has started studies on the question of glutamate synthesis during conidiation. Since one possible source of glutamate might be via glutamate dehydrogenase (NADP-dependent), conidia from the double mutant *bd, am* were examined for their glutamate content. He has found ∿ 2 fold lower levels of glutamate

*fellow of the Swiss National Research Fund

in these conidia and a much slower rate of germination on minimal media. Conidia from the double mutant *am, en-am* had even lower levels of glutamate, and an even slower rate of germ tube formation. This latter double mutant is thought to be blocked in two pathways of glutamate synthesis (Table 2). Although it is encouraging to see these effects on glutamate levels and on germination, one cannot definitively state that it is these pathways alone that are responsible for the glutamate buildup in conidia. There is a definite possibility that inbalances in nitrogen metabolism caused by these mutations inhibit the pathways that are actually responsible. Obviously, a series of carefully planned and executed isotope tracer experiments will be necessary in order to ascertain the source of the conidial glutamate pool before any definitive conclusions can be reached.

At the time that the high glutamate level was reported (Schmit and Brody, 1975, 1976) it was noted that conidia also contained low levels of certain basic amino acids, i.e., arginine, ornithine, compared to mycelia. It was also noted that a mutant strain blocked in ornithine catabolism was aconidial. It was speculated at that time that these amino acids might be stored in mycelia in order to support conidiation. It is interesting to extend this speculation a little further now and propose that the stored arginine and ornithine in the mycelia may be one of the major sources for the high levels of glutamic acid in the conidia. The levels of these three amino acids in conidia or mycelia as well as the levels of certain other molecules are listed above (Table 1). The levels of the siderochromes, which are rich in ornithine, are listed in this table (Horowitz *et al.*, 1976), as is the level of ergothioneine (Fahey, personal communication). Ergothioneine is known in *Neurospora* to be synthesized from histidine (Ishikawa and Melville, 1970). The interesting generality which emerges from these data is that mycelia contain high levels of certain basic amino acids which are precursors to compounds found in high levels in conidia. It is known that mycelia sequester the bulk of their ornithine and arginine inside a vesicle (Davis, 1975; Weiss, 1973), and that the bulk of the lysine and histidine are also associated with this vesicle (Weiss, personal communication). It is tempting to propose that this vesicle is a reservoir of precursors for compounds needed during conidiation. This proposal is illustrated as Fig. 4.

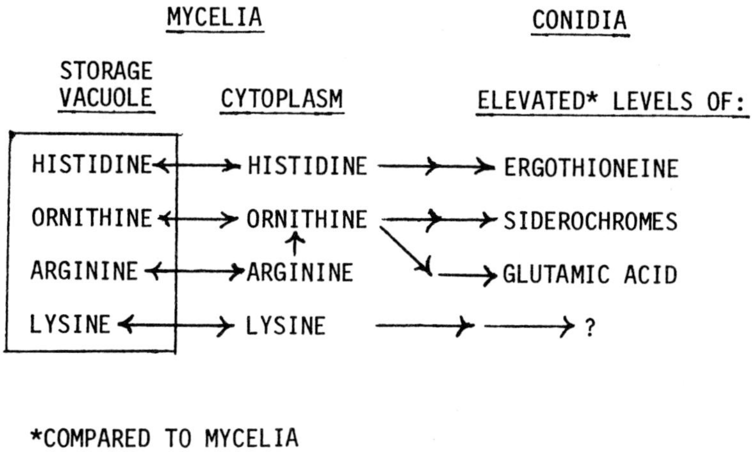

Fig. 4. Schematic relationship between compounds stored in the mycelial vacuoles and those found in high levels in conidia.

It is also known that catabolism of some of these stored amino acids occurs under nitrogen, but not carbon, starvation (Legerton and Weiss, 1979), thus making a possible connection between nitrogen starvation, catabolism of these amino acids, and the conidiation process.

There are some studies which suggest that the level of glutamine plays a role in the mobilization of these stored amino acids (Vaca and Mora, 1977; Weiss, personal communication), and it would be interesting to see if a decreased level of glutamine in conidiating regions and aerial hyphae are involved in the conidiation process as well. Direct measurements of glutamine levels would be one necessary item, as would the effects of mutants blocked in glutamine synthesis, glutamine metabolism, and mutants unable to derepress enzymatic functions under nitrogen starvation conditions. It should be pointed out that many other areas, i.e., purine metabolism, etc. are also affected by nitrogen starvation, so that one should not focus just on amino acid catabolism. In this regard, it should be noted that there are reports suggesting that low levels of guanine nucleotides are necessary for bacterial sporulation (Lopez *et al.*, 1979). It would appear that sporulation will be like many other areas of development biology, in that it will be difficult to sort out the necessary events from associated events.

Another experimental approach to studying conidiation is to look amongst the biochemical mutant strains for conidiation defects. These strains were isolated on the basis of a deficiency in a particular area of metabolism. Although these mutations may not have been shown to be structural gene mutations, at least the area of metabolism is localized and one has an area on which to focus. The opposite genetic approach of just selecting for developmental mutants (aconidial) often comes to a dead end, since it is not clear how to approach biochemical problems with these strains in a rational way. With this in mind, Dr. Guignard has been analyzing strains blocked in the conversion of ornithine to glutamate, *ota* mutants, and has found that these strains form conidia poorly (15-20% of normal). This developmental defect is associated with the ornithine non-utilizing trait in all of the isolates tested from a cross of *ota* x *bd*. This *ota* mutation has no serious or discernible effects on vegetative growth but does have a unique morphology when grown on a slant, which, as far as is known, has not been described in other biochemical mutants. Two other interesting mutants (UM-700 and UM-723) which are aconidial, were also selected on the basis of a defect in ornithine catabolism (Davis and Mora, 1968). The exact deficiency is still unknown in both of these strains, but it is interesting that exogenous glutamate or a complete medium did not "cure" the conidiation defects in any of these strains. The finding of conidiation defects associated with deficiencies in ornithine catabolism is precisely the type of example mentioned a few lines above, in that it suggests many possible experiments in one area. For instance, one can look at the levels of ornithine in conidiating regions and conidia of an *ota* strain, for the source of glutamate in the few conidia that are formed, at the levels of ornithine transminase in conidiating cultures and aerial hyphae, etc. One could also try to isolate temperature-sensitive *ota* strains, and to screen existing aconidial strains for defects in ornithine catabolism. At this time, it is not clear how the *ota* mutation affects the conidiation process, but some of the possibilities can be mentioned. One suggestion is that an intermediate in the conversion of ornithine to glutamate is required for conidiation. Another possibility is that an accumulation of ornithine outside of the vesicle inhibits conidiation. Other guesses as to mechanism have to do with some regulatory effects of the mutation which prevent both ornithine transaminase synthesis or activity and the induction of conidiation. Clearly, more experimental evidence is required to understand further these initial findings.

C. DISCUSSION

1. *Conidiation.*

The genetic and biochemical approaches towards understanding conidiation in *Neurospora* are obviously just the initial studies of this phenomenon. As mentioned above, most of the important experiments are yet to be performed, and possibly the important questions are even yet to be framed. However, the existing lines of evidence do yield a tentative framework for future studies even if alternate explanations finally emerge for the present observations. Some of the alternate explanations that can be considered for our observations are as follows: the catabolism of the basic amino acids could be just a consequence of nitrogen deprivation and not a necessary part of conidiation; the defects in ornithine metabolism (*ota*, etc.) could just affect conidiation by some indirect means, i.e., a pleiotropic effect, and not directly affect a key part of the conidiation mechanism. Although our evidence to date does not rule out these alternate explanations, or prove our proposition, it should be pointed out that it is quite consistent with the idea that the catabolism of the basic amino acids are an important process in conidiation. It should also be mentioned that many other processes will be important steps in conidia formation, and that the timing between the catabolism of the basic amino acids and the formation of glutamate will have to be coordinated with these other biochemical steps. In this regard, one might not expect all of the aconidial or poor conidiating mutants to be blocked in basic amino acid metabolism, but to be blocked in other areas as well, and the finding of aconidial mutants that are not deficient in ornithine catabolism will not disprove the proposal. At this point, it is worthwhile to indicate that there are many morphological mutants of *Neurospora* which grow slowly and conidiate poorly, if at all (Dicker *et al.*, 1969). These strains are not as interesting as those which are only blocked in the conidiation process. These so-called "phase-specific" mutants will be the ones worth pursuing.

One other facet of conidiation worth discussing is the apparent paradox of how conidia can simultaneously contain both high levels of an enzyme, like glutamic acid decarboxylase, and high levels of its substrate, glutamate. Several suggestions can be made as to the nature of this phenomenon. First, conidia may contain an inhibitor of the enzyme. Secondly, a necessary cofactor for the enzyme may be sequestered away somehow, or even the substrate may be in a vesicle and physically separated from the enzyme. A third possibility, which is not mentioned as often, has to do with the differential effect of drying on the activity of enzymes. This explanation proposes that during conidia formation, the relative water content declines slowly, and most of the enzymes become immobilized due to the lack of water. If the enzymes involved in glutamate buildup, i.e., ornithine transaminase, etc., are less susceptible to drying, then they will continue to convert compounds to glutamate during this period of dryness, but the glutamate will not be metabolized any further. Upon hydration, all of the enzymes will become active again and the glutamate will be catabolized. Although this scheme is easy to consider since it avoids the necessity of proposing inhibitors for every one of the many reactions involved in glutamate degradation, it would be probably difficult to test experimentally.

2. *Germination.*

The process of conidial germination obviously requires a large number of different biochemical steps. Many of these steps have been described by other contributors to these proceedings, as well as elsewhere. What seems to be lacking so far in all of these reports is some feeling for the independence or interdependence of these multiple processes. It is not clear whether metabolic or genetic blocks in one process affect a second process, and what sequence of events precedes another sequence. At present the information available appears as independent sets of observations, with little effort given to identifying the effects of one variable on all of the other processes. For this reason, one can only make rough guesses as to the major landmarks in the germination process. Perhaps it would be helpful to have a series of temperature-sensitive and cold-sensitive mutant strains blocked at various points in germination, and then to disseminate these widely to a variety of laboratories? It might also be helpful to compile a list of germination-specific inhibitors or conditions and to employ these alone or in combination with the mutants. These inhibitors should have, if possible, few side effects so that their mode of action will be more clear. Perhaps several inhibitors affecting the same general process would be more realistic since they presumably would have a different spectrum of side effects. Inhibitor-resistant mutants are also a way of testing the side effects. In this regard, we have been exploring recently two conditions for phase-specific inhibitors. One of these is isonicotinic hydrazide, which is known to inhibit glutamic acid decarboxylase, and unfortunately other pyridoxal-phosphate containing enzymes. Another possibility we have been investigating is the incubation of conidia at 40°C at a low density. This prevents germination almost completely although once germinated, growth can proceed at this temperature readily. Since the inclusion of a fatty acid, palmitic acid, reverses the temperature block of germination, one suspects an involvement of some membrane in this phenomena.

Finally, it seems worthwhile to point out that some of the approaches and observations on fungal spore germination and formation have also been employed and occurred in bacterial systems. For instance, the enzymes involved in the conversion of ornithine to glutamate rise $\approx$ 10 fold during sporulation of *Bacillus subtilis* (Ramaley and Bernlohr, 1966) and some strains of bacilli requires glutamate for maximum sporulation, but not growth (Buono *et al.*, 1966), etc. Perhaps we should keep closer track of our colleagues studying bacterial spores, just in case there are more analogies than are immediately apparent.

ACKNOWLEDGEMENTS

The author appreciates his colleagues, Roland Guignard, Robert Fahey, Steve Mikolajczyk and Joseph Schmit for their communication of unpublished results, and/or helpful suggestions on this article. The author's work has been supported by the National Science Foundation.

REFERENCES

Ando, Y. (1979), Spore lytic enzyme released from *Clostridium perfringens* spores during germination, *Journal of Bacteriology* 140, 59-64.

Aronson, J.N., Doerner, J.F., Akers, E.W., Borris, D.P. and Mani, M (1975), γ-Aminobutyric acid pathway of glutamate metabolism by *Bacillus thuringiensis*. *In* "Spores VI" (Ed. P. Gerhardt, R.N. Costilow and H.L. Sadoff). pp. 404-410. American Society for Microbiology, Washington, D.C.

Beck, D.P. and Greenawalt, J.W. (1976 a), Biogenesis of mitochondrial membranes in *Neurospora crassa* during cellular differentiation: ultrastructural changes accompanying differentiation, *Journal of General Microbiology* 92: 97-110

Beck, D.P. and Greenawalt, J.W. (1976b), Biogenesis of mitochondrial membranes in *Neurospora crassa* during cellular differentiation: changes in oxidative phosphorylation and synthesis of mitochondrial phospholipids, *Journal of General Microbiology* 92, 111-119.

Beck, D.P. and Greenawalt, J.W. (1977), Composition and synthesis of cellular lipids in *Neurospora crassa* during cellular differentiation, *Journal of Bacteriology* 131, 188-193.

Beever, R.E. and Dempsey, G.P. (1978), Function of rodlets on the surface of fungal spores, *Nature* 272, 608-610.

Beever, R.E., Redgwell, R.J. and Dempsey, G.P. (1979), Purification and chemical characterization of the rodlet layer of *Neurospora crassa* conidia, *Journal of Bacteriology* 140, 1063-1070.

Brody, S. (1977). Differences in chemical structure between *Neurospora crassa* conidial and mycelial proteins. *In* "Abstracts of Second International Mycological Congress" (Ed. H.E. Bigelow and E.G. Simmons). Vol. A-L, p. 69. University of South Florida, Tampa, Florida.

Buono, F., Testa, R. and Lundgren, D.G. (1966), Physiology of growth and sporulation in *Bacillus cereus, Journal of Bacteriology* 91, 2291-2299.

Charlang, G. and Williams, N.P. (1977), Germination-defective mutant of *Neurospora crassa* that responds to siderophores, *Journal of Bacteriology* 132, 1042-1044.

Clutterbuck, A.J. (1978). Genetics of vegetative growth and asexual reproduction. *In* "The Filamentous Fungi" (Ed. J.E. Smith and D.R. Berry). Vol. 3, pp. 240-256. Edward Arnold, London.

Collins, D.M. and Wilson, A.T. (1972), Metabolism of the axis and cotyledons of *Phaseolus vulgaris* seeds during early germination, *Phytochemistry* 11, 1931-1935.

Dabrova, N., Taxer, S.S. and Howard, D.H. (1976), Germination of *Candida albicans* induced by proline, *Infection and Immunity* 13, 830-835.

Dancer, B.N. and Mandelstam, J. (1975), Production and possible function of serine protease during sporulation of *Bacillus subtilis, Journal of Bacteriology* 121, 406-410.

Davis, R.H. (1975), Compartmentation and regulation of fungal metabolism : genetic approaches, *Annual Review of Genetics* 9, 39-65.

Davis, R.H. (1979), Genetics of arginine biosynthesis in *Neurospora crassa, Genetics* 93, 557-575.

Davis, R.H. and Mora, J. (1968), Mutants of *Neurospora crassa* deficient in ornithine-δ-transaminase, *Journal of Bacteriology* 96, 383-388.

Dicker, J.W., Oulevey, N. and Turian, G. (1969), Amino acid induction of conidiation and morphological alterations in wild type and morphological mutants of *Neurospora crassa*, *Archives of Mikrobiologie* 65, 241-257.

Eberhart, B.M. (1977), Inhibition of germ tube formation in *Neurospora*, *Journal of Bacteriology* 130, 516-520.

Ernst, V., London, D.H. and London, I.M. (1979), *In situ* phosphorylation of the α subunit of eukaryotic initiation factor two in reticulocyte lysates inhibited by heme deficiency, double stranded RNA, oxidized glutathione, or the heme-regulated protein kinase, *Proceedings of the National Academy of Science* 76, 2118-2122.

Fahey, R.C. (1977). Biologically important thiol-disulfide reactions and the role of cyst(e)ine in proteins: an evolutionary perspective. *In* "Protein Cross-linking, Part A" (Ed. M. Friedman). pp. 1-30, Plenum, New York.

Fahey, R.C., Brody, S. and Mikolajczyk, S.D. (1975), Changes in the glutathione thiol-disulfide status of *Neurospora crassa* conidia during germination and aging, *Journal of Bacteriology* 121, 144-151.

Fahey, R.C., Di Stefano, D.L., Meier, G.P. and Bryan, R.N. (1980), Role of hydration state and thiol-disulfide status in the control of thermal stability and protein synthesis in wheat embryo, *Plant Physiology* 65, 1062-1066.

Fahey, R.C., Mikolajczyk, S.D. and Brody, S. (1978), Correlation of enzymatic activity and thermal resistance with hydration state in ungerminated *Neurospora* conidia, *Journal of Bacteriology* 135, 868-875.

Farach, M.C., Farach, H. and Mirkes, P.E. (1979), Control of development in *Neurospora crassa:* nutrient requirements for conidial germ tube emergence, elongation, and metabolic activation, *Experimental Mycology* 3, 240-248.

Fincham, J.R.S., Day, P.R. and Radford, A. (1979). "Fungal Genetics" University of California Press, Berkeley and Los Angeles.

Horowitz, N.H., Charlang, G., Horn, G. and Williams, N.P. (1976), Isolation and identification of the conidial germination factor of *Neurospora crassa*, *Journal of Bacteriology* 127, 135-140.

Ishikawa, Y. and Melville, D. (1970), The enzymatic α-N-methylation of histidine, *Journal of Biological Chemistry* 245, 5967-5973.

Kosower, N.S., Vanderhoff, G.A. and Kosower, E.M. (1972), Glutathione VIII. The effects of glutathione disulfide on initiation of protein synthesis, *Biochimica Biophysica Acta* 272, 623-637.

Legerton, T.L. and Weiss, R.L. (1979), Mobilization of sequestered metabolites into degradative reactions by nutritional stress in *Neurospora*, *Journal of Bacteriology* 138, 909-914.

Linko, P. and Milner, M. (1959), Enzyme activation in wheat grains in relation to water content. Glutamic acid-alanine transaminase and glutamic acid decarboxylase, *Plant Physiology* 34, 392-396.

Loo, M. (1976), Some required events in conidial germination of *Neurospora crassa*, *Developmental Biology* 54, 201-213.

Lopez, A., Marks, C. and Freese, E. (1979), The decrease of guanine nucleotides initiates sporulation in *Bacillus subtilis*, *Biochimica Biophysica Acta* 587, 238-252.

Matsuyama, S. Nelson, R. and Siegel, R. (1974), Mutations specifically blocking differentiation of macroconidia in *Neurospora crassa*, *Developmental Biology* 41, 278-287.

Mirkes, P.E. (1977), Role of the carbon source in the activation of ribonucleic acid synthesis during the germination of *Neurospora crassa* conidia, *Experimental Mycology* 1, 271-279.

Nelson, D.L. and Kornberg, A. (1970), Biochemical studies of bacterial sporulation and germination. XVIII. Free amino acids in spores, *The Journal of Biological Chemistry* 245, 1128-1136.

Niederpruem, D.J. and Hunt, S. (1967), Polyols in *Schizophyllum commune*, *American Journal of Botany* 54, 241-245.

Pendyala, L. and Wellman, A.M. (1977), Developmental-state-dependent adenine transport in *Neurospora crassa*, *Journal of Bacteriology* 131, 453-462.

Ramaley, R.F. and Bernlohr, R.W. (1966), Postlogarithmic phase metabolism of sporulating microorganisms, *Archives of Biochemistry and Biophysics* 117, 34-43.

Rosenberg, G. and Pall, M.L. (1978), Cyclic AMP and cyclic GMP in germinating conidia of *Neurospora crassa*, *Archives of Microbiology* 118, 87-90.

Rosenberg, G. and Pall, M.L. (1979), Properties of two cyclic nucleotide-deficient mutants of *Neurospora crassa*, *Journal of Bacteriology* 137, 1140-1144.

Rousseau, P. and Halvorson, H.O. (1973), Macromolecular synthesis during the germination of *Saccharomyces cerevisiae* spores, *Journal of Bacteriology* 113, 1289-1295.

Schmit, J.C. and Brody, S. (1975a), *Neurospora crassa* conidial germination: role of endogenous amino acid pools, *Journal of Bacteriology* 124, 232-242.

Schmit, J.C. and Brody, S. (1975b), Developmental control of glucosamine and galactosamine levels during conidiation in *Neurospora crassa*, *Journal of Bacteriology* 122, 1071-1075.

Schmit, J.C. and Brody, S. (1976), Biochemical genetics of *Neurospora crassa* conidial germination, *Bacteriological Reviews* 40, 1-41.

Schmit, J.C., Fahey, R.C. and Brody, S. (1975). Initial biochemical events in germination of *Neurospora crassa* conidia, *In* "Spores VI" (Ed. P. Gerhardt, R.N. Costilow and H.L. Sadoff). pp. 112-119, American Society for Microbiology, Washington, D.C.

Serna, L. and Stadler, D. (1978), Nuclear division cycle in germinating conidia of *Neurospora crassa*, *Journal of Bacteriology* 136, 341-351

Setlow, B. and Setlow, P. (1977a), Levels of oxidized and reduced pyridine nucleotides in dormant spores and during growth, sporulation, and spore germination of *Bacillus megaterium*, *Journal of Bacteriology* 129, 857-865.

Setlow, B. and Setlow, P. (1977b), Levels of acetyl coenzyme A, reduced and oxidized coenzyme A, and coenzyme A in disulfide linkage to protein in dormant and germinated spores and growing and sporulating cells of *Bacillus megaterium*, *Journal of Bacteriology* 132, 444-452.

Shimizu, M., Egashir, T. and Takahama, U. (1979), Inactivation of *Neurospora crassa* conidia by singlet molecular oxygen generated by a photosensitized reaction, *Journal of Bacteriology* 138, 293-296.

Singh, R.P., Setlow, B. and Setlow, P. (1977), Levels of small molecules and enzymes in the mother cell compartment and the forespore of sporulating *Bacillus megaterium*, *Journal of Bacteriology* 130, 1130-1138.

Spedding, D.J. and Wilson, A.T. (1968), Studies of the early reactions in the germination of *Sinapis alba* seeds, *Phytochemistry* 7, 897-901.

Stuart, W.D. and Woodward, D.O. (1977), New class of ribonucleic acid in *Neurospora* associated with the outer cell envelope, *Journal of Bacteriology* 129, 395-399.

Vaca, G. and Mora, J. (1977), Nitrogen regulation of arginase in *Neurospora crassa*, *Journal of Bacteriology* 131, 719-725.

Weber, D.J. and Ogawa, J.M. (1965), The specificity of proline in the germination of spores of *Rhizopus arrhizus*, *Phytopathology* 55, 262-266.

Weiss, R.L. (1973), Intracellular localization of ornithine and arginine pools in *Neurospora*, *The Journal of Biological Chemistry* 248, 5409-5413.

Yanagita, T. (1957), Biochemical aspects on the germination of conidiospores of *Aspergillus niger*, *Archiv für Mikrobiologie* 26, 329-344.

MICROCYCLE CONIDIATION

J.E. Smith, J.G. Anderson, B. Kristiansen,
A. Al-Rawi and A.G. Yahya

Department of Applied Microbiology, University of Strathclyde, Glasgow G1 1XW, Scotland

INTRODUCTION

Microcycle conidiation in filamentous fungi has been defined as the recapitulation of conidiation following conidial germination without an intervening phase of mycelial growth (Smith *et al.*, 1977). The advantage of this form of conidiation for studying the biochemical and physiological aspects of the transition from the vegetative to the conidiating state of fungi are now being fully developed with several microcycle systems.

From an historical point of view surface cultivation techniques were originally the only means of studying the asexual conidiating process of filamentous fungi. Although these techniques still offer the best environmental system for spore production they are basically not suited to gaining information on the biochemical aspects of the process. The use of submerged cultures for sporulation studies has been widely practiced in recent years. In this way more homogeneous mycelial cultures are possible although special inducing conditions were often required to overcome a natural inhibition of sporulation that characterises submerged culture development. The use of batch and continuous fermenter systems in the study of fungal conidiation has undoubtedly created new dimensions to understanding this complex type of differentiation (Smith, 1978). Many environmental parameters now known to be involved with microbial development have been considered in the context of fungal sporulation and interesting interpretations derived. However, it must not be overlooked that most submerged mycelial cultures will be heterogeneous in morphological form and any subsequent analysis can only be a summation of various physiological states.

In part, some of these problems can be overcome by imparting some degree of synchronous control of the developmental processes. Thus, synchronous sporulation has been attempted and to some extent achieved with *Aspergillus niger* (Anderson and Smith, 1971a), *Achlya* sp (Griffin and Breuker, 1969) and *Penicillium digitatum* (Zeidler and Margalith, 1972) using either nutrient changes or cultural manipulations. By using a medium replacement technique Anderson and Smith (1971a) were able to achieve a good separation between the various developmental stages of conidiation in *A. niger* which subsequently allowed biochemical analysis of the developmental phases to be made (for reference see Smith, 1978). Studies using chemostat cultivation systems (Ng *et al.*, 1973; Robinson and Smith, 1976; Larmour and Marchant, 1977) have confirmed that for conidiation to occur in filamentous fungi there must be a check in the normal unlimited pattern of vegetative growth. This change can be promoted by a limitation of growth rate and any environmental factor which can influence growth rate may also influence conidiation. Thus the fundamental role of growth rate in determining the nature and extent of conidiation in filamentous fungi is of primary importance. The microcycle conidiation techniques that have been developed can offer major advantages over the conventional mycelial systems that have been discussed above. In particular, the higher proportion of vegetative tissue associated with normal sporulation (giving higher masking effect biochemically) is greatly reduced allowing clearer differences to be detected between the phases and where thermic variation has been used to control the microcycle there is much greater synchrony of conidial development. This is particularly well demonstrated by the fermenter controlled systems for *Aspergillus niger* and *Paecilomyces varioti* developed by Smith and co-workers.

INDUCTION OF MICROCYCLE SPORULATION

Microcycle sporulation processes have been recorded and described for several fungi in response to different environmental conditions. However, temperature has been shown to give the most dramatic control of the process. In practice, microcycle conidiation results in the production of a morphologically distinct unit comprising the parent conidium connected to the spore bearing apparatus (which may also arise from an abbreviated germ tube) and the second generation conidia.

Induction not utilising temperature treatments

There are many examples in the literature which demonstrate that sporulation can occur immediately after spore germination with an apparent absence or extreme limitation of the normal vegetative phase. Cultural conditions which lead to particular complete inhibition of hyphal development following germination have been most often encountered. Thus Roten and Bashi (1969) concluded that the direct formation of secondary conidia in *Alternaria* spp and *Stemphylium botryosum* was induced by various environmental factors which inhibited the vegetative development of the mother conidium. A somewhat similar situation was believed also to exist in *Geotrichum candidum* (Park and Robinson, 1969).

Inhibitory conditions created by spore crowding can result in microcycle sporogenesis. Spores borne directly on the germ tube were found in *Melanconium fulgineum* by overcrowding of germinating spores (Timmick *et al.*, 1952), while autoinhibitors associated with conidia have been considered by Lingappa and Lingappa (1969) to be the causative determinant inhibiting mycelial growth and initiating the development of secondary conidia in *Glomerella cingulata*. Sporulation from germ tubes of *Penicillium notatum* conidia when growing in a medium in which they had originally been produced was considered as further evidence of germination inhibitors creating microcycle type of sporulation (Hadley and Harrold, 1958).

Induction of microcycle conidiation by nutrient limitation has been recorded with several fungal species. Thus 'iterative germination' in which conidia directly form fertile conidiophores has been described in *Helminthosporium speciferum* when the conidia are germinated in water in the absence of external carbon source (Mangenot and Reisinger, 1976). Under less defined conditions a similar germination pattern has been described for *H. sativum* as 'precocious sporulation' (Boosalis, 1962) and on acetate as single C source, "conidiation prématurée", had been observed in *Neurospora crassa* (Turian, 1961). Zeidler and Margalith (1973) reported microcycle conidiation in *Penicillium digitatum* when conidia germinated in media containing certain amino acids (glutamate and serine) as the sole nitrogen source. Microcycle conidiation primarily occurred with those conidia that did not germinate during the early stages of cultivation. The microcycle formation of chlamydospores and conidia of strain 244 of *Claviceps purpurea* was considered to be due to a pleiotropic mutation intervening with the normal differentiation of the cells (Pažoutová *et al.*, 1978). Direct sporulation from zoospores of *Blastocladiella emersonii* (Hennessy and Cantino, 1972) when growing under starvation conditions in dilute phosphate buffer was termed 'lag phase sporogenesis'. Microcycle sporulation has been observed in a number of *Streptomyces* spp (Carvajal, 1947) and in bacteria (Vinter and Slepecky, 1965).

Induction by temperature-nutrient manipulation

The initial studies demonstrating the microcycle conidiation concept in *Aspergillus niger* were obtained by using a combined temperature and nutrient manipulation (Anderson and Smith, 1971b, 1972). Although the nutrient content of the medium was important the critical determining factor was the supraoptimal temperature (44°C) which caused conidial enlargement by spherical growth but inhibited germ tube formation. When the temperature was later dropped to 30°C there was a direct production of one or more fertile conidiophores from the enlarged parent conidium - the giant cell - without any accompanying vegetative filamentous growth. However, by minor changes in the thermal treatment (41°C) it was possible to produce giant cells which subsequently produced both conidiophores and limited vegetative growth at the same time (Anderson and Smith, 1972). Limited studies at Strathclyde have also demonstrated microcycle conidiation with *A. flavus*, *A. giganticus*, *A. oryzae* and *A. glaucus* and giant cell formation with *Rhizopus stolonifer*. All these systems depended on thermic variation.

Following the *Aspergillus* studies several variations of the system have been developed to induce forms of microcycle conidiation in *Penicillium urticae* (Sekiguchi *et al.*, 1975a, b,c), *Paecilomyces varioti* (Anderson *et al.*, 1978) and microcycle macroconidiation (Cortat and Turian, 1974) and microcycle microconidiation (Rossier *et al.*, 1977) in *Neurospora crassa*. With the exception of microcycle microconidiation in *N. crassa* where plate culture was used, all the other systems have been developed using submerged agitated conditions. Each system can be clearly shown to be composed of two stages, viz. (a) *Stage 1* - elevated temperature, spore enlargement and restriction of hyphal growth; (b) *Stage 2* - at a lower temperature and in the presence of an appropriate nutritional environment to produce the conidiation phase of the particular system. For each organism there is a precise temperature, duration of each stage, and a specific nutritional status of the medium. For *Neurospora crassa* - the difference in temperature between Stage 1 and Stage 2 is 21°C for *Aspergillus niger* 14°C, for *Paecilomyces varioti* 8°C and for *Penicillium urticae* 2°C. An approximately similar time is required to achieve microcycle conidiation in each of the systems (between 24-48 h after incoculation of conidia) except with microcycle microconidiation in *Neurospora crassa* which requires 24-48 h for stage 1 and a further 96 h for stage 2.

With the exception of microcycle microconidiation in *N. crassa*, where development occurred in a non-nutrient (phosphate buffer) medium, all the other systems had a medium composition which would permit normal vegetative growth in the absence of temperature limitation. The role of CO_2 in microcycle conidiation is not yet clearly understood but studies with the *Aspergillus* system would suggest that it is playing an important morphogenetic role. Giant cells produced by fermenter conditions will tend to germinate after 20 h at 44°C if 5% CO_2 is not added to the air supply. Whether or not CO_2 metabolically produced might play some role in the induction of microcycle conidiation in other species has not been reported. Microcycle microconidiation in *Neurospora crassa* occurs solely at the expense of endogeneous conidial reserves since both stages occur on non-nutrient medium. With *Penicillium urticae* stage 1 occurs in a nutrient rich medium and stage 2 in a nitrogen poor medium. There is no nutrient limitation which microcycle conidiation in *Aspergillus niger* and *Paecilomyces varioti* and in microcycle macroconidiation in *Neurospora crassa*. Glutamate is an essential medium component to give maximum conidiation during stage 2 with *Aspergillus niger* and *Paecilomyces varioti*.

An important aspect of most microcycle systems has been the increasing synchronous control of the developmental process. Again microcycle microconidiation in *Neurospora crassa* is the exception and lacks synchronous development during germination and subsequent growth throughout stages 1 and 2. With the other systems temperature manipulation has been found to maintain a separation of the morphological events which comprise each stage. Synchrony persists because of the severe restriction on vegetative growth resulting from exposure to elevated temperature and in some cases to the suboptimal nutrient conditions which have been imposed.

The age of the conidium may well be an important factor in determining synchronous development. In many of the microcycle systems that have so far been developed the spore is a phialoconidium. Thus in normal subaerial development, chains of spores are produced which will vary considerably in age due to the nature of formation. Thus when a spore inoculum is prepared it will usually contain spores ranging in age from several days to several hours. There must exist considerable biochemical differences between such spores and in particular in the levels of endogeneous and exogeneous spore germination inhibitors. Thus a 'normal' spore inoculum is clearly a relatively asynchronous population. What therefore can be done to create a more synchronous system? Studies at Strathclyde extensively examined spore germination of *Aspergillus niger* using methods which encompassed both induction synchrony and selection synchrony. Repeated washing, up to 5 times in sterile distilled water, was found to yield the most synchronised population with respect to uniformity in 95% of the population during spherical growth phase (Deans, 1978).

The removal of water-soluble germination inhibitors was believed to be happening during the washing routine. Attempts to characterise and isolate this inhibitory compound(s) were unsuccessful. Although comparable results could be obtained with continuous and discontinuous sucrose gradients, Ficoll 400 and CsCl gradients, the complexity of these procedures, along with problems in maintaining sterility, precluded their routine use for the large numbers of conidia used. All other methods of induction synchrony, with the exception of dimethylsulphoxide (DMSO), severely limited giant cell formation. Since DMSO treatment involved application of an external chemical agent whose precise action on the spores was not known, it was decided that simple washing would reduce the risk of any metabolic aberrations. The 5-wash routine was adopted as a standard method for cleaning and synchronising conidia for *Aspergillus niger*. It is surprising to note that few, if any, developmental studies with fungal spores have fully considered the need for germination synchrony. On the present study it has been found that improving the synchrony of giant cell formation greatly facilitates subsequent synchrony in later stages of microcycle conidiogenesis. The synchronised pattern of microcycle conidiation in *Aspergillus niger* is shown in Fig. 1.

DEVELOPMENTAL CHANGES DURING MICROCYCLE CONIDIATION

Conidium enlargement (Stage 1)

Most forms of microcycle conidiation show an initial increase in conidial diameter, associated with a period of supraoptimal temperature. This is seen particularly with the Aspergilli systems (Anderson and Smith, 1972; Deans, 1978). Under normal conditions of germination *A. niger* conidia will undergo on average a 2x swelling before germ tube outgrowth. However, when this process is carried out at 44°C germ tube formation is completely inhibited while enlargement of the conidia continues resulting in the formation of large spherical cells approximately 20-25 μm in diameter. Spore surface area increases 30x. This extensive increase in size is a growth process accompanied by large increases in macromolecular synthesis and a net increase in dry weight (Kuboye *et al.*, 1976; Deans, 1978).

The high temperature used in the *Aspergillus niger* system appears to inhibit the apical growth process and subsequent hyphal development. However, the growth process in itself

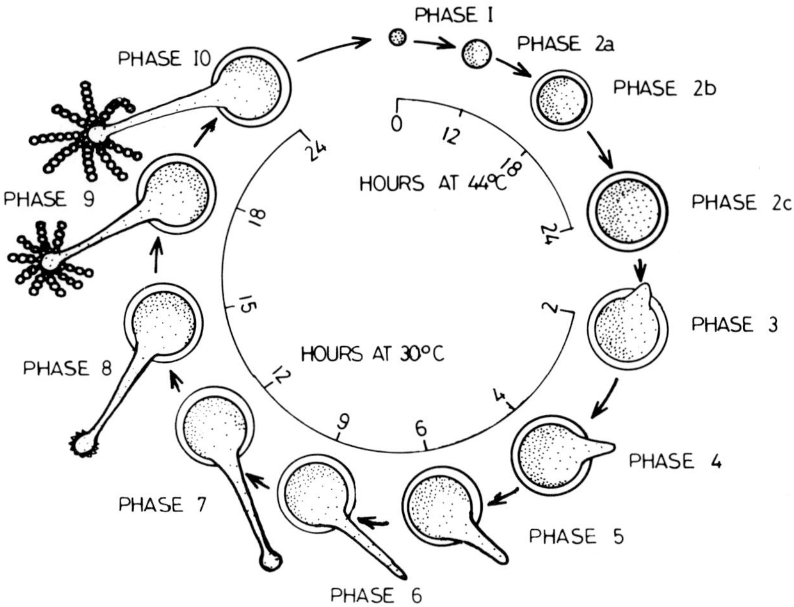

Fig. 1. Microcycle clock for Aspergillus niger

is not stopped but continues by way of uniform wall deposition resulting in spherical cells of variable size and thickness. When such cells are exposed to lower temperatures (30°C) the normal apical outgrowing process can occur. Thus the high temperature effect is not permanent and most probably indicates a temperature sensitive metabolic step.

Ultrastructural studies have been carried out during conidium enlargement in *Penicillium urticae* (Sekiguchi *et al.*, 1975a) *Aspergillus niger* (Deans *et al.*, 1980) and *Neurospora crassa* (Rossier *et al.*, 1977). There was no evidence to indicate that the high temperatures which were severely restricting development were causing any ultrastructural damage. Ultrastructural organisation was typical of most fungal cells although nuclear numbers increased progressively to *c* 16 after 24 h at 44°C (Yahya, 1979). The giant cells contained numerous mitochondria and abundant ribosomes. The cell wall attained a thickness of up to 2 µm in the fully developed giant cell. This thickened cell wall presented many problems with respect to stain preparations and sectioning. Dictyosome-like structures were observed in the immediate vicinity of the growing wall and vesicles were seen to 'bleb' off from the dictyosome plates and move to the plasmalemma.

The ultrastructural appearance of the walls of enlarged conidia of *N. crassa* (Rossier *et al.*, 1977) and *P. urticae* (Sekiguchi *et al.*, 1975a) again imply that isotropic wall growth is occurring at this stage. Heat-treated micro- and macroconidia of *N. crassa* showed electron dense, vacuolar and perinucleolar dense materials respectively (Ton That and Turian, 1978).

Progressive vacuolations observed in the enlarged conidia of *A. niger* and *P. urticae* at later stages in the microcycle process (particularly with nutrient exhaustion) probably reflects degeneration and cellular ageing. Extensive microbody formation was characteristic of these stages (Smith *et al.*, 1976).

Conidiophore outgrowth and spore formation (Stage 2)

In all microcycle systems, excluding the Aspergilli, the outgrowths which develop from the enlarged conidia are morphologically similar to normal germ tubes. Such outgrowths exhibit a limited amount of apical extension before conidiogenesis. With *N. crassa* microcycle macroconidiation (Cortat and Turian, 1974; Ton That and Turian, 1978) the outgrowths convert to simple conidiophores producing proconidia from the tip by basifungal budding. In *N. crassa* microcycle microconidiation (Rossier *et al.*, 1977) apical growth of the outgrowth stops and is replaced by lateral growth giving rise to hooks which finally become phialides bearing microconidia. In microcycle conidiation with *P. urticae* phialides are produced at the tips of the outgrowth. Ultrastructural analysis of this form of phialoconidiogenesis shows no difference from normal subaerial formation (Sekiguchi *et al.*, 1975c). Extensive size reduction occurs with the conidiogenous apparatus of microcycle *Paecilomyces varioti* (Anderson *et al.*, 1978). The outgrowths become typical bottle shaped phialides producing chains of normal conidia.

With *A. niger* and other Aspergilli a thick conidiophore stalk is produced which is characteristic of the species. In the microcycle system the giant cell is apparently performing the function of the *Aspergillus*-type foot cell (Anderson and Smith, 1972; Smith *et al.*, 1977). In microcycle conidiation there is a considerable reduction in size of the conidiophores when compared to subaerial conidiation and whereas in the latter case the conidiophore head possesses both metulae and phialides, only phialides are produced in the vesicle of the microcycle structures. Conidia produced by the microcycle technique are similar in size to normal type conidia but lack the characteristic black pigmentation

The conidia can undergo repeated microcycle conidiation without any change in the ability to revert to a normal developmental pattern.

It is not uncommon for a giant cell of *Aspergillus niger* to produce more than one conidiophore and sometimes even as many as five have been recorded. The substantial biomass involved in these outgrowths clearly indicates that the conidiating units, i.e. giant cell and conidiophore(s), must be continuously taking nutrient from the environment and not relying totally on the re-utilisation of reserves in the giant cell.

Ultrastructurally the conidiophore outgrowth is derived from the inner cell wall layer of the giant cell (Deans *et al.*, 1980). Associated with the outgrowing conidiophores were numerous microvesicles which persisted in the apically extending conidiophores. The conidiophore cell wall thickness, in median sections along the conidiophore length, were no longer of the dimensions of the giant cells, but rather of that found in hyphae.

At a late stage of conidiophore development septum formation occurred near the point of conidiophore outgrowth, occasionally this was in the form of a double septum. Such septum formation does not appear to occur during subaerial conidiophore development. It should be noted that early septum formation occurred at the junction of the enlarged conidium and the conidiating outgrowth in both microcycle *Penicillium urticae* (Sekiguchi *et al.*, 1975b) and *Neurospora crassa* (Rossier *et al.*, 1977).

KINETICS OF MICROCYCLE CONIDIATION IN *ASPERGILLUS NIGER*

Microcycle conidiation can be routinely achieved in a batch fermenter system (Kuboye *et al.*, 1976), slightly modified by Deans (1978).

For kinetic studies samples (10 ml) were removed every two hours from the fermenter and giant cell and conidiophore development measured microscopically.

Kinetics of giant cell growth

There have been only a limited number of reports on growth kinetics of spores. Trinci (1971) reported that during the first 4 h after inoculation *Rhizopus stolonifer* sporangiospores increased linearly in diameter (7.6 - 9.8 μm) before germination occurred. However, the relatively small increase in diameter and short observation time did not give sufficient indication of linear growth.

The growth of the giant cells of *Aspergillus niger* is

shown in Fig. 2, where D is the giant cell diameter and t is the fermentation time. The straight line obtained indicates that the following kinetic model can be used for the growth of the giant cells:

$$\frac{dD}{dt} = \mu_D D \qquad - (1)$$

where

D = giant cell diameter (μm)
μ_D = specific growth rate (h^{-1})

The subscript D is introduced to indicate that the specific growth rate is related to the diameter and not the dry weight which is the standard practice in fermentation studies.

The value of μ_D is given by the slope of the straight line in Fig. 2.

$$\mu_D = 0.075\ h^{-1}$$

Pirt (1966) has derived the following expression for growth of mycelial pellets:

$$\frac{dr}{dt} = \alpha w \qquad - (2)$$

where

r = pellet radius
α = specific growth rate
w = width of the peripheral growth zone of the pellet after the pellet has exceeded a certain diameter

In the giant cell it would be expected that the whole cell would be active in the growth process, i.e. the certain diameter referred to by Pirt (1966) has not been exceeded. Equation (1) therefore is the form of equation (2) when $w = r$.

Thus the giant cells of *Aspergillus niger* exhibit exponential growth and have a specific growth rate, based on cell diameter, of 0.075 h^{-1}. The kinetic model proposed represents a special case for a model previously employed to describe growth of mycelial pellets.

Kinetics of conidiophore growth

The growth of conidiophores is illustrated in Fig. 3 where

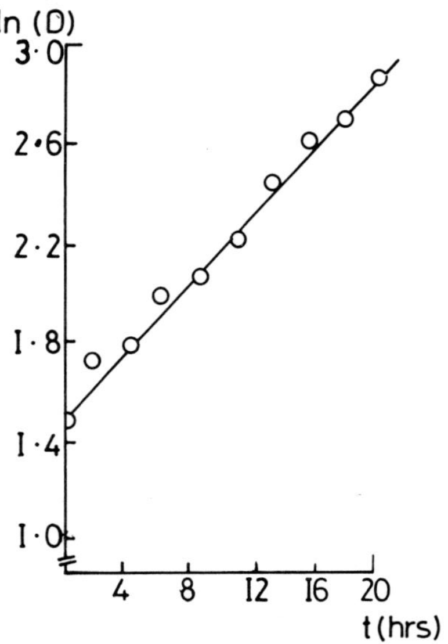

Fig. 2. *Giant cell growth where D = giant cell diameter. Each point represents the average of 200 readings.*

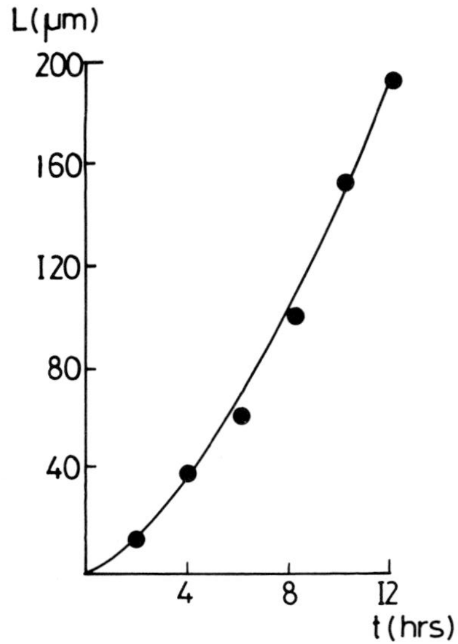

Fig. 3. *Growth of conidiophores where L = length of conidiophore. Each point represents the average of 200 readings.*

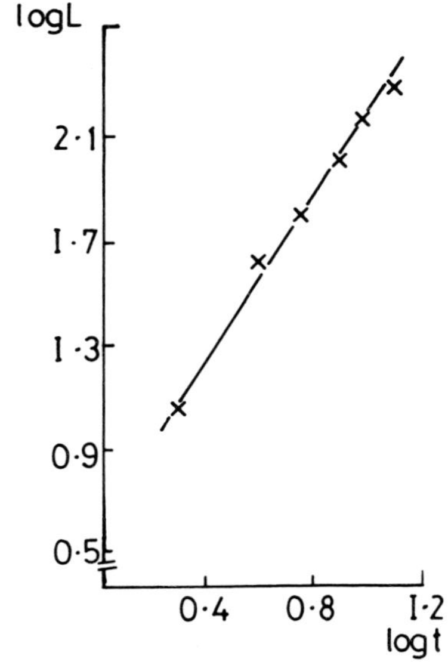

Fig. 4. *Growth of conidiophore. Log L as log t, where L = length of conidiophore, t = time.*

L is the length of the conidiophore in μm. From Fig. 4 is appears that the model describing the growth of the conidiophore is of the form

$$L = At^n \qquad - (3)$$

where

L = conidiophore length (μm)
A, n = constants

From Fig. 4 and equation (3) the values of A and n are found to be

A = 4.0
n = 1.55

If the conditions for promoting growth of the conidiophores are not met, germ tube growth will occur. It has been shown for a number of fungi, including *Aspergillus niger*, that germ tubes may grow exponentially in length (Trinci, 1969, 1971).

A kinetic model for the growth of germ tubes will therefore be of the form

$$\frac{dLg}{dt} = \mu_L \; Lg \qquad - (4)$$

where

Lg = germ tube length
μ_L = specific growth rate based on length

For conidiophore growth, equation (3) and the values obtained from A and n can be used to give

$$\frac{dL}{dt} = BL^{0.35} \qquad - (5)$$

where

B = constant

A comparison between equations (4) and (5) shows that the germ tubes are growing at a faster rate than the conidiophores. This is only to be expected considering the smaller diameter of the germ tubes, giving a high surface to volume ratio. With a surface active in absorption of nutrients a faster linear rate will be exhibited.

PROTOPLAST FORMATION AND REGENERATION FROM MICROCYCLE CONIDIATION

Protoplasts have been obtained from giant cells and conidiophores of microcycle *Aspergillus niger* (Davis *et al.*, 1977; Yahya, 1979), and from normal *Aspergillus niger* mycelium using a lytic enzyme system derived from *Trichoderma harzianum*. Giant cell protoplasts were only released by the complete digestion of the cell wall and always in the presence of added commercial chitinase enzymes. The inability of *T. harzianum* lytic enzymes alone to cause sufficient digestion for protoplast release (whereas rapid release occurs with normal mycelium) suggests that chitin plays a more dominant structural role in giant cell wall architecture. Protoplast production decreased with decreasing giant cell age again indicating a changing wall biochemistry. Whereas protoplast size from giant cells was reasonably constant, production from outgrowing conidiophores varied considerably. The first protoplasts were mainly derived from conidiophore tips and were 7-10 μm in diameter. Such protoplasts only rarely had vacuoles. With increasing time of incubation larger protoplasts (12-23 μm) were produced from the sides of the conidiophore walls and contained several large vacuoles. The largest protoplasts (25-42 μm) and the last to be produced, were derived from the point of junction between the giant cell and conidiophore outgrowth (Fig. 5)(Yahya, 1979).

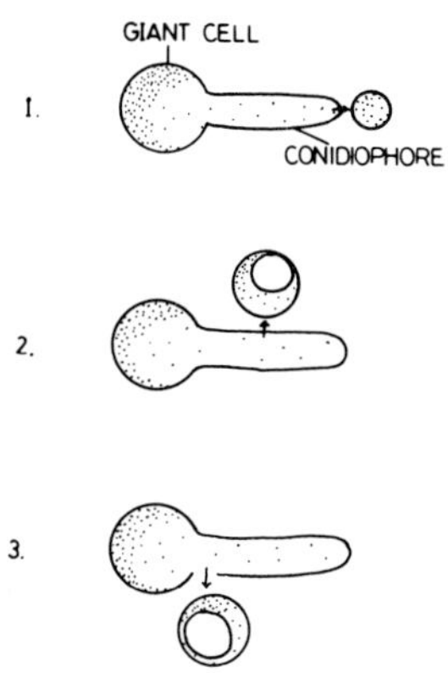

Fig. 5. Sites of protoplast release from outgrowing conidiophores.

When protoplasts were prepared from giant cells at 40°C (the optimum temperature for formation) they were subsequently unable to show any form of regeneration at 30°C. Many combinations of chemical compounds were tried without success.

However, when giant cell protoplasts were produced at 30°C a good level of regeneration occurred. Fewer protoplasts were produced at 30°C than at 40°C but a higher percentage could regenerate. It is possible that some essential part of the protoplast, probably the plasma membrane is damaged during release at 40°C. Because of the good synchrony of giant cells and the formation of only one protoplast per cell, the resultant populations are quite homogeneous. In contrast, mycelial protoplasts are very variable in size and often are enucleate. Giant cell protoplasts produced simple wide germ tubes after 7 h incubation; no conidiophores were ever observed to develop.

Conidiophore protoplasts were able to regenerate at 30°C and to produce single germ tubes after 2 h which grew steadily without branching. Again there was no conidiophore development with these regenerations. Protoplast formation from conidiophores resembles protoplast formation from mycelium with development occurring through wall pores producing protoplasts of variable size and vacuolar complexity. In contrast, giant cell protoplasts occur only with complete dissolution of the cell wall, with one protoplast being formed per giant cell. This will offer a unique homogeneous population for physiological studies.

CELLULAR ANALYSIS DURING MICROCYCLE CONIDIATION IN *ASPERGILLUS NIGER*

Dry weight: Dry weight steadily increased during the entire cycle (Fig. 6). The early studies by Kuboye *et al.* (1976) had indicated a phase of autolysis during the late stage of conidiogenesis. However, current work uses an increased glucose level (15 g l^{-1}) which ensures that an excess of glucose is present throughout the cycle. The highest level of dry weight increase occurred between 20-32 h which coincided with maximum wall formation and conidiophore outgrowth (Yahya, 1979).

DNA and RNA levels and nuclear numbers: The pattern of DNA and RNA levels are shown in Fig 7. The relationship between DNA concentration and nuclear number during giant cell formation is given in Table 1. Spores of *Aspergillus niger* are normally uninucleate (3.2×10^{-8} µg DNA cell $^{-1}$) and nuclear number increased steadily during giant cell formation with 16 nuclei being observed on average after 24 h at 44°C. Fluorescent microscopy using 4,6-diamidino-2-phenylindole (DAPI) as a DNA binding agent gave visual confirmation of the nuclear patterns (Yahya, 1979).

It was not possible to determine whether nuclear division

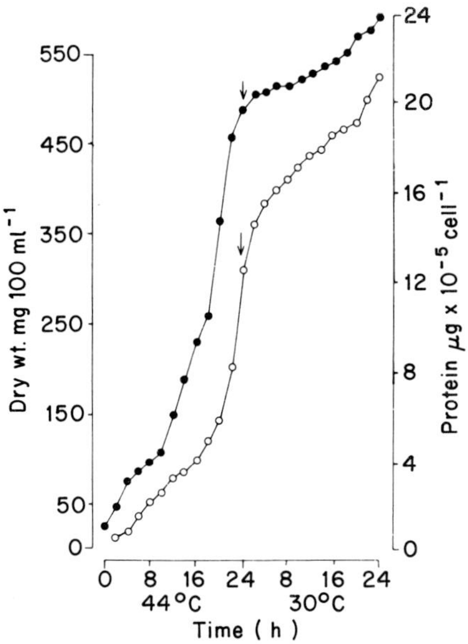

Fig. 6. Changes in dry weight (●) and protein concentration (o) during microcycle conidiation of Aspergillus niger

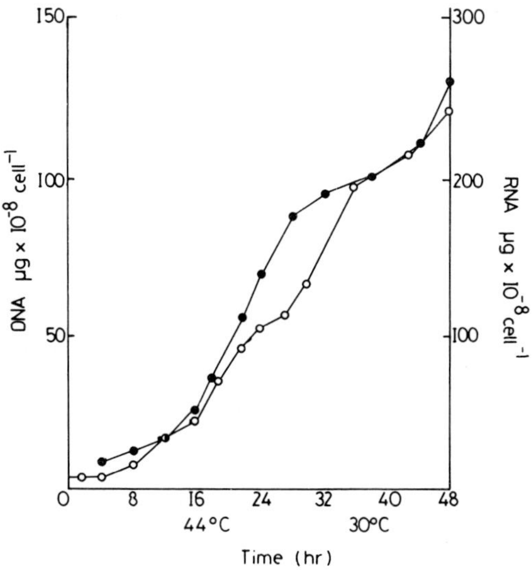

Fig. 7. Changes in DNA (o) and RNA (●) during microcycle conidiation of Aspergillus niger

in each giant cell was occurring synchronously. The considerable depth of the cell, together with the extreme thickness of the cell wall, made direct observation difficult.

TABLE 1

Relationship between DNA levels and nuclear number during giant cell formation in Aspergillus niger

Time (h)	Relative DNA level ($x10^{-8}$ μg)	Number of nuclei
1 - 4	3.2	1
5 - 8	6.9	2
8 - 10	10.0	3
10 - 12	14.0	4
12 - 16	18.5	6
16 - 19	28.0	11
19 - 21	38.0	14
21 - 23	44.5	14

Using the original method of inducing microcycle conidiation in *Aspergillus niger* (Anderson and Smith, 1971b, 1972) pulse labelling using ^{14}C leucine and ^{3}H uridine have indicated that the maximum rate of protein and RNA biosynthesis occurred during giant cell development (Duncan *et al.*, 1978). Fractionation of the RNA using acrylamide gel electrophoresis indicated that label had been incorporated into 1.294×10^6D and 0.84×10^6D ribosomal RNA and into 4S and 5S RNA fractions. Evidence for the presence of mRNA was also obtained by the isolation of polyribosomes from the giant cells.

Protein: Total cellular protein achieved maximum levels of increase between 20-24 h at 44°C to 4 h at 30°C (Fig. 6). The increase in protein level observed at and beyond 16 h at 30°C can be related to the demands of conidiospore formation.

Cell wall chemistry: The chemical structure of the cell wall has been determined throughout the microcycle of *Aspergillus niger* measuring six major components i.e. protein, lipid, α and β-glucan, chitin and galacto-mannan polymer (Deans and Smith, 1980a).

The lipid and protein levels in the cell wall are shown in Fig. 8. Cell wall lipids were low during the initial stages of giant cell formation but began to increase about the 12 h stage and continued through to the 6 h stage of conidiophore outgrowth. There was then a decrease over the next 12 h with a further increase during conidia formation. Cell wall protein increased steadily during the first 15 h incubation and then decreased to the end of the giant cell stage. Protein then increased throughout conidiophore extension achieving a maximum during conidiospore formation. The

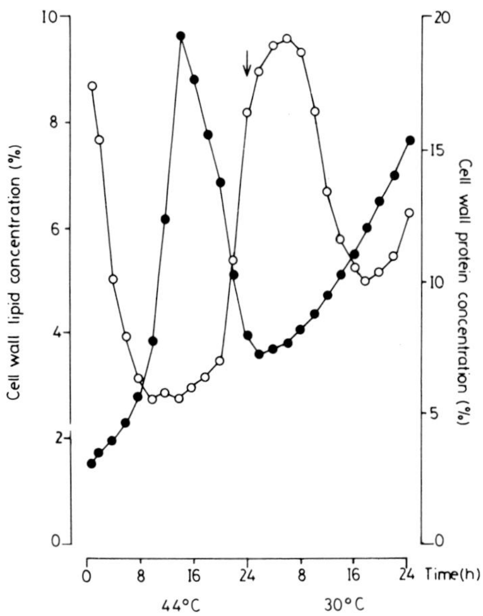

Fig. 8. Lipid (○) and protein (●) levels during microcycle conidiation of Aspergillus niger

varying pattern of cell wall protein during the microcycle may be related to the changing activities of specific wall localised enzymes (Deans and Smith, 1980b).

Chitin levels increased steadily throughout the entire cycle achieving a maximum level (34%) of total cell wall material at the final stages of spore formation (Fig. 9). The microfibrillar nature of chitin within fungal cell walls has led to the consideration that chitin gives both structural rigidity and shape to the organism.

The major α-glucan detected in the giant cell wall was the S-glucan, a straight chain α(1-3) glucan (73%). Nigeran, an alternating α(1-3) and α(1-4) polymer remained relatively constant at 20% throughout. During the early stages of giant cell formation α(1-3) glucan was synthesised in the cell wall while in the later stages of conidiophore outgrowth and development there was a net decrease in levels (Fig. 10). β-glucan increased throughout the microcycle consistent with the proposed cell wall matrix role of this polymer (Fig. 11). The galactomannan polymer fraction of the cell wall seldom exceeded 10-14% of the total wall material.

Cell wall related enzymes: The major enzymes involved in cell wall synthesis and degradation, i.e. α-glucanase, β-glucan synthase, β-glucanase, chitin synthase and chitinase

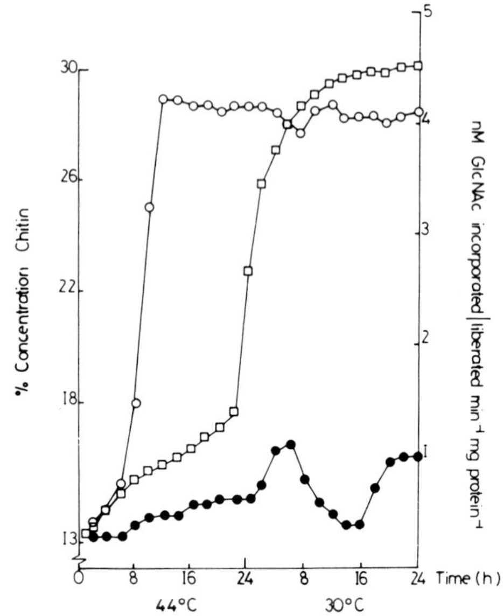

Fig. 9. Changes in chitin (□), chitin synthase (○) and chitinase (●) during microcycle conidiation of Aspergillus niger

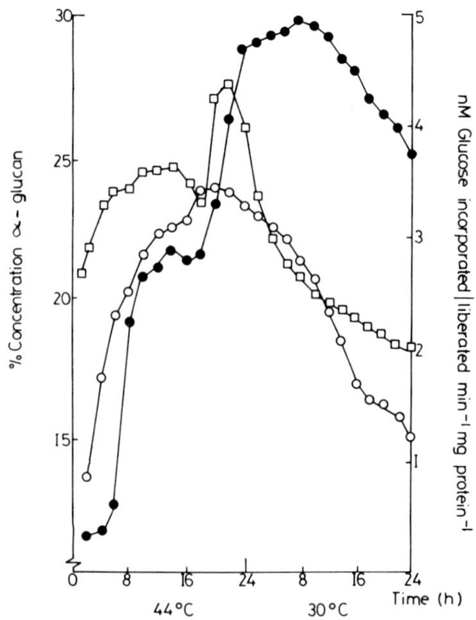

Fig. 10. Changes in α -glucan (□), α -glucan synthase (○) and α-glucanase (●) during microcycle conidiation of Aspergillus niger

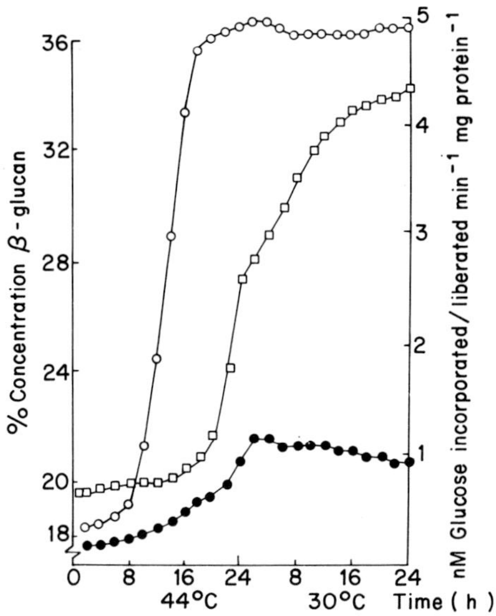

Fig. 11. Changes in β-glucan (□), β -glucan synthase (○) and β -glucanase (●) during microcycle conidiation of Aspergillus niger

have been subcellularly localised and assayed (Deans and Smith, 1980b).

Chitin synthase level increased rapidly during the early stages of giant cell formation and remained consistently high for the remainder of the cycle. In contrast, the generally low activity of chitinase had small peaks of activity corresponding to conidiophore formation (Fig. 9). During the microcycle the chitin synthase complex displayed distinct functional changes - the enzyme from the giant cell stage showing features normally associated with yeast-type chitin synthase - a zymogenic form - whereas conidiophore chitin synthase was similar to normal hyphal chitin synthase. The prolonged life experienced by the giant cell chitin synthase may help to explain the massive thick wall of the giant cell, while the activity present in the conidiophore which does not retain the thickened cell wall structure, supports a different pattern of chitin synthase activity.

β-glucan synthase increased throughout the microcycle while the activity of the glucanase was low at most points of the cycle. In contrast, α-glucan synthase showed an early high activity during giant cell development then this activity decreased. Co-incident with the decrease there was a marked increase in the activity of the α-glucanase (Fig. 11).

Thus there was evidently some re-utilisation of α-glucan

moieties or rechannelling of glucose monomers towards the end of giant cell incubation and into conidiation. Thus α(1-3) glucan may function as a carbohydrate storage compound to be used during the early stages of conidiophore outgrowth.

THE INFLUENCE OF METABOLIC INHIBITORS ON MICROCYCLE CONIDIATION OF *ASPERGILLUS NIGER*

Differentiation requires changes in cellular metabolism which will involve the establishment of new metabolic pathways to produce the cellular component necessary for the new developmental state. Such metabolic changes may well involve the formation of new enzymes. By studying the pattern of biosynthesis of RNA and protein, information relevant to enzyme synthesis may be obtained.

Recent studies by Deans and Smith (1979) have considered the effect of several inhibitors of protein and nucleic acid biosynthesis and mitochondrial functioning on conidial germination and microcycle conidiation of *Aspergillus niger*. The results showed that normal conidial germination was highly susceptible to most of these inhibitors with all compounds exerting similar dosage response patterns although the individual compounds were effective at different concentrations; vegetative growth or giant cell formation at 44°C was the least susceptible whereas conidiation at 30°C was intermediate in response.

Addition of actinomycin D, chloramphenicol or cycloheximide at any time up to 15-16 h of giant cell formation inhibits conidiophore outgrowth (Deans and Smith, 1979) while addition of these inhibitors after this period did not result in inhibition of conidiation. These studies further implied that transcription of information necessary for conidiation was occurring as early as 8 h at 44°C, but that expression was not able to occur until the permissible temperature of 30°C was achieved.

The diameter of the giant cell of *Aspergillus niger* must achieve a minimum size before being able to support conidiophore development. Giant cell diameter less than *c* 15-16 μm both in inhibitor experiments and in non-inhibited controls (achieved by removing giant cells at different time intervals to 44°C) rarely produced normal conidiophores but instead developed more vegetative mycelium. This does not appear to be a requirement of the other types of microcycle conidiation. For normal conidiation Axelrod (1972) has demonstrated that with surface and liquid growing cultures of *Aspergillus nidulans* (Eidam) commitment to conidiation was achieved only when a critical amount of biomass or colony growth had occurred.

When a low concentration of actinomycin D (20 μg ml^{-1}) was added at zero time of giant cell formation there was no subsequent conidiophore development but all giant cells germinated at the lower temperature of 30°C to produce vegetative mycelium. The later the addition of actinomycin D the more vegetative development was suppressed and the more conidiophore development proceeded (Duncan *et al.*, 1980). Thus temperature alone does not appear to suppress the capacity for vegetative development, as was originally considered (Anderson and Smith, 1972). Vegetative mycelial growth appears to be suppressed only in conditions which favour conidiophore development, i.e. a regulatory mechanism operates which suppresses vegetative growth when the genes for conidiophore development are being expressed.

REFERENCES

Anderson, J.G. and Smith, J.E. (1971a). Synchronous initiation and maturation of *Aspergillus niger* conidiophores. *Transactions of the British Mycological Society 56*, 9-29.

Anderson, J.G. and Smith, J.E. (1971b). The production of conidiophores and conidia by newly germinated conidia of *Aspergillus niger* (microcycle conidiation) *Journal of General Microbiology 69*, 185-197.

Anderson, J.G. and Smith, J.E. (1972). The effects of elevated temperatures on spore swelling and germination in *Aspergillus niger*. *Canadian Journal of Microbiology 18*, 289-297.

Anderson, J.G., Aryee, V. and Smith, J.E. (1978). Microcycle conidiation in *Paecilomyces varioti*. *FEMS Microbiology Letters 3*, 57-60.

Axelrod, D.E. (1972). Kinetics of differentiation of conidiophores and conidia by colonies of *Aspergillus nidulans*. *Journal of General Microbiology 73*, 181-184.

Boosalis, M.G. (1962). Precocious sporulation and longevity of conidia of *Helminthosporium sativum* in soil. *Phytopathology 52*, 1172-1177.

Carvajal, F. (1947). The production of spores in submerged cultures of some *Streptomyces*. *Mycologia 39*, 426-440.

Cortat, M. and Turian, G. (1974). Conidiation of *Neurospora crassa* in submerged culture without mycelial phase. *Archives of Microbiology 95*, 305-309.

Davis, B., D'Avillez-Paixao, M.T., Deans, S.G. and Smith, J.E. (1977). Protoplast formation from giant cells of *Aspergillus niger*. *Transactions of British Mycology Society 69*, 207-212.

Deans, S.G. (1978). Cell wall ontogeny during microcycle conidiation of *Aspergillus niger*. PhD Thesis, University of Strathclyde, Glasgow.

Deans, S.G. and Smith, J.E. (1979). Effect of metabolic inhibitors on microcycle conidiation of *Aspergillus niger*. *Transactions British Mycological Society 72*, 201-206.

Deans, S.G. and Smith, J.E. (1980a). Cell wall composition during microcycle conidiation of *Aspergillus niger*. Manuscript in preparation.

Deans, S.G. and Smith, J.E. (1980b). Changes in cell wall synthesis and lytic enzymes during microcycle conidiation of *Aspergillus niger*. Manuscript in preparation.

Deans, S.G., Gull, K, and Smith, J.E.(1980). Ultrastructural changes during microcycle conidiation of *Aspergillus niger*. *Transactions British Mycological Society 74*,(in press).

Duncan, D.B., Smith, J.E. and Berry, D.R. (1978). DNA, RNA and protein biosynthesis during microcycle conidiation in *Aspergillus niger*. *Transactions British Mycological Society 71*, 457-464.

Duncan, D.B., Berry, D.R. and Smith, J.E. (1980). Selection inhibition of conidiophore development in *Aspergillus niger* by actinomycin D. *FEMS Microbiology Letters 7*, 333 - 335.

Griffin, D.H. and Breuker, C. (1969). Ribonucleic acid synthesis during the differentiation of sporangia in the water mold *Achlya*. *Journal of Bacteriology 98*, 689-696.

Hadley, G. and Harrold, C.E. (1958). The sporulation of *Penicillium notatum* Nestling in submerged liquid culture. *Journal Experimental Botany 9*, 408-417.

Hennessy, S.W. and Cantino, E.C. (1972). Lag-phase sporogenesis in *Blastocladiella emersonii*: induced formation of unispored plantlets. *Mycologia 64*, 1066-1087.

Kuboye, A.O., Anderson, J.G. and Smith, J.E. (1976). Control and autolysis of a spherical cell form of *Aspergillus niger*. *Transactions British Mycological Society 67*, 27-31.

Larmour, R. and Marchant, R. (1977). The induction of conidiation in *Fusarium culmorum* grown in continuous culture. *Journal General Microbiology 99*, 59-84.

Lingappa, B.J. and Lingappa, Y. (1969). Role of autoinhibitors on the mycelial growth and dimorphism of *Glomerella cingulata*. *Journal General Microbiology 56*, 35-45.

Mangenot, F. and Reisinger, O. (1976). Form and function of conidia as related to their development. *In The Fungal Spore* (Ed. D.J. Weber and W.H. Hess) pp. 789-847, New York: John Wiley.

Ng, A.M.L., Smith, J.E. and McIntosh, A.F. (1973). Conidiation of *Aspergillus niger* in continuous culture. *Archiv für Mikrobiologie 88*, 119-126.

Park, D. and Robinson, P.M. (1969). Germination studies with *Geotrichum candidum*. *Transactions British Mycological Society 54*, 83-92.

Pažoutová, S.R., Řeháček, Z. and Pokorný, V. (1978). Microcycle sporulation in the *Claviceps purpurea* 244.*Folia Microbiologica 23*, 376-378.

Pirt, S.J. (1966). A theory of the mode of growth of fungi in the form of pellets in submerged culture. *Proceedings of the Royal Society B 166*, 367-373.

Robinson, P.M. and Smith, J.H. (1976). Morphogenesis and growth kinetics of *Geotrichum candidum* in continuous culture. *Transactions British Mycological Society 66*, 413-420.

Rossier, C., Ton-That, T.C. and Turian, G. (1977). Microcycle microconidiation in *Neurospora crassa*. *Experimental Mycology 1*, 52-62.

Roten, J. and Bashi, E. (1969). Induction of sporulation of *Alternaria porri* sp *solani* by inhibition of its vegetative development. *Transactions British Mycological Society 53*, 433-439.

Sekiguchi, J., Gaucher, G.M. and Costerton, J.W. (1975a). Microcycle conidiation in *Penicillium urticae*: an ultrastructural investigation of spherical spore growth. *Canadian Journal of Microbiology 21*, 2048-2058.

Sekiguchi, J., Gaucher, G.M. and Costerton, J.W. (1975b). Microcycle conidiation in *Penicillium urticae:* an ultrastructural investigation of conidial germination and outgrowth. *Canadian Journal of Microbiology 21*, 2059-2068.

Sekiguchi, J., Gaucher, G.M. and Costerton, J.W. (1975c). Microcycle conidiation in *Penicillium urticae:* an ultrastructural investigation of conidiogenesis. *Canadian Journal of Microbiology 21*, 2069-2083.

Smith, J.E. (1978). Asexual sporulation in filamentous fungi. *In The Filamentous Fungi* (Ed. J.E. Smith and D.R. Berry) *3*, 214-238. London, Edward Arnold (Publishers).

Smith, J.E., Anderson, J.G., Deans, S.G. and Davis, B.(1977). Asexual development in *Aspergillus*. *In: Genetics and Physiology of Aspergillus* (Ed. J.E. Smith and J.A. Pateman) pp. 23-58. London: Academic Press.

Smith, J.E., Gull, G., Anderson, J.G. and Deans, S.G. (1976). Organelle changes during fungal spore germination. *In The Fungal Spore: Form and Function* (Ed. D.J. Weber and W.M. Hess), pp 301-354. New York: John Wiley & Sons.

Timmick, M.B., Barnett, H.L. and Lilly, V.C. (1952). The effect of methods of inoculation of media on sporulation of *Melanconium fuliginum*. *Mycologia 44*, 141-149.

Ton That, T.C. and Turian, G. (1978). Ultrastructural study of microcyclic macroconidiation in *Neurospora crassa*. *Arch. Microbiol. 116*, 279-288.

Trinci, A.P.J. (1969). A kinetic study of the growth of *Aspergillus nidulans* and other fungi. *Journal of General Microbiology 57*, 11-24.

Trinci, A.P.J. (1971). Influence of the width of the peripheral growth zone on the radical growth rate of fungal colonies on solid media. *Journal of General Microbiology 67*, 325-344.

Turian, G. (1961). L'acétate et son double effet d'induction isocitratasique et de différenciation conidienne chez les *Neurospora. C.R. Acad. Sci. Paris 252*, 1374-1376.

Vinter, V. and Slepecky, R.A. (1965). Direct transition of outgrowing bacterial spores to new sporangia without intermediate cell division. *Journal of Bacteriology 90*, 803-807.

Yahya, Abdul-Ghani (1979). Physiological studies on microcycle conidiation of *Aspergillus niger*. PhD Thesis, University of Strathclyde, Glasgow.

Zeidler, G. and Margalith, P. (1972). Synchronised sporulation in *Penicillium digitatum* (Secc). *Canadian Journal of Microbiology 18*, 1685-1690.

Zeidler, G. and Margalith, P. (1973). Modification of the sporulation cycle in *Penicillium digitatum. Canadian Journal of Microbiology 19*, 481-483.

EPILOGUE

The last afternoon session was devoted to a general discussion of topics covered during the Symposium and to an assessment of proposals for future research. Here are some ideas and excerpts mainly derived from contributions by the session chairmen.

Dr. Fuller emphasized the need for a better understanding of the nature and function of various organelles such as vesicular elements and plasmamembrane-related lomasomes and plasmalemmasomes. Interestingly, the latter structures have not been visualized with the recently introduced freeze-substitution technique. More emphasis should be devoted to the possible significance of intrahyphal and intrasporal compartmentation. This is exemplified by the spatial separation of mitochondria and the ribosomal nuclear cap in the zoospores of Blastocladiales and the loss of the barrier accompanying inception of germination. Combined studies at the ultrastructural and biochemical levels are urgently needed.

Ultrastructural reorganization at sporulation implies not only intracytoplasmic, spatial changes but also a precise timing of sporogenic events : this is of special concern for the surface structures which typify the spores formed (smooth, dry or wet, spinulose, verrucose, etc.). Attention will still have to be paid to the formation and function of septa in relation to the problem of spore secession. There are sharp differences between septation processes depending on whether one deals with phialides, arthroconidia, ascospores or basidiospores. It might also be rewarding to isolate and characterize specialized membranes such as the forespores from sporulating yeasts, or the endoplasmic reticulum and the Golgi dictyosomes. Again, care should be taken to avoid possible artifacts by including the biochemical, biophysical, histochemical and ultrastructural techniques available today.

Dr. Madelin stressed the need for concentrating on developmental problems. This could be achieved by devoting more effort towards the study of relatively simple systems of conidiogenesis. On a broader scale more emphasis should be placed on consequences that studies of this kind might have upon problems of fungal pathogenicity, hyperparasitism and lichenization.

According to Dr Dutta fungal genes are now better understood thanks to the new techniques advanced by molecular biologists. Fungal genomes seem to be strikingly different from other eucaryotes in that short (200-400 base pairs) repetitive DNA sequences do not exist, suggesting different kinds of genomic controls.

Nuclear controls have to be understood as the result of interaction between the genome and the cytoplasm with the aim of identifying master control switches for the sporogenic processes. Mutational analysis has led to a better understanding of the problem of competence for sporulation and has indicated the presence of genes that control the choice between the asexual and sexual pathways of the fungal life cycle.

More knowledge of actual numbers of genes participating in the sporogenic processes is needed and work in this area has recently been intensified. Cloning of tightly clustered structural and operator genes of the regulatory circuits will help to unravel regulatory processes. Such studies have been initiated with structural genes of the quinic acid catabolic pathway. Increasing attention should also be paid to the role of intervening sequences in fungal genes associated with morphogenesis.

Dr. Weber drew attention to the fact that expression of the vegetative and sporogenic information relies upon the synthesis of many proteins and an intensive turnover of these macromolecules during the transition from one phase to the other. This remake of the protein pattern necessarily involves proteases as shown during yeast ascosporogenesis. In Blastocladialian fungi late protein synthesis intervenes to ensure zoosporangial papilla formation as a last step of differentiation. Sophisticated methods should be developed in attempts to isolate control proteins not only of the biosynthetic processes associated with sporogenesis but also of those leading from germinated spores to the formation of infection structures in phytopathogenic fungi.

In his remarks Dr. Sussman points to the need of relating environmental controls more closely to the ecology of the respective organisms. In the attempts to understand more clearly the activation processes required for germination of many types of spores (mainly of sexual origin), there still appear some divergent opinions. On the one hand, the diversity of activators known, including heat, acetate, urea and other chemicals is explicable on the basis of an effect on proteins. On the other hand, lipids, or at least lipoprotein membranes of spores of *Phycomyces*, may be the site of the effect of the numerous activation treatments for these spores. What is noteworthy is that the most detailed biochemical studies of the constitutively dormant ascospores of *Neurospora* and the spores of *Dictyostelium discoideum* and *P. blakesleeanus* still have yielded no agreement on the mechanism of the central effect of activation. Experiments with physico-chemical parameters like phase-shifts of lipids and activation energies of proteins are performed with systems that appear as too mixed, so that an average across all proteins and lipids is obtained. Reciprocally, the use of values obtained from pure proteins, like pepsin, does not represent the situation for the naturally occurring lipo- and glyco-proteins and lipids whose role in activation may be central. If the field is to progress, extraction of membranes and analyses of their protein and lipid contents will have to be undertaken. Preceding this step, perhaps, is the need to isolate 'properly' the organelles, starting with the mitochondria, from dormant spores to study them separated from their natural environment. Again, both ultrastructural features and biochemical processes should be pursued concomitantly in such studies.

Dr. Bartnicki-Garcia's remarks highlighted some of the points raised during the session on biochemical events and controls. Growth processes during germination resume as early as the so-called swelling stage, partly in form of cycloheximide-sensitive protein synthesis. This stage should, therefore, more correctly be termed stage of spherical growth. During these stages growth must be more clearly defined on the basis of linear versus logarithmic (exponential) increase in biomass. The next stage of growth where a switch from the multidirectional to a polarized, unidirectional, pattern occurs remains refractory to a more detailed analysis. Obviously, the new wall formation requires active polysaccharide synthesis and in many cases chitosomes are concentrated near the front of the polarized germ tube. Polysaccharide composition changes in a still unclear fashion during spore formation and is often interrelated with melanin synthesis and the deposition of external spore layers. In the maturing spore lipids accumulate as energy sources. Their possible role in the regulation of various processes including those involving membrane lipids deserves increased attention. The sporopollenins, while adding mechanical and chemical resistance to spores might also be of ecological significance not hitherto investigated.

Dr. Gottlieb's summary statements illuminated the fact that metabolism with its regulatory devices pervades the whole of formation and germination processes. In pathogenic fungi, it is known that protrusion of the germ tube through the spore wall is controlled by endogenous inhibitors. Next, the formation of infection structures (appressoria) occurs in response to environmental stimuli which act on a still elusive cytoplasmic system for transmitting the signals from the cellular surface to the nucleus (onset of DNA replication). That respiratory enzyme systems reappear in the early phases of germination was already well-documented but current progress is now concerned with the assembly of the subunits into the functional enzymes (cytochrome oxidase, ATP-ase, etc.) inside outgrowing conidia. From a teleological point of view, biochemical switches intervening at spore formation should provide the reducing power required for their germination. In such perspective, glutamate synthesis appears as a critical requirement met by reoriented biosynthetic pathways at conidiogenesis of *Neurospora*.

Microcycling techniques, as first employed for the *Aspergillus niger* system, allow a maximal reduction of the vegetative growth phase. With the use of such systems one may hope to unravel the nature of the factors crucial to the transition from the vegetative to the reproductive stages, as Dr. Smith pointed out in his remarks.

The final round of discussion was concerned with the future of these meetings. It was noted that despite additional prospective topics such as industrial, ecological (spore dispersal and survival) parasitic, or symbiotic aspects major emphasis should always center around spore formation and spore germination.

SUBJECT INDEX